# The Handbook of
# Design Management

# THE HA

# OF

# MANA

# NDBOOK
# DESIGN
# GEMENT

Edited by
Rachel Cooper, Sabine Junginger
and Thomas Lockwood

with
*Richard Buchanan, Richard Boland
and Kyung-won Chung*

Oxford • New York

English edition
First published in 2011 by
**Berg**

Editorial offices:
First Floor, Angel Court, 81 St Clements Street, Oxford OX4 1AW, UK
175 Fifth Avenue, New York, NY 10010, USA

Berg is an imprint of Bloomsbury Publishing Plc.

**Library of Congress Cataloging-in-Publication Data**

A catalogue record for this book is available from the Library of Congress.

**British Library Cataloguing-in-Publication Data**

A catalogue record for this book is available from the British Library.

ISBN      978 1 84788 488 6
e-ISBN    978 1 84788 490 9 (institutional)

Typeset by Apex CoVantage, LLC, Madison, WI, USA.

Printed in the UK by the MPG Books Group

www.bergpublishers.com

# CONTENTS

## PART IV: INTO A CHANGING WORLD

    Future of Design Management                                        495
    *Rachel Cooper, Martyn Evans and Alex Williams*

31  Working the Crowd: Crowdsourcing as a Strategy
    for Co-design                                                      512
    *Mike Press*

32  On Managing as Designing                                           532
    *Richard J. Boland, Jr.*

    Conclusions: Design Management and Beyond                          539
    *Rachel Cooper and Sabine Junginger*


    INDEX                                                              547

# ILLUSTRATIONS AND TABLES

## FIGURES/ILLUSTRATIONS/DIAGRAMS

## TABLES

# LIST OF CONTRIBUTORS

**Anuja Agarwal**, Associate Professor at We School, Mumbai, specializes in business design, innovation and interaction design. She is a postgraduate in computer applications and is currently pursuing her research in the area of 'creativity, design thinking and innovation in management education'. She aims to bring design thinking and social sciences into mainstream management education in order to create the socially sensitive business leaders of tomorrow. Anuja spearheads a number of research projects on rural India, which incorporate understanding of economic and social needs to further analyse the opportunities for the various industry sectors to reach out to the grassroots and benefit the rural population.

**Leonard Bruce Archer** CBE (1922–2005) was a chartered mechanical engineer and later Professor of Design Research at the Royal College of Art, where he was made an Honorary Fellow upon his retirement in 1988. He helped found the Design Research Society in 1966, was appointed its first President from 1992 to 2000, and was presented with a Lifetime Achievement Award by the Society in 2004. He was also a Member of the Design Council from 1972 to 1980. He has been considered a champion of research in design and helped establish design as an academic discipline.

**Sir Misha Black** (1910–77) was an industrial designer, interior designer and architect. In 1933, he joined Charles and Henry Bassett and Milner Gray in forming the forerunner of the Industrial Design Partnership (later called the Industrial Design Unit) in London. This was one of the first multidisciplinary design consultancies in Britain. He later joined the Ministry of Information and was given the job of principal exhibitions designer. He also became involved in setting up a new design group, the Design Research Unit, in 1943. In 1959 he was appointed the first Professor of Industrial Design at the Royal College of Art, a highly influential role in design education that he held until retirement in 1975. Black was also a founder member and later president (1959–61) of the International Council of Societies of Industrial Design (ICSID), which first met in London in 1957; a Fellow of the Chartered Society of Designers, and winner of the Minerva Medal, the Society's highest award; and between 1974 and 1976 Black was President of the Design and Industries Association. He was knighted in 1972.

**Richard Boland** is the Elizabeth. M. and William. C. Treuhaft Professor of Management, Department Chair, and Professor of Information Systems at Case Western Reserve University, Ohio. He undertakes qualitative studies of individuals as they design and use information and is interested in how people make meaning as they interpret situations in an organization, or as they interpret data in a report. His primary focus is on how managers and consultants turn an ambiguous situation into a problem statement and declare a particular course of action to be rational. He has approached this in a variety of ways, including symbolic interaction, metaphor, cause mapping, frame shifting, language games and exegesis. He is a Senior Research Associate at the Judge School of Business at the University of Cambridge, as well as a Visiting Fellow at Sidney Sussex College.

**Brigitte Borja de Mozota**, a pioneer in design management, received her PhD in design management from Université Panthéon Sorbonne in 1985. She is an author of several books and participates in international research networks and journals. She is an expert for international industry bodies, including the European Office for the Harmonization of Internal Markets (OAMI), Design Management Europe, and the International Council of Societies of Industrial Design (ICSID). She is currently Director of Research at Parsons Paris School of Art and Design and a professor at Université Paris Ouest. She teaches design management to MBA students at ESSEC and Audencia. She is a Design Management Institute Life Fellow.

**Margaret Bruce** is Professor of Design Management and Marketing at Manchester Business School, University of Manchester. She carries out research in design management, fashion and luxury marketing, social media and lifestyle consumption with current projects focused on consumption in a period of austerity and the implications for retailers and producers. Her interests in social media relate to the role of new media specialists in the design and communication process, brand loyalty and promotions using social media, and how young adults use social media. Best-selling books include *Design in Business* (2002) and *Fashion Marketing* (with Tony Hines, 2007). Professor Bruce is the Retail Chair of the Chartered Institute of Marketing (CIM) and a Senior Fellow of CIM and is a Trustee of Noise, a charity dedicated to the employment of young creatives. She holds Honorary Professorships at the University of the Arts London, ICN University of Nancy 2, France and the Xi'an Institute of Science and Technology, China.

**Ralph Bruder** was appointed Professor for Ergonomics in Design at the University of Essen in 1996. He founded the Institute for Ergonomics and Design Research at University of Essen in 2001. From 1998 until 2002 he was Dean of the Faculty for Design and Arts at University of Essen. In April 2002, he was named the first president of the Zollverein School of Management and Design. He has been President and Managing Director of the Zollverein School of Management and Design since 2003 until 2006. Since January 2006 he has been Professor at the TU Darmstadt in the Faculty of Mechanical Engineering and Director of the Institute of Ergonomics at the TU Darmstadt.

**Cai Jun** is Professor in the Department of Industrial Design, Academy of Arts and Design, Tsinghua University. He has helped to establish the leading position of the department in China with his colleagues since 1980s. Concentrating on research on design strategy and innovation development in industry, he was consultant to the establishment of Lenovo design and Beijing Industrial Design Centre (BIDC) in the 1990s. He has received both international and domestic design awards from Norway, Hong Kong and Mainland China, has published twenty papers on design management and collaborated with Nokia, LG, Herman Miller, GE, Lenovo, A. O. Smith and Motorola for research projects since 1999. He was the main organizer of the 2009 D2B – Tsinghua International Design Management Conference.

**Kyung-Won Chung** has been the Chief Design Officer/Deputy Mayor of Seoul Metropolitan Government during his leave of absence from the Department of Industrial Design at KAIST since 2009. He served as the President and CEO of the Korea Institute of Design Promotion (2000–3), an executive board member of ICSID (1995–9), and an advisory council member of DMI since 2002. He has been awarded the Silver Tower Industrial Medal (2003) and the Presidential Prize (1999) from the Korean Government. He received a doctorate from Manchester Metropolitan University (1989), MID from Syracuse University (1982), and MFA and BFA Degrees in Industrial Design from Seoul National University.

**Rachel Cooper** is Professor of Design Management at Lancaster University, where she is Chair of the Lancaster Institute for the Contemporary Arts and Co-Director of Imagination Lancaster, a centre for research into products, places and systems for the future. She has authored several books in the field of Design and more than 200 research papers, and she is currently commissioning editor for an Ashgate series on *Design for Social Responsibility*. She is editor of *The Design Journal* and President of the European Academy of Design; and was on the Arts and Humanities Research Council until 2010. She has undertaken several advisory roles to national and international universities, government and non-governmental organizations. Her research interests cover design management, design policy, new product development, design in the built environment, urban regeneration, design against crime and socially responsible design. Professor Cooper is currently co-investigator of 'Urban Futures', a five-year EPSRC-funded research project.

**Angela Dumas** consulted and wrote on design and management predominantly in the 1980s. She has been head of the Centre for Design Management at the London Business School, and Senior Associate at the Judge Institute of Management Studies at the University of Cambridge, where she led a research team dedicated to developing new techniques for managing innovation and concept design. She was formerly Research Director at the Design Council where, with a small team, she developed a new programme of research initiatives in design.

**David Dunne**, BComm (University College Dublin), PhD (University of Toronto) teaches at the Joseph L. Rotman School of Management, University of Toronto. His research focuses on how designers approach problems and the implications of this way of thinking for management education. He has published widely on this and other topics. He is a former marketing/advertising manager with Unilever and Young & Rubicam. He teaches design in collaboration with design schools in the US and Canada; his teaching has been recognized through numerous awards, including the University of Toronto President's Award and the prestigious 3M National Fellowship. He leads executive programs and acts as a consultant for major clients in North America, Asia and Europe.

**Gerard Endenburg** is founder and former director of the Sociocratic Centre Netherlands, located in Rotterdam. He is currently an honorary professor of organizational learning at Maastricht University. Gerard Endenburg was educated as an electrical engineer and holds a doctoral degree from the University of Twenty. He has pioneered the development of sociocratic circular organizing in Europe and elsewhere. Endenburg is also former CEO of Endenburg Elektrotechniek in Rotterdam.

**Martyn Evans** is Senior Lecturer in Design at Lancaster University. As a trained product designer, his research interests explore the approaches designers use to consider the future – in particular the ability of designers to envision potential social, cultural, technological and economic futures. He was co-investigator in 'Design 2020', an AHRC/EPSRC-funded research project looking at potential futures for the UK design industry. He has secured and supervised a number of government-funded knowledge transfer initiatives within the area of design and new product development. With extensive experience of leading undergraduate and postgraduate design curricula, he is currently Course Director for MA Design: Management and Policy and external examiner at a number of UK institutions.

**Michael Farr** became editor of *Design* magazine in 1952 and went on to become Chief Information Officer at the Council of Industrial Design (COID). He wrote *Design in British Industry* (1955), which criticized the division between designer and manufacturer, and also an early book on *Design Management* (1966), often considered as the first comprehensive literature on the subject. He taught design management in the UK and Hong Kong.

**Gorm Gabrielsen** is Professor of mathematics and statistics at Copenhagen Business School. He has conducted research in forensic psychiatry, criminology and design. He has taught statistics at all levels and also participated in quantitative and methodology courses for PhD and master studies. His role in research is bringing statistical methods and mathematics to conceptual research work. He has conducted numerous experimental studies in a variety of fields, including district psychiatry and murders, co-branding, country of origin. In this context Gabrielsen has developed new ways of measuring cognitive and emotional elements that are partly unconscious and contradictive.

**Peter Gorb** graduated from Cambridge University with an MA in history, and from Harvard University with an MBA in business studies. He was Associate Professor in Design Management at London Business School, a subject he pioneered, and also ran the Institute of Small Business there. He has produced a number of publications on design management and education reform, as well as two anthologies of poetry. Within business, he has been deputy chief executive of two large UK public companies: Northgate Group and the Burton Group. A Fellow and former Council Member of the Royal Society of Arts, Peter founded its national Education For Capability movement, a major influence on the development of vocational education in schools. He also holds The Society's Bicentenary Medal for 1988.

**David Hands** is based within the Faculty of Arts, Media and Design, Staffordshire University, UK. He is a senior lecturer in design and innovation and also Award Leader for the MA Design Management programme. His main areas of research interest include design auditing; design against crime and design briefing. He has written extensively on design management with his recent book *Design Management: Vision and Values* (2009) investigating the many different facets of design management in both theory and practice.

**Bill Hollins** undertakes consultancy for Direction Consultants (www.directionconsultants.co.uk), which he started with his wife in 1985. He has a doctorate in design management from Strathclyde University and now specializes in service design. He has been actively involved in British Standards on design management and the UK Design Council website on Service Design. His first book on service design management was published in 1991 and his fifth book was published in 2006. He has over 120 other publications. Bill also teaches at various universities and has worked in twenty-five countries. He is a qualified referee and can often be found yelling on the terraces of teams in the lower divisions.

**Ulla Johansson** is the Torsten and Wanja Söderberg Professor of Design Management and Director of the Business and Design Lab, a cooperation between the School of Design and Crafts and the School of Business, Economics and Law at Gothenburg University, Sweden. Her research interests cover a variety of areas: gender, methodology, irony and organizations, critical management and design management. In design management she is specifically interested in the relationships between the designer and other professions and how the designer seems to be somewhat of a catalyst for change in many organizations. She has published six books and a number of articles. She is actively involved in the Swedish and Scandinavian networks for design and design management research and is a reviewer for the European Academy of Design. She has supervised doctoral students in both design and management.

**Frans Joziasse** holds an MBA in design management from the University of Westminster (London) and lectures/teaches at several universities in Europe and the US and at the Conferences of the Design Management Institute on strategic design management issues. Having graduated as an industrial design engineer from Delft University in 1986, he founded his own industrial design consultancy in Rotterdam and subsequently he co-founded PARK, an international design management consultancy, in 1998. Joziasse has been cited for numerous awards for design excellence by the Gute Industrie Form in Hanover (Germany). He has research interests in organizational change, innovation within multinational organizations, creative networks, design and corporate strategy and local/global consumer behaviour/trends.

**Sabine Junginger** explores the links and relationships between the human activities and tasks of designing, changing, organizing and managing. She connects human-centred design theories and methods with theories and methods in organizational change, product development, design management and systems design. She develops methods and tools with the aim of inviting, engaging and enabling managers in public and private organizations to develop their organizations' design capabilities and to innovate products and services around people. Her publications include several book chapters on this topic as well as journal articles in academic design journals (*Design Issues* and *The Swedish Design Journal*) and in business journals (*Journal of Business Strategy* and *Design Management Review*). Sabine was among the first two people who were awarded a PhD in design by Carnegie Mellon University. Her dissertation looked at the role of design in changing public organizations.

**Yu-jin Kim** has been Assistant Professor of the Department of Media Image Art and Technology at Kongju University in Korea since 2008. As a member of evaluation committee of Web Award Korea, her research interests are design management, Internet service design, digital contents and interface/interaction design, and others. She was involved in many design strategy projects for leading corporations including L.L. Bean, American Greetings, Nokia, Sprint and others during her stay in the Design

Continuum in Boston as a design strategist in 2005. She has received degrees of BSID (1999), MSID (2001) and a doctorate (2008) from KAIST.

**Lucy Kimbell** is a researcher and designer. From 2005 to 2010 she was Clark Fellow in Design Leadership at Saïd Business School, University of Oxford, where she taught an MBA elective in design leadership and researched designing for service. For over fifteen years Lucy has been involved in the design of software and public services. In her art practice she creates installations and performances that blur the boundaries between social science, art and design.

**Philip Kotler** is the S. C. Johnson Distinguished Professor of International Marketing at the J. L. Kellogg Graduate School of Management, Northwestern University. He received his Master's Degree at the University of Chicago and his PhD Degree at MIT, both in economics. He carried out postdoctoral work in mathematics at Harvard University and in behavioural science at the University of Chicago. Professor Kotler is the author of *Marketing Management: Analysis, Planning, Implementation and Control*, the most widely used marketing book in graduate business schools worldwide. He has also authored or co-authored dozens of other leading books on marketing. He has published over a hundred articles in leading journals, several of which have received best article awards. Apart from his teaching and writing Kotler has worked as a consultant to many of America's top companies, and has travelled extensively throughout Europe, Asia and South America, advising and lecturing to many companies about how to apply sound economic and marketing science principles to increase their competitiveness. He has also advised governments on how to develop and position the skill sets and resources of their companies for global competition.

**Tore Kristensen** is Professor of strategic design at Copenhagen Business School. He has conducted studies in economics of design and the use of physical space both at work and in the home. Consumers' transformative experiences due to arts, science exposure, food and music are another active area of research. His research concerns co-branding and country of origin. Design experiments may enable insights into various aspects of how individual users of design experience, such as cognitive, affective and economic effects. Such effects are often ill understood and respondents are often unable to tell about their experiences. Kristensen teaches strategic design, consumer experiences and communication management.

**Thomas Lockwood** is the President and member of the Board of DMI, the Design Management Institute, a non-profit educational organization based in Boston, Massachusetts. He is one of the few people in the world with a PhD in design management and is recognized as a thought leader in the area of integrating design and innovation into business, and building great internal design organizations. Lockwood is a design advisor to corporations and to countries, a visiting professor at Pratt Institute in

New York City, a frequent design award juror, and has lectured and led brand and design workshops in over twenty countries.

**Angela Meyer** is a business consultant, designer and educator, with experience taking design into organizational contexts and making design accessible and effective for business and government clients. Angela holds a masters degree in design from Carnegie Mellon's School of Design. She has a strong interest in the intersection of design and management and is currently a Visiting Executive at the Darden School of Business at the University of Virginia. She was previously a consultant at 2nd Road, in Sydney, Australia, where she worked with large organizations in the areas of design transformation, strategy design, and service design.

**Mark Oakley** taught Management at Aston Business School. He was one of the early researchers who worked in the Managing Design Initiative in the 1980s and edited the first *Handbook of Design Management* (1990). His research focused on product design and management. Oakley saw the link between design practices and the role of design in the organization as key to design's ability to contribute to an organization's business success. He published significant papers on design management education (1988) and was the authority on design management during the 1980s.

**James Pilditch** (1929–95) was a design consultant and writer who played a significant role in promoting the importance of design in business. He wrote numerous books including *The Silent Salesman* (1961), *The Business of Product Design* (1966), *Talk about Design* (1976) and *Winning Ways* (1987), many of which have become standard works in higher education. In 1959 he founded Package Design Associates (later Allied International Designers) and steered the company into becoming one of the first integrated marketing consultancies. He was also active in many organizations including The Design Council, the Royal Society of Arts and the London Business School. He was honoured by professional bodies in Britain and the United States and in 1983 he was appointed CBE for his contribution to design.

**Mike Press** is Professor of Design Policy at Duncan of Jordanstone College of Art and Design, University of Dundee, where is he also Associate Dean of Design. His books include *The Design Agenda* and *The Design Experience* (both co-authored with Rachel Cooper), which have been translated into a number of languages and are established texts in the field of design management. Research interests and publications also include digital processes in craft design, and the role of design in social innovation.

**G. Alexander Rath** was Principal of Tah & Associates in Chicago when he co-authored *Silent Design* with Philip Kotler in 1984.

**Davide Ravasi** is Associate Professor of Management at Bocconi University. His research interests include the management of design and designers and the interrelations of organizational identity, image and culture. He is associate editor of the *Journal* *of Management Studies*. His works have appeared in the *Academy of Management Journal, Organization Science, Journal of Management Studies, Journal of Business Venturing, Industrial and Corporate Change and Long Range Planning*, among others.

**A. Georges L. Romme** is Professor of Entrepreneurship and Innovation at Eindhoven University of Technology (TU/e). He is currently also Dean of the Industrial Engineering and Innovation Sciences Department of the TU/e. Georges Romme obtained a master degree in Economics from Tilburg University and a doctoral degree in Business Administration from Maastricht University. His current research addresses design methodology, innovation, high-tech entrepreneurship and organizational learning. Georges Romme also serves as a senior editor for the journal *Organization Studies*.

**Uday Salunkhe**, Group Director of Welingkar Institute of Management Development & Research (We School) with campuses in Mumbai and Bengaluru, (India), is an engineer with a management degree in 'operations' and has pursued his doctorate in 'turnaround strategies for sick companies'. He has extensive experience both in academia and the corporate world. His passion for innovation and a design thinking approach powered by his leadership has impelled him to pioneer programs in business design and innovation, e-business, retail, healthcare and legal management in India. He has also initiated linkages with foreign universities and innovation firms. He is actively associated with universities and educational authorities at the national and SAARC level to bring about reforms in management education.

**Daniela Sangiorgi**'s research focuses on service design. In her most recent work she focuses on the applications and implications of 'personalization' and 'participation' in the design and supply of services. In doing so she is interested in working with and within service organizations to explore resistance to and capacity for change, with a particular focus on the education and health sectors. She is interested in the emergence of new service models and in the practice and theory of service innovation.

**Ileana Stigliani** is a research associate with Design London and the Innovation and Entrepreneurship Group of Imperial College Business School. She holds a PhD in Business Administration and Management from Bocconi University, Milan. Her dissertation focused on the role of artefacts and aesthetics in product design and was based on a ten-month ethnographic research in a world-leading design consulting firm located in Boston, MA. Her research interests include the emergence of service design as a design discipline, the role of artefacts in designers' creative processes, designers' aesthetic knowledge and knowing, and the collaborations between business firms and design consulting firms.

**Lisbeth Svengren Holm**, professor in fashion and design management at the Swedish School of Textiles (THS), University of Borås, and head of the Fashion Market Studio at THS. She is part of research projects with a focus on design and innovation, design and strategic development, and communication. She is also a guest researcher at Lund University Department of Design Sciences. Prior to this she was at Stockholm University School of Business, head of the Market Academy as well as responsible for research and education at Swedish Industrial Design Foundation. She has published several articles within the field of design management and a recent book about the interface between design, technology and marketing, together with Ulla Johansson.

**Paolo Tombesi** studies challenges of design management in large complex buildings and holds the Chair in Construction at the University of Melbourne, Australia. A former Fulbright Fellow, he has a PhD in architectural practice and regional development from UCLA. He has lectured at various universities in Europe and North America, has been on the board of the *Journal of Architectural Education*, *UME* and *Construction Management and Economics* and has worked as a consultant for state and federal governments as well as the Royal Australian Institute of Architects on research and heritage issues. In 2005, he was one of the recipients of the Institute's research award, the Sisalation Prize. The resulting book, *Looking Ahead: Defining the Terms of a Sustainable Architectural Profession*, was published in 2007.

**Alan Topalian**, Principal of Alto Design Management, has researched the management of design since 1976. 'Firsts' from this research program included the analysis of design leadership, board-level responsibility for design, and what constitutes corporate design management; the teaching of designers and managers together in the classroom; and detailed corporate design management case studies. When Visiting Professor of Design Management at Middlesex University Business School, he introduced pioneering mandatory modules into all its MBA programs. The Design Leadership Forum that Alan established in 2002 brings together business executives and design professionals to explore issues and experience, particularly relating to leading through design.

**Surya Vanka** is Principal Manager of User Experience at Microsoft Corporation, and oversees best practices and engineering standards to create high-quality user experiences for Microsoft's customers. He has worked as designer and manager on several products during his more than ten years at Microsoft. His mission is to put the users rather than technology at the centre of the development process for all of Microsoft's products. Surya was associate professor of design at University of Illinois at Urbana-Champaign and was a Fellow at the prestigious Center for Advanced Study. He is the author of two books on design, has lectured on design in over twenty countries, is a frequent keynote speaker worldwide and is widely published. His work has appeared in numerous publications including a Design Council UK global design study, *Form*, *ID Magazine*, *WIRED*, *Interactions*, BBC Radio, National Public Radio, and Channel 15 Television. He regularly teaches design, research and innovation courses in Asia, Europe, and the Americas.

**Bettina von Stamm** is hugely passionate about understanding and enabling innovation, particularly in large organizations. To follow this passion she has set up the Innovation Leadership Forum (ILF, www.innovationleadershipforum.org). The six legs on which she stands are teaching, researching, running a networking group, writing, workshops and seminars, and public speaking. She also very much enjoys the role of a 'catalyst' in large organizations, speeding up the understanding and creation of innovative organizations. In terms of academic background, she has a first degree in architecture and town planning and an MBA as well as a PhD from London Business School. In addition to numerous articles she has published two books, one for general managers and those who want to start on their 'understanding innovation' journey – *The Innovation Wave* (2002) – and the text book *Managing Innovation Design and Creativity* (second edition, 2008); the latter is also available in simplified Chinese and shortly also in Arabic. She has published a further book, *The Future of Innovation*, which was conceived together with her Russian colleague Dr Anna Trifilova.

**Jennifer Whyte** studies the organizational challenges of design innovation in the delivery of large projects and programmes, such as Heathrow Terminal 5, the London 2010 Olympics or the Building Schools for the Future Programme. Her work has delivered new indicators for design quality, techniques for studying visual practices, and technologies and methods for digital engineering. She is a Reader in Innovation and Design in the School of Construction Management and Engineering at the University of Reading, UK; an Advanced Institute of Management (AIM) Fellow and Director of the Design Innovation Research Centre. She has published a book and more than twenty journal articles, including articles in *Design Studies, Building Research and Information* and *Organization Studies*. She has also advised to the UK government on design and creativity in the firm.

**Alex Williams** is actively engaged in design management research, supervises a range of international design research students and has led a range of related projects, including an AHRC-funded 'Design 2020' project; an RCUK-funded Design Research Summer School held recently in Shanghai, a British Council Prime Minister's Initiative project addressing design student employability in China and an EU-funded Asia Link programme, which developed international curriculum in design management. He instigated and is current chair of the 'Design 2 Business' International Design Management Conference Committee.

**Jill Woodilla** PhD is a visiting professor at the School of Business, Economics and Law, University of Gothenburg. Formerly she was Associate Dean and Associate Professor of Management at the Welch College of Business, Sacred Heart University, USA. Jill's ironic perspective provides her with a critical view of the multiple realities of any situation. Her current scholarly interests include the theoretical underpinnings of design management, varieties of organizational discourse, and innovative pedagogy for teaching design management. She pays particular attention to ways in which people in management find it difficult to understand fully the design process and designerly way of thinking. She enjoys collaborating with others in the research process and helping emerging scholars as they navigate the writing and publication process.

# ACKNOWLEDGEMENTS

We were inspired to take on the task of producing this book by the rise in the number of colleagues whose work has implications for the practice, interpretation and theory of design management. There are, of course, too many to fit within one volume. We can thus only offer a glimpse into a vibrant scholarly field and want to point out that, of course, there are many more works and researchers whom we wish we could have included in this edition. We are grateful for their continued efforts to advance the education and practices of design management.

We would like to acknowledge in particular the support of the community of authors who have contributed to this volume. We appreciate their patience with us as editors and their diligence in dealing with our editorial queries and requests. We would also like to mention our consultant editors, Richard Buchanan, Richard Boland and Kyung-won Chung, whose work and thinking has informed the development of this book. A special note of thanks goes to the widow of James Pilditch, Anne Pilditch, who kindly granted us permission to reprint the chapter 'Into a Changing World'. We are equally indebted to the lawyers of Tiffany & Co. and to the University of Pennsylvania Press, who worked with us to include the chapter by Sir Misha Black. We are further grateful for the support of staff at the University of Brighton's Design Archives, who maintain the vads.ac.uk archives. This remains a treasure for design researchers.

Finally, we could not have delivered this manuscript without the guiding support and encouragement of Tristan Palmer at Berg and most importantly, Nicky Sarjent, our support at Lancaster, who dedicated her time to helping us produce this final manuscript. A project like this takes a toll and one big thank you goes out to our families who supported us throughout this task. Thank you.

*Rachel Cooper, Sabine Junginger and Thomas Lockwood*

# FOREWORD

This comprehensive addition to the literature on design management provides an overview of the subject since the early 1970s and a review of the present-day state of the discipline.

My own interest in the contribution that design can make to the management world began in the 1960s and became my main preoccupation when I moved from a management job to teach at London Business School. During my years there I identified a number of key issues that needed to be understood if the chasm of misunderstanding between the design and the management worlds was to be bridged. They can be classified as follows:

- The practical contributions that design brings to management.
- The financial measurement of those contributions.
- The clarification of some myths about design.

The first of those practical contributions is visual literacy. To be able to see and reproduce what you see is a design-honed skill without which no manager can operate effectively.

The second is a care and concern for products. Designers are often obsessed with the products with which they work. Managers, all too often, have little feeling for them.

The third, and most important is a way of thinking about problems that gives precedent to finding out *how* over finding out *why*. It is a route to knowledge favoured by designers who like to get things done. Too many managers, influenced by academia, opt for the opposite route, which favours understanding over action.

Measuring the financial contribution of the design function to business performance is vital if design is to be appreciated. Product design can be measured as part of the gross margin figure in the profit-and-loss account of the enterprise. Furthermore it can usually be shown to have a higher financial contribution than either the marketing or production functions of the business.

Environmental design costs are measured as fixed assets on the balance sheet. The cost of communication design is found in the list of expenses that determine net profit. All designers need to learn some financial accounting in order to demonstrate their importance to the management world.

And so to deal with some misconceptions about design:

- Don't leave design to designers! The more managers are involved in design the more they will appreciate its value.
- Design can't be seen! *Design is a planning process for artefacts;* it is not the artefacts themselves. This is in fact the definition of design. And whilst we are about definitions: *design management is the management of that process.*
- Design is not (or at least not always) a creative process and although designers are usually creative people, they should never deny creativity to others. Accountants, for example, are very creative!

Finally, some long-term advice to young and ambitious designers. Leave your profession! Accountants learnt years ago to swap professions as a way to get on in the world, so when the opportunity occurs for a management job, take it. Your design skills will help to equip you to manage. There are many examples of designers succeeding in top management jobs.

These are exciting times for design management. Recently, I attended in London the largest and most international conference that the Design Management Institute has run. Bill Hannon who founded the DMI, and I, its first Honorary Fellow, were presented to the conference. We used to be known as the godfathers and later the grandfathers of design management. Perhaps archaeological specimens is now a more apt description in a profession that, as this new book demonstrates, has as strong a past as it has a future.

*Peter Gorb*

# General Introduction: Design Management – A Reflection

RACHEL COOPER AND SABINE JUNGINGER

Design is anarchic; it doesn't fit easily a structural model.

*Roger Guilfoyle*

Design. Management. Management of design. Design and management. Putting two words together has generated forty years of lively debate. Researchers and practitioners have tried to specify what is good about design (or what is 'good' design), why design is important, how we can manage design and why it should be managed to begin with. Resistance towards the idea that design can be managed persists in some quarters but new word combinations are emerging and they shift the meaning of this relationship yet again.

In this introduction we trace some of the drivers and the perspectives of protagonists who influenced design management in the contexts in which they have operated. Contexts in design management are as important as in any other field of research and practice. They allow us to situate the questions that are being posed and the answers that are provided. Perhaps most obviously, there exists no generic history of design management. Attempts at historical timelines of design management have only recently been undertaken (cf. Best 2006; open-sources such as Wikipedia 2010). All the current genealogies of design management show a heavy bias towards developments in the UK, the US and continental Europe. This is not least because the origins of design management continue to be mostly examined in the context of the Industrial Revolution, which triggered new means of production and new ways of managing people, processes and materials. Is it an accident that much of the work on this book took place in Manchester, UK, the cradle of the Industrial Revolution?

We, too, find ourselves succumbing to the overwhelming abundance of historic materials accessible from these Western, mostly capitalist and certainly developed countries. We do this in the consciousness that many stories of design management are waiting to be told. Our contributors from India, China and Korea provide further evidence of the need to broaden the cultural and geographical boundaries of design management studies. Furthermore, if we understand designing and managing to be interconnected activities, the history of design management will have to include such achievements as the building of the pyramids (cf. Isler 2001) or the operations of the Arsenal of Venice (cf. Stovall 2010). While we call on all researchers to fill these gaps quickly, we begin with the resources we have access to and these predominantly originate from Europe, the UK and the US. Our story of the origin of design management therefore begins with developments in Europe.

## THE ORIGINS OF DESIGN MANAGEMENT

The origins of design management have their roots long before Fredrick Taylor established his scientific principles of management. In terms of combining the activities of designing and managing in one individual, Josiah Wedgewood, an English master of pottery, stands out. Born in 1730, he married the beauty and the art of designing ceramic pottery with a set of management skills to form a world-renowned business. When he died, in 1795, he provided a first case study for management and design (cf. Falk 1961). The beauty of products was one of the original drivers for designers to engage with management and business. 'Nothing Need Be Ugly' was thus the slogan and the title for 'A proposal for the foundation of the Design and Industries Association' developed in 1915. This proposal subsequently led to the formation of the Design and Industries Association (DIA) in the UK.[1] Pointedly, the proposal was distributed among businessmen who had just visited the special exhibition of German goods of successful design at the Goldsmith's Hall in London (DIA 2010). The proposal pointed to the necessity of the various stakeholders in industrial production to work together: 'What is needed at the present time is the gathering together of all the several interests concerned with industrial production into a closer association; an association of manufacturers, designers, distributors, economists and critics.'

The initiators of the UK Design and Industries Association were astutely aware of the simultaneously evolving Arts and Craft Movement led by William Morris, the association states on its webpage today (DIA 2010): 'They all felt the need for both industry and designers to be motivated, not merely by the pursuit of commercial profit but by some sense of social purpose and responsibility to provide consumers (a term not yet coined) with well-designed and well-made products.'

The DIA thus pursued aims similar to those of the German Werkbund, which was founded by a group around German architect Hermann Muthesius in 1907. Ironically, Muthesius' design knowledge and understanding came from 'studying all things related to design' during his seven year long stay in England from 1896 to 1903 (Schneider 2000). Designer Peter Behrens was also a founding member of the German Werkbund, he began to work as a consultant for the company AEG in 1907. His role in shaping the overall corporate identity, buildings, posters and actual products of AEG make Behrens the archetypical design manager of the early twentieth century. This new design manager had senior level influence on the company he worked with. His position was that of a consultant, rather than that of an employee. Although the Bauhaus also straddled issues of design and management, the next major developments occurred in the 1950s when design consultancies entered the picture. The UK saw the rise of consultants, such as Terence Conran, who formed his design consultancy in 1955, soon to be surrounded by now familiar names in the consultancy industry worldwide, including Wally Olins and Michael Wolf and Michael Peters. In the US, Raymond Loewy, Henry Dreyfuss and Walter Dorwin Teague and Paul Rand made an impact (Cooper and Press 1995: 30).

The 1950s also saw the rise of the in-house designer. Several individual designers had found ways to engage with senior management and were in a position to shape the visual corporate identities of large and influential companies like Braun, where Dieter Rams developed a distinctive product aesthetic and design from which he generated ten principles of design that continue to be used by design and management today. Their individual work pointed to the importance of design not only to industries but to individual organizations as well. For instance the iconic CBS 'eye' logo; created by designer William Golden in 1951 remains CBS's logo today and although it has undergone slight modifications and updates, the original design from 1951 is the visual icon of the CBS brand to this day (Doordan 2000). These early design managers aided the companies they were part of, or their clients, in gaining visibility and recognition through distinct visual appearances and through their aesthetic product form language.

The first time, interest in the relationships between design and business took centre stage in the US came about through the vision of Walter Paepcke, an American industrial philanthropist with German roots who had made his fortune through packaging. Two thousand people attended his first Aspen event in 1949, 'Goethe's Bicentennial', for which he convinced philosopher Albert Schweitzer to hold the convocation (AIGA 2009). Athough not yet honed in on the issues of design and management specifically, this changed two years later when Paepcke established the International Design Conference in Aspen (IDCA). Paepcke saw the purpose of the IDCA as being to bring together designers, artists, engineers, business and industry leaders. In June 1951, 250 participants from around the world convened for the first time on the theory and practice of design. The title of this first IDCA conference, 'Design as a Function of Management,' highlighted a new understanding of the relationship between designers and managers and attracted designers, business and industry leaders. However, the term design manager as such did not exist during these days and the IDCA in

1951 can be considered a first explicit attempt at organizing these individual efforts at design management and working towards an infrastructure that is conducive to these activities.

The reasons for this renewed interest in design management by 1950, in the UK, the US and also in Germany could be explained by a renewed industriousness after the Second World War, which stimulated production and eagerly sought to develop new markets. But something else seemed to have happened in the war years. Facing a shortage of traditional manufacturing materials because, for example, metals and steel were reserved for war production, the US government turned to designers to explore new ways of use for materials that could substitute the scarce materials. Arthur Pulos (1988) documents how this kind of experimentation led to glass dishes and the now famous glass coffee jug without which today's coffee makers would be incomplete. Discovery and invention for the common good brought design and designers a new recognition and pointed to different roles of design in the organizational context. Rather than delivering a specific product, designers were encouraged to generate new possibilities for materials and products. Their results would foster new industries, establish new companies and create badly needed jobs while delivering new and innovative consumption goods. Postwar design in Germany and in the UK had a very different role in rebuilding large devastated areas. Moreover, in Germany, designers turned their energies to envisioning and proposing possible ways of living after many had lost all: friends and family, house and bed, religion and belief.

Design management as a term, concept and description of a discrete design activity emerged in 1964, when the Royal Society of Arts in the UK reached out to a wider audience through its 1964 Presidential Prize for Design Management. Not surprisingly, design management soon started to seep into the consciousness of both academics and practitioners. 'Design' then was understood to apply specifically to the production of products, industrial goods, graphic items and interior designs. In the meantime, the architectural profession used 'design' as a verb to describe parts of their activity. Essentially in the UK and in the US, this separation reverberates through today's educational practice: design management continues to be applied to architecture and indeed in project management in construction, yet the wider concerns of design management as a scholarly discipline and professional practice were adopted more quickly by manufacturing-related design disciplines as they came to appreciate the business dimensions in which they operated.

More important, perhaps, is the fact that design management interested only few management scholars through the 1960s. Most were still grappling with the emergence of marketing, strategy and new product development. The topic of design management and its problems found articulation in design journals, in design magazines and at specific conferences.

## DESIGN MANAGEMENT DISCOURSE: IMPORTANT JOURNALS AND MAGAZINES; INFLUENTIAL CONFERENCES AND SEMINAL BOOKS

In the UK, the discourse around design management began in earnest during the 1960s. However in terms of publications, there were no scholarly journals specifically aimed at design management in the 1960s and indeed no significant scholarly journals were dedicated to design. The *Design* magazine had a wide readership and often fulfilled the role of informed commentary on design and industry. For instance in the February issue of *Design* in 1965, an article by Maurice Jay appeared under the theme of design management titled 'Furniture: autocracy versus organisation.' Jay argues that:

> The furniture industry has certain features that make it deserving of separate consideration in any study of design/management relationships. First, the products of the industry have virtually no mechanical component and their design, therefore, can be wholly the responsibility of one man. While designs must be worked out with the production engineer or works manager, there is no need for compromise between industrial designer and engineering designer and the former – often an amateur – is frequently, by tradition, the managing director himself. (Jay 1965)

Jay goes on to discuss issues of manufacturing scale, or the cost of distribution, the challenge of fashion and the need for maintaining control through family businesses rather than turning themselves into public companies. Which of course, many companies did with the emergence of the company Habitat led by Designer Conran, or later IKEA. This is still a privately owned organization.

In the June issue that same year, Dennis Cheetham writes under the title 'Design management: four views on design decision-making' (Cheetham 1965). Cheetham based his article on interviews with executives from the four firms that had won medals in the Royal Society of Arts Presidential Medals for Design Management. The winners included Terence Conran of Conran, Geoffrey Gilbert managing director of Jaeger and Co., Leslie Julius of S. Hille & Co Ltd and Jasper Grinling, managing director and design director of W. & A. Gilbey Ltd. The theme of the articles was one of the need for 'design detail' to run through everything. In what might be interpreted and labelled as design thinking today, Cheetham quotes Leslie Julius: 'Once you get involved, you find your designer's way of thinking and looking at things rubs off on yourself. There's a sense of conviction running through things – now it runs through the whole staff from top to bottom.'

The theme of design leadership is revealed in Cheetham's interview with Jasper Grinling, who also saw power as the essential ingredient of design policy: 'I report to the board of directors on all matters relating to Gilbey design. The design director has got to be on the board, he must be trusted by his board, and of course he must bear in mind all the time sound commercial principles.'

The articles that appeared in *Design* magazine in 1965 tended, as Michael Farr recollects in an article that year, 'to have concentrated largely on reporting the ways in which design is being treated as a management problem in specific industries or in individual firms' (Farr 1965). In this article Farr provides a first formal definition of design management: 'Design management is the function of defining a design problem, finding the most suitable designer, and making it possible for him to solve it on time and within a budget. This is a consciously managed exercise which can apply to all the areas where designers work.' He goes on to explain why the designer should not be tasked with management and other management functions should focus on their core skill; therefore there is a need for a design manager, where design is core to the business then that person should be on the board and 'where companies need occasional but intense design activity then a consultant design manager' is necessary. Farr and other writers of his time are mostly concerned with the competitive value of design and the need to manage the design resource more effectively.

In a 1969 paper, Bruce Archer reflected on the demands of 'Design and Management for the 70s' (Archer 1969). He presented a version of this to the Society of Industrial Artists and Designers (in the UK). He put his focus on 'industrial design in relation to industrial management' but much of the paper is taken up by the similarities of concern between the designer and the manager but the crossover is not recognized. He identifies several great challenges for industrial management, including the cost of labour, the scale of manufacturing occurring overseas, the need for British industry to increase scale and runs, the challenge of technology and the lack of entrepreneurs and innovators in the UK. In a prophetic paragraph he describes many of the struggles that continued over the following forty years and, indeed, which some companies are still grappling with today:

> If managements are going to achieve better vertical integration of the innovative effort then there will certainly have to be much more management involvement on the part of designers than has been normal in the past. That is to say, – designers must become more directly and more deeply involved in problems of investment, problems of the marketplace, and problems of value, price and cost as well as with the problems of function, pro auction and aesthetics with which they are already familiar. And since management itself is becoming so much more skilled in operational research – that is to say, in the use of mathematical techniques in decision making – and since management itself is placing so much more reliance on data and data processing, then the designer will have to become sufficiently knowledgeable in these subjects and sufficiently skilled in these arts to be able to conduct effective dialogues with managements and technologists. If there is going to be a greater emphasis

on high technology, then, of necessity, designing is going to become more and more a multi-disciplinary affair. It will be necessary for physicists and industrial designers and statisticians and psychologists to work in single teams getting out of high technology new ideas which will be launched as finished products and which will work and be acceptable in a short time. Will the designer be fit to face the world or the 1970s? Can he adapt himself to the highly professional, multidisciplinary team: to the vertical integration of industrial innovation? Is he willing and able to provide the link – the calculable link – between absolute value, marginal value, price, premium and cost without losing control of those intangible, cultural values of which he must remain the guardian? If the answer were No, he might well find himself displaced by the machinery of consumer research, his memory relegated to the pages of the comic books alongside the ghost of the old-time inventor. But I don't think he will be. The design profession sees progress as its business.

During the 1970s, *Design* magazine published less on design management *per se* and focused much more on the broader application and use of design, in relation to new products or new materials, innovation and technology. Writers who had contributed to the early series, like Michael Farr and James Lilith, went on to influence young academics with their book publications in the field.

In the US the approach to and interpretations of design management were very similar, according to J. Roger Guilfoyle, the former editor of the US magazine *Industrial Design* (*ID*). In a recent email discussion, Guilfoyle explained that 'over time, we did numerous articles in the general subject area of design management' (email message to Rachel Cooper, 19 August 2010). The magazine reported on Beckman Instruments, CBS Laboratories, Herman Miller, IBM, Steelcase, Westinghouse, Braun, Brionvega, Danese and 'a myriad of others'. Interestingly, the focus of many of the early articles was on 'product planning'. Later, the shoe was on the other foot and the focus shifted to 'design management'. The publications on product planning range from March 1956 to the beginning of 1967. These early authors include Don Wallace ('Shaping American Products: Design and Craftsmanship in Large Industries', March 1956), Jane Fisk Mitararchi, editor at the time ('How Industry Uses Design', October 1956), Richard Latham, Robert Tyler and George Jensen ('The Process of Product Planning', October 1956), Hugh B. Johnston ('IBM System of Design Coordination', March 1957), Maude Dorr ('Case Studies: How American Corporations Use Design Abroad', February 1963), who was then one of the staff associate editors, and finally Robert Malone, editor-in-chief ('Product Planning in Corporations', May 1963). In 1967 there was a shift towards design management, which ran as a theme through 1971. Authors here included E. W. Seay ('Design Management 2: Westinghouse', 1967), William J. Hannon, then editor-in-chief ('Design Management', April 1969) and J. Roger Guilfoyle ('Design Management', September 1971).

In his email, Guilfoyle (email message to Rachel Cooper, 19 August 2010) recalls that

the thrust of most articles in *Industrial Design* during the twenty year period from 1954 to 1975 . . . was on design consultants. The evolution of industrial design in America, which began with a handful of individuals in the twenties – prominently, Bel Geddes, Deskey, Dreyfuss, Loewy, Teague – was configured into large consultancies after the war. *Industrial Design* magazine's thrust was to support and report on the profession. That profession then was essentially the independent consultancy. Certain companies, such as Herman Miller, whose products were 'design,' grew differently, but relied heavily on outside designers. And although Teague, father and son, designed cars as Loewy famously did, the auto industry had strong inside styling departments . . . When Dreyfuss left his partnership in 1967, he became an 'über design consultant' advising the top executives of his four major clients (American Airlines, AT&T, Bankers Trust, Deere) on which designers to hire for what projects. The argument had already been joined. J.C. Penney and other mass marketers of products had internal design departments, which supervised the implementation of the brand. Some innovation took place. Often, it was modifying something generic. These internal design departments did not always enjoy the same access to the top executive level as consultants traditionally had. Design management guided these internal departments whose effectiveness was measured primarily by strenuous cost analysis. The measure was Profit and this trumped Design effectiveness in the same way Design Management superseded Product Planning. It wasn't just semantics, but an essential philosophical divergence.

Interestingly, in Guilfoyle's (1971) paper on design management, he described the work of the Centre for Advanced Research in Design. This was an autonomous subsidiary of Container Corporation of America (sponsored by Walter Paepcke). This centre, as design director John Massey said, offered clients a total communication service, plus marketing, research and development facilities. Massey also described how the Centre turned into an 'almost integral part of the client's operation. This enables us . . . to anticipate solutions rather than react to a problem' (Guilfoyle 2010).

Magazines like *ID* and *Design* provided critical insights into the world of design, the practice of designers and their use by industry. It was not until 1979 that *Design Studies*, the first scholarly journal in design, began. Its focus was on engineering, architecture and design theory. However, an early paper on engineering design management in French industry appeared in 1983 (Pitts 1983). But it was not until Peter Gorb published 'The Business of Design Management' in 1986 (Gorb 1986) that the focus shifted to the wider notion of design management. In this piece, Gorb discusses the increasing recognition of the importance of teaching design to managers, discussing the main blocks to this, one of which is the confusion of what design management is as a subject area. He goes on to explain the multiple meanings of design management under five headings: design office management, educating designers for management, educating managers for design, design project management and design management organization.

The *Design Management Journal* appeared in Fall 1989. Over the next twenty years it remained one of the few design-management-specific resources for design managers, related professionals, and a small circle of academics. The *Design Management Journal* is striving for a balance between offering practical industry insights and reporting on academic research. In his keynote article Tom Peters (1989) refers to design as 'only secondary about pretty lumpy objects and primarily about a whole approach to doing business, serving customers and providing value'. As a well known management consultant Peters was able to point to the relevance that design has to the issues of the day, listening to the customer, customization, quality and perception, linking design to each. However he does indicate one of the main barriers to the uptake of design management amongst MBAs:

> We all know that everybody who graduated with an MBA won't go to the bathroom unless they can find a measurement. Well, don't complain; invent a measurement, invent a value-added measurement. Invent a perception measurement. Designers have to translate their concerns into the facts and figures that are part of the decision-making process. And it can be done.

Peters emphasizes the need for passionate leadership 'because design is about obsession'. In what may seem surprising today, he (in 1989) already refers to Steve Jobs at Apple and Akio Morita at Sony.

This first issue of the *Design Management Journal* is notable for its renowned contributors. Aside from management expert Tom Peters, we find management scholar Henry Mintzberg, industrialists Paul Cook from Raychem and Herbert Kohler representing the Kohler Company. Both the healthcare and the service industries are present with Angela Emrick from Mount Carmel Health and Tim Cook, the Service Design Strategist for Cook Company. In addition, several marketing scholars, including Arvand Phatack and Rajan Chandron from Temple University and Roland Krapfel from the University of Maryland contributed essays. Leading academic strategists Kim Clark and Takahiro Fujimoto from the Harvard Business School highlight the importance of Design Management by adding an article of their own. These familiar and respected names instantly established the *DM Journal* as a serious publication and also gave credence to the topics surrounding design management.

Indeed, the first few years of the journal have significant gravitas and much of the thinking still espoused today appears in the early issues of the journal. Robert Hayes (1990) of Harvard Business School, for instance, provided a keynote article on 'Design has become the missing link in US competitiveness.' He distinguishes between design as a facilitator; design as a differentiator; design as an integrator, and design as a communicator. His principle argument was that organizations need good design but good executives and more importantly good teams and co-operation between the many disciplines involved.

As one of the side effects of the efforts to attract a broader audience to the *DM Journal* and to engage more closely with the practice sector, the level of the contributions could not keep up with that of the first issues. In its stead, a proliferation of practice took form, providing insights from practitioners who often do not have the time to reflect in a deeper way. Ten years ago, in 2000 the Design Management Institute sought to remedy this situation by relaunching the *Design Management Journal* as a refereed academic journal to develop the theories in the field whilst a sister publication, the *Design Management Review* was founded to retain the practitioner insight. However, the *DM Journal* still struggles with the demands of an academic publication. In contrast, the *Design Management Review* enjoys popularity among practitioners. A review of the articles from 1(1) (Fall 1989) to 21(4) (Fall 2010) illustrates the way design management has engaged with management issues of the day, from corporate identity, customer focus, marketing, strategy in the early 1990s through user centred design, innovation and creativity, brand strategy in the late 1990s to a deeper engagement with managing design in the organization, measuring value, developing frameworks for design strategy to transforming organizations through design. Since mid-2006, more articles have concerned themselves with design-led policies and their implications for design management. The editorial introduction to Volume 18, Number 4 in Fall 2007 (DMI 2007) addresses this shift in focus:

> Design can make an impressive contribution to business success; the challenge is fitting it into the big picture. Contributors to this issue offer insights on establishing a design-focused culture and stimulating innovation in general, as well as on marketing individual product and service innovations. They articulate the unique wisdom designers bring to problem solving. They spell out metrics for calculating the impact and value of design. They critique the roles of design and technology as catalysts in the creative process, and they illustrate important lessons in case studies as such companies as Microsoft, Dell, and KitchenAid.

In summary, the dominant contributions to the *Design Management Review* continue to be written from a practitioner's perspective. They cover the salient topics in day-to-day design in the context of a demanding and fast-paced business world. With that, matters of branding, social media and lowering environmental impact of products remain popular. Also evident in these articles is a continuous search for the measure and metrics of design that we can use to assess its contribution to business, society and the environment in general or to social media and brands in particular. But the changes in the role of design in the organization are being noted as well. In a recent issue, for example, Eckersley and Alexis (2010) point to the rise of a new generation of business professionals who 'see the world differently from their predecessors, and who view the gap between business and design less as a problem and more as a curious anachronism.'

Today, papers that address the relationships between design and management appear in many design publications (cf. *Design Journal, Design Studies, Design Issues*). The term 'design management', however, is seldom explicitly applied. For many, the term has become both too narrow and specific (i.e., relating to project management) whereas others struggle with its many interpretations (cf. DMI 1998). On the other side of the spectrum, management theorists and practitioners continue to show interest and contribute to the body of literature periodically. Their works can be found in the *Journal of Business Strategy*; the *Journal of Product Innovation*; more recently in publications by the Academy of Management (cf. *The Academy of Management Journal; The Academy of Management Education and Learning*) and the influential *Harvard Business Review*. A remarkable number of non-design journals have released special issues on design matters over the past five years. These include, for example, *Organization Studies*, the *Journal of Applied Behavioural Sciences, Organization Science* and *the Journal of Business Strategy*.

## INFLUENTIAL CONFERENCES

Another set of influential texts that have shaped the discourse of design management have their roots in lectures and conferences. Often, the initiative for such events has come from business and management schools. The six Tiffany-Wharton Lectures on Corporate Design Management, sponsored by Tiffany & Co – yes, the same as in 'Breakfast at Tiffany's' – and held at the University of Pennsylvania in 1973, were collated into a book in 1975: *The Art of Design Management: Design in American Business* (Schutte 1975). Peter Gorb hosted *Design Talks* and *Design Management* at the London Business School in the 1980s and disseminated the contents of these two workshops through two namesake publications. Both have a special place in the classic literature on design management today. Following this pattern are recent conferences and workshops, for example the Convergence Workshop at the Weatherhead School of Management, part of Case Western Reserve University in Ohio.

The Design Management Institute organized the *Design Management Research and Education Conference* throughout the 1990s. The most recent conference took place in 2008 in Paris. Selections of these academic papers were published in the *Design Management Journal* but no official proceedings exist. Other international design conferences, for example those run by the Design Research Society, the European Academy of Design, place design management amongst a much broader design focus. Only relatively recently has a wider international perspective on the links between designing and managing, management and design generated a new interest in specific design management conferences among academics. Currently, several universities are working on concepts for academic design management conferences that fit the shifts and the changes in the field over the past decades. Likewise national and regional initiatives have developed networks. For instance, the UK Design Council has backed networks of design management academics to initiate workshops. The 2010 *Design for Social Business Workshop* in Milan is another example of this trend. Another development, long awaited and anticipated, is the emergence of events and conferences in Asia and other parts of the world. The *Design to Business* (D2B) conference in Shanghai has shown once more that there is growing awareness of this topic among Asian business managers and designers alike.

## SEMINAL TEXTS AND BOOKS

A major effort to gather the loose ends of the design management discourse was undertaken by Noon and Warner in 1988. Their annotated bibliography of product design management (Noon and Warner 1988) cites over 600 relevant papers, reports and books. The topics covered range from new product development, marketing, engineering and production, processes and team management. In this abundance of materials, it is worthwhile to focus on a few seminal works that address core issues of design management. Already, we have mentioned three books that have shaped the ideas and concepts of the relationships between design and management – *Design Talks* (Gorb and Schneider 1988); *Design Management* (Gorb 1990) and *The Art of Design Management: Design in American Business* (Schutte 1975). This is the place to describe their contributions. Our start, however, will be the work of earlier researchers.

**Michel Farr** wrote prolifically on design management and claimed that his company Michael Farr (Design Integration) Ltd was the first independent design management consultancy to operate in the UK and Europe. Farr's perspective was that design management was about defining the design need, sourcing the design expertise and managing the design process in the organization. He published *Design Management* in 1966 because he believed that there was a grossly inadequate number of experienced design managers in industry in the UK and that managers were 'muddle headed about design'. He set out a list of the delusions about design: (1) designing is easy, (2) designing is concerned with appearance only, (3) designers are artists alien to industry, (4) one designer can do all types of designing, and (5) design decisions can be made by everyone; to these he gave very sound and evidenced responses. In the following chapters, he elaborated on the reasons for designing, which for him included competition, distribution systems, market developments, social changes, invention, variety reduction and diversification. The remainder of this monograph sets out the tasks in design management from planning a design programme, selecting designers, developing the brief and managing the design process. All of which were illustrated with case studies. Much of this work by Farr has great resonance today and the guidelines are equally relevant. He argued that at that point design management as a conscious activity in industry was only just beginning; he argued for the design manager as an individual who can 'be a neutralist in every specialist's empire and yet a committed entrepreneur where the product was concerned'.

**James Pilditch** who set up Allied International Designers Limited (AID) in 1959 was another voice that championed design beyond the boundaries of the design consultancy. Like many of the writers and practitioners during the 1950s, 1960s and 1970s, his focus was very much around product or industrial design. In a book *The Business of Product Design* he wrote with Douglas Scott, a principal designer from AID, he argued that the pace of change was driving the need for design to work with manufacturers;

> . . . against a background of global communications , searchers into space and massive computation, many settled ideas about design are irrelevant. Throughout the western world industry now competes on a scale never seen before. In every industrial society factories face the need to sell the vast output made possible – and necessary – by modern technique. Manufacturers and designers alike must plan for these new conditions. Admittedly, the pace of change is hard to grasp. But already events have made it clear that new skills are needed to compete in the inevitable struggle for markets. (Pilditch and Scott, 1965)

Pilditch and Scott discuss mass manufacturing and the need for differentiation, through good organization and product planning, market research, ergonomics and production analysis, maintenance and design. They describe what industrial designers do and how to find them, and take the product from idea through to production. They also illustrate how design can exploit new materials and processes and cut costs.

Roughly ten years later, in 1976, Pilditch offered a broader perspective on design in *Talk about Design*. In a series of essays he argued that 'the tasks designers now are no longer peripheral, but central to corporate survival'. He suggested that designers can go into new countries and markets with marketing men to help them turn local resources into producing goods the wealthier countries can use. This was set very much in the context of internationalism, that there was no longer a need to be 'near a source of power, a source of raw material or near convenient communications' because we have new ways of creating energy and substituting primary materials and we can command communications anywhere. He argues that the hunger for knowledge would alter the designer's job beyond recognition especially in the face of innovation as a 'dynamic of a changing world'. He stated that the designer's role, 'increasingly, will be that of thinker, planner and co-ordinator'. That by 'de-specialisation, therefore, the abandonment of at least some of the traditional skills that the designer can fulfil his role in the future'. Pilditch rarely discussed design management as a role or activity. Yet in writing directly to business he was arguing for the value of design to industry, the need for it to be managed appropriately but also for designers to recognize their contribution.

This early work of Pilditch's touched on the designer as a big thinker, the relationship of design to marketing and corporate identity, to innovation and also to designing for people. Indeed, in 1969, he discussed the problems of the environment by looking at business, design and pollution. Pilditch's later work in 1987 *Winning Ways* was much more a book about management written by a designer but again arguing the case for design at the centre of the organization. Here again we were in a crisis in relation to the decline of manufacturing in the UK, and Pilditch uses the book to describe how to design and manufacturing products within the context of a well managed company. Here he espouses the need for creativity in everyone. He advises the use of designers because of their imagination but also to create an environment that 'stimulates people to develop and exercise their imagination' to produce creative teams.

The works by Farr, Pilditch and Scott, among others, are the context in which design management regained momentum in the mid-seventies, especially in the UK and the US. **Peter Gorb**, who was a friend of James Pilditch and therefore not only familiar with but also in a dialogue with Pilditch's views, produced two books based on the seminars at the Design Management Unit at London Business School. In *Design Talks* (1988), Peter Gorb and Eric Schneider present the first set of papers from the Design Management Seminars. In his own contribution, Gorb argues for a classification of design into Product design, environmental design and information design, (he later added corporate identity design). Such a classification, Gorb suggests, offers a way for organizations to make use of design and more importantly, to quantify it in business terms. Gorb also suggests that each cannot be managed in the same way – identity needs to be managed as a central resource, whilst product design, being 'operational in style and directly relevant to performance', is most likely to be managed by line managers. Information design, because it covers packaging and advertising, will follow the organizational pattern of business through the hands of brand managers, buyers. Finally environment design in terms of building would be handled by consultants and maintenance through facilities managers.

In the second set of essays, published under the title of *Design Management*, Gorb (1990) offers his own definition of Design Management:

> The effective deployment by line managers of the design resources available to an organisation in the pursuance of its corporate objectives. It is therefore directly concerned with the organisational place of design, with the identification of specific design disciplines which are relevant to the resolution of key management issues, and with the training of managers to use design effectively.

In Gorb's view, once managers had learned to appreciate design then deploying design as an effective resource could be left in their hands. He also indicated that the design profession had not yet dismantled its frontiers enough to move from designing into managing as other disciplines, such as computing or marketing, had done. And in relation to marketing the aspect of consumer focus and non-price factors was emerging as a design-related factor through the later talks. Gorb's work had great influence in management and in design. For him, design 'pervades every aspect of business world and contributes significantly to business profitably at every level.' He also understood that designing and managing are closely related activities with similar methodologies: 'The methodology of the designer is a better model for use by managers than the methodology of the scientist or the scholar and the manager needs to learn how to use it' (Gorb 1992).

**Alan Topalian** was another early advocate of design management, working from 1971 as a design consultant. In his seminal text *The Management of Design Projects*, Topalian (1980) provides a framework for the management of design, the work was prompted by a survey of managers and designers. The work focused on how to manage design for managers. Topalian identified, as others have, that there are different types of design and design projects and therefore there should be different approaches to managing them. He also introduced the notion of the design audit, the dimensions of the design process and the design project team. He differentiates between the design project management role and that of design responsibility – 'the design manager's principal role is to organize; the design responsible's principal role is to provide leadership. Of course the roles of design manager and design responsible are not mutually exclusive' (Topalian 1980: 83). He suggested in 1980 that the responsibility for any companies design activity can never be passed to an outsider and that ultimate design responsibility always rests with the chief executive. Interestingly he also suggested that unlike the role of design responsible, the role of design manager need not always be held by a company employee. This text is perhaps the first most very practical guide for design managers, in which Topalian outlines quite precisely a framework (Topalian 1980: 88–89, Table 6.2) of tasks for corporate design managers.

In 1975, an important book resulted from the Tiffany-Wharton Lectures at Wharton School at University of Pennsylvania. **Thomas Schutte** (1975) in his introduction to *The Art of Design Management* stated 'Even in the most progressive schools of management, design sensitivity or 'the way things look' rarely finds its way into the classroom'. Something perhaps a little naïve to say in relation to view on the value of design being espoused at the time, which was a little deeper than aesthetics, and even in relation to the backer of the programme, Walter Hoving, Chairman of Tiffany & Co, who thought the challenge was foreign competition outdesigning American business. Schutte also believed that there were developments such as the increasing specialization of the American corporation into finance, operations, marketing with little room for design education, the lack of design awareness in large corporations, the lack of innovation in management education and the concern for efficiency. Hoving (Foreword in Schutte 1975) was convinced that 'if a new corporate conscience towards the role and implementation of good design and aesthetics were to be generated within the American business community, it had to begin with a leading business school'.

The marketing dimension entered into design management literature in the late 1970s and early 1980s. **Christopher Lorenz**, at that time management editor with the *Financial Times,* became known as an authority on design and strategy. His seminal book *The Design Dimension* links to the work of marketing scholars Levitt and Kotler. He employs case studies to illustrate the relationship between design marketing and technology. Lorenz (1986) discusses the globalization of products and markets, referencing the success of Japanese management. He suggests that this should provide extra stimulus for the emergence of product design but warns that if globalization is managed badly 'Design would be pushed back to the dark-ages of skin-deep styling.' He saw, alongside the necessary relationship between marketing and design for 'imaginative marketing and product strategy', that emerging technologies in products and processes provided a further set of opportunities. The value of designers for Lorenz was not only their fundamental skills but their ability to synthesize or cross-pollinate from one field to another closely associated with their abilities of imagination and visualization. Lorenz's core argument was that the design dimension was no longer an optional part of marketing and corporate strategy but should be at the very heart.

**Mark Oakley** considered the field from the perspective of product design and saw a lack of ability to manage design within companies and an absence of designers from the board level. A management academic, Oakley whose text *Managing Product Design* (1984) described what design was, and provided a much-used spiral model of the product design process that included the formulation stage, evolution stage, transfer stage and reaction stage, which he used to illustrate problems that arise in practice, Oakley covered issues relating to the development of design policy, audits of design, planning products, organizing and locating design in the organization. He also went into considerable depth in relation to generating briefs and specifications. Oakley's perspective was very much from the product, production perspective and whereas he recognized the threats to industry these were related to external competition, legal and political trends, diminishing resources and changes of attitude, fashion and trends, however where he saw the major changes to the role of the designer was the growing complexity and the development of cross national teams and the increasing role of CAD. Oakley saw the pivotal for designers as a pivotal role in forming the bridge between the market and the design and the production engineers.

In parallel to these arguments there was a body of work related to developing the professional practice of designers, reflected in the rise of books addressing issues of how to run a design business. Seminal works on this topic include *The Professional Practice of Design* by Dorothy Goslett in 1961 (one of the few women contributing to design management at that time), *The Business of Design* by Ian Linton in 1988 and *Design Protection* by Dan Johnston in 1989. These were targeted towards design students and practising designers, to enable them to engage more professionally with their clients, to communicate on a similar if not equal footing.

A watershed occurred in 1990 with the publication of Mark Oakley's *Design Management: A Handbook of Issues and Methods*. Here we have quite clearly moved on from texts by managers and designers, articulating the value of design to managers, or the value of management to designers, to papers trying to develop some theoretical insight into what design management was, addressing issues and methods of design management from a corporate and business perspective oriented towards developing goods for markets and appearances for corporations, 'the achievement of good results from design'. This first handbook of design management revealed a group of active scholars from which grow a mature discipline and formed the basis for a raft of further publications on design management and related topics. In his introduction to the book, Oakley talked openly about the difficulties of putting together an edited volume on design management. It seems that many people took note and shied away from following in his footsteps. Instead, a series of single and co-authored books had international influence and have since shaped discussions and the research in the field. Among these count two co-authored publications by **Rachel Cooper** and **Mike Press**. *The Design Agenda – A Guide to Successful Design Management* was written as an introduction to design management in 1995. This book provides an overview of different design disciplines and highlights the importance of incorporating design in all aspects of corporate life. In contrast, *The Design Experience*, released in 2003, looked at the changing role of design in response to changes in consumer behaviour and a rising focus on customer experience. *The Design Experience* sought to shift the focus from managing businesses and projects or honing in on specific design processes towards design management activities that keep the experience of people in mind.

Another major book in this era is *Design Management: Using Design to Build Brand Value and Corporate Innovation* by **Brigitte Borja de Mozota**. Published in 2003, it demonstrates how management theories and models can relate to the business of design. It has become a mandatory reading for students interested in issues of marketing and design strategies. Borja de Mozota is also credited with setting up one of the first doctoral programmes in design management.

The most designerly presentation on the topic was put forth by **Kathryn Best** in 2006. *Design Management* has quickly gained popularity among practitioners and students in the field. Complemented with many colourful and visually attractive images and diagrams, Best's book has filled the need for a publication on design management that is easily accessible and serves to familiarize the novice with key concepts, basic historical developments and examples from design practice.

In addition to these internationally known works, is also important to point to a wide range of research that took place in the Nordic countries throughout the 1990s. Finland and Sweden especially maintained an interest in design management issues, methods and practices, which are documented in a series of smaller print publications. We defer to the chapter by Ulla Johansson and Jill Woodilla in Part IV, which offers an overview of the history and developments of design management in Scandinavia.

In conclusion to this brief review we can say that the discourse on design management as it took place and takes place in journals, magazines and conferences has a scattered, unstructured and disorganized history, yet the issues remain of interest to practitioners, that is to design professionals, and business managers. For instance, a range of national, regional and local professional conferences are often held around brand, innovation or product development, although few of these are in a position to advance our insights into designing and managing because they frequently take place away from any ongoing academic discourse.

On the academic side, the discourse on design management is picking up again, as core issues of design and management are being re-examined and explored in light of new design and management concepts. Design management as such has steadily woven its way in and out of mainstream design and management publications and conferences. One of the ways in which it has tried to remain relevant is by picking up emergent themes of the day. This has allowed design management to capture the imagination of business leaders for a moment but it has also meant that it neglected its own research into design and almost abandoned its roots. A lot of energy was spent on fitting design into the management paradigms and with aligning design processes with those that were established and accepted in management. We will take a closer look at how design management sought to fit existing management paradigms and how it is now moving into managing change. In this context, it will be interesting to see the role of education in design management.

## DESIGN MANAGEMENT EDUCATION

In the UK, support for the development of design management education came from an initiative in the 1980s by Prime Minister Margaret Thatcher. She held a seminar in 1982 at 10 Downing Street on product design and market success, attended by representatives of the design industry, education and government. Thatcher explained her aim 'to tap the thoughts of successful men and women who are committed to the idea that good design is the cornerstone of successful business' (Thatcher 1982). The resulting **Managing Design Initiative** had two objectives. Firstly, it aimed to improve the emphasis on design education at all levels and secondly, it sought to increase top managements' awareness of good design.

Mark Oakley, at that time, saw the pivotal role for designers as forming the bridge between an organization and the market, the design and the production engineers, and thus became one of the lead researchers. He undertook a study on the role of design in industrial performance. The findings reflected the breadth of the topic as it had emerged over the century. He identified eight relevant areas of research and education. These stretched from 'design in an economic and business context' to 'the nature of design work', from 'design and product strategies' to 'design policy making', from 'researching design and product requirements' to 'managing design projects' and finally from 'elements of design work' to 'evaluating design work and legal aspects of design work' (CNAA 1984). As part of this research project, Oakley was charged with developing curriculum content for design management and related courses.

Following the Managing Design Initiative, the **Council for National Academic Awards** (CNAA) went on to fund three initiatives dedicated to the exploration of design management. The first initiative supported the development of postgraduate programmes on managing design. The second initiative funded an evaluative study in collaboration with the Fulbright Commission by north American academics and industrialists. The third initiative came later and led to undergraduate programmes in managing design and programs for in-company training. These activities flourished while the CNAA funding was available but very few programmes continued after their original funding had run out. Nonetheless, these three initiatives did produce a new generation of design management academics in the UK. As a consequence, new design management programmes continued to be developed at undergraduate and postgraduate level. In response to the Fulbright review findings, which pointed to a lack of sufficient resources for design management education, further research and publications sought to put these resources in place.

The CNAA initiative fostered several new courses and program in design management in the UK. One of these was situated at the Royal College of Art, then as now considered to be one of the leading design education institutions in the world. Naomi Gornick had successfully set up a postgraduate programme in design management, which required students to spend a significant amount of time working with design managers in the industry. A number of those graduates have since gone on to very successful careers with companies such as BA, BAA, etc. Nevertheless in the early 1990s, the RCA felt it was no longer a relevant programme for them. Gornick moved on to Brunel University where she founded and directed the MA Design, Strategy and Innovation Programme but several other universities shared RCA's view on design management. The London Business School, Westminster University and Oxford University, for example, all silently withdrew from design management education. It is worth asking what prevents design management building a sustainable legacy that could thrive in either management or design schools – and survive long after funding money has dried up.

In light of the problems of design education in higher education, the activities of professional associations moved into the foreground. The American **Design Management Institute** emerged as a new driver of matters relating to design management. A singular person was the catalyst for this: William (Bill) Hannon. Hannon was an industrial designer who had experienced corporate challenges before he created his own role as corporate design manager in the 1950s (Tobin 2010). Throughout the 1960s, he ran Boston's largest design consultancy and held a chair in the Division of Design. In 1976, the Massachusetts College of Art launched the Design Management Institute following a successful conference for corporate design managers. The Design Management Institute soon became an independent organization in its own right. Peter

Lawrence led the DMI as its first president until 1984, followed by Earl Powell. Thomas Lockwood, the president to 2011, followed in the footsteps of Earl Powell. The DMI presents the biggest independent professional organization dedicated to design management in the world. It continues to offer workshops and professional development seminars to designers and managers.

Another initiative, 'The Competitive Edge, The Role of Design in American Corporations' was undertaken in the early 1980s and led by Colin Clipson, the Director of the Architecture and Planning Research Laboratory at the University of Michigan. This was perhaps one of the most influential projects worldwide in terms of raising awareness for design management. It produced four reports: A *Business Design Index* – a database of books and periodicals and other materials on the role of design in business; The *Business/Design Issues* – a report on key business/design issues supported by research out by the competitive Edge Team; *The Business/Design Cases* – a series of detailed case studies on companies such as Citicorp, and Hewlett-Packard; and lastly *Business/Design Promotion* – a report on an international survey of design promotion organizations (Clipson 1985). This significant body of work certainly stimulated the debate that ensued throughout the 1980s and 1990s in terms of what is design management.

The lack of resources for Design Management Education was noted by Earl Powell at DMI in the mid-1980s and so in collaboration with Harvard Business School, DMI set out to commission and develop fourteen detailed case studies, five from Europe (Bacho, Braun, ERCO, Nautech and Philips), five in Japan (Canon, CKD Corporation, Sharp, Sony, and Yamaha); and four from the US (Black & Decker, Dictaphone, Digital Equipment Corporation and Texas Instruments). Research was undertaken by DMI staff, led by Karen Freeze Director of Research at DMI and included others, such as Christopher Lorenz and Lisbeth Svengren. In his introduction to the work Robert H Hayes, Professor of Management at Harvard Business School refers to an explosion of industrial competition around the world and states 'New ways of doing things, of dealing with employees, making products, and serving customers are being experimented with and adopted. Putting these new approaches into practice often requires a whole new management mindset. Now where is the impact of changing the traditional mind-set more evident than in the area of design and particularly the management of design.' The resulting cases were exhibited under the title *Designing for Product Success* in six cities in Europe and the US and were use as Harvard teaching cases (Triad Design Project 1989).

## PRACTICES AND METHODS IDENTIFIED WITH DESIGN MANAGEMENT

One of the key objectives of design management as a field of practice and research was to provide methods to bring design to the consciousness of management. Initially, it was thought that design could achieve this by speaking the language of management. Management at that time meant decision making, controlling processes, containing cost and ensuring profit. Of the techniques and practices now closely linked with the practices and methods of design management, the influence of this management interpretation are evident. These practices include stage gate processes for product development; standardizations and best practice models; design audits and models for evaluating an organization's design competencies.

*Stage-gate processes.* One of the most obvious places to align design practices with management concepts was the area of product development. In this context, stage-gate processes offered ways to articulate and regulate the design activities in a development project (cf. Cooper and Kleinsmidt 1979). Stage gate processes were seen as formal processes that could be learned and applied to new product development (Griffin 1997).

*Standardizations.* Another effort sought the standardization of design management practice. One further initiative that deserves to be mentioned was the development in the UK of the British Standard in Design Management Systems, initiated by Topalian, this resulted in a series of guides (British Standards Institution 1994, 1996. 2001, 2005, 2008a, 2008b, 2008c). These standards arose out of practice rather than theory, and filled the gap in practice rather than scholarly theoretical reflection and the development of testable models. They are operation guidelines and don't push the boundaries of the theory or indeed test the suggested approaches to managing design.

*Design audits*: frequently mentioned in general design management rhetoric but never articulated transparently, one of the first attempts to describe an approach to Design Audits was in Cooper and Press (1995). Whereby the audit must be undertaken of the physical outcomes of design as well as the organizational attitudes to design, illustrating that audits can be both qualitative and quantitative. This has been followed by research and publications focused on specific contexts such as the work of Fleetwood (2005) who illustrates in his work in New Zealand that New Zealand businesses must become design enabled and design audits by research can provide a knowledge base for building such businesses. In addition, Moultrie et al. (2006) looked at design audits for SMEs. James Moultrie has gone on to develop a specialism in the measurement of design capability at both at firm and national level. Although measurement of design has hitherto received little critical attention and many attempts to do so have been criticized by those who believe it is unquantifiable.

*Models for evaluation.* The tools developed here include the Design Ladder by the Danish Design Center (Ramlau and Melander 2004) and the Design Management Staircase by Design Management Europe (Kootstra 2009). The Design Ladder assigns a company to one of four levels of design maturity based on their attitude towards design. The Design Management Staircase, developed and used by Design Management Europe (DME) as a self-assessment tool for organizations allows an organization to

rank itself from one to four along two dimensions: First, on the place it assigns design management (project level; functional level or across organization) and second,
on how the organization uses design management. This second dimension allows for a
deeper organizational analysis of design management: is it used as a process (level 1);
does the organization have design management expertise (level 2); are resources made
available to design management (level 3) and is the organization overall aware of the
benefits of design management (level 4)? The design management staircase focuses on
how an organization values and uses design management across the organization for
traditional design problems (product development, branding, corporate identity, corporate design). Like the Design Ladder, the Design Management Staircase emerged
as a tool in the context of traditional product development – that is, product design.
But both models point to the different roles, functions, methods and practices of design managers in situations where they have to make a case for design and in situations
where an organization perceives its corporate strategy and its organizational design as
a product in development.

## DESIGN MANAGEMENT: DESIGNING, CHANGING, ORGANIZING AND MANAGING

Looking at the discourse, the resulting practices and methods thus far, we observe
confusion about the boundaries of design management, how it addresses problems
of the organization and what purpose it fulfils within the field of design itself. The
language and the arguments put forth range from discussions of the value of design
in general, the role of design in the organization, and the location of design within
the organization, to efforts at defining design management and setting boundaries of
practice by identifying methods and by assigning tasks. Yet, what seems to be confusing and at times schizophrenic can be explained in strikingly simple ways. Design
is undertaken by an individual or groups and usually takes place in the context of
some form of organization, whether that be a private or public company, a community or voluntary group. The nature and purpose of one organization may be different from that of another. As a result, issues of organizing and managing are part of
the activities of designing. We argue that early contributors like James Pilditch were
acutely aware of this. This is one reason why *Talk about Design* (1976) remains one
of the most profound books on design, business and management. Throughout this
book, Pilditch talks about 'design,' not 'design management' as a comprehensive activity that is closely linked with the activities of managing and organizing. While he
later addresses design management specifically, in this early book, he does not feel a
need to create a new subcategory for design and its role in the organization. This puts
Pilditch close to the work of Roger Falk (1961) who, fifteen years prior, highlighted
design as part of the 'art of management' in *The Business of Management*. Falk considered management as 'a subject, which, at the very heart of its nature, strikes at the

heart of social behaviour' (Falk 1961:12). Falk therefore provides another bridge into design – that of change.

As in the 1960s and 1970s, our time appears to be full of upheavals, uncertainties – and change. Change and design, as Herbert Simon (1969) observed, go hand in hand. People who engage in designing always find themselves involved with aspects and issues of change. Yet, what exactly changes, can be changed, should be changed or could be changed depends on the kind of design approach people apply as much as on the very reasons, hopes and aspirations for that particular design outcome. Changing, in turn, is an activity that cannot be accomplished without organizing or reorganizing people, tasks, structures and resources in such a way that they align with these reasons, so that the potential hopes and aspirations have a chance to reach fulfilment. All of these activities require various levels and forms of management. But if the activities of designing, changing, organizing and managing are deeply interconnected, then there is another side to design management that Pilditch, Falk and Oakley foresaw and that is now gaining currency in management and business schools. That side looks to the transferability of design concepts, practices and methods to managerial and organizational problems as well as to the possibilities design practice holds for strategy and policy formulation and implementation. We are back at discovering management as a subject at the very heart of social behaviour. Today, we find that designing and managing are less the polar opposites they are often made out to be. Instead, we find that managing organizations involves uncertainties and incomplete information and poses challenges akin to 'wicked problems' – a category of problems that defy traditional decision-making and problem-solving approaches.

With this in mind, we can organize the discourse of design management around three paradigms: The first paradigm concerns the concepts and ideas of design practice, how it is to be done and what it should be concerned with. The second paradigm refers to notions surrounding methods of design management, the characteristics of a manager and the matters that can be managed. The third paradigm comprises principles and practices, concepts, methods and matters of design as a general human undertaking. We might describe it as the paradigm of design as general capability. Much of the discourse on design management from 1960 to the beginning of the second millennium documents a struggle with the first two paradigms, professional design and management.

On the one hand, the field found its *raison d'être* in explaining the value of professional design to an organization and thereby making the case for employing and utilizing designers. The arguments the field put forth concerned the attractiveness and the need for beautiful products, product innovation and product differentiation in the market. In parallel, design management concerned itself with developing methods and practices to structure the design process, introduce predictability and cost control. This emphasized skills in writing a design brief and managing the design and delivery process. It focused on the role of design in solving market-related design problems. Design problems in this time frame constituted those kinds of problems that concerned products, brands and services in the commercial context. As a consequence, the tasks for design management aimed at resolving those issues within an organization, which interfered with the work of design professionals. In that sense, design management was interested in all aspects of an organization but only insofar as it directly affected the way in which they touched on the objectives and resources of specific design projects. The objectives of individual design projects were to create beautiful and functional products and services that could form a brand and delivered a good return on investment to the bottom line. Questions centred on how design contributes to product development and innovation, to corporate communication and brand and, more recently, to services in both the public and private sector. For design management, this meant to scope and manage each design project to existing strategic goals of an organization. To achieve this, design management called for design advocates in top and senior management positions and to be installed on company boards. Design competency from the perspective of the design profession was to be present in top management, on company boards, within individual design leaders (either in-house or external design consultants) as well as within the design teams and as part of cross-disciplinary teams.

Whilst the concern expressed in relation to design management was how to deliver this design contribution effectively, whether it was through more design-aware managers, through design consultants who could influence senior managers or through a new profession of design management. There was also a debate as to whether design management is an internal or external function. The way in which organizations respond to this argument for design and design management in an organization is often related to the business sector they operated in, their size and their maturity. As Walker (1990:43) argues as an organization matures it becomes more ambitious in its use of design, it uses a wider variety of specialism, spends more on design and therefore needs to manage it more effectively. Table I.1 summarizes some of these points of contention.

However, the landscape and the context in which companies operate have shifted since the early 1990s. This has introduced a third paradigm to the discourse of design management. While design will doubtlessly maintain its authority in influencing a product's appearance through styling and in orchestrating the context in which it is sold or the manner in which it is packaged, design management today is a challenge and an opportunity for all organizations – even for those that do not manufacture and sell goods or services. We have seen the shift in the sector towards looking at design management in relation to services and public policy. Indeed, many businesses are currently honing in on those characteristics they share with non-profit organizations. This shift away from the manufactured good and the reflection on some of the competitive models has helped to 'unpack' design in the organization – that is to free it from the limiting realms of specific design specialties and to be valued as a general organizational capability.

**TABLE I.1: The two principal paradigms shaping the discourse of design management from 1960 to 2000**

| Function | Design Practice Paradigm | Design Management Paradigm |
| --- | --- | --- |
| Adds value through … | Aesthetics, product innovation, differentiation, | Interpreting the need, writing the brief, selecting the designer, managing the design and delivery process |
| Solves problems of design relating to … | Products, brands, services | All aspect of design in the organization, but principally products, brands and service |
| Develops and fosters design competency among … | Top management, board members, design leaders, design consultants, design team, cross disciplinary teams | Top management, board members, senior management, design management consultants |
| Achieves objectives of … | Designing products and services that are beautiful, functional, create a brand and make a profit for the organization | Managing design to deliver strategic goals |

Oakley (1990: 7) observed that designers are often concerned about issues beyond the product, such as for example, 'a desire to improve the environment, to elevate the public taste in art and design – or even to help bring about social and political changes'. Since then, economic and social effects of globalization such as climate change and an increasing inability to deal with the garbage we produce, a projected shortage of resources, an aging population, etc. have put into question management models that purely aim for profitability. Even the gross domestic product curve (GDP), which has guided economic growth for the past decades, is now being challenged in its value as indicator for human progress. Increasingly, alternative indicators are held against it, which show that material and quantitative growth have not resulted in an increase in life quality for most people.

We now have a new generation in the management arena, picking up on these issues. Roger Martin (2009), Dean of the Rotman School of Management at the University of Toronto, has turned the attention of managers to the mental aspects of design work. Design thinking, he suggests, is one of the key success factors for businesses in the future:

the most successful businesses in years to come will balance analytical mastery and intuitive originality in a dynamic interplay that I call design thinking. Design thinking is a form of thought that enables movement along the knowledge funnel and firms that master it will gain a nearly inexhaustible, long-term business advantage. The advantage, which emerges from the design-thinking firms' unwavering focus on the creative design systems, will eventually extend to the wider world.

Not surprisingly, with the management arena showing an interest in design there arise new opportunities for design consultants to benefit from their insights. Many design consultants have since tried to clarify the potential of design thinking and their role in delivering design thinking to the organization. Tim Brown (2009) CEO and president of IDEO a global design consultancy, too, promotes the idea of design thinking as a tool for businesses. Brown's interpretation of design thinking reiterates Pilditch in 1976, who demanded a more comprehensive approach to design for the organization. In today's language, we would say that he calls for an integrated approach to product, communication, information and service design.

In a further push for a more intrinsic role of design within organizations, Richard Boland and Fred Collopy from the Weatherhead School of Management at Case Western Reserve University in Ohio explore the possibilities of *Managing as Designing* (Boland and Collopy 2004). Their premise is that managers should act not only as decision makers, but also as designers. They remind us that although decision making and designing are inextricably linked in management action, managers and scholars have concentrated mostly on decision making and neglected those aspects that concern design. Drawing on examples of decision-making and leadership in architecture, art, and design, they establish that 'the design attitude' applies to problems of management. In that same volume, Richard Buchanan (Boland and Collopy 2004: 54) suggests that:

> the popular understanding of design tends to reduce it to self-expressive artistic activity associated with the appearance of graphic communication, industrial products, interior spaces and buildings. The prospect of bringing this kind of design into the business of managing organizational life seems at best metaphoric and at worst frivolous.

Buchanan explains that this is not the way in which leading designers regard design. In his opinion, they see design as 'a deeply humanistic and intellectual activity that focuses on the creation of practical, effective products that serve human beings in all aspects of their lives.' We have therefore come full circle and are back to assess the value of design to people, societies, environments and organizations. It was, of course, James Pilditch who observed in 1976 that the issues of meaning and value are gaining weight and are introducing a new kind of rationality into probability calculations of success or failure. Design, he summarized, is all about people and it is people that support or abandon an organization.

The argument remains that environmental, societal, economic and political challenges are about how people and organizations approach matters of change. Because the skills and methods that constitute design are useful in responding to the challenges facing us today, designing is now being recognized as a general human capability. As such, it can be harnessed by organizations and apply to a wide range of organizational problems. We can summarize the key characteristics of this third paradigm, designing as a general capability, as follows:

- A focus on human problems.
- An advanced concept of the nature of a 'product' and of how 'things' come into being.
- A comprehensive and systematic design approach that allows for emergence.
- A wide range of inclusive and participatory and/or collaborative design methods.
- Situating designing in the contexts of changing, organizing and managing.

It is clear from these characteristics that most people could develop this capability. Design management, as a consequence of paradigmatic shifts in design and in management, is reorienting itself to fit new circumstances and new demands. Table I.2 shows how this new paradigm operates in relation to the previous two paradigms. The effects of this reorientation are already felt, as design management is regrouping into two overlapping strands of research and practice. On the one hand, we have design management with regard to creating products, places, communications, which involves managing the process, the people, and promoting the role of design in the creation of a physical contribution to an organization's strategic goal. On the other hand, we have design management with regard to creating an organizational capacity to adopt and use the design approach as part of an overall response to change and external challenges.

These strands can and do operate separately but as is clear from the chapters in this book (as well as from history) that they do weave and interact with each other. The existence of these two strands is one of the strongest indicators that design requires careful management. The design manager of the future will need to know how, when and where these two strands link up. Hands-on experience, theoretical knowledge, a grasp of the past and of the technology of today, will allow the design manager to address the challenges of tomorrow.

## THE HANDBOOK

It is clear from our introduction that design management serves as an anchor for those who wonder about the complex relationship of design and management and the relevance and consequences of these different facets in all design fields, including place, product and service design, strategy and policy design.

The timing for an updated *Handbook on Design Management* seems right. While a range of authors (cf. Topalian 1980; Wolf 1993; Cooper and Press 1995; Borja de Mozota 2003; Press and Cooper 2003; Kern and Kern 2005, 2009; Best 2006; Hands

**TABLE I.2: Three paradigms of design management in 2010**

| Function | Design Practice Paradigm | Design Management Paradigm | Design Capability Paradigm |
|---|---|---|---|
| Adds value through . . . | Aesthetics, product innovation, differentiation, | Interpreting the need, writing the brief, selecting the designer, managing the design and delivery process | Humanistic, comprehensive, integrative, visual approaches |
| Solves problems of design relating to . . . | Products, brands, services | All aspect of design in the organization, but principally products, brands and service | change in environment, society, economy, politics and organizations |
| Develops and fosters design competency among . . . | Top management, board members, design leaders, design consultants, design team, cross disciplinary teams | Top management, board members, senior management, design management consultants | Every area of the organization |
| Achieves objectives of . . . | Designing products and services that are beautiful, functional, create a brand and make a profit for the organization | Managing design to deliver strategic goals | Delivering sustainable organizations in the context of societal and global wellbeing |

2009) have pursued the topic and the problems of design management from various angles, there exists no current work that tries to connect these individual efforts. Various volumes on design management have attempted to frame the national discourse (for example, see Hase et al. 2006–7). These isolated efforts fall short of advancing our insights into the purpose, practice, theories, research and methods of design management today.

We are, of course, only too conscious of the precedents set by scholars already mentioned. There is the awareness that this handbook can hardly scratch the surface of a large number of works in our field. Nonetheless, by bringing together researchers and practitioners from China, India, Korea, Germany, Italy, the United States, Canada, France, the Netherlands, Sweden, Denmark and the UK, we are hopeful that the handbook is beginning to tell the new story of design management as it unfolds today.

The degree to which design management can hope for longevity and sustainability remains to be seen and, in fact, remains a question reflected on throughout this volume. To answer this question, we have gathered a selection of current thoughts and present understanding from leading researchers who span the boundaries of design management in one way or another. Today, design management constitutes a field of practice and research. Over twenty years of undergraduate and postgraduate design management programmes have contributed to this. A range of recognized professional design management consultancies have emerged. A professional design manager is now common in many private and increasingly in public organizations. And yet, as this volume demonstrates, significant work on design and management continues to be done by boundary spanners and people who are not part of the mainstream design-management discourse. Or are they? Where or what are the boundaries of design management? We hope that this handbook will contribute to this ongoing debate.

The handbook is divided into parts and, although there is no chronological or conceptual order, we would recommend that all readers who want to understand where design management is today start with Part I. This part reprints a selection of the influential papers published since the early 1960s on design management or, more specifically, design in relation to management. Reading this part will provide the genealogy for the subsequent parts; it also illustrates the design/design management dilemma discussed above, which continues throughout the remaining text. From Part I onward, the sections cover specific aspects/domains of design management and therefore can be approached in any order depending on your specific interest. For instance, Part II – 'New Educational Perspectives for Designer and Managers' – looks at the approaches to education both in the West and in emerging countries such as India. Part III, Design, management and the organization is the largest section; as such there are subdivisions. Obviously this is the area where much of the current research and interest remains, the subdivisions address how design is embedded in the organization, how it informs strategy or is informed by it and finally how we evaluate design and lead design. Part IV brings to the fore the changes in design management and alternative approaches being taken in the light of the current global challenges. This handbook, then, provides another line in the sand – an overview of the subject – and offers ideas for possible development and new research directions.

## NOTE

1. The initiators of the Design and Industries Association included Lord Aberconway, Chairman of the Metropolitan Railway, Kenneth Anderson of the Oriental Steam Navigation Company, Frank Brangwyn, the artist, Fred Burridge, Principal of the Central School, B. J. Fletcher, Principal of the Leicester School of Art, St John Hornby of W. H. Smith and *Imprint*, John Marshall of Marshall & Snelgrove, James Morton of Sundour Fabrics, Frank Pick of the London Underground, Gordon Selfridge, Frank Warner, the silk manufacturer and H. G. Wells. It also carried the signatures of the seven members of the working committee.

'Very quickly, the committee was invited to a meeting with Sir Hubert Llewellyn Smith and Sir Cecil Harcourt Smith, the permanent secretary of the Board of Trade and the director of the V & A' (DIA 2010).

# REFERENCES

AIGA (2009) 'Aspen Design Summit.' AIGA, www.aiga.org (accessed 8 October 2010).

Archer, B. (1969) 'Design and Management for the 70s', *Design Journal* 246: 66–7.

Best, K. (2006) *Design Management: Managing Design Strategy, Process and Implementation.* Lausanne: AVA Publishing.

Boland, R. and Collopy, F. (eds) (2004) *Managing as Designing.* Stanford, CA: Stanford University Press.

Borja de Mozota, B. (2003) *Design Management: Using Design to Build Brand Value and Corporate Innovation.* New York: Allworth Press.

British Standards Institution (1994) *BS 7000-3 Design Management Systems. Guide to Managing Service Design.* London: British Standards Institution.

British Standards Institution (1996) *BS 7000-4 Design Management Systems. Guide to Managing Design in Construction.* London: British Standards Institution.

British Standards Institution (2001) *BS 7000-5 Design Management Systems. Design Management Systems. Guide To Managing Obsolescence.* London: British Standards Institution.

British Standards Institution (2005) *BS 7000-6 Guide to Managing Inclusive Design.* London: British Standards Institution.

British Standards Institution (2008a) *BS 7000-1 Design Management Systems. Guide to Managing Innovation.* London: British Standards Institution.

British Standards Institution (2008b) *BS 7000-2 Design Management Systems. Guide to Managing the Design of Manufactured Products.* London: British Standards Institution.

British Standards Institution (2008c) *BS 7000-10 Design Management Systems. Vocabulary of Terms Used in Design Management.* London: British Standards Institution.

Brown, T. (2009) *Change by Design: How Design Thinking Transforms Organisations and Inspires Innovation.* New York: Harper Business.

Cheetham, D. (1965) 'Design Management: Four Views on Design Decision Making', *Design Journal* 198: 62–9.

Clipson, C. (1985) *The Role of Design in American Corporations: The Competitive Edge.* Washington DC: National Endowment for the Arts.

CNAA (1984) *Managing Design – An Initiative in Management Education* [Report]. Sponsored by the CNAA, DTI and the Design Council. London: CNAA.

Cooper R. and Kleinsmidt, E. (1979) 'Industrial Design in Canada: Its Role and Strategic Impact', *Design Canada*, April. Cited in: Noon, P. and Warner, T. (1988) *Product Design Management: an Annotated Bibliography.* Aldershot: Gower.

Cooper, R. and Press, M. (1995) *The Design Agenda – A Guide to Successful Design Management.* Chichester: Wiley & Sons.

DIA (2010) *Design and Industries Association*, www.dia.org.uk (accessed 8 October 2010).

DMI (Design Management Institute) (1998) '18 Views on the Definition of Design Management', *Design Management Journal* 9(3): 14–19.

DMI (Design Management Institute) (2007) 'Editorial.' *Design Management Review*, 18(4). Available at: www.dmi.org/dmi/html/publications/journal/articlelist_d.jsp?itemID=JNL-V18N4 (accessed 8 October 2010).

Doordan, D. (2000) 'Design at CBS.' In D. Doordan (ed.). *Design History: An Anthology.* Cambridge, MA: MIT Press.

Eckersley, M. and Alexis, J. (2010) 'Meet the New Hybrid Designers', *DMI Review* 21(2): 54–62.

Falk, R. (1961) *The Business of Management.* Harmondsworth: Penguin Books.

Farr, M. (1965) 'Design Management. Why is it Needed Now?' *Design Journal* 200: 38–9.

Farr, M. (1966) *Design Management.* London: Hodder & Stoughton Ltd.

Fleetwood, R. (2005) 'Design Audit by Research. Building a Knowledge Base for Competitiveness by Design.' *Joining Forces,* Conference held at the University of Art and Design Helsinki, September 22–24.

Gorb, P. (1986) 'The Business of Design Management', *Design Studies* 7(2): 106–10.

Gorb, P. (ed.) (1990) *Design Management: Papers from the London Business School.* London: Architecture Design and Technology Press.

Gorb, P. (1992) 'Design Management Education: A Personal Retrospective', *Design Management Journal* 3(3): 19–22.

Gorb, P. and Schneider, E. (eds) (1988) *Design Talks!* Aldershot: Ashgate Publishing.

Goslett, D. (1977) *The Professional Practice of Design.* London: Batsford.

Griffin, A. (1997) 'PDMA Research on New Product Development Practices: Updating Trends and Benchmarking Best Practices', *Journal of Product Innovation Management* 14(6): 429–58.

Guilfoyle, J. (1971) 'Design Management', *Industrial Design* 18(7): 60–5.

Hands, D. (2009) *Design Management – Vision and Values.* Worthing: AVA Publishing.

Hase, H., Hinz, K. and Schnackenberg, H. (eds) (2006–7) *Design Management*, Volumes 1–3. Berlin: International Design Zentrum.

Hayes, R. (1990) 'Design: Putting Class into "World Class"', *Design Management Journal* 1(2): 8–15.

Isler, M. (2001) *Sticks, Stones and Shadows: Building the Egyptian Pyramids.* Norman: University of Oklahoma Press.

Jay, M. (1965) 'Design Management: Furniture Autocracy versus Organisation', *Design Journal* 194: 49–54.

Johnston, D. (1989) *Design Protection: A Practical Guide to the Law on Plagiarism for Manufacturers and Designers.* London: The Design Council.

Johnston, H. (1957) 'IBM System of Design Coordination', *Industrial Design*, March.

Kern, U. and Kern, P. (2005) *Designmanagement – Die Kompetenzen der Kreativen.* Hildesheim: Olms Verlag.

Kern, U. and Kern, P. (2009) *Designplanung: Prozesse und Projekte des wissenschaftlich-gestalterischen Arbeitens.* Berlin: ProBusiness Verlag.

Kootstra, G. (2009) *The Incorporation of Design Management in Today's Business Practices; An Analysis of Design Management Practices in Europe.* Rotterdam: Design Management Europe (DME), ADMIRE Program.

Linton, I. (1988) *The Business of Design.* Wokingham: Van Nostrand Reinhold.

Lorenz, C. (1986) *The Design Dimension: The New Competitive Weapon for Business*. Oxford: Basil Blackwell.

Martin, R. (2009) *The Design of Business: Why Design Thinking is the Next Competitive Advantage*. Boston, MA: Harvard Business Press.

Moultrie, J., Clarkson, P. and Probert, D. (2006) 'A Tool to Evaluate Design Performance in SMEs', *International Journal of Productivity and Performance Measurement*, special edition on Performance in Design and Manufacture, 55(3/4).

Noon, P. and Warner, T. (1988) *Product Design Management: An Annotated Bibliography*. Aldershot: Avebury Gower.

Oakley, M. (1984) *Managing Product Design*. London: Weidenfeld & Nicolson.

Oakley, M. (1990) *Design Management: A Handbook of Issues and Methods*. Oxford: Basil Blackwell.

Peters, T. (1989) 'The Design Challenge', *Design Management Journal* 1(1): 8–14.

Pilditch, J. (1976) *Talk about Design*. London: Barrie & Jenkins.

Pilditch, J. (1987) *Winning Ways*. London: Harper & Row.

Pilditch, J. and Scott, D. (1965) *The Business of Product Design*. London: Business Publications Ltd.

Pitts, G. (1983) 'Engineering Design Management in French Industry', *Design Studies* 4(3): 158–60.

Press, M. and Cooper, R. (2003) *The Design Experience*. London: Gower Press.

Pulos, A. (1988) *The American Design Adventure 1940–1975*. Cambridge, MA: MIT Press.

Ramlau, U. and Melander, C. (2004) 'In Denmark, Design Tops the Agenda', *Design Management Review* 15(4): 48–54.

Schneider, U. (2000) 'Hermann Muthesius and the Introduction of the English Arts and Crafts Garden to Germany', *Garden History* 28(1): 57–72.

Schutte, T. (ed.) (1975) *The Art of Design Management: Design in American Business*. New York: Tiffany & Co.

Simon, H. (1969) *The Sciences of the Artificial*. Cambridge, MA: MIT Press.

Stovall. S. (2010) 'Recreating the Arsenal of Venice: Using Experiential Activities to Teach the History of Management', *Journal of Management Education* 34(3): 458–73.

Thatcher, M. (1982) 'Take it from the Top', *Design* (May): 3–4.

Tobin, J. (2010) 'A Man Ahead of His Time: Bill Hannon and the Founding of DMI', *Viewpoints* (June): 1–2.

Topalian, A. (1980) *The Management of Design Projects*. London: Associated Business Press.

Triad Design Project (1989) *Designing for Product Success*, Exhibition catalogue. Produced in conjunction with the Triad Project. Boston: Design Management Institute, The Design Council.

Walker, D. (1990) 'Design Maturity: The Ladder and the Wall', in M. Oakley (ed.), *Design Management: A Handbook of Issues and Methods*. Oxford: Basil Blackwell, pp. 43–6.

Wikipedia (2010) 'Design Management', *Wikipedia,* http://en.wikipedia.org/wiki/Design_management (accessed 8 October 2010).

Wolf, B. (1993) *Design-Management in der Industrie*. Wetzlar: Anabas Verlag.

# PART I

# Traditions and Origins of Design Management

# Editorial Introduction

SABINE JUNGINGER AND RACHEL COOPER

In a departure from a traditional handbook, the first part of this handbook contains seven previously published works by authors who have made significant contributions to design management as a field and as a practice. Our decision to begin with the past in a book that is intended to be about the present and the future is based on a number of observations that we have made in our teaching, in our research and in our work with practitioners who concern themselves with design management. One notable finding is that fewer and fewer people are familiar with these early works. The reasons for this may be that much design management writing published in the 1960s, 1970s and 1980s is difficult to access today. It tends to be out of print and only selected libraries hold one of the remaining copies. These, in turn, are seldom called for since their existence is no longer widely known. This explains why we recently received a first edition of James Pilditch's *Talk about Design*, published in 1976, which was as pristine as it was when it came off the printing press – the pages still stuck together. The possibility exists that this problem is greater outside the US and the UK, because many of the early prolific writers were associated with these two countries.

We see a need to bring these important texts back into the consciousness of current scholars, practitioners and students. We believe that one can hardly grasp the changes and shifts that have occurred both in design and in management without revisiting some of the original thinking. Can we grow the field of design management without looking back? What changes in design and in management have occurred since then? Which questions still have relevance? As we are scanning the current landscape of research, practice and education, we find that many of the issues and themes of the contemporary design management discourse are variations on themes and concepts that were introduced decades ago by these early scholars. What does this mean? Are we asking the same questions over and over? Have we made any progress? How? Where? These are just some of the questions that come to mind and for which it is worthwhile to find answers. Reprinting these works is also a way for us to pay homage and to express our gratitude to the achievements of some of the design management pioneers.

There is, of course, the uncanny sense of bias in our selections, as they seem to convey that research into design management only took place in the UK, to some extent in the US and otherwise nowhere else. This is hardly the case. Nonetheless, we are just now beginning to embark on a journey of discovery that we hope will reward us with an even richer picture of the concepts, practices, methods and principles of design management than the one we can paint at this point in time. In making selections, one always has to omit someone or something. There are many outstanding works that we could and would have liked to add to this particular section. But, alas, the goal was to produce a handbook, not a classic reader in design management or its history.[1] If anything, our limited selection demonstrates the need for such a book. With the pieces presented here, we hope to provide students, researchers and practitioners interested in the bigger picture of design management some pointers as to where they can look for further materials.

We begin with a piece by **Bruce Archer**, published by *Design Magazine* in 1967. In 'A Place for Design in Management Education?', Archer argues for management programmes to include education about matters of design. For those readers interested in the developments of design education in business and management schools, this first chapter offers a historic background to Part II (next part of this handbook). We continue with **Michael Farr's** answer to the question 'Design Management – Why is it needed now?' In this essay from 1965, Farr offers an early definition of design management and a description of the tasks of the design manager. Farr positions design as a 'competitive advantage'. As will become obvious in Part III, design and the activities of designing today continue to circle around aspects of competitive advantage. This piece, too, was originally published in the *Design Magazine* (1965). Farr was the first to call for design management as a specialization of design. In this piece, he provides a first definition of what the position of a design manager encompasses.

Our third classic reading is currently enjoying renewed popularity both among students of management and of design. The discussion of 'Silent Design' by **Peter Gorb** and **Angela Dumas** marks a significant shift for design and management, as it begins to look into the question of who is actually engaging in design activities. This question has gained new relevance with the foray of design in public organizations, public services and even in governments. Today, many people remain unaware that they are (a) dealing with a design problem and (b) actually making things, shaping all sorts of human interactions which influence their own environments. Several essays in Part IV touch on different aspects of the ubiquity of the practice of design.

We are delighted to include 'The Designer Manager Syndrome' by **Sir Misha Black**, a chapter from one of the first books on design management that originated in the

US and which is based on the six Tiffany-Wharton lectures on the Art of Corporate Design Management hosted by the University of Pennsylvania in 1973. Walter Hovington edited the book for Tiffany & Company in New York in 1975. In this chapter, Black refers back to Moholy-Nagy (1947) to suggest 'design is an attitude of mind'. He then observes that, for design to have a greater role in business, a design attitude 'must permeate management'. The development of design attitude in organizations poses one of the key contemporary challenges, both for managers and designers.

**Mark Oakley** counts among the most prolific and relentless writers on design management. His interests in the problems surrounding managing and designing ranged from educating managers (Oakley 1986) to the influence of design in industrial and economic achievement (Oakley 1985). We have decided to reprint his insightful analysis of 'Organizing Design Activities', an excerpt from his book 'Managing Product Design', published in 1984. This chapter is mandatory reading for anyone trying to come to terms with some of the core challenges designers face while working with or within organizations. This includes the location of design within a particular organizational structure, which has great influence on the authority and power of design. It also includes organizational and managerial resistance to change. Oakley does not limit the organization of design activities to these two aspects. Nonetheless, power and resistance to change remain two important challenges for today's designers, who engage with ever more complex problems in ever more complex organizations.

In the context of marketing and strategy, the work by **Philip G. Kotler** stands out. We decided to include the paper on 'Strategic Design – A Powerful but Neglected Tool', which he co-authored with **Alexander Rath** and which was first published in the *Journal of Business Strategy* in 1984.[2] For anyone wondering about the origin of the term 'design thinking', it might be worthwhile noting that Kotler and Rath used 'design thinking [that] goes into product development work' as a criterion to measure a 'corporation's design sensitivity and design management effectiveness'. In regards to design and marketing, 'Humanistic Marketing' (Kotler 1987) offers another fundamental reading. Unfortunately, we could not include it in this edition.

**James Pilditch**, one of the grandfathers of design management, will have the final word in this section. We have included the opening chapter from his classic work *Talk About Design*, which was first published in the UK in 1976. 'Into a Changing World' provides insights into the hopes for design management at the time. More importantly, though, we can gather that design here is bigger than a product or a company. Social problems come into the fore, the need for designers to conceive of products that are 'good' not for their own egos but for the people they design them for. In this sense, Pilditch already foreshadows one of the most significant shifts that have taken hold in design and management over the past years. No longer is it sufficient to worry about how to make something, although some design managers refuse to acknowledge this. Instead the question begins with why we should make this in the first place. Foremost of all, design is concerned with change. And this is a theme that is driving the current discourse on design management – be it in the form of innovation, transformation or organizational change. It is noteworthy that Pilditch was familiar with the work by Peter Drucker, another management scholar whose work is highly relevant in the context of designing and managing. Drucker, says Pilditch, considered marketing and innovation to be the 'only worthwhile occupations of management.' Pilditch offers this wry response: 'Design, by the way, runs all through both' (marketing and innovation) (Pilditch 1976: 10).

## NOTES

1. A history of design management would also need to include the key developments and paradigm shifts within the field of management. Mauro Guillén's *Models Of Management*, for example, offer a helpful background of the developments in management thought and practice that link directly to the role designers and design can have in an organization (Guillén 1994. *Models of Management*. Chicago, IL: University of Chicago Press).
2. The publication date on the *Journal of Business Strategies* website cites the year 1993 as the year of publication, whereas Kotler himself cites the same edition of the journal (vol. 5, issue 2, pages 16–21) but from 1984.

## REFERENCES

Kotler, P. (1987) 'Humanistic Marketing: Beyond the Marketing Concept', in A. Fuat Firat, N. Dholakia and R. Bagozzi, *Philosophical and Radical Thought in Marketing*. Lexington, MA: Lexington Books, pp. 271–88.

Moholy-Nagy, L. (1947) *Vision in Motion*. Chicago: Paul Theobald & Co.

Oakley, M. (1985) 'The Influence of Design on Industrial and Economic Achievement', *Management Learning* 23(4): 3–13.

Oakley, M. (1986) 'Bringing Design into the Management Curriculum', *Management Education and Development* 17(4): 352–62.

Pilditch, J. (1976) *Talk about Design*. London: Barrie & Jenkins.

## RECOMMENDED READING

Guillén, M. (1994) *Models of Management*. Chicago, IL: University of Chicago Press.

# 1

# A Place for Design in Management Education?*

## L. BRUCE ARCHER

*There is now a steady flow of trained designers into industry. But are these designers being skilfully managed so that they can produce the most effective results? An examination of management courses shows that little time is given in most of them to the management of creative design work and of innovation generally. In 1965, the Council of Industrial Design raised the problem of design management in a joint meeting at the* Financial Times *between management consultants and industrial designers. Since then, the CoID has joined some of the marketing conferences of leading consultants, and has taken part in a few management courses.*

*In the following article, L. Bruce Archer argues the case for including design as a subject in management courses and comments on the reactions to this view of some principals and course directors at selected business management schools. In a subsequent article, J. Noel White, deputy director, CoID, who is a council member of the National Marketing Council, will discuss the practical aspects of design management which concern industrial organisations. These aspects include the relationship of a company's innovation policy to its product development programme; the organisation of an efficient and speedy flow of information from market and technical research into the design department in a usable form; and the appropriate location of the design department in a company's management structure. If business schools are to include design management in their curricula these questions must first be answered in broad outline.*

The principal object of any normal company with shareholders is to make a profit. Indeed, many other institutions, such as nationalised industries, even though they have *no* shareholders, must conduct themselves *as if* they were in business to make a profit.

*Source*: Reprinted from Archer, L.B. (1967) 'A Place for Design in Management Education?', *Design Journal* 220: 38–43. Courtesy of the University of Brighton Design Archives, www.brighton.ac.uk/designarchives/.

A manufacturing company makes its profit by manipulating materials into more valuable forms – that is, by producing artefacts which command a higher value in exchange than their intrinsic or manufactured (and marketed) cost. One could argue further that, to be commercially viable, the one irreducible quality that a product must have – whether beautiful or ugly, efficient or inefficient, durable or transitory – is a value in exchange which is greater than its manufactured and marketed cost.

But what constitutes value in exchange? A product is worth what people are willing to pay for it. The qualities which a purchaser seeks and values may be summarised as utility, singularity and/or emotivity. Perhaps one should also add availability. The term 'utility' covers concepts such as the need for food, the convenience of a lawnmower and the earnings from a machine tool. The term 'singularity' embraces meanings such as the scarcity of antiques, the uniqueness of a painting, and the individuality of a dress. The term 'emotivity' describes the beauty of a textile, the covetability of a jewel and the status symbolism of a motor car.

And what constitutes cost? This is the outlay on materials, processes, labour, power, transport, promotion, research and development, plant, premises, and the servicing of capital – all of which may be discerned in the product as the quality of economy. Where a product exhibits appropriate degrees of economy, plus the qualities which constitute value, then a profit is made. Moreover, where this profit is made by the conversion of *raw materials* into more valuable forms, especially where the quality of utility is present, new wealth is created – to the benefit of the user and the community

FIGURE 1.1  The design profession has been singularly inarticulate about design policy and design management/Dr S.K. Manstead, deputy principal, Ashridge Management College, Berkhamsted. Photograph by Eugene Ankeny.

at large, as well as of the producer. The accounting of the effects of variations in these qualities is the technique of logistics, a recognised marketing and management tool. The provision of these qualities is design.

But to determine more precisely which qualities are needed in the design of a product, questions of strategy and tactics must be considered. If we have conceived a radically new idea, how far should we go in perfecting and testing it before going on to the market? Should we introduce our new designs at the *avant-garde* end of the market, taking as big a mark-up as the market will stand, and recovering our research, design, development and tooling costs as quickly as possible? Or should we sell to the mass market at the lowest possible price, writing down the costs over a long period, and leaving few loopholes to competitors?

These are marketing and management questions. But the answers both affect and are affected by the configuration of the design – its function, efficiency, styling, costing and construction. The means and the ends are thus intimately related. And because of this, the function of design in a manufacturing company must be a primary responsibility of management and cannot be dismissed as being of secondary importance.

This argument would suggest that the management of design should occupy a significant place in general management education. But does it? To try to find out, I carried out an inquiry which showed that the problem is only beginning to be tackled in a few management training centres. But before going on to describe the inquiry in more detail, it is necessary to look at recent developments in the training of both managers and of designers.

FIGURE 1.2 British businessmen show little thirst for knowledge about design/Dr Arthur Earle, principal, London Business School. Photograph by Eugene Ankeny.

It is certainly relevant to the present argument that the same two decades which have
seen engineering and industrial designers attempting to formalise their professional organisation and training have also seen similar changes in the organisation and training of
management. Managers have perhaps advanced more rapidly towards the development
of a science of management than designers have towards a science of design. Certainly
students of rational methods in design find themselves leaning heavily on techniques
borrowed from management science. Teachers of management are at least two years
ahead of teachers of design in raising the level of instruction to that of a valid university
subject. Even so, the business schools have far from completed their development.

Ever since the end of the Second World War, management in British industry has
been accused of amateurism, timidity, inefficiency and backwardness. In 1963, these
criticisms were focused by the Robbins and Franks reports, and the cost of rectifying
the situation was calculated by the Normanbrook report the following year. At that
time, many full-time and part-time courses of instruction in management techniques
were already in existence – organised mainly by technical colleges, professional bodies
and management consultants, and consisting mainly of short courses. The same year,
the Confederation of British Industry, the British Institute of Management and the
Foundation for Management Education raised £5 million which, with the promise of
£2½ million of Government money, was to be devoted, among other things, to the
development of graduate business schools on the American model. This gave a great
fillip to these courses, as well as stimulating much jostling among the universities for
a share of the money.

A research fellowship is needed on design
decision making in management/
A. L. Minkes, director, Graduate Centre for
Management Studies, Birmingham.

FIGURE 1.3   A research fellowship is needed on design decision making in management/A.L. Minkes,
director, Graduate Centre for Management Studies, Birmingham. Photograph by Eugene Ankeny.

It so happened that these events roughly coincided with renewed activity on the design side, stimulated by the Coldstream and Feilden reports on the re-organisation of art-based and engineering-based design education respectively, and by the Robbins report which recommended that the colleges of advanced technology should be turned into universities.

This then is the background to the inquiry, which suggests that the time has arrived when design could be linked much more directly to the management chain. The inquiry, which was in two parts, set out first to establish the facts, and second to collect opinions based on interviews with principals and course directors at selected management training centres.

In the first part of the inquiry, DESIGN wrote to all the organisers of courses on management subjects who could be identified, and asked them for prospectuses. Out of several hundred prospectuses received from 45 centres, 140 courses were selected as dealing with subjects within which reference to some aspect of design might seem appropriate. These ranged from 'The organisation of industry and commerce', a three year honours degree course at the Faculty of Social Sciences, University of Edinburgh, to 'Operator training (skill development method)', a one day course at P. E. Consulting Group Ltd., Surrey.

For the purposes of this survey, five aspects of design were considered to be relevant: aesthetics, ergonomics, design for function, design for marketing, and design for production. It was assumed that a management student would not need to acquire skill in the techniques of designing, but that he would be expected to gain some understanding of the effects of at least some aspects of design on marketing and production, and of the principles of organisation of the design function. Since attitudes to this would depend a great deal on the special slanting of each course, all those examined were divided into four classes:

1. Management courses, including business administration, industrial administration and management science.
2. Marketing courses, including sales management, industrial marketing, advertising, market research and management of new product development.
3. Industrial engineering courses, including works management, personnel management, work study and ergonomics for engineers.
4. Production engineering courses, including production planning, process control, quality control, and value analysis.

Every course syllabus was then examined and a judgement made as to which, if any, aspects of design *should* have been referred to and which, if any, were *in fact* referred to. The results were depressing. Only 35 of the 140 courses appeared to refer to design at all, and of these only seven to any significant degree. In general, it was clear that design was regarded as an activity for backroom boys – important, perhaps, but separate.

This attitude was painfully driven home by the follow-up interviews. These formed the second part of the inquiry, and were intended to discover whether or not there were any special reasons for the low rating of design as a management interest, and whether or not any great changes were in store. On the advice of J.F. Sinclair, director of the Foundation for Management Education, and Mrs Molly Adams, of the British Institute of Management, it was decided to visit the Graduate Centre for Management Studies, Birmingham; the London Graduate School of Business Studies; the Manchester Business School; the Administrative Staff College, Henley-on-Thames; and the Ashridge Management College, Berkhamsted.

## WHAT THE SCHOOLS HAD TO SAY

Ashridge Management College was the only one visited which has actually given instruction in design as a management tool. The principal, Dr Christopher Macrae, fully acknowledges the importance of the subject, and the CoID has co-operated with the college in arranging lectures and discussions. I spoke to Dr S.K. Manstead, deputy principal, who said, 'I think we must distinguish between the situation as it is and the situation as we would like it to be. As of now, design is dealt with as a significant subject in our longer marketing courses. But we regard design as a general management responsibility, not just marketing. However, it enters our general management courses mainly in the shape of value analysis, which is an extremely useful tool, but only part of the story. The problem is that the design profession has been singularly inarticulate about design policy and design management. We recognise the need for expertise in this field but it is still not very easy for us to know where to find it.'

On a previous occasion, W.G. McClelland, director of the Manchester Business School, had talked to James Noel White, deputy director of the CoID. At that time, McClelland had expressed doubts about industrial design as an appropriate subject for management teaching at university level, unless and until it could be shown that there was an established corpus of knowledge and well documented case studies – a consideration, incidentally, that is currently occupying a number of minds at the Royal College of Art, which is having to turn itself into a university.

During the present inquiry I spoke on the telephone to Ken Simmonds, professor of marketing at Manchester, and he confirmed McClelland's earlier view. 'The trouble is,' he said, 'there is a tremendous amount for our students to assimilate, and not enough time to put it across.'

## MANAGEMENT TECHNIQUES

At the University of Lancaster, to which I was referred by Professor Simmonds, Professor Raymond Lawrence took almost exactly the same line. He added that in his view the propositions put forward were concerned with the marketing function.

Marketing people prepare design briefs, and design people carry them out. In Birmingham, however, A.L. Minkes, director at the Graduate Centre for Management Studies, was intrigued by the idea that management techniques were being used by designers to resolve design problems, and said that he would welcome an initiative on the part of the CoID in setting up a research fellowship on design decision making in management.

At the London Business School, Dr Arthur F. Earle not only spoke of design in strategic terms, but was able to discuss product design, house style and the design of environments with obvious knowledge and experience. In the second year syllabus of his two-year master's degree programme, which began only a few months ago, subjects such as entrepreneurship and innovation, product design and development, product formulation to meet consumer needs, tactical aspects of product and marketing management, the build-up of the product line, and decisions on branding, packaging, container and label design are listed. Perhaps his experience as managing director of Hoover Ltd had something to do with it.

Dr Earle admitted, however, that his view was not widely shared. 'On the whole, British businessmen show little thirst for knowledge about design,' he said. 'Although I know that any designer worth his salt will treat design as an all-in problem, most people think of engineers as inside designers and industrial designers as outside designers, and no more.'

J.P. Martin-Bates, principal of the Administrative Staff College, Henley-on-Thames, also had well formulated ideas on design decision making as a management function. He always strives to get the design side represented in each course, although some sessions are stronger than others in this respect. Design, along with development and research, is always the subject of supporting studies, and also receives attention in the examination of technical and economic change in marketing. But he, too, finds it difficult to achieve real depth of understanding of design as a matter of fundamental concern to the production and commercial sides.

## A MEASURE OF SYMPATHY

It can be seen that in the leading business schools, at least, there was a measure of sympathy for the view that design and designer have a role in management and policy making. What of business itself? In its survey *Attitudes in British Management*, published in 1965, Political and Economic Planning quotes many examples of narrow attitudes to design (the survey regards design as a part of research and development). Among other things, the survey concludes, 'There was in many firms an unsatisfactory relationship between the R & D, production and sales or marketing departments. . .', and 'It has become almost a cliche to say that much of British industry is too dominated by the production side, but the interviews confirmed that it is still, nonetheless, true.'

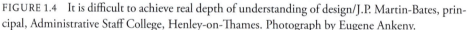

FIGURE 1.4   It is difficult to achieve real depth of understanding of design/J.P. Martin-Bates, principal, Administrative Staff College, Henley-on-Thames. Photograph by Eugene Ankeny.

The inescapable conclusion from both the paper study and the direct approaches is that there are still great gulfs of misunderstanding, not only between the management profession and the design profession, but also between the engineering and the industrial design branches of the design profession. Industrial designers are generally regarded as aestheticians, unversed in the needs of product marketing and business strategy. Engineering designers are generally regarded as technicians, unversed in the needs of marketing, business strategy, and human factors.

Perhaps when the British business schools recognise that the management of design is a primary rather than a secondary responsibility of management, they will proceed to do something about it. If they do so, the design side will be expected to produce the established corpus of knowledge and the well documented case studies which McClelland said would be demanded. And unless the design profession and the design schools exert themselves a good deal more than they have so far, that corpus of knowledge and technique will be found wanting. Bridges are best built from both banks of a divide.

# 2

# Design Management: Why Is it Needed Now?*

## MICHAEL FARR

*The articles in the design management series have so far concentrated largely on reporting the ways in which design is being treated as a management problem in specific industries or in individual firms. This article begins a new series in which the emphasis will shift to take in expert views on the way certain specialised skills could and should be used by industry to solve design management problems. The first article in the new series is taken from Michael Farr's book on design management, which will be published by Hodder and Stoughton Ltd towards the end of this year.*

The types of design now needed by industry and commerce are daily growing more complex. No longer can the manufacturers of most products or the suppliers of most services rely on rule-of-thumb methods each time they make a fresh attack on their markets. To take two examples: the producer of a gas range is not *selling* a cooker, he is marketing leisure in a prestige casket; the supplier of ironed shirts is selling not a laundry service, but pride in appearance. These are typical of the forces which motivate the market place and so condition the types of design required. From scientific research leading to technological development, there are further forces which affect design, requiring the exploitation of new and frequently untried methods and materials. And within the companies themselves there are yet other, often conflicting, reasons why products and services should be changed, and hence re-designed.

The design skills needed to meet modern requirements are also becoming more complex, and their practitioners more specialised. No longer can one designer do all types of designing. Designers' training, experience and opportunities are tending

*Source*: Reprinted from Farr, M. (1965) 'Design Management. Why Is It Needed Now?', *Design Journal* 200: 38–9. Courtesy of the University of Brighton Design Archives, www.brighton.ac.uk/designarchives/.

to become less diffuse and more profound. Larger markets for products and services mean greater capital outlay on tools, plant and promotion before they are tackled. The penalty for starting off with the wrong design is growing more severe. Picking the right designer for the job, and making certain that he is also a person who works well with a company's own team, has become a crucially important task.

For these and other reasons, which this series of articles will attempt to indicate, design management is becoming an integral part of profitable trading which commerce and industry can no longer ignore.

## DEFINITION

Design management is the function of defining a design problem, finding the most suitable designer, and making it possible for him to solve it on time and within a budget. This is a consciously managed exercise which can apply to all the areas where designers work.

It seldom does. Industry and commerce generally are still confused and sceptical about the whole business of designing. The problem has many roots. One to be eradicated is mistrust of something new. Another is fear of being fleeced by a horde of idle artists. And yet another is lack of accord in the management of an organisation, which inevitably leads to stifling indecision. These, and other rooted objections to recognising design as a valid and viable commercial tool that must be kept in trim, are frequently caused by nothing more serious than lack of practice.

Lack of practice, that is, in design management. When his public desires a change – or at least appears to be ready to accept one – the manufacturer who wants to supply it must think in terms of design. For many it will be an unfamiliar thought process because the majority of factories are not producing new designs every month, and in some cases not every year. They have no need to. Lack of familiarity with the design process can best be gauged by comparing it with manufacturing methods or the practices of the selling staff. In both cases the key to efficient and profitable operation lies in continuity; in the handling of the minimum number of variables in the greatest possible quantities. A smoothly operating production line is the manufacturer's daily concern. So, too, is the sales policy. Neither would do well if it were not constantly appraised and improved. Both would become inefficient and uneconomic if the designs they handled were *also* subject to continuous improvement.

Companies need works directors and sales managers to ensure that their business operates efficiently. They need design management when they need new designs.

An underlying premise of this series is that if designers are good at designing they should not have the time to spare to manage the ramifications of their design projects, regardless of whether or not they are also good managers. A good designer has been trained and experienced in designing, and is, therefore, best used (and better fulfilled) as a designer. The same may be said for managing directors, works and sales managers, and other executive staff who have defined duties in any business and are most effective when they concentrate on these duties. Each of them, including the designer, makes a particular contribution to the process of transforming a product, and each sees the outcome of the process from a specialist's viewpoint. The design manager comes in when the need for his unbiased co-ordinating services is felt, and for the majority of companies his is not a full time job.

## TASKS OF THE DESIGN MANAGER

But the design manager, as such, is still rare in industry. His job, in brief, is to investigate, from the *designing* point of view, the requirements for a new product; set a time and budget for the design development period; find and brief the designer (or team of designers); set up and operate an easily understood network of communication between all parties concerned in the new product; and be responsible for the co-ordination of the project until the prototype reaches the production line and the designing of packaging and supporting printed matter is complete. At these latter stages the design manager supplies all required information to those responsible for product marketing, sales promotion, advertising, publicity and public relations.

Throughout any design project, there is the tension of conflicting forces – the marketing manager may be in a hurry for results, while the works director does not want his existing, profitable operations disturbed – and in countless cases this tension has succeeded in spoiling or shelving the designer's work. It is the design manager's job to resolve these forces, preferably by anticipating them, so that, to put it simply, the designer can give of his best and the company can receive it.

When the amount of designing handled makes it necessary for the design manager to be on the staff of a company, he should have a status that allows him to be on equal terms with the works and sales managers. He should be directly responsible to the managing director. This is desirable because the work he handles will not be confined to either product design or packaging and publicity design. The total design policy of the company would be his responsibility, and it could embrace its house style, office design and furnishing, and exhibitions, as well as the products and their promotional aids. Usually it is only the managing director who can decide on all these matters.

## AT BOARD LEVEL

This equal status with other senior executives could mean that the design manager is, in fact, on the board of directors. For companies where designing is habitually extensive and planned far ahead, the design manager on the board should be able to bring an original and vital dimension to his colleagues' discussions. Of course, he can be equally valuable where the board is primarily concerned with day-to-day matters. The presence or absence of the design manager on the board can be justified differently for each company; what matters is his direct contact with the managing director, who can give him a mandate to question the motives of those who might unwittingly obstruct a design project.

For companies which need occasional but intensive design activity for their products, a design management consultant could be called in temporarily. In many cases he would act as if he were a member of the company's senior staff for the duration of his assignment. Again, he should be directly responsible to the managing director. Often the consultant design manager would find that, owing to his initiative or his client's, it would be desirable for him to extend his original terms of reference in order to co-ordinate all the company's other design requirements. In such cases he would probably be retained on a more long-term basis.

However employed, the design manager needs knowledge, specific working methods and skill. His tasks lie in problem solving, planning, briefing, communications and co-ordination. They occur with every project, but their content is never the same. Assuming he is a consultant, his knowledge must be of a large variety of freelance designers in this country and abroad. He must *know* them, and know what they do and how they set about it. He must know what the trends are over the whole field of industrial design, and be a good critic of them. He should be familiar with the different types of specialised services that designing frequently needs, such as consumer, market and industrial research, and ergonomics. In addition, he must have a working knowledge of the main industrial processes and the general characteristics of leading materials and finishes in all the areas in which he offers his services.

The staff design manager, for all but a few companies, would obviously not need such extensive experience, but it would be important for him to be alert to all the design trends that could conceivably affect his company's business. This means that he should not be closeted within its offices all day. Above all, both types of design manager need vision: the capacity to see how people from different disciplines are likely to interact in order to produce something new and worthwhile.

## METHOD

The design manager's normal working methods are to be described later in this series. He should be fully conversant with the 'technology of design', with the various tools that can solve his problems (such as statistical method), and with those that can help to run his programmes (such as network analysis). His skill cannot be defined in any way which is meaningful in a general context, but it can be said that if he practises without a really keen intuitive sense of what is wanted and how to get it, he will not practise for long. Sorting out the relevant information for a designer's brief is one of the design manager's most difficult tasks.

Often the design manager is the only person who can appreciate the potential abilities of a young and diffident designer. By breaking down the design problem into manageable stages, by carefully phasing the integration of the designer with the company's own staff, and by giving more encouragement than is normally forthcoming to the novice in industry, the design manager can realise for his company's profit some original talents that would otherwise be lost altogether – or lost to a competitor. Much designing brilliance, fresh from art school, runs to waste for lack of a perceptive *entrepreneur*.

Experienced or inexperienced, the designer is the specialist who performs what ultimately matters to the company. He wants to be able to perform it to the best of his ability. Only in so far as he helps in this definitive process has the design manager got a *raison d'etre* in commerce and industry. He has to show that he is the best person to absorb the various and often conflicting requirements of the company's executives, and to interpret them to the designer in a consistent manner. He will not do this successfully by relying solely on re-editing, re-writing and re-issuing reports. He will, during the course of any job, do it mainly by conversation in which his *rapport* with the designer is a factor of fundamental importance.

What, then, are the reasons for publishing a series of articles on design management? There are many, but lying at the root of all of them is a need for industry to become more competitive, particularly competitive abroad. Design is a unique factor in competition. Skilful management of designers and designing, therefore, becomes imperative. If better returns are to be earned on the capital invested in the new materials, processes, plant and marketing systems, then what is required is a more precise application of the right designing skill to their products. In this country and throughout the world there is a grossly inadequate number of experienced staff design managers, let alone consultants. Perhaps this series will go some way to show that there is an opening for more of them.

3

# Silent Design*

## PETER GORB AND ANGELA DUMAS

*This paper describes the outcomes of a one-year pilot research study and outlines the routes for the two-year wider study to follow.*

*The research was prompted by the growing interest in the UK in design and its contribution to business performance, and the need to replace anecdote about 'best practice' in organizing and utilizing design, with information about more 'general' practice.*

*After defining design as 'a course of action for the development of an artefact' and suggesting that design activity pervades organizations, the paper describes the methodology used to examine how design is organized. Using matrices to explore the interaction of design with other business functions the report suggests that 'silent design' (that is design by people who are not designers and are not aware that they are participating in design activity) goes on in all the organizations examined, even those which have formal design policies and open design activities.*

*It is the scope and nature of 'silent design', and its conflict and/or co-operation with formal design activity, which will form the basis for the hypothesis on which the wider investigation will be built.*

This is an interim report on a research investigation into the *organisational place of design*. The research, which is funded by the Leverhulme Trust, is expected to take three years to complete. For operating purposes it was divided into two parts: the first part, a pilot study over one year, was intended to develop propositions for examination in the wider part of the research. This interim report discusses the outcome of that study, which has just been completed. However, before discussing the pilot study, it is worth restating the overall objectives of the research, and describing the context in which it is taking place.

*Source*: Reprinted from *Design Studies*, 8(3), Gorb, P. and Dumas, A. (1987) 'Silent Design', 150–6, with permission from Elsevier.

In the last decade, and particularly since 1982, there has been a growing interest in Britain in the contribution that design can make to business profitability. The government, through the Department of Trade and Industry, has given significant financial support to the promotion of this proposition. Through the Design Council it has funded a design consultancy scheme designed to help smaller industries to use design more effectively. It has asked the larger industries formally to commit themselves to supporting the importance of design, and is conducting a national advertising campaign. A Government Minister is charged with the national responsibility for design.

In the field of education the subject of design management has emerged as a new field of study in both design and business schools. The teaching of design at the MBA level was pioneered at London Business School in 1976 and is now firmly established here. In 1985 the Council for National Academic Awards published a report on design management which has led to initial work at a postgraduate level in five polytechnic management studies departments[1]. In-house training in the subject is beginning to happen in industry. In 1986 design appeared on the agenda of the conference of the European Foundation for Management Development, for the first time.

Britain's leadership in promoting the importance of design is being watched and emulated both in Europe and the USA. An outcome of this growing interest has been a key discussion on how best to organise and utilise the design activity.

## THE RESEARCH OBJECTIVES

It is perhaps a natural outcome of the interest shown in a new subject that what has been published so far about the organisation of design has been largely anecdotal and almost exclusively concerned with best practice. This information has proved of great value in motivating others and as a starting point for investigation. However, like all best practice it may be relevant only to the organisation concerned and indeed may be only temporary.

Our research objective is to discover what constitutes *general* practice. We have set out to examine all those aspects of the business where design is utilised and to identify how the enterprise organises itself to make best use of design. We are also concerned to identify strategies surrounding the implementation of design if and where they exist.

In order to achieve this, the general objectives of the research were to identify:

- the design understanding of managers; their view on the relevance and scope of design in their organisations;
- the operational role of design: how it relates to problem solving and decision making;
- the assignment of responsibility and accountability for the various aspects of design in organisations;
- resource commitments by way of people and funds, and the ways in which the performance of these resources is measured.

It was also hoped as an outcome of the research to establish a database which could be used by subsequent research into issues of design organisation.

## THE SCOPE OF DESIGN

Design is a process. It is perhaps necessary to affirm this rather obvious fact in view of the common confusion between process and product which takes place when definitions of design are attempted. Furthermore, in defining design, we also need to recognise that external appearance, style, colour and other aesthetic and subjective considerations with which design is commonly associated constitute only part of the design process. Design is also concerned with use, with marketing and production considerations and a wide range of technical and engineering resources and requirements. However, above all it is concerned with a methodology.

> Everyone who designs devises a course of action aimed at changing existing situations into preferred ones. The intellectual activity that produces material artefacts is no different fundamentally from the one that prescribes remedies for a sick patient or the one that devises a new sales plan for a company or a social welfare policy for a state. Design, so construed, is the core of all professional training: it is the principal mark that distinguishes the professions from the sciences. (Simon 1982)

This quotation proposes a methodology of design which Herbert Simon differentiates from the methodology of science. He also points out the very wide-ranging nature of that methodology which has application well beyond his own concern with artefacts. His comment illuminates one main reason why confusion exists over the place of design in organisations, and why there is a need to identify and later specify some landmarks in the process that is design.

Simon limits his definition of design to artefacts, that is, man-made things; although he is also concerned with systems of artefacts, and how the individual artefacts within such a system relate to each other. We have adopted that limitation. However, we are also concerned with and how people in organisations understand and make various contributions to the planning, strategies and goals surrounding those artefacts and systems of artefacts.

Our working definition of design is therefore:

> a course of action for the development of an artefact or a system of artefacts; including the series of organisational activities required to achieve that development.

It is important to point out that this definition is not so exclusive as to encompass merely the activities of the professional designer, a limitation which we abandoned at an early point in our investigation. If this definition seems narrow it is worth emphasising that artefacts pervade industrial organisations. They comprise the products which a manufacturing organisation makes and sells or a retailing organisation buys in order to sell, or the products used by service business to provide its service. They also embrace those artefacts like buildings and equipment which go to make up the physical environment which constrains or enables the organisation to achieve its purposes. In this context we include the work of the architects, engineers, interior designers and space and environmental planners. Finally they also cover the artefacts which make up the information systems through which the organisation communicates its purposes to its various audiences (e.g. employees, shareholders, customers); and include everything from annual reports to advertising material.

This means that although we begin with a simple definition, it has to span numerous activities usually planned and managed in different parts of the organisation in different functional departments.

## THE PILOT STUDY

Because of the inevitable complexity of identifying an activity which cuts across traditional organisational lines we decided on a pilot study which would, we hope, not only throw up issues for wider study, but also alert us to any flaws in our method of enquiry. In addition we decided to cover a number of industry sectors to see if comparisons between sectors could be made.

Accordingly, the pilot study was undertaken within four industry sectors: electronics and apparel in the manufacturing sector and retail and transport in the service sector. Sixteen firms in total were investigated, four in each sector. Unstructured interviews were conducted with a range of people involved (as the organisation saw it) with the design process. In some organisations, more people were interviewed to add greater depth to the pilot study.

The method we used to undertake the investigation was the completion of a matrix. Along the horizontal axis of the matrix were placed the main areas of artefacts in which design operates and which have been outlined above. These are the 'products', the 'environments' and the 'information systems', each of them being subdivided into appropriate categories.

Down the vertical axis were shown seven levels of involvement in detail from 'shallow to deep'. A description of each of these together with the original matrix is shown in the appendix (Figure 3.1). Using this matrix it was hoped to plot resources for a commitment to design within each organisation, and thereby identify and characterise the various organisations and business sectors in terms of their management of design.

## DEVELOPING THE MATRIX

From the information we received during our interview programme we realised that our original matrix could be expanded, as an analytical tool. Our task was to examine the integration and interaction of the design process, through the activities of individuals. However, we also began to see a need to identify separately contributions by professional designers. Accordingly we developed two new matrices each dedicated to producing a clearer picture of one aspect of the data. The original matrix was retained, with one main modification, the ability to record 'professional' design activity. Now known as Design Matrix 1, it operates as a reference map for all design activity. Of the two new matrices, Design Matrix 2 concentrates upon interaction between functional areas and Design Matrix 3 on activity by individuals. All these matrices are shown in the Appendix (Figures 3.1–3.4).

## OUTCOMES

As an outcome of our work with the matrices, we were able to make the following two statements:

1. Design activity appears to be widely dispersed throughout the organisation.
2. Design is very interactive, and cuts through many traditional functional areas.

Perhaps more significantly, we were able to make a third statement.

3. Design activity is frequently not classified as such within organisations, nor does there appear to be any consistency of classification.

This statement was achieved in the course of the interview procedure, during which it became apparent that we needed to interpret certain information in order to complete the matrices.

The matrix development had led us to the realisation of a 'covert activity' in all the organisations we had investigated. This covert activity was clearly an important element in all transactions affecting both individual goals and motivation and ultimately, therefore, the central goals of the organisation. We could also begin to see why organisations were experiencing difficulties implementing a design programme.

## SILENT DESIGN – THE ARGUMENTS

It can be argued that a great deal of design activity goes on in organisations which is not called design. It is carried out by individuals who are not called designers and who would not consider themselves to be designers. We have called this 'silent design'.

The aims and intentions behind the design activity of an individual cannot simply be subsumed under 'design' if his [or her] job description, title and his own intentions are not perceived by him and others as having design as a central activity. To assume that if the job entails design, it should be undertaken by a professional designer. Indeed within his particular business context his set of decisions might be more appropriate than those of the designers.

The degree to which the 'silent designer' is aware of his design role needs to be understood better. It is also of great significance in the interaction between the 'silent designers' and the 'professional designers'.

In our major investigation we plan to explore the 'silent' design issue. To develop a framework for this investigation we have posed the following questions:

1. How widespread is 'silent design'?
2. How does silent design relate to overt design, where that function exists in an organisation?
   – Is it productive – are there conflicts and how are they resolved?
   – Does the amount of each affect these issues?
3. Should silent design be made overt? Is this even possible?
4. If not is there an optimum balance?

## THE MAJOR INVESTIGATION

With the establishment of a focus for the major investigation the authors plan to modify the form of the questionnaire as it was set out in their original research proposal. They plan to reduce slightly the number of organizations to be investigated on a broad front in order to make room for a few in-depth and more detailed studies. These in-depth studies will deal with the problems involved in asking direct questions about 'silent design'. They also expect the in-depth studies to uncover some customs and habits of the silent designer and the wider questionnaire to provide an organizational context in which these customs and habits exist.

## CONCLUSION

The pilot study has indicated that design activity pervades organizations and that it is dispersed, interactive and frequently undertaken by people who would not recognize that their job involves design. For the time being we are naming this phenomenon 'silent design'. During the major investigation, the authors will look at organizations with and without formal design policies and resources.

Whilst most of the organizations already studied do have formal design policies, the initial study has found that 'silent design' exists in these organizations as well as those that do not. In the organizations that do have a formal and declared design policy, it is interesting to note words and phrases used to describe aspects of maintaining them. The words 'discipline' and 'control' and the phrases 'top management commitment is vital' and 'custodian of design' were all used by most of them more than once. This suggests a degree of unease, not surprising in a relationship as ambiguous and unclear as the design and management relationship appears to be. It is likely that the increasing promotion of design by government and other agencies will persuade more organizations to take on board a policy toward design.

In identifying the existence of 'silent design', the need for more knowledge on the interaction between 'covert' and 'overt' design activity is clearly vital. Without it one could postulate that a rush toward the introduction of design policies and practices might inadvertently demolish long-standing and successful 'silent design' activities. The need for caution is clear. One must be careful not to hamper success in utilizing design to increase profits.

## APPENDIX

The methodological approach for the pilot study originated from a working paper written for the general manager who had little background in design. Embodied in the paper is a conceptual framework with which to consider design activity within an organization. A simple matrix accompanied the paper and provided a device to explore the scope and depth of commitment to design within an organization. The matrix is illustrated in Table 3.1. Scope for design is allocated to the horizontal axis under three areas of activity: products, environments and information. Depth of commitment is on the vertical axis.

The working paper describes the vertical axis 'as a guide to establishing the level of involvement at all design stages'. Seven stages make up a scale which starts 'shallow' with an 'auditing' process where direct involvement is limited to evaluation, and finishing 'deep' where an organization directly implements design (probably by manufacturing). The stages are described from the paper as follows.

**TABLE 3.1: The original matrix**

| A Design Implementation Matrix | Products | | Environment | | | Information | |
|---|---|---|---|---|---|---|---|
| | Finished products | Components | Buildings | Machines | Equipment | Internal | External |
| Audit | | | | | | | |
| Advise | | | | | | | |
| Plan | | | | | | | |
| Specify | | | | | | | |
| Supervise | | | | | | | |
| Demonstrate | | | | | | | |
| Implement | | | | | | | |

- *to audit.* An attempt should be made to audit every product with the view to ensuring that it adheres as closely as possible to the design principles of the organization.
- *to advise.* It will certainly be useful prior to or during an audit procedure to offer advice on design modifications which will help this happen.
- *to plan.* Better still, as part of the general planning process for all products, attempts should be made to establish design planning guidelines.
- *to specify.* As a more rigorous planning procedure for certain products it will be desirable to establish actual specifications for the design of those products.
- *to demonstrate.* As a way of demonstrating the effectiveness of a specification, it is sometimes useful to design directly and make either a model of the product, or the first of a production run, or a fully completed detail. (An example might be the first of a chain of shops for which a specification has been established.)
- *to implement.* The deepest level of involvement: actually undertaking the implementation yourself.

Employed as an analytical tool, the original matrix was our first formal step in developing a picture of the use of design in organizations. It also permitted those organizations who were considering developing the strategies for design to understand how broad the scope for design could be.

In dealing with the operational and organizational issues as they interact with the design process, the literature on complex and developing organizations has provided some useful definitions which have assisted in clarifying our task. One such is a working guideline of the roles of purposes of people within an organization, taken from Lawrence and Lorsch (1969).

An organisation is the coordination of different activities of individual contributors to carry out planned transactions with the environment.

Lawrence and Lorsch also describe an organization as 'a system of differentiated units which require integration and the view of the individual contributor as a complex problem solving system himself' (1969). This in particular is appropriate to the issues that emerged in the organizations that were talked to during the pilot study.

In all cases the authors asked the organizations to direct them to those people they considered to be the most active participants in design projects. This was to enable the authors to gain insight into the way the organization and the individual conceptualized their activity over design. On analysing the data received by using the original matrix, the authors discovered that the matrix itself needed development.

The two issues they found themselves unable to cover with the original matrix were firstly, the activity of the individual as a contributor to the design process, and secondly, the integration of the various design activities.

This resulted in the development of three new matrices. In using these the authors did not attempt to alter significantly the terminology or the structure, merely to arrive at a working tool to record and analyse the two issues described above.

These three matrices are shown. The inclusion in Matrices 1 and 2 of a separate category for 'input by professional designer' was one of the final amendments and reflected the growing awareness of the need to differentiate between activity by 'professional designers' and all other design activity.

The word design can be, and often is, utilised legitimately in many different activities. Consequently, the word 'designer' cannot refer solely to those groups of individuals whose education and training overtly equips them to operate as professional designers. With a separate category one can record in matrices 1 and 2 the activity of an individual who by education and training is qualified to produce design work. This differentiates between professional designers and all those individuals who are found to be active agents operationally in the design process irrespective of the degree of cognisance of their activity.

Design Matrix 1, 'Involvement of steps within artefacts' (Figure 3.1), is the closest to the original matrix. It allows activity to be recorded in the seven steps of involvement across the 11 artefact categories. It is important that it is not used to make comparisons and it operates as a map for the design activity recorded in the other matrices.

The second matrix was employed to record relationships at or between activity points. This is a very complex matrix as its lengthy title implies, 'Interaction between artefacts – involvement of steps'. Each of the seven steps on the vertical axis of the first matrix uses one of these matrices. Taken together they can be used as a three-dimensional model on the contour map principle, with cumulatively high spots of end activity forming the peaks of the map. Figure 3.2 shows the basic matrix.

However, each of the seven steps of this matrix must first be considered separately. This allows design activity to be considered in relationship to the eleven artefact categories. Figure 3.3 shows an example of the step 'specify' in an apparel manufacturing organization. Each square on the matrix is subdivided into two triangles. In certain squares both sides of the square are filled in; others only on one side. This is done to suggest the likely direction of the activity. 'Research and development' have an activity with 'sourcing', since R&D will test a fabric that has been sourced and only if the test is successful will the source be utilized. The major responsibility in the (specify) activity resides in R&D. However, it can also been seen that there is an 'input by professional designer' activity which also has responsibility toward 'sourcing'. With 'process' (under the product category) and 'equipment' (under the environment category) the direction of activity is shown as more balanced. In this instance there is also activity *toward* the 'input by professional designer' column, suggesting that he would be affected by activity in the two categories rather more than he would directly affect them.

This matrix series allows us to look at the relationship of design activities. However, the activity of the *individual* as a contributor is only implied. The third matrix

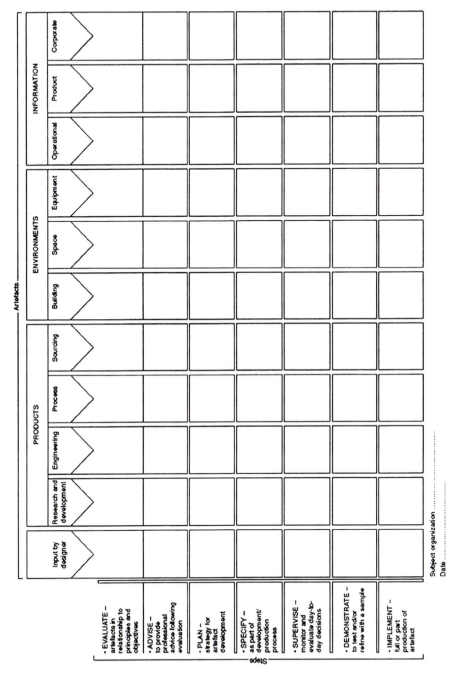

FIGURE 3.1 © *Design Studies,* 1987.

'involvement of personnel within steps' allows one to record an individual's activity and is illustrated in Figure 3.4. Here each of the seven steps is represented, while the eleven artefact areas are not. Each square on the grid is subdivided into twelve rect-angles. By referring to the personnel key one has a layout of the people involved.

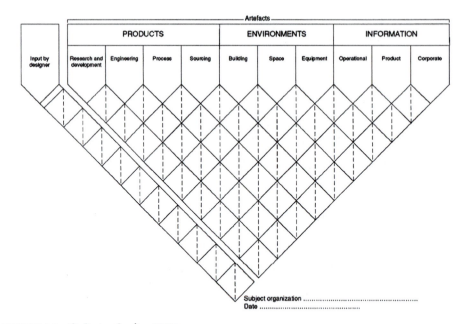

FIGURE 3.2    © *Design Studies,* 1987.

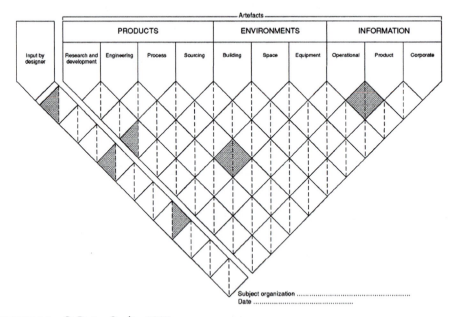

FIGURE 3.3    © *Design Studies,* 1987.

What the matrix does not in its present form illustrate is whether individuals operate as part of a team or singly or whether the activity is constant or sporadic.

Further adaptations of the matrix would enable us to quantify this and much other
information about the design activities of individuals and groups. For example one

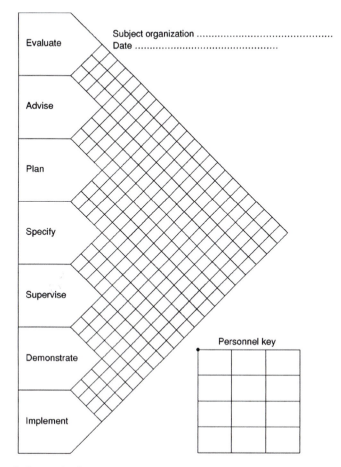

Subject organization .............................................
Date .................................................

Evaluate

Advise

Plan

Specify

Supervise

Demonstrate

Implement

Personnel key

FIGURE 3.4 © *Design Studies,* 1987.

may wish to know the rate at which they work, whether the design activity is continuous or intermittent.

However, before attempting to extend the matrix it would be clearly advisable to determine which kinds of design activities have priority in the eyes of the people participating in the design process. The questionnaire (to which the main text refers) is intended to provide priorities of this kind in our explanations of the place that design occupies in organizations.

## REFERENCES

CNAA (1985) Managing Design, An Initiative in Management Education, Report of a projected sponsored jointly by CNAA, Department of Trade and Industry and The Design Council.

Simon, Herbert, A. (1982) The Sciences of the Artificial, The MIT Press.

Lorsch, Lawrence (1969) Developing Organizations, Diagnosis and Action, Addison Wesley.

4

# The Designer and Manager Syndrome*

## SIR MISHA BLACK

*Two worlds have to be bridged – that of the executive in management who creates design policy and the designer who creates design. Management and designers have important responsibilities which go beyond the reaches of the business firm and the studio. Designers and management ought to share their concern for the environment and structure of society – a concern for 'man as a living, mutating, organizing, and dying entity'. Design is an attitude of mind and must permeate management and be more than a painting for the president's dining room or mural in the cafeteria. Lack of design understanding and awareness in our society is due to our world-wide educational system. 'Visual literacy and the cultivation of aesthetic discrimination are sacrificed to undue concentration on the development of intellectual faculties.'*

*A behavioural understanding of the social symbolism and acceptance of products are key social and marketing research tools for a designer and manager who must plan for future products.*

There is a chasm between talking about merchandising and being responsible for selling, between reading about surgery and actually inserting the scalpel, between being concerned about design and turning a blank sheet of drawing paper into instructions for manufacture. I am sure that many of the recent efforts to throw some bridges across the abyss which separates those who determine business policy from those who, as designers, give it physical form have helped somewhat to a degree, but we still

*Source:* Reprinted from Black, M. (1975) 'The Designer and Manager Syndrome', in T. F. Schutte (ed.) *The Art of Design Management – The Tiffany-Wharton Lectures on Corporate Design Management.* New York: Tiffany & Co., pp. 41–55. Courtesy of Tiffany & Co. and the University of Pennsylvania Press.

remain gazing at each other with the inbred suspicion of sparring partners. My intention is, therefore, to at least partially expose the potentialities and limitations of my
profession of industrial design.

We both face problems which did not greatly worry our ancestors. The purpose
of business, as the executive once saw it, was purely to make money; the duty of the
designer was to produce beautiful objects and environments. If he lowered his sights
to enable industry to increase its profits, he was almost certainly, he imagined, pros-
tituting his art. We are wiser, in some ways, than our grandfathers even if our clearer
vision has exposed new problems reaching to the horizon. The late 20th century attitude to business will be discussed later, but let me first explain what I believe should
now be the concern of the designer. To be a designer, as I understand this ill-defined
and much misused word, is to be conscious of, and accept, some responsibility for the
physical form of our world; to be continuously aware of the shape, size, color and texture of those parts of our environment which are man-made, of the interrelationship
of component parts, whether they be static or in motion, which produce a single object or a system; to be prepared to distinguish between those objects and relationships
which are aesthetically acceptable and those which fall below our personal standards.
'Design,' as Maholy Nagy said in the late 1920s, 'is an attitude of mind,' to which I
would add, 'and the capacity for tactile and visual discernment.'

But concern for the condition of our environment and the capacity (or assumed capacity) for aesthetic discrimination do not in themselves produce a designer. Concern
without the capacity for implementing change is the role of the consumer as critic;
to work as a designer, technical skills and experience are essential requirements. The
need to master complex techniques separates designers into fields of specialization –
architecture, engineering, graphic communication, the design of products for the industries based on the ancient crafts, and into many sub-divisions of these principal
fields. To be a designer, in the sense that I am now using this generic term, requires
not only the skills, but also the willingness to deploy them for the improvement of
the environment, rather than its desecration. To this extent design is, or should be, a
moral act undertaken within the constraints of the political, economic and social systems. It is a practical art, and as such different from the arts of painting, sculpture, literature and music, which can transcend the immediate present and open windows to
ecstasy. The fine arts, as Modrian has said, enable those who look at, or listen to, their
manifestations – to be conscious of 'the union of the individual with the universe.'
They have been described by your American philosopher Susanne Langer as 'the creation of forms symbolic of human feeling.'

Design operates at a more mundane level; its concern is with man as a living, mutating, organizing and dying entity. It is here that the business executive's interests
and mine, as a designer, coincide. The executive is involved in the structure of society
and so am I. His desire is to improve the physical condition of millions of our fellow
human beings who inhabit the world and the quality of our lives, and this equates

with the ambition of all designers who have escaped despair and enervating pessimism. Business is becoming a profession, design is achieving professional authority. It is now necessary to establish a basis for understanding and co-operation so that the business executive can harness design skills to further our common purpose. I have deliberately used the word 'harness,' as the initiative must come from management. A designer without a client is as impotent as an actor declaiming to an empty theatre. The first step in improving design standards is for management to decide that it wishes to do so. From this board room decision much good can flow, both in financial returns to the company and for society as a whole.

I have already suggested that design is an attitude of mind. This must permeate management if more is to be achieved than the purchase of some paintings for the President's dining room and the commissioning of a mural for the staff cafeteria – admirable though such patronage may be. Design is inevitably an aspect of many facets of business organization. All business executives are committed to the employment of designers, the only variant being that some are conscious of what they are doing and others somnambulistic. Design exists in the company letter heading, its trade symbol, the livery of its delivery vans, its factory and administration buildings, the furniture and tableware in its staff and executive dining rooms, in the products of its factories and the containers in which they are marketed. It is impossible to be in business without making a commitment to design; even a stockbroker has a corporate identity, while those who manufacture products or provide services are dependent on buyers' reactions for their existence.

The techniques for employing and co-operating with architects and graphic designers by enlightened management are well understood. I regret only that good intention does not more commonly achieve exemplary results. This is the fault of educational systems throughout the world, a world in which visual literacy and the cultivation of

FIGURE 4.1   Large-scale industrial catering. The staff restaurant for BP Limited. Designers: Design Research Unit, London.

aesthetic discrimination are sacrificed to undue concentration on the development of intellectual faculties. To compensate for their hereditary cataract, management should, during this intermediate period, look for wise counsel to support or question their aesthetic predilections. Such counsellors now exist in the United States and in Great Britain but they should be selected with the care and attention equal to that which characterizes management selection of chartered accountants and lawyers.

The role of the third type of designer required by production industries is less well understood. The function of the industrial designer is still too often envisaged as a luxury which can be employed or discarded at the whim of management. The industrial designer's specialization is two-pronged. Its first manifestation is based on anthropology and ergonomics. It is concerned with the relationship of the users or operators to artifacts or machines, be they simple domestic or office appliances, complex machine tools, agricultural equipment or control systems. It is the aspect of engineering design that determines whether a hand tool is properly shaped and balanced to ensure maximum efficiency in operation, whether the refinement of kitchen appliances is reasonably related to the capacity of housewives to understand their intricacy, whether the control system of an automobile contributes effectively to safe and enjoyable driving. The need for specialization within the product or system development team as well as the complexity of design development necessitates the devolution of specific industrial design responsibility in creating the anonymous team which is now usually the generator of new products and systems.

FIGURE 4.2   Design for Control – Collaboration between marine architects, equipment engineers and industrial designers. Design Consultants: Design Research Unit, London. Photograph by John Maltby Ltd.

The education of industrial designers enables them to participate in the creative processes which are essential to product and system development. This arises partially from the structure of their curricula, which exclude the depth of study in engineering science required of the mechanical engineering undergraduate, so as to provide time for divergent thinking and developing conceptual attitudes to product and system innovation. To these is added the advantage of studying industrial design in colleges of art and design where there is the inevitable tension engendered between students of the fine arts and of the useful arts. The argument and counter argument by which the artist and the designer attempt to defend their different but related activities heighten self-criticism and set standards of personal achievement and responsibility which are crucial in developing creativity.

The second specialization of the industrial designer is his overt concern with aesthetics, with the formal qualities of objects, with shape, texture and color, and with the visual and tactile relationship of the component parts of machines and products. Separated from mechanism and structure this becomes 'styling,' which aims only to encourage sales irrespective of social need; considered as a refinement of the mechanism and structure of industrial products it becomes 'style,' as much an attribute of product design and systems of engineering as it is of literature and music. This second specialization is not the exclusive prerogative of the industrial designer. Engineers and business executives are concerned with the total visual impact of the products for which they are responsible just as the industrial designers are – the difference is one of degree. The industrial designer has the advantage of this being one of his specializations, a task to which he is consciously devoted and for which he has been specially educated. In this way he can be usefully teamed with design engineers whose satisfaction is dependent more on the efficiency of mechanisms than on visual and tactile qualities.

Style is the signal of a civilization. Historians can date any artifact by its style, be it Egyptian, Grecian, Gothic, Renaissance, Colonial American or Art Nouveau. It is impossible for man to produce objects without reflecting the society of which he is a part and the moment in history when the product concept developed in his mind or was the creative outcome of a group sharing common attitudes and technical capacity. In this sense, everything produced by man has 'Style,' but this can be debased and perverted when factors other than the achievement of excellence become the dominant motivating forces. In its pure sense, style is 'an aesthetic sense based on admiration for the direct attainment of a foreseen end, simply and without waste. Style in art, style in literature, style in science, style in logic, in practical execution have fundamentally the same aesthetic qualities, namely attainment and restraint … with style the end is attained without side issues, without raising undesirable inflammations … style is the ultimate morality of the mind.' In his Presidential address to the Mathematical Association in 1916, A. N. Whitehead admonished his audience not to 'bother about your style, but solve your problems, justify the ways of God to man.'

We must attempt to relate this purist attitude to the design of industrial products as a whole, which seldom measure up to the Whitehead definition of style. The world is littered with products from the engineering industries disguising their mechanical efficiency (and sometimes inefficiency) with symbolic forms and decorative embellishment, and which have only a marginal relationship to engineering necessity and manufacture. These products are usually the outcome of aesthetic decisions having been made by managers or engineers who are unskilled in making such a judgment. Subjective aesthetic design problems are constantly posed in all design projects which allow for alternative solutions, as do all those which are not conceived at the frontiers of knowledge and mathematically determined. If there are five such decisions to be made during a design development program, and if each discloses ten alternative solutions, this can produce 100,000 design variations. If the number of decisions to be made and the possible alternatives increase, the number of possible design variants quickly reaches astronomical figures. The problem is compounded by the fact that some industrial products consciously need to serve symbolic as well as practical needs. This is exemplified by the automobile, but is equally apparent in the design of domestic appliances, office machinery and, to a lesser extent, in machine tools and agricultural equipment. The motoring correspondence of the British GUARDIAN, on 10 July 1972, described the Ford Capri 3000 E: 'The Ford Capri has a powerful looking bonnet, a racy tail and squashed up seating just like connoisseurs' cars that make sacrifices for speed. The Capri's sacrifices are for style. The bonnet is only two-thirds full of engine and the stubby rear makes the boot small … Ford, traditionally value conscious, has produced a stylish and successful car, and the 3000 on its own terms has the panache of a car twice its price.'

So long as people are not cheated by superficial design into believing that stylish elegance compensates for engineering negligence (and this is clearly not the case in the Capri, which is excellent value for money), I see no cause for puritanical objections to dressing up the ordinary with the glamour of the extraordinary. It differs only in materials and technique from making the visual best of one's personal appearance. The desire to obtain positive aesthetic pleasure and social status from the objects used by man is as old as mankind itself. A concern with the shape and decoration of man-made products is an endemic characteristic of the human race; it stretches from the decoration of neolithic pots to supporting the structure of 19th century beam engines by classical Grecian columns, from the decoration of Saracen scabbards to the form and livery of the Japanese Tokaido express train. The present need is not to disparage this aspect of style but to ensure that the formal decisions are appropriate to the object – to decide which artifacts and systems should be negative and self-effacing and which may proudly and aggressively acclaim the social and symbolic implications of their mechanical purpose. The motor car, high speed trains and television sets fall into the expressive, symbolic category; machine tools, electronic equipment, refrigerators and hospital equipment have willingly accepted a discreet anonymity in which their formal qualities are the outcome mainly of operational efficiency and economy in pro-duction. Some products which are initially expressive recede to a negative acceptance without social overtones and then later burst out again as positive social symbols. The telephone is an example of this see-saw process. In the early 1900's the telephone was both a utility and an indicator of social status; by 1914 it was hidden under dolls with crinoline skirts; by the thirties, it was accepted as a piece of domestic, commercial and industrial equipment which warranted no greater attention or commendation than its technical efficiency; by the 1970's it has again attracted symbolic, social attitudes as the availability of new types of instruments provides the opportunity for the type of personal choice and decision which can provide positive aesthetic pleasure and indi-cate social status. Already there are different models offered by the British Post Office. Computers are in the early stages of an aggressive-recessive cycle. They are still a source of pride to their owners but soon to be relegated to commonplace acceptance compa-rable with the boiler room and the air conditioning plant, now of interest only to the specialist. The automobile has moved in years from brash spaceman exuberance to a more subdued and mannerly concern with ground speed images.

In a competitive society, the need for manufacturers to be aware of the move-ments in public taste and its effect on sales is an essential factor in marketing strategy. The problem is compounded as attitudes towards products and systems change at ac-celerating speed. For a manufacturer planning a new product requiring design devel-opment, the tooling and initial production program may spread over 18–24 months

FIGURE 4.3    Industrial designers in collaboration with engineers: a train for London Transport. Con-sultant designer: Misha Black of Design Research Unit, London. Photograph by John Maltby Ltd.

and is no longer assured of public acceptance unless he can foresee what form will be acceptable in years' time by the targeted section of the market. The capacity to sense movements in public taste does not require clairvoyance. It requires market research (which usually operates negatively by eliminating catastrophically wrong decisions and may sometimes indicate useful lines of technical development) as well as a capacity for instinctive comprehension based on a knowledge and understanding of movements in painting and sculpture; an awareness of the attitudes of the most creative and experimental architects and designers; and of changing social values. What is happening one year in the studios and experimental workshops will inevitably later influence mass markets. The difficulty is in the discrimination between the mainstream and eccentric excursions into shallows, and ensuring that the time scale is correct. The industrial designer, by his association with the fine arts during his education, and his continued interest in their influence on public taste, is a useful ally of both management and the production team when forward decisions must be made which will influence marketing success or failure.

The need to prognosticate social acceptance, to decide whether products have moved from aggressive to recessive positions, is not only an aspect of our capitalist society. It has proved to be a requirement in the Soviet Union where public reaction to consumed products has also become selective now that the requirements of minimal existence are more easily satisfied. The U.S.S.R. has established its All Union Research Institute of Industrial Design (V.N.I.I.T.E.) to ensure that Soviet manufacturers become aware of the volatile and sometimes seemingly irrational movements in public taste on which the domestic and export sale of industrial products partially depends. The other Eastern European countries are equally conscious of the need to employ the specialized abilities of industrial designers to ensure that their engineering products reflect the style of our century and that their consumer products combine aesthetic satisfaction with efficient utility.

In this field of man–artifact relationships, the designers still operate largely subjectively – by hunches which they would find difficult to describe or justify, except by results. The considerable volume of research work on design method has been concerned with practical problem solving, and tends to gloss over aesthetic problems as not being a suitable subject for objective analysis. There is a need for more theoretical and case study of the morphology and typology of man-made objects. Until more research work is undertaken we can do little better than follow the advice of Palladio who, in the 16th century, wrote: 'Although variety and things new may please everyone, yet they ought not to be done contrary to that which reason dictates.'

Reason dictates to management that they should seek specialized design skills to augment their own abilities and experience. This can be done by establishing an effective design office within the company organization, by the management of consultant designers, or by a combination of both. The latter is normally the most fruitful method in large scale industry. The company design office is in close daily liaison with those other departments which utilize or affect its work: with research, product planning, production engineering, sales and publicity. The essential modifications to products which must be made at short notice, the need to watch deviations from the established house style, and the opportunity for educational work within the organization all fall naturally within the province of the inside design office. The consultant makes periodic visits to advise and criticize. He brings to one industry his experience in many ways; not only can he talk with the company designer on terms of professional equality but he can also talk to management with a freedom not normally enjoyed by employed staff.

Whatever organizational permutation may best meet the needs of a specific industrial or marketing group, one irreducible fact remains – the quality of the design program is absolutely dependent on the capacity of management to appreciate its potentiality. Management is an invisible participant in every design project. The most frequent comment in any design office is: 'It's no use, *he* will never accept it – you couldn't possibly persuade him to have it.' Hence a brilliant innovation is sadly filed away in favour of a mediocre solution which will offend no one, and will make only a modest contribution instead of a leap forward. Fortunately, this is not the inevitable rule. There are too many brilliant design innovations in American and British industry to justify despair. Yet, for every innovation which sees the light of manufacture and marketing, a dozen are neglected for the lack of management confidence in the capacity of their designers, and for the lack of confidence of the designers in the vision of their directors or clients.

The essential quality of design is creativity: the capacity for predicting what technology will make possible and what people will desire and need a year or two ahead. As Swift wrote: 'Vision is the art of seeing things invisible,' and the invisible future will be predicted by designers who have been trained to make and then test creative hypotheses. For designers to work effectively, they need the sympathetic understanding of management. I am sure that the new breed of managers will provide the oyster shell in which designers can be the essential irritant.

Around 1881, when Joseph Wharton established his now world-renowned business school, Queen Victoria wrote: 'Beware of artists, they mix with all classes of society and are therefore most dangerous.' Designers are concerned only with the practical arts, and the danger is thus diminished, but they are amongst those who are willing to accept a responsibility for the look and feel of our environment. And, if this involves action which disturbs those who wish only to conserve the social and environmental patterns of the past, then Queen Victoria was right and shall continue to be so.

But we are no longer isolated. The council of the Confederation of British Industries has recently accepted a report from one of its committees which reads: 'A company, like a natural person, must be recognized as having functions, duties and moral obligations that go beyond the pursuit of profit and the specific requirements of legislation ... profit alone is not the whole of the matter.' The moral obligation includes a respect for our man-made environment and the elements of which it is comprised. The new generation of business executives and their natural allies the designers, can ensure that the industries under their control will improve the physical world and not degrade it. As Camillo Olivetti said more than 50 years ago: 'A good business is one which *also* makes money.'

I have so far talked about design as though it were a straightforward respectable activity undertaken by men and women who differ only in their technical knowledge and specialization from the business executives with whom they must collaborate. Viewed from one position this is an accurate image of the designer's persona. When design is an ordering process that changes the inventors' or development engineers' lash-up into a marketable product, or concerns itself with industrial or civic tidiness and good manners, it does not differ greatly from any other professional occupation. But this is only one facet of design which has achieved exaggerated importance in our society, because only those specifically trained to see are able to perceive the physical world and suggest how it might be improved. While the world remains sick, all who are not motivated solely by greed must combine to construct crutches so that society can continue to exist while it heals itself. Our democratic societies are, however, not monolithic. They provide opportunities for immediate manifestations of the human spirit even if development of the body politic as a whole must be intolerably slow. There are moments, even in the most staid of businesses, when a new product or service can be perceived, when a new headquarters building or factory can be planned, when an outmoded corporate image is ripe for complete reassessment. These are moments of excitement in which all those involved may share. But they are also the moments when professional competence alone is insufficient. For innovation, creative management must be linked with design creativity, when the visionary capacity of the designers subsumes their daily plain competence. Designers with the capacity for innovation are thin on the ground in all countries – but they do exist, and a place should be found for them within the structure of industry and business. Such designers may be uncomfortable companions; they may be motivated by forces which do not answer to the bridle of normal social behaviour; they may well be the odd man out in the necessarily tidy industrial hierarchy, but the odd ball often bounces higher.

Design when it is creative is not a tidy affair: it is a search for perfection in an imperfect world. But business executives will tolerate its nonconformist intolerance, will discover how to harness creativity to their practical needs – then a partnership can be established which will make their working lives more exciting (if less ordered and comfortable) and may make the products and services which they control contribute to a future which will make our present, in retrospect, appear to have been sadly inadequate.

# 5

# Organising Design Activities[*]

## MARK OAKLEY

## INTRODUCTION

The discussion of the Beta Engineering case in the preceding chapter was intended to highlight the factors which may influence the outcome of design projects. Prominent amongst those factors was the question of organising for design activities. Senior management must consider very carefully the special features of design work and design departments; it is then necessary to decide how best to structure such departments and how they may work with and relate to other parts of the company.

Several different structures may be encountered in practice and some of these are discussed in this chapter. Some companies organise themselves in the Beta fashion with separate design units headed by a senior manager to whom is delegated (or abandoned) part or all of the responsibility for product design and strategy decisions. At the extreme are companies whose design activities are diffused throughout the organisation. In this case, top management may or may not set policy guidelines or seek to ensure some degree of conformity in design results.

Some companies identify a role for 'product champions' to push new product projects through any organisational or procedural barriers which may be encountered. In such companies, a 'task force' approach is quite common with small groups of engineers, designers and others assigned to specific projects. Other companies may prefer to use a more broadly based method, perhaps built on one or more large units within which there are a number of projects. Individuals move in or out of these projects depending upon the need for their particular skills. It is clear from the available literature

[*]*Source*: Reprinted from MANAGING PRODUCT DESIGN by Mark Oakley, 1984, published by Weidenfeld & Nicolson, an imprint of the Orion Publishing Group, London: 49–63.

that there is little agreement about the precise circumstances in which each method should be applied, although it is possible to derive guidance in broad terms about selecting a suitable approach.

Whatever organisational structure is adopted, leadership and management will be required. Some experts see a need for a special kind of 'design manager' who, amongst other functions, is able to bridge the knowledge and language gap that frequently exists between ordinary managers and designers.[1,2]

Hence, the purpose of this chapter is to examine these organisational questions and to try to give an idea of the advantages and disadvantages of the alternative structures that may be adopted. The special demands placed upon those who manage design are compared with the demands placed upon managers in other functions. Following from this, it is possible to draw some conclusions about the personal characteristics that should be sought in those who manage design activities.

## CONCEPTS OF CHANGE

All managers are confronted by change – if no change ever occurred then there would be no need to have managers at all. If markets were static or totally predictable, no-one would need to take decisions about production quotas. If customers' tastes and loyalties never changed then no new products would be needed. But, of course, in the real world change is an ever-present fact of life. For millions of different reasons – some natural, some man-made, some accidental, some deliberate – all companies must continually adjust their activities to compensate for changes which are taking place. In the same way that changes in the market dictate a need for product design, the act of designing new products, or improving old ones, gives rise to some degree of change within the company. This change may be slight, as when modifications are made to minor components, or it may be major, as in the case where a factory must be totally re-organised and re-equipped to accommodate a new product.

Regardless of its extent however, many organisations will resist change right up to the point where survival of the organisation is seen to be at stake. In the normal course of events, preoccupation with current activities may leave little scope for any real innovation. Many firms demonstrate this attitude by paying lip-service to their product design activities whilst proclaiming official doctrines of innovation. Such firms may, for example, encourage new product ideas, only to find consistently that none of them meets the stringent criteria laid down in advance.[3] Similarly, many firms effectively eliminate change by oscillating between support and resistance – an 'on-again, off-again' approach to design and innovation.

Not infrequently, a myopic concentration on the manufacture of existing products means that new products are never successfully developed. Greiner[4] is critical of this attitude to new methods and products. He reinforces the opinion that successful change does not begin until strong environmental and internal pressures 'shake the power structure at its very foundation'. A more recent publication,[5] appearing in the depths of the worst economic recession for 50 years, still finds it necessary to challenge managers to consider the implications of failures to adjust to change and urges them to be receptive to innovation.

According to Twiss,[6] resistance to change is more often a feature of older companies than younger ones. He has found that as companies reach maturity, strategies become more defensive and fewer projects lead to new products which depart substantially from current practice. He believes that the reason for this often lies with the 'Chief Technologist' who is normally found to be the creative force in a young company but is less effective as the company evolves and a management team is built up around him. This means that in the mature company it becomes necessary to design some formal approach in order to cope with change.

An American study by Lynton[7] has examined large and small organisations and found that each may face circumstances in which they can no longer deal with change by intuitive means. It may be thought that an organisation's size is the primary factor in handling change. But Lynton has found that a decision to formally cope with change is not so much related to the size of the organisation, as to the degrees of uncertainty in technology, markets and the environment. The necessary redesign of the organisation into one which can accommodate higher degrees of uncertainty is invariably more difficult in older companies but is not markedly so in larger companies as opposed to smaller ones. New designs always involve some change and for this reason they may be resisted. After all, whilst the modern company is often built around the production process which is (usually) rational and standardised, design is not always seen as rational and can be disruptive to those affected by it. Bright[8] offers 12 reasons (Table 5.1) why an innovative change such as a new design may be resisted by the employees of a company.

It is interesting that the book in which the list first appeared was published some 20 years ago – and it is 60 years since the pioneering work of Elton Mayo and others[9] drew attention to the importance of developing good human relations within companies. Yet still the lessons do not seem to be generally understood. It is true that there are companies apparently able to adjust to change with the support and commitment of the whole workforce. But there seem to be many others which are quite unable to cope because their employees, including managers, fear change and consequently block the progress which is necessary for survival. It would be misleading to generalise about underlying causes, but since management is given (or assumes) responsibility for managing, this is where the basic problem must be sought. Managers who cannot tolerate uncertainty and seek to avoid change, or who are unable to communicate with their subordinates or see no need to do so, are not suited to the task of managing product design or similar change-inducing activities.

**TABLE 5.1: Reasons for resisting an innovative change**

1. To protect social status.
2. To protect an existing way of life.
3. To prevent devaluation of capital invested in an existing facility.
4. To prevent a reduction of livelihood because the change would devalue the knowledge or skills presently required.
5. To prevent the elimination of a job or profession.
6. To avoid expenditure such as the cost of replacing existing equipment.
7. Because the change opposes social customs, fashions and tastes and the habits of everyday life.
8. Because the change conflicts with existing laws.
9. Because of rigidity inherent in large or bureaucratic organisations.
10. Because of personality, habit, fear, equilibrium between individuals or institutions, status and similar social and psychological considerations.
11. Because of tendency of organised groups to force conformity.
12. Because of the reluctance of an individual or group to disturb the equilibrium of society or the business atmosphere.

Not only is change itself important, so is the rate of change in a company. A firm has to change at a sufficient rate and in an ordered manner to meet the conditions imposed from outside. A rate which is too high leads to chaos; one which is too low may end in bankruptcy. Also, in order to achieve the goal of prosperity which is the reason for the existence of most companies, the effects of competition demand that each subsequent design change has to be done a little better and a little more profitably than the preceding one. It is disturbing that so few organisations seem prepared for this challenge. Many are engulfed in systems which perpetuate conformity, precedent and procedure, and reaction to crisis, continues to be the primary model of adjusting to change.

The problem is huge (and in its entirety is quite outside the scope of this discussion) involving issues of governmental policies, the attitudes and activities of trade unions, divisions of wealth and power and the availability of resources such as fuel and materials. However, as already noted, one major factor is the style of management which is practised and encouraged in design and other innovative activities. Efficiency in recognising the need for new products and the careful screening and selecting of ideas will all be in vain if the management of the company is not suited to the special requirements of new product work. The nature of these requirements can be better understood if the special features of design departments are compared with those of other departments, particularly production with which design is often closely linked, since both are often considered to be the 'technical' parts of the business and to have

**TABLE 5.2: Comparison of organisational features of production with those of product design**

| Features of Production | Features of Design |
| --- | --- |
| 1. Rational, standardised, predictable | 1. Irrational, novel, unpredictable |
| 2. Operations accurately timed | 2. Accurate timing of activities usually impossible |
| 3. Long runs of identical products | 3. Activities frequently changing |
| 4. Creativity and initiative not developed in workforce | 4. Highly creative personnel essential |
| 5. Work closely controlled – essential for profitability. Risk eliminated | 5. Profitability related to skill, change, judgment, intuition, risk taking, etc. |

similar methods of operation. Table 5.2 summarises the main organisational features which distinguish design and production.

Comparison with other functions such as marketing or finance will also show differences of emphasis; for example, marketing's need for short lead times on new products compared with design's concern to spend as long as possible on projects to achieve the best results. In the light of these differences, we need to consider both the organisation of the design unit itself, and its relationship with the rest of the company.

## ORGANISATION OF PRODUCT DESIGN UNITS

A number of years ago Burns and Stalker[10] analysed the organisational aspects of a sample of firms involved with the design of new products; their work remains important today. They observed that within these companies there were organisational styles ranging from what they termed 'mechanistic' which were very formal, hierarchical, bureaucratic and inflexible, to styles which they termed 'organic' which were informal, based on teams and tended to 'shape' themselves to the problems being tackled.

They concluded that mechanistic systems work satisfactorily only where conditions are relatively stable – flow line production departments for example, or other situations where close control of highly specialised work is essential. Mechanistic forms of organisation are not likely to prove satisfactory when applied to design units, which need flexibility in many respects. Here organic systems are more appropriate and, as Burns and Stalker observed, such systems improve the prospects of success for new products. Table 5.3 lists some of the features that may be found in organic systems. Managers responsible for product design departments need to consider how they can promote 'organic features'.

Promoting such features may be a delicate matter, especially within those firms which are otherwise organised along precise and inflexible lines. Even in situations where they do not have the opportunity to develop ideal systems, managers should

**TABLE 5.3: Typical features of organic systems**

1. Unifying theme is the 'common task' – each individual contributes special knowledge and skills – individual's tasks are constantly re-defined as the total situation changes.
2. Hierarchy does not predominate – problems are not referred up or down, but are tackled on a team basis.
3. Flexibility – jobs not precisely defined.
4. Control is through the 'common goal' rather than by institutions, rules and regulations.
5. Expertise and knowledge located throughout the organisation not just at the top.
6. Communications consist of information and advice rather than instructions and decisions.

be aware of the actions they can take to assist creative work so that organisationally desirable features predominate. Whitfield,[11] discussing a range of issues associated with creativity and innovation, shows that design teams are likely to be most effective where:

- All members make a full contribution; co-operation is accepted as the way of achieving the best result.
- Short term leadership tends to rotate according to the immediate needs of the job.
- Decisions are made by the people who are best informed on the subject. These are not necessarily the most senior present.

It is important to understand that mechanistic and organic styles are categories of organisation which are unlikely, in practice, to be found in 'pure' forms. There are many intermediate stages between these categories and most design units will exhibit features of both styles, with the 'right balance' depending on the nature of the company, the industry and the projects being undertaken. In all cases, the best policy for the design manager is to be vigilant for the evolving of mechanistic characteristics and then take steps to prevent them becoming dominant within the design operation. The kind of features that need to be checked include:

- Increase in the number of levels of supervision.
- Control by detailed inspection of work methods rather than by evaluation of results.
- Communication consisting of instructions and decisions rather than exchange of advice, consultation or information.
- Confrontation of win-lose nature rather than collaboration.
- Insular attitude of top management with sense of commitment to past decisions.

**TABLE 5.4: Differences in managerial roles**

| The Traditional Manager | The Design Manager |
|---|---|
| 1. Experience-based know-how. Education as one-time activity. | 1. Managerial knowledge based on recurrent up-dating. |
| 2. Technical and analytical skills emphasised. | 2. Skills needed to deal with ambiguity, complexity and conflict. |
| 3. Expectation of continuity of organisational experiences. | 3. Ability to adapt to unpredictable new events. |
| 4. Standard operating procedures guide decisions. | 4. Decisions augmented by environmental and other inputs. |
| 5. Inward perception emphasises internal, issues, competition. | 5. Inward/outward perception includes societal problems. |
| 6. Importance attached to stable relationships. | 6. Temporary relationships tolerated. |
| 7. Assumes rational organisational behaviour. | 7. Rationality seen as subjective. |
| 8. Task orientated. | 8. Goal orientated. |
| 9. Action oriented to keep physically busy. | 9. Combines periods of reflection with action. |
| 10. Individualistic approach to specialised problem solving. | 10. Interdisciplinary team approach to complex problem solving. |

The role of the design manager is clearly a crucial one in promoting successful results. In many respects the qualities required in order to be effective in this job are substantially different from those traditionally exhibited by managers. The main emphasis must be on the design manager's ability to deal with change and ambiguity;[12] Table 5.4 summarises the main differences.

## LOCATION OF DESIGN WITHIN THE FIRM

It is not always easy to decide where design activities should be located within the firm. As already stated, product design is often considered to be a technical activity of a kind similar to production. However, because of the fundamentally different natures of the two functions, giving control of design to production may result in failure. This may happen either because organisational conditions inappropriate to design are imposed or simply because production has resistance to new products (because of the disruption involved). This resistance may take the form of constant rejections of new designs, refusal to supply information and help, or just general obstruction – all while paying lip-service to the need for new products. These attitudes may be particularly acute in long-established firms where design work has been limited previously to improvements and modifications. Nevertheless, in a great many companies design departments, are to be found as part of the production set-up, under the control of a production or technical manager, as in Figure 5.1.

It will be noted that design is effectively isolated from the highest level decision maker, the chief executive. Information and directives may be 'filtered' by production to the extent that design work is reduced to a mundane level. Modifications to existing designs and extensions to current ranges of products may well be achieved, but more ambitious projects are likely to be stifled. Any work that is undertaken will be subject to pressure to ensure that production considerations are given paramount attention – at the expense of other requirements including those of the customer. Several of the case studies in this book highlight the limitations caused by treating design as a component of production. Very similar problems may arise if design is organised as a part of one of the other functions such as marketing.

Some companies seek to avoid these problems by establishing design as an independent department equal in status to the other major functions, as shown in Figure 5.2.

This may well lead to improved design performance because with direct linkage to the highest decision level, projects can be monitored and encouraged. Radical projects may be attempted and some success anticipated. However, some drawbacks will remain. The design manager (or design director) may have to bargain with other

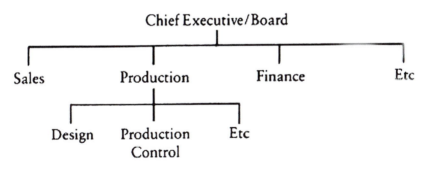

FIGURE 5.1   Design as part of production.

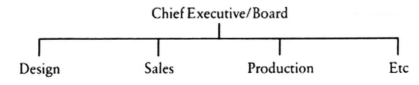

FIGURE 5.2   Design as an independent function.

departmental bosses for cash and resources. Because design is a long-term activity it may seem to senior managers in other functions that it is just a wasteful consumer of the income which they create. Boardroom politics may work to the disadvantage of the design function; power is often related to the size of financial budgets. Sales directors or finance directors whose scale of operation may be many times greater than that of design, might acquire or assume correspondingly greater influence in decision-making. A further disadvantage with an independent design unit is that it may well be viewed with suspicion by the rest of the firm. The fact that it *is* separate from other functions may cause speculation about its relevance to company objectives. This usually manifests itself when a new product is ready to be handed over at the end of a design exercise. The product may be rejected because it is seen as an intrusion from outside – sometimes referred to as the 'not invented here' syndrome.

Some companies try to overcome this problem by directing new product operations through a steering committee which represents all major functions. Others appoint 'project champions' whose job it is to push the new product through all barriers and solve any problems which may arise. Unfortunately, project champions are not often given the authority which is a necessary for them to be effective. Companies sometimes delude themselves into believing that a bright young manager will be able to achieve the progress that has been denied to others simply by force of personality. Frequently, the problem is not one of leadership at all, but of availability and allocation of resources. For example, if the design unit needs to use a particular manufacturing process for test purposes and that process is fully occupied with production work designated as top priority by the Board, then no amount of dynamic design management will ensure success.

Despite these reservations, the steering committee approach as summarised in Figure 5.3, does offer a number of advantages. The most important is the potential ability of the committee to ensure both the relevance of new design activities and the acceptability of design results.

The intimate involvement of the committee members, drawn from all major areas of the company, should promote well-integrated design, in theory at least. In practice, the method may be less successful. As well as the usual problems associated with management by committee – compromise, indecision and procrastination – individual managers may still succeed in undermining projects which they find undesirable. These managers may find that conflict arises out of their dual roles (as departmental

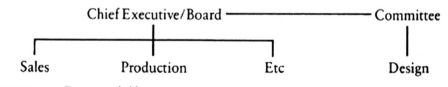

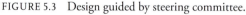

FIGURE 5.3   Design guided by steering committee.

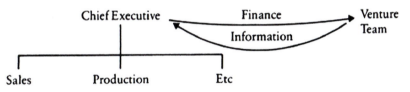

FIGURE 5.4   The venture concept.

managers and as steering committee members) and that this causes the taking of inappropriate decisions.[13] The use of a 'project champion' in place of a committee may
avoid many of these problems, although selecting the right person for the job assumes
crucial importance. Ideally, the champion needs to have all the qualities of the change-
responsive manager listed in Table 5.4. In addition, he must be able to understand the
technical aspects of the project and do his best to deliver an acceptable new product
within the deadlines set. Whilst satisfying colleagues elsewhere in the firm, he must
always see the completion of the project as the main priority and he must inspire his
team towards this goal.

In order to encourage this approach, some firms move one step further still and, except for a 'financing link', effectively set free the design project so that it can grow and
mature independently of the main organisation. This concept of organising for new
products is sometimes referred to as the 'venture method'. It is especially attractive for
traditionally organised companies which find it difficult to manage design within its
normal boundaries. By setting up design projects as 'mini-businesses', a high degree of
success may be achieved.

## ORGANISING FOR DESIGN OUTSIDE THE COMPANY

Up to now, the discussion has been based upon the assumption that design work is
being carried out within the company. Of course, this is not always the case; for a variety of reasons, design may have to take place outside the company. Projects may be
handed over to a specialist outside design consultancy when suitable resources do not
exist within the firm, or where they are already fully stretched by other projects. Typically, this may happen when a firm is attempting to move into a new product area of
which it has little direct technical experience; rather than build up a new design group,
it may decide to go outside, at least as far as the 'mark one' version of the product is
concerned. Once the new concept has established itself in the market, it may then be
appropriate to set up an internal design operation.

Other firms may use outside design consultancies as concept generators. In order
to get around any creative blocks within the company the initial design work may be
done outside followed by a shortened development programme of testing and refinement inside the firm. Finally, a few companies make a practice of sub-contracting all
of their design activities, usually in the expectation of achieving very high standards in

design results coupled with short project durations. This approach is most commonly found amongst small companies, but at least one major European manufacturer, with an outstanding reputation for the quality and design of its products, also follows this policy and does not directly employ any designers.

In such circumstances, the role of the design manager as a co-ordinator becomes highly important. His main concerns are no longer to do with motivation and leadership – he is now responsible for ensuring the continuing relevance of the externally based design work and for monitoring progress against time and cost budgets. In this respect, the necessary administration may become extremely tedious as Topalian[14] shows in his discussion of the documentation required to keep track of projects – job cards, job files, progress reports etc. Also the design brief assumes even greater importance than in the case of in-house design work where ambiguities or problems may be readily resolved. Once a brief has been issued to an outside design group the scope for amending errors is much reduced. Indeed the brief is likely to form part of the contract between the two parties so it may need to be drawn up with legal precision. The design manager is responsible for this and he or she should also be the focus of communications. If a problem does arise, the design consultant should not have to telephone around the client's company in order to find the answer to a question.

In addition to the advantages already mentioned, one attractive feature of commissioning design work outside the company is that it is often easier to abort unsuccessful projects. The whole business is less personal than where emotionally-saturated design projects inside the company remain in continued existence, even as their prospects become ever more bleak, because everyone has a personal stake and no-one has the heart to 'pull the plug'. As each contract period with the outside designer comes to an end, a cool assessment may be made of the real progress and prospects before a further stage is embarked upon.

As well as the design brief, it may be necessary to provide the outside designer with a copy of the company's design manual, if one exists. This is the best way of conveying technical information such as approved colours, typefaces and logos, all of which are necessary in maintaining the corporate identity of the company. Another form of outside design work is where non-specialist managers in very large organisations have to commission design work on a local basis; for example, in large retailing chains, banks, building societies, and large, multi-plant manufacturing companies. Clearly, it would be impossible for a centrally-based design manager to personally co-ordinate all design projects throughout his company. Instead the work must be delegated to local staff and the design manager's responsibility is to provide technical back-up. In this case, the design manual may provide step-by-step instructions on how to hire designers, assess costings and budgets, prepare a brief, evaluate prototypes or 'mock-ups', choose between alternative design solutions, and so on. Only when a problem arises which is outside the scope of the design manual need reference be made to the design manager.

The design manual should be comprehensive enough to indicate when these problems may occur and advise referral within its step-by-step format.

## CONCLUSION

In this chapter, an attempt has been made to highlight the organisational features associated with design work. The most important point is that to design new products, or to modify existing ones, always involves some degree of change. Whatever the scale of this change, reaction to it within the company may well be negative and sometimes will be sufficiently intense to kill off a design project before its outcome can be properly evaluated. This calls for special design management skills to maintain a good working relationship between the design unit and the rest of the company. The form of management practised within the unit is also important and some guidance was given about features to be developed and those that should be discouraged. Finally, the management of design work sub-contracted outside the company was considered; here it was seen that administrative skills are most important.

Together with the issues raised in the preceding chapters of this book, it will now be apparent that the responsibilities of design management can be very wide indeed. As well as a responsibility for planning for design and preparing a product strategy, design management is also concerned with 'picking the winners' – evaluating product ideas – and selecting the right designer for the job within or, from outside the company, or putting together the best design team. The next chapter discusses the venture concept at some length and provides a case study illustration of some of the benefits and limitations of the approach. An interesting insight into the way design relates to other functions in a company is provided by a recent report[15] on the organisation of design in several different companies within the footwear industry. Also, while not dealing specifically with design organisation, Blakstad's case studies[16] highlight more of the problems.

## REFERENCES

1. Farr, M., *Design Management.* Hodder and Stoughton, 1966.
2. Topalian, A., 'Designers as Directors.' *Designer,* February 1980.
3. Schon, D.A., 'The Fear of Innovation' in *Uncertainty in Research Management and New Product Development,* Ed. Hainer R. M. et al. Reinhold 1967.
4. Greiner, L.E., Patterns of Organisational Change, *Harvard Business Review* pp 119–130, May/June 1967.
5. *Innovation in Manufacture,* Institution of Production Engineers, London 1982.
6. Twiss, B.C., *Managing Technological Innovation,* 2nd edition, Longmans, 1980.
7. Lynton, R.P., Linking an Innovation Sub-system Into the System, *Administrative Science Quarterly,* pp 398–415, Sept 1969.

8. Bright, J.R., *Research, Development and Technological Innovation,* Irwin, 1964.

9. Roethlisberger, F.J. and Dickson, W.J., *Management and the Worker,* Harvard University Press, 1939.

10. Burns, T. and Stalker, G.M., *The Management of Innovation,* Tavistock Publications, 1966.

11. Whitfield, P.R., *Creativity in Industry,* Penguin, 1975.

12. Basil, D.C. and Cook, C.W., *The Management of Change,* McGraw-Hill, 1974.

13. Mueller, R.K., *The Innovation Ethic,* American Management Association, 1971.

14. Topalian, A., *The Management of Design Projects,* Associated Business Press, 1980.

15. Oldham, S.W., *Design and Design Management in the UK Footwear Industry.* The Design Council, London 1982.

16. Blakstad, M., *The Risk Business,* The Design Council, London 1979.

## 6

# Design: A Powerful but Neglected Strategic Tool*

## PHILIP KOTLER AND G. ALEXANDER RATH

*Design is a potent strategic tool that companies can use to gain a sustainable competitive advantage. Yet most companies neglect design as a strategy tool. What they don't realize is that good design can enhance products, environment, communications, and corporate identity.*

In this era of intensifying global competition, companies are searching for ways to gain a sustainable competitive advantage in the hope of protecting or improving their market positions. A great many industries are characterized by intense service and/ or price competition that only succeeds in driving down everyone's profits to an unhealthy level. One of the few hopes companies have to 'stand out from the crowd' is to produce superiorly designed products for their target markets.

A few companies stand out for their design distinctiveness, notably IBM in computers, Herman Miller in modern furniture, and Olivetti in office machines. But most companies lack a 'design touch.' Their products are prosaically styled, their packaging is unexciting, their information brochures are tedious. Their marketers pay considerable attention to product functioning, pricing, distribution, personal selling, and advertising, and much less attention to product, environment, information, and corporate identity design.

Many companies have staff designers or buy design services, but the design often fails to achieve identity in the marketplace.

*\*Source*: Reprinted from Kotler, P. and Rath, G. A. (1984) 'Design: A Powerful but Neglected Strategic Tool', *The Journal of Business Strategy* 5(2): 16–21. © Emerald Group Publishing Limited all rights reserved.

The following real (though disguised) example is typical of many managers' attitudes toward design:

Steven Grant, an entrepreneur, visited one of the authors and described a device he was developing called the Fuel Brain, which monitors room temperature and controls the heating and air circulation functions of oil furnaces. When asked whether he would use professional design services to assist in this venture, he said there was no need. His engineer was designing the product. His next door neighbor was designing the logo. His marketing officer was designing a four-page brochure. The Fuel Brain would not need any fancy packaging, advertising, or general design work, because he felt that the product would sell itself. Grant believed that anyone with an oil burning furnace and a desire to save money would buy one. A year later, upon being recontacted, he sadly explained his disappointment in the sales of the Fuel Brain.

One only has to look at current U.S. products in many product categories – kitchen appliances, office supplies, air conditioners, bicycles, automobiles, and so on – to acknowledge the lack of good design. Yet its potential rewards are great. Consider the dramatic breakthroughs that some companies have achieved with outstanding design:

- In the stereo equipment market, where several hundred companies battle for market share, the small Danish company of Bang & Olufsen won an important niche in the high end of the market through designing a superbly handsome stereo system noted for its clean lines and heat-sensitive volume controls.
- In the sportscar market, Datsun endeared itself by designing the handsome 240Z. For most buyers before 1976, the 240Z was a dream car at an affordable price, around $4,000–$6,000. The latest copy is by Mazda, which coupled innovative pricing with the 240Z design, capturing a large share of the sportscar market with its first offering, the RX7.
- In the hosiery market, Hanes achieved a dramatic breakthrough in a mature market by using creative packaging design and modern packaged goods marketing techniques, catapulting the L'eggs division to the position of market leader. The L'eggs boutique (in-store display) used information design effectively, pulling consumers from other stores and brands. Design was a key component in the marketing strategy and created instant product recognition for the brand.
- In the kitchen furnishings market, Crate & Barrel selects products for its retail stores that meet good standards of material, finish, form, and color. Most of the products are Italian and Finnish. The look has become so well entrenched that many consider it to be the standard in kitchen furnishings. Crate & Barrel also designed environments to promote traffic and used seconds of expensive products as loss leaders. Once again, good design is used as an element in a marketing strategy.

Well-managed, high-quality design offers the company several benefits. It can create corporate distinctiveness in an otherwise product- and image-surfeited marketplace. It can create a personality for a newly launched product so that it stands out from its more prosaic competitors. It can be used to reinvigorate product interest for products in the mature stage of its life cycle. It communicates value to the consumer, makes selection easier, informs, and entertains. Design management can lead to heightened visual impact, greater information efficiency, and considerable consumer satisfaction.

This article aims to help company strategists think more consciously and creatively about design leadership and to help company marketers work more effectively with designers. It addresses the following questions:

- What constitutes effective design?
- What keeps executives from becoming more effective design managers?
- How can a corporation's design sensitivity be measured?
- How can the interface between marketers and designers be improved?

## WHAT CONSTITUTES EFFECTIVE DESIGN?

The term 'design' has several usages. People talk about nuclear plant design and wallpaper design even though the two emphasize different design skills – those of functional versus visual design. Design also appears in the description of higher priced products, such as designer jeans and designer furniture.

Certain countries – notably Italy, Finland, Denmark and Germany – are often described as being outstanding in design. These countries use design as a major marketing tool to compete in world markets. Even here, design connotes different qualities depending on the country: Italian design is artistic, Finnish design is elegant, Danish design is clean, and German design is functional.

Design is also used to describe a process. Pentagram, the noted British design firm, sees design as a planning and decision-making process to determine the *functions* and *characteristics* of a finished product, which they define as something one 'can see, hold, or walk into' [1]. Our definition of design is as follows:

Design is the process of seeking to optimize consumer satisfaction and company profitability through the creative use of major design elements (performance, quality, durability, appearance, and cost) in connection with products, environments, information, and corporate identities.

Thus, the objective of design is to create high satisfaction for the target consumers and profits for the enterprise. In order to succeed, the designers seek to blend creatively the major elements of the design mix, namely performance, quality, durability, appearance, and cost. These elements can be illustrated in the problem of designing, for example, a new toaster:

- *Performance*. First, the designer must get a clear sense of the functions that the target consumers want in the new product. Here is where marketing research come in. If target consumers want a toaster that heats up rapidly and cleans easily, then the designer's job is to arrange the features of the toaster in a way that facilitate the achievement of these customer objectives.
- *Quality*. The designer faces many choices in the quality of materials and workmanship. The materials and workmanship will be visible to the consumers and communicate to them a certain quality level. The designer does not aim for optimal quality but affordable quality for that target market.
- *Durability*. Buyers will expect the toaster to perform well over a certain time period, with a minimum number of breakdowns. Durability will be affected by the product's performance and quality characteristics. Many buyers also want some degree of visual durability, in that the product doesn't start looking 'old hat' or 'out of date' long before its physical wearout.
- *Appearance*. Many buyers want the product to exhibit a distinctive or pleasing 'look.' Achieving distinctive style or form is a major way in which designed products, environment, and information can stand out from competition. At the same time, design is much more than style. Some well-styled products fail to satisfy the owners because they are deficient in performance characteristics. Most designers honor the principle that 'form follows function.' They seek forms that facilitate and enhance the functioning of the object rather than form for its own sake.
- *Cost*. Designers must work within budget constraints. The final product must carry a price within a certain range (depending on whether it is aimed at the high or low end of the market) and designers must limit themselves to what is possible in this cost range.

Consumers will form an image of the product's design value in relation to its price and favor those products offering the highest value for the money. Effective design calls for a creative balancing of performance, quality, durability, and appearance variables at a price that the target market can afford. Design work needs to be done by a company in connection with its products, environments, information, and identity.

## WHAT KEEPS EXECUTIVES FROM BECOMING MORE EFFECTIVE DESIGN MANAGERS?

According to one estimate, over 5,000 U.S. companies have internal design departments and many others use outside design consultants. There are eight industrial design consulting firms with over ten employees, as well as numerous smaller ones [5]. In spite of the availability of design services, many companies neglect or mismanage their design capabilities. The reasons are design illiteracy, cost constraints, tradition-bound behavior, and politics.

## Design Illiteracy

Some designers charge that U.S. managers are largely illiterate when it comes to design. According to RitaSue Siegel:

> For the past 20 years American industry has been run by managers. They are trained in business schools to be numbers-oriented, to minimize risks and to use analytical, detached plans – not insights gained from hands on experience. They are devoted to short-term returns and cost reduction, rather than developing long-term technological competitiveness. They prefer servicing existing markets rather than taking risks and developing new ones [5].

Although this is stereotyped thinking, it represents a widespread view that many designers have of the people who run America's corporations.

## Cost Constraints

Many managers think that good design will cost a lot of money, more than they can afford. Using Skidmore, Owings & Merrill to design a new warehouse will be expensive. But bad design can cost even more money. Actually good design does not have to be expensive. Many companies have found that having an internal designer or outside design consultant on retainer pays for itself many times not only in avoiding costly errors but in creating a positive image for the company.

## Tradition-bound Behavior

Tradition-bound behavior is also a barrier to effective design management. A catalog format is very hard to change; and a product design or a company name is even harder to change. Salespeople will argue that their customers will be confused by name, product, and calalog changes. Managers prefer to stick with the original design instead of exposing their tastes to critical judgment.

For example, after Pillsbury bought Green Giant Foods, several suggestions were made that a facelift was in order. Pillsbury asked Leo Burnett, the Green Giant's agency, to look into this, but after initial creative development, the agency gave up because no one would commit to backing the new designs.

*Politics*

Company politics play a role in every firm. Some executives might oppose a proposed design simply because they want to block another group. Politics surface in creative reviews, budget meetings, and strategy planning sessions.

## HOW CAN THE INTERFACE BETWEEN MARKETERS AND DESIGNERS BE IMPROVED?

If a company recognizes the need for more and better design work, then a two-way process of education must occur. Marketers must acquire a better understanding of the design process and designers must acquire a better understanding of the marketing process.

Marketers need to be aware of the split in the design community between the functionalists and the stylists. The orientation of the functionalists is based on putting good functional performance, quality, and durability into the design. The orientation of the stylists is to put good outer form into the design. Functional designers are normally responsive to marketing research and technical research, while stylists often resist a marketing orientation. The stylists prefer to work by inspiration and tend to pay less attention to cost. Fortunately, few designers are at the extremes, and most are willing to pay some attention to market data and feedback in developing their designs.

Marketers also often split into the same two camps. Some marketers, notably those in the salesforce, often plead with the designers to add 'bells and whistles' to the product to catch the buyers' attention and win the sale. They press for features and styling that are eyecatching, even though they might not contribute to good design and performance. Other marketers hold that the key to customer satisfaction and repeat sales is not simply attracting initial purchase but providing long-term product-use satisfaction. These marketers are more interested in supporting the incorporation of good performance, quality, and durability characteristics into the product. They point to the success of Japanese automobiles as based not on style leadership so much as the consumer belief that Japanese automobiles offer better quality, durability, and useful features. So marketers also need to get their act together when they work with designers and make recommendations as to what counts most in the consumers' mind.

A common management mistake is to bring designers into the new product development process too late or to bring in the wrong type of designer. There are eight stages in the new product development process:

1. Idea generation
2. Screening
3. Concept development and testing

4. Marketing strategy
5. Business analysis
6. Product development
7. Market testing
8. Commercialization

Typically, the designer is invited in at stage 6, product development, when the prototype product is to be developed. Designers, however, should be brought in earlier, preferably in the idea generation stage or at least the concept development and testing stage. Designers are capable of producing ideas that no customers would come up with in the normal course of researching customers for ideas. And, during the concept development and testing stage, designers might propose intriguing features that deserve investigation before the final concept is chosen.

## Design Philosophy

Each company has to decide on how to incorporate design into the marketing planning process. There are three alternative philosophies. At one extreme are design-dominated companies which allow their designers to design out of their heads without any marketing data. The company looks for great designers who have an instinct for what will turn on customers. This philosophy is usually found in such industries as apparel, furniture, perfumes, tableware, and so on.

At the other extreme are marketer-dominated companies which require their designers to adhere closely to market research reports describing what customers want in the product. These companies believe designs should be market-sourced and market-tested. This philosophy is usually found in such industries as packaged foods, small appliances, and so on.

An intermediate philosophy holds that designs need not be market-sourced but at least should be market-tested. Consumers should be asked to react to any proposed design because often consumers have ways of seeing that are not apparent to designers and marketers. Most companies espouse the philosophy that designs should be market-tested even if not market-sourced.

Here is how one firm, Atmospheres, develops its designs for bank retail environments:

The designers at Atmospheres construct settings of bank interiors and test them on small focus groups of bank customers. Customer responses to different layout arrangements, textures, furniture, etc., help Atmospheres gain insight into customer perceptions and preferences. Based on customer responses, the designers then develop a design proposal for the bank. The design package is tested with another focus group to refine and verify the effectiveness of the design. The final version is presented to management with evidence of the degree of interest and satisfaction of the bank's customers in the proposed design.

This rhythm between the visual conceptions of the designer and the consumers' reactions to proposed designs represents the essence of market-oriented design thinking. It neither inhibits the designer from coming up with great ideas nor allows bad design ideas to be accepted without testing.

---

### How a Corporation's Design Sensitivity and Design Management Effectiveness Can Be Measured

Companies need to review periodically the role that design plays in their marketing program. At any point in time, company management will have a certain degree of design sensitivity. A design sensitivity audit (Exhibit 1) consists of five questions that will indicate the role design plays in the company's marketing decisionmaking. A design management audit (Exhibit 2) asks five more questions that rank how well management uses design. Each question is scored 0, 1, or 2. A corporation's design sensitivity will range from 0 to 10, and its design management will also range from 0 to 10. Companies with a combined design sensitivity and design management effectiveness rating of anywhere from 14 to 20 are in fairly good shape. Those scoring less than 8 should examine whether they are missing a major opportunity by not making more use of design thinking in their marketing strategy.

---

**EXHIBIT 1**
**Design Sensitivity Audit**

1. What role does the company assign to design in the marketing decision process?
   (0) Design is almost completely neglected as a marketing tool.
   (1) Design is viewed and used as a minor tactical tool.
   (2) Design is used as a major strategic tool in the marketing mix.

2. To what extent is design thinking utilized in product development work?
   (0) Little or no design thinking goes into product development work.
   (1) Occasionally good design thinking goes into product development work.
   (2) Consistently good design thinking goes into product development work.

3. To what extent is design thinking utilized in environmental design work?
   (0) Little or no design thinking goes into environmental design work.
   (1) Occasionally good design thinking goes into environmental design work.
   (2) Consistently good design thinking goes into environmental design work.

4. To what extent is design thinking utilized in information design work?
   (0) Little or no design thinking goes into information design work.
   (1) Occasionally good design thinking goes into information design work.
   (2) Consistently good design thinking goes into information design work.

5. To what extent is design thinking utilized in corporate identity design work?
   (0) Little or no design thinking goes into corporate identity design work.
   (1) Occasionally good design thinking goes into corporate identity design work.
   (2) Consistently good design thinking goes into corporate design work.

---

**EXHIBIT 2**
**Design Management Effectiveness Audit**

1. What orientation does the design staff follow?
   (0) The design staff aims for high aesthetic ideals without any surveying of the needs and wants of the marketplace.
   (1) The design staff designs what marketing or consumers ask for with little or no modification.
   (2) The design staff aims for design solutions that start with an awareness of consumer needs and preferences and adds a creative touch.

2. Does the design staff have an adequate budget to carry out design analysis, planning, and implementation?
   (0) The budget is insufficient even for production materials.
   (1) The budget is adequate but typically cut back during hard times.
   (2) The design staff is well budgeted, especially on new product development projects.

3. Do managers encourage creative experimentation and design?
   (0) Creative experimentation and design are discouraged.
   (1) Designers are occasionally allowed creative freedom, but more typically they have to design within tight specifications.
   (2) Designers have creative freedom within the limits of the project parameters.

4. Do designers have a close working relationship with people in marketing, sales, engineering, and research?
   (0) No.
   (1) Somewhat.
   (2) Yes.

5. Are designers held accountable for their work through post-evaluation measurement and feedback?
   (0) No.
   (1) Designers are accountable for cost overruns in the production process.
   (2) Design work is evaluated and full feedback is given to the designers.

FIGURE 6.1

## CONCLUSIONS

While every corporation buys and uses product, environmental, information, and identity design, very few have developed a sophisticated understanding of how to manage design as a strategic marketing tool. Design has been defined as a process that seeks to optimize consumer satisfaction and company profitability through creating performance, form, durability, and value in connection with products, environments, information, and identities. Strong design can help a company stand out from its competitors. The best results can be achieved by training general managers, marketers, salespeople, and engineers to understand design and designers to be aware of and understand the functions of these people. Design ideas should at least be market-tested, and preferably be market-sourced or stimulated by market survey data. As other strategic marketing tools become increasingly expensive, design is likely to play a growing role in the firm's unending search for a sustainable competitive advantage in the marketplace.

## REFERENCES

1. P. Gorb, ed., *Living by Design* (1979), pp. 7-8.
2. J.P. King, 'Robots Will Never Be Practical Unless Products Are Designed for Them.' *Industrial Design*, Jan./Feb. 1982, pp. 24–29.
3. P. Kotler, 'Atmospheres as a Marketing Tool,' *Journal of Retailing*, Winter 1973–1974, pp. 48–64.
4. See 'Architecture as a Corporate Asset,' *Business Week*, Oct. 4, 1982, pp. 124–126.
5. R.S. Siegel, 'The USA: Free to Choose,' *Design*, Jan. 1982, p. 24.

## ACKNOWLEDGEMENTS

The authors, Kotler and Rath, wish to acknowledge Lawrence Salomon, Director of Graduate Studies, School of Art and Design, University of Illinois at Chicago; Jane Bell, President, Atmospheres, Boston; and Barbara Braverman-Davis, Design Consultant, Brookline, Mass., for their help.

# 7

# Into a Changing World*

## JAMES PILDITCH

*In February 1974 the European head of one of the world's largest American corporations visited me. 'I guess you wonder why I'm here', he said. 'Our world has changed. It's a whole new ball game. We need people now who are specialists in new situations, or at adapting existing resources to suit new situations … and that is how I understand you to be.' His words, if flattering, were a good definition of what designers ought to be.*

*Five years before, in April 1969 when this first paper was written, the world looked different. The present was a continuation, albeit accelerating, of the past. Designers were struggling to define their professionalism. At a meeting of the Society of Industrial Artists and Designers I warned of greater preoccupations to come.*

Victorian painting, so R.H. Wilenski once wrote, is made up of what he called 'emotive fragments' – the ruined battlement, the dead tree lit by a pale moon obscuring (but not wholly) an ancient plough or, better by far, a marble goddess coy in her nakedness.

Rue de la Paix, Somme, Bauhaus, Bobby Charlton, automation, Apollo 8, Harold Wilson (there's an emotive fragment for you) – our world is full of such charged words that trigger off emotions of joy, prejudice, fear, nostalgia.

In this country few words provoke more alarm than those relating to the future. Computer, redundancy, mass marketing, technocrat, merger, teaching system, Rayner Banham's 'triumph of software', cybernetics, ultrasonics, on and on they roll. All 'A OK', 'all-systems-go' words we relish or abhor. Coined for the most part on another

*Source: Reprinted from Pilditch, J. (1976) *Talk About Design*. London: Barrie & Jenkins, pp. 3–14. Courtesy of Anne Pilditch.

continent (a fact not without significance) these emotionally packed words relate to a world only lately being made apparent to us.

Will the future be, as Churchill once described something else, 'made all the more sinister by the lights of perverted science' as some among us fear? Or might we hope for a steady improvement in the quality of our lives? This uncertainty lies at the root of the decaying, stalling, equivocal attitudes we notice all about us.

I sometimes wonder whether fear of the future isn't like a wasp's sting. The sting is 1/32nd of an inch long. The other 5 inches is our imagination.

Design is – and always has been – concerned with tomorrow. Whether the time scale is long or short, designing a city centre or a package, one is always thinking about and influencing future activity.

I risk this commonplace to shift the rubble of muddled emotion that so easily hampers clear thought. In defence of the futurists I want to say something else. It is this: many people suppose that if one embraces the future, grasps at it with hope, plans for its consequences, one is scornful or neglectful of the past. This is just as false as it is to assume that previous experience will not be relevant to the future.

Two contradictory thoughts occur to me. So great is the so-called explosion of knowledge and capability (I read somewhere that the amount of knowledge in the world is doubling every three years) that the world we are entering is, in some ways, unlike the one we went to sleep in last night. But equally, present and future are inseparable. Decisions we take today always have future implications.

Prophesying is difficult, as Robert Storm-Peterson once said, especially where the future is concerned. What one can say, however, is that it will be impossible to plan for the next decade if we do one of two things: either see it as a wholly natural continuation of the world we grew up in, or if we apply to it the criteria of the present.

You perhaps remember the thought expressed in Tennessee Williams *The Milk Train Doesn't Stop Here Any More* that our lives are all memories. The title of his story I uttered a moment ago is now a memory. So are most of the standards we apply to our work today – shaped, as they were, to suit some past event. In other words we really must think fresh if we hope to cope with the decade to come.

Internationalism is at the heart of all developments in the years ahead. For all of us it will be hard, even impossible, to plan effectively without accepting this view of a wider world.

Look at a few of the forecasts to see the kind of world we are entering. From this we can try to see what the implications might be for design and designers.

The place to start is not outer space and all that holds (a Swedish scientist told me 8000 people will be on the moon by 1990) but closer to home. Start with the basics of feeding people. The importance of this to designers will become clear in a moment. Poor people in undeveloped countries are multiplying much faster than relatively rich people in developed countries. And the gap between the two is widening alarmingly. The growth of population in undeveloped countries between now and the year 2000 will be greater than the existing world population. A million new mouths will have to be fed each week.

In this time the number of people in the so-called developed world will increase by 60 percent. In the undeveloped world by no less than 150 percent.

This might be alright if the undeveloped were able to work and earn. But by 1985 four out of ten people in Africa, Asia and Latin America will be under 15 years old – almost twice as high a proportion as in Europe and America.

So, more and more people; the rich getting richer and the poor staying poor. By 1985 we in Europe will earn thirty times as much as Indians, nearly six times that of Latin Americans. North Americans will have 60 times the income of Indians, ten times the income of South Americans.

By 1985, too, the undeveloped areas of the world will have more than four-fifths of the people in the world – and they will be producing less than one fifth of the world's goods.

I don't want to be a bore with figures – already I feel like the woman who whispered to a friend 'I won't go into any more details, I've already told you more than I know' – but there's one more set I must give you. It concerns growth of income. Think of the disparity of wealth already in the world today. Realise that between 1965 and 1985 the EEC is expected to increase per capita wealth by 82 percent, EFTA by 77 percent. Then think that in the same time Indian income will grow only 39 percent. The gap is widening.

I dwell on this because our present East–West problems pale before – what shall we call it – this North–South peril. The gap is like the crack in a creaking, straining rock being torn apart at a boiling volcano's peak.

Couple this with the increasing self-determination of many emerging countries and it is not hard to see that it will be a matter of practical politics, no less than humanity, for us all to do more about the backward countries.

To some extent, at any rate, we may expect the natural drives of self-interested business to be deflected into more social activity in a wider world.

What can designers do?

What designers can do is to go into these countries together with marketing men to help them turn their local resources into producing goods the wealthier countries can use. And to do so in ways that suit local ability. What is the alternative? Are less developed countries to be condemned to remain agrarian? If not they must learn to compete with all comers for the favour of Western consumers. This is partly a matter of design. The marketing man can suggest what should be designed. But only designers can design it. It seems to me a waste of time to pour money into countries so they can make more and more baskets nobody wants.

There is no reason to suppose undeveloped countries must follow the same tortuous path as we have to reach our present state. They can leapfrog generations of ideas. The emergent African airline doesn't buy bicycles. It buys jets. Local demand leaps from nothing to transistors. A conclusion is that the emergent nations can represent good markets for Western goods, and so widen the wealth gap still further.

We've been talking about wealth. In the next 15 years business in Europe and America will more than double. By then the Common Market and EFTA will be half as big again as all North America is today.

The consequences of this are many. For one thing, people who read with concern Servan-Schreiber's *Le Defi Americain* should note that American firms will have bigger markets outside the US than in it. Already the tough anti-trust laws are forcing US business to seek growth beyond its natural boundaries. I find it hard to think that American business will not become much more powerful everywhere, and that the next few years won't see the full blooming of a large number of huge, wholly international corporations.

These fast-rising standard of living and population are pressures to build the multinational corporation. They are pressures for internationalism, international laws and the decay of the nation state.

No less a person than George Ball, former Secretary to the US Treasury, has said that business is at least twenty-five years ahead of Government in its understanding of this new international world. National boundaries, as any businessman could tell any politician, are ceasing to have any meaning except as obstructions. We should note this as we strive for more national standards. The more we create our own national standards, the harder it will become to accept international ones – unless we think internationally now.

Certainly designers should think in these terms.

This decline in what they call 'national interest values' was forecast by Khan and Wiener in their book *The Year 2000,* a detailed and serious attempt to study the next 31 years. We'll pick one or two other broad trends they outline as we go along.

There is another basic point about internationalism. It is that many of the traditional reasons for locating industry are ceasing to be significant. To be near a source of power, a source of raw material or near convenient communications, all these are falling before the ability to create new forms of energy, to substitute for primary materials, to command communications anywhere.

The original source of an industry or company has long since been proved irrelevant and even restrictive. That a company started in London or Amsterdam or Wichita is seen increasingly as an historical accident. The *multi*national corporation, an extension of the international company, deliberately strives, by staffing and location, *not* to have a national bias. Working for ITT in Brussels, I've had meetings with Americans, English, Germans, a Norwegian, a Dane – chosen not as national representatives but because they are good at their jobs.

With modern communications and with rational computer analysis, companies will locate resources where it makes most sense. And that often means close to their customers. They will strive harder than ever not to be foreign mercenaries bleeding local wealth, but partners in local growth wherever they operate.

It is fashionable to talk about exports. In this word lies the worst fallacy of international thinking.

The whole idea of home and export is essentially out of date. Speaking at the European business school at Fontainebleau a few months ago, Dr Gross, the best thinker about business I've ever heard, said, and get this, 'You cannot remain an important factor in business by sticking to the export philosophy'.

I wonder whether we haven't missed the real significance of what Marshall McLuhan has called the 'electric speed' of modern communications. It is *not* that the traditional centres of an empire, political or industrial, can control outposts more effectively. If that were so, perhaps the Commonwealth would be stronger than ever. No, the new fact is that *anywhere can be the centre.* For a few bitter days Dallas was the focal point of the world. As instantly it can be Vietnam, Tokyo, Brussels or the sun-scorched deck of a ship in the Pacific.

The dynamic companies of the world see this clearly enough. The best aren't bothered about exports; they want to be where it makes most sense, near their customers. Of them all IBM is the outstanding example of the emergent multinational corporation. Successfully crossing borders, putting research centres on the dusty mountains behind Nice, or in Scotland, locating production on a 'product by plant' basis, responding to the new realities of modern production, communication and control.

In other words, business and life in the 70's *is* international. It is also big. We've heard often enough that in a few years a handful of firms (is it 300?) will command 80 percent of world business. True in detail or not, the tendency is clear. What has happened is that a few firms have got the point of what is happening, and they are the ones that are spurting ahead.

An underlying point is internationalism, as I have tried to say. Two others are highly relevant to designers everywhere. They are related.

First is the move away from production as the key to success and towards the ability to define and solve problems. In what has been called our 'techno-dynamic' society, of highly saturated markets, the new need is to create one's own markets. And the way to do this is to define a human want or need and satisfy it. The great growth companies like IBM, Polaroid, 3M and others are all ones that have created their own markets. Coupled with this, intricately, is the need to innovate.

At once it is necessary to say that innovation is not the same as R&D. One is concerned with pure research, the other with getting new things into society. It is worth reflecting that in a report published by the US Department of Commerce two years ago, Britain was cited as a country that spends almost as much on a per capita basis as the US on R&D but achieves nothing like the same practical, marketable results. We should ask why. This report was the outcome of a study group set up to study the problems of innovation. Among many interesting findings was this: companies that had as they reported 'committed themselves to innovation as a way of life' grew, on average, between five and twelve times as fast as the national average.

Here we see another widening gap, a gap more controllable because it is a gap in the use of brains – a gap, above all, in outlook.

It is a truism to say we live in a world of change, and that the rate of change is accelerating beyond credence. But it is not at all as trite to suggest that designers are in the centre of such change. They should be. And to get there I'd advocate far *less* concern with conforming to canons of good taste as an occupation and far *more* involvement with using new resources to solve human and industrial problems. We are talking about shaping the future by our own fantasy, our own planning.

We've glanced at how forward-looking industry is shifting its focus from the ability to produce to the ability to define and solve problems. But this means having knowledge. The current cliche is that growth companies are 'knowledge-orientated', and this raises pressing questions for designers. The rush and rash of new ideas, new processes, materials, methods presents us, as designers, with a stark choice: keep up or fall behind. Designers will find they need more and more knowledge to keep up. How will they acquire this knowledge? Are we to be content with local knowledge? Are we to ignore developments in the US, Japan, Germany, Sweden and elsewhere? Of course not. One of the consequences of the 'small world' we've all heard forecast will be the need for knowledge of how the Japanese build roads, or what the French labelling laws will become.

This hunger for knowledge will alter the designer's job beyond recognition. First is this need to recognise innovation as the dynamic of our changing world. And with it to accept that the rate of change is such that we will inevitably find ourselves reckoning with inventions not yet made. Think of the many installations laid down in the last twenty years that have failed to perform adequately because this certainty was ignored. The failure to understand growth rates is still the central characteristic of official planning. Designers must do better.

The second line of thought, bearing in mind the explosion of knowledge, is that the nearer designers come to genuine problem-solving the less they can specialise. As the level of competence rises in industry so the range of skills necessary to solve problems is seen more clearly. The constellation of skills required varies from job to job. We cannot hope to be master of them all. Today we need a computer analyst, tomorrow a warehouse specialist, the day after an ergonomist. The old idea of the designer as a kind of Leonardo turning his hand to designing aeroplanes between paintings, the 'name' designer as a maitre, fount of all wisdom, is no longer credible or welcome. The designer's role, increasingly, will be that of thinker, planner and co-ordinator of diverse skills.

It is by de-specialisation, therefore, the abandonment of at least some traditional skills (no matter how regretfully) that the designer can fulfil his role in the future.

I have to touch on the magic word 'marketing'. We cannot stay alive ourselves or understand our clients if we fail to understand what marketing means. Peter Drucker has written 'marketing and innovation are the only two worthwhile occupations of management'. Design, by the way, runs all through both. We're all familiar with the shift towards the consumer that has been going on a long time. Salvatore Teresi, an Italian professor whose lectures have been likened by a British diplomat to 'conducting Verdi without an orchestra', puts it if not succinctly then at least forcibly. He said 'don't make love to your products. Make love to the consumer'.

What are the problems of *our* market, as designers? Can we hope to serve modern business when we don't know how the German furniture industry differs from the Italian, how they sell paint in Sweden, or where industrial strength lies in Belgium?

Ignorance of markets is no new subject. It is not for me to add to the list of hair-raising stories, but there is one that underlines the point. A design office in Paris once designed a symbol for a Dutch company based on a letter Q for quality. Trite, you might think as an idea, but possibly acceptable. The only snag: in the Netherlands quality is spelt with a K (kwaliteit). It may be fair to assume that a local designer anywhere has more local knowledge. But the essence of international success must be to combine with that local knowledge – and service – the new sophistication made possible by relatively large resources and by a variety of experience in any field.

While advocating internationalism, it is only fair to note that recent political events have thrown up an interesting reversal of this concept. It seems that the more complex problems become, the more centralised decision-making must be, the more people on the peripheries want to participate in these decisions or, indeed, have them all to themselves. Who could say whether nationalism is declining or becoming more powerful a force? 'Participation', as President de Gaulle called it, is the general cry. How this will affect us is hard to say. It seems to be very unlikely that design will become more local, except in the sense that design competence is fanning out throughout Europe and the world. But what may be thought is that designers (like industry) will need to give more thought to local sensibilities, probably to the extent of having nationals serving each market.

Earlier we touched on the 'shrinking world' concept. Among many implications is the obviously competitive one. New York will be three hours away. US competition in the design profession will doubtless increase, so will the impact of US client organisations, and the spread of their attitudes not only to designers but to life. The Continent too is coming closer. The popular European joke about 'Fog in the Channel, Continent cut off', is becoming a nostalgic memory, along with monogrammed cabin trunks and the Orient Express. Unquestionably, modern European attitudes to life will influence us too. Which force will be the stronger? I have no answer, only a question. But in any event we may expect competition among designers and the design requirement of business to alter.

Among social changes of gale force, is the move predicted by Khan and Wiener towards what they call a more 'sensate' hedonistic society – bringing a collapse in the traditional values of work, thrift, authority and so on. We saw it expressed if not coherently (for an essence of the new attitude is an annihilation of order) then at least powerfully in *Hair*. Do you remember that poignant line that says so much? 'Sex isn't love, it isn't even fun anymore'. Even freedom turns bitter to the taste.

Yet while this mood may be glimpsed everywhere, it may be wrong to assume that the acceptance of such a trend is the same universally. Crosscurrents and contradictions are extreme. One may expect that in any society – ours, or that in France or Norway or Spain, we must try to understand the attitudes that exist among the groups we are aiming at, on behalf of our clients. This is a complication of internationalism. In other words, Carnaby Street may knock the 'copains' dead, but leave German bankers less than ecstatic.

From this spill a number of thoughts. One, that of designing for one's markets, rather than oneself. Two, receiving briefs that cover the extent of market hoped for, rather than simply the domestic scene. Traditionally companies create goods to suit their own markets which they then try to sell elsewhere. The limitations are obvious. Designers must press for proper international briefs, if products or whatever else we design are to have international currency.

Third, the extension of this idea is that one can readily imagine that the greater market for a product lies outside the home country. Population speaks for itself. It is therefore entirely possible that we will find ourselves recommending designs that suit the majority outside the homeland more than they suit people within it.

A fear springs from this. What a cigarette firm has nonsensically called 'cis-Atlantic' and television aims at – 'mid-Atlantic', in other words the emergence of a smooth bland design which, springing from no definable origins, lacks character. Is this what internationalism and multi-national marketing means? It may happen, perhaps already is. To the extent it drives out our very own, indigenous Odeon Hemel Hempstead style I think I'm all for it. Mid-Atlantic design might be better than three-quarters of the articles that adorn our High Streets today.

The real point of the continued rise in imports has not been grasped by manufacturers.

In the last few months I have bought a record player and a lamp from Denmark, carpets from Spain and Germany, a lamp from France, chairs from the United States (or designed there anyway), a clock from Italy as well as British goods. I buy these things because, in my view, they're good, well designed and value for money.

Today the initiative for design is not coming only from the United States, but as much from Japan, Germany, Italy, Denmark and this country too. The hope of internationalism must be that good things, wherever they come from, can flourish because markets are larger. No longer need a design award be the 'kiss of death', as we have all heard it described.

Good design is available more widely. Dim, conservative or autocratic manufacturers no longer enjoy the captive markets into which they can pour whatever they choose. In other words, there's more competition and more choice, both marvellous benefits of international marketing.

How many designers, no less than industrialists, will go on being 'production-orientated' rather than 'marketing-orientated'? That is to say, more concerned with producing what *they* conceive to be good design than with assessing what their clients want and need. How many will go on offering their simple speciality without linking it to others that may, together help solve clients' problems better? How many will go on thinking of their city or their country as the beginning and end of their professional world, failing to understand the homogeneity of Europe as their natural zone, failing to appreciate that their clients must act internationally?

Having access to larger markets, good design can become a decisive factor in international business. Rising standards of education, of literacy, of the exchange of ideas, will tend to raise mass standards rather than lower them. So I am hopeful that universally the acceptance of good design will become normal.

This is the real point for us. International competition means rising standards. Diffused and dissipated as it has become, there's little doubt that the invasion of Scandinavian furniture pushed up standards of British furniture.

Among criteria for commercial success in the seventies good design will loom large. That's where we come in. For years designers have been urging industry to get with it. Now that industry – bigger, more international than ever before – is responding, we must ask 'Are we ready to serve their new needs; have we the competence, the resources, the *outlook* they need?' But note that the requirement is for *good* design, not necessarily for *our* design. The key to the success of the design profession in the next decade is for it to be good at its job, able to create the things people prefer. Never mind preoccupation with professional status – leave that to the lawyers who have built a fortress of tangled laws and custom no foreigner can enter. Let us concentrate on international competence with universal integrity.

# PART II

# New Educational Perspectives for Designers and Managers

# Editorial Introduction

SABINE JUNGINGER AND RACHEL COOPER

The very categories "design student" and "management student" should not be regarded as immutable. The possibility should not be excluded of radical shifts in the current boundaries of the two disciplines.

*William Callaway, 'New Directions in Design and Management Education'*

What are the links and boundaries between design and management education? This section focuses on design education in the context of management studies to provide a glimpse into the changing relationship between designing and managing as it is taking hold in business and management schools across the globe. We put this education section at the beginning of the book, as we find that the contributions in this area demonstrate important shifts in management and in design alike. Our focus on business and management schools rather than on design schools is a conscious choice.

This choice is based on several observations. For example, we found that design schools interested in exploring the links between design, management and organizations nowadays tend to develop programmes in collaboration with existing and preferably established management or business schools and, as a consequence, we find innovative design-related changes in these cases. There are earnest efforts under way by educators in management to study the possibilities that design offers in improving business practices and in achieving desirable results. For many of these scholars, the activities, principles and practices of designing present a much-needed tool for future leaders that adds value and complements existing approaches. A recent article in the *Academy of Management Learning and Education* provides testimony to how designing and managing are viewed in a new light. There, Ken Starkey and Sue Tempest (2009) argue that the 'design challenge facing business schools is to create a more holistic view of management and management education, suited to our complex and difficult times'. In short, we currently detect significant motion in management scholarship and in management programmes with regard to exploring design practices and conducting research into the principles and values of designing.

There is, however, another and perhaps more important reason to focus on design education in business and management schools. This reason relates directly to design research: if design is being integrated into managerial education, design researchers need to understand how, when and why design comes into play. In other words, we need to ask which aspects of designing are being taught and explored (and which are not!), what kind of products are being designed by whom, how and for what end. Are these programmes geared, for example, to look to design as 'delivering products that work well with clients' or does a programme treat designing as 'a creative process that enables ideas to come to life' – that is, are they geared towards traditional manufacturing or towards the growing service industry (Tether 2008). Is there a covergence of design and management and, if so, how is this manifested?

There have been prior efforts to 'find a place for design in management education'. Bruce Archer was instrumental in this in 1967 (see Part I). We have also already mentioned a range of educational initiatives. Not surprisingly, Mark Oakley addressed the issue of design in the management curriculum in a journal paper 1986. At no time before, however, did the arts and humanities expand their roles in the management curriculum as quickly and as saliently as they do presently. The current developments differ from those earlier ones in that management programmes themselves are critically reflecting on the kinds of skills and knowledge they need to impart to future leaders. Creativity and the ability to navigate uncertainty and complexity have emerged as key abilities for successful managers. The paradigms under which management and design have operated for decades are in question. This is accompanied by other developments – for example by a different outlook on life and work in general. Two new key terms are wellbeing and work–life balance. They express an orientation towards creating work environments built around concepts of sustainability, not only in environmental but also in social and individual terms. The *organization man*, vividly brought to life by Whyte (1956), has become a rare individual and is about to be extinct. Job biographies and life flows today demand flexibility, mobility and internationalism as much as entrepreneurial skills, all of which thrive on abilities for critical thinking, emotional intelligence and the ability for lateral thinking – the marks of design.

Design management has always struggled with the balance of educating managers about design on the one hand and educating designers in matters of business on the other. It had long been accepted that managers and designers represent two radically different personality types. One, the manager, was driven by order and routine. The other, the designer, was driven by experiments rejecting structure and repetition (Walker 1990).

Today, we witness a renewed eagerness among management scholars and students alike to embrace a design attitude (Michlewsky 2008). The design attitude, that is, the skilful and purposive use of design, is turning into a valuable if not essential asset for leaders. The chapters in this section reflect on this idea from very different cultural and national perspectives. It emerges that the motivations behind these different efforts at bringing design education into the management fold are quite different. In India, global competitiveness is one of the key drivers, but so is a rediscovery of its culturally unique wealth of design and management history. Design has thus become a means to express pride in India's national distinctiveness. In Germany we sense an effort to move beyond the Bauhaus and the HfG Ulm, which remain the undisputed sources for many successful products and brands made in Germany, yet which are now in danger of stifling efforts to broaden the realm of design practices, education and design research. For Canada, a focus on design thinking and design methods has created an opportunity to show leadership in management education. The motivations in the UK are not quite clear as the emphasis swings from design to management and back. If anything, the current economic situation will make very clear where design stands in relation to management. There are, of course, many other countries that deserve mentioning in this section: Turkey, Poland, Israel, Hungary, Mexico and Brazil immediately come to mind. The four chapters included in this section therefore only offer us a glimpse into the upheaval of the management world and the promise design seems to hold.

To begin with, **Anuja Agarwal, Uday Salunkhe and Surja Vanka** discuss 'New Approaches to Design and Management in India'. They begin with the evolution of management education in India and go on to show the current change management education is undergoing. They then turn to the development of design in India and trace the nation's history of design education. Before this background, Agarwal, Salunkhe and Vanka draw up the new relationship design and management are entering in India.

In the next chapter, **David Dunne** argues for user-centred design and design-centred business schools. Dunne is interested in a model of management education that is based on the ideas and methods of user-centred design. He suggests that managers could think like user-centred designers and derives from that the requirements of a new educational curriculum for management. He illustrates the possibilities of such an approach with an exemplary design-based business course he has developed at the Rotman School of Management in Canada.

'Mutual inspiration and learning between management and design' is also a topic for **Ralph Bruder** and the title of his chapter. Bruder begins with a general reflection on what he calls the 'cross-relations' between design and management before he goes on to reflect on the challenges that are part of teaching management *and* design. He then discusses a concept for teaching business design, which the Zollverein School of Management and Design was meant to implement. As founder and former president of this German educational experiment, his case study shares lessons learned. Lessons that offer both a cautionary note and a message of hope for future such undertakings.

**Lucy Kimbell** concludes this section with an intentionally provocative piece. Her interests are not in improving on management education. Instead, she wants to create better futures. Thus her 'Manifesto for the M(B)A in Designing Better Futures' calls for an enquiry-based curriculum that goes beyond the creation of better objects and instead enquires into future practices.

## REFERENCES

Michlewsky, K. (2008) 'Uncovering Design Attitude. Inside the Culture of Designers', *Organization Studies* 29: 373–92.

Starkey, K. and Tempest, S. (2009) 'The Winter of our Discontent – The Design Challenge for Business Schools.' *Academy of Management Learning and Education* 8(4): 576–86.

Tether, B. (2008) 'Service Design: Time to Bring in the Professionals?' In L. Kimbell and V. P. Seidel (eds). *Designing for Services – Multidisciplinary Perspectives: Proceedings from the Exploratory Project on Designing for Services in Science and Technology-based Enterprises, Saïd Business School*. Oxford: University of Oxford, pp. 7–8.

Walker, D. (1990) 'Two Tribes at War?' In Oakley, M. (ed.). *The Handbook of Design Management*. Oxford: Blackwell.

Whyte, W. H. Jr. (1956) *The Organization Man*. Garden City, NY: Double Day.

## RECOMMENDED READING

Callaway, W. (1990). 'New Directions in Design and Management Education', in M. Oakley (ed.). *Handbook of Design Management*. Oxford: Blackwell.

CNAA (1984) *Managing Design: An Initiative in Management Education*. London: Council for National Academic Awards.

CNAA (1987) *Managing Design: An Update*. London: Council for National Academic Awards.

Oakley, M. (1986) 'Bringing Design into the Management Curriculum', *Management Education and Development* 17(4): 352–62.

# 8

# New Approaches to Design and Management in India

ANUJA AGARWAL, UDAY SALUNKHE
AND SURYA VANKA

India, with its rapidly changing economy, is at an interesting juncture. It is a diverse country, economically, socially and politically, with a large population. There is a lot left to be done for the country in terms of economic and social welfare, infrastructure development and mapping resources for growth and development. India has the youngest population in the world, which can be educated and trained to take up these challenges. The fast rise of the software and the manufacturing sectors in India along with blooming retail industry, advanced healthcare services, textile and fashion enterprises, also means that India needs a lot of appropriately educated young people to convert these challenges into business opportunities. For these reasons, India continues to invest in management and in design education.

## EVOLUTION OF MANAGEMENT EDUCATION IN INDIA

According to Sougata Ray and Anup Sinha:

> Management education in India has a history dating back to the late 1940s. Later, the Indian government took initiatives to promote management education and set up the Indian Institute of Management. Subsequently, as management education gained prominence, all major universities started to offer degrees in management. Privately owned business and management schools have proliferated during the last two decades, offering programmes both at undergraduate and postgraduate levels. (Ray and Sinha 2005)

The number of business schools, including the MBA-granting departments of universities, is now estimated to be over 3,000.

The historical concept of an Indian Institute of Management (as opposed to an Institute of Business) was to indicate that the task of' 'managing' was not restricted to a commercial business organization alone. Management was equally essential for public administration, running non-profit organizations and small establishments with some focused purpose. Recognizing this plurality of needs and the impossibility of offering everything for everybody, management education in India developed along the lines of providing a general management education with a combination of knowledge and skills that would provide the foundations for creating effective and efficient managers. The graduates of these institutes have the potential to work in any organization.

Management education has the potential to influence the country's growth and development. But with changing times a need has arisen for a radical change in management education. The emerging unarticulated needs and realities of people in India demand new approaches.

Our environment has changed from market boom in 2006–7 to a slowdown in 2007–9. In such times we need leaders who will embrace ambiguity. In line with our vision to produce the leaders of tomorrow, we need to look at relevant management education that will equip them to navigate the future in this rapidly changing environment. We need creative leaders to find answers to difficult questions (Rowe 2004).

Second, with innovation fast becoming the buzzword in corporate circles and in an era of shared common resources – be they technology, processes, products or information, even the common knowledge pool – what is it that differentiates one business from another and how can a management institute best be able to act as a catalyst for growth in this new environment? There needs to be new thought as to what we impart as management education today.

Inspiring innovations in business and management techniques need the integration of multiple disciplines. According to Johansson (2004) 'operating within a field will generate ideas along a particular direction but when you step into an intersection, you can combine concepts between multiple fields and generate ideas that leaf in new directions'. An integrated disciplinary approach, bringing design, management, technology and social sciences together, will nurture a breed of managers who are trained not only to manage but who can also innovate by looking at the human, social and psychological needs of the users of their products and services and by improving profits in their line of business.

With this thought, the We School[1] in Mumbai (acclaimed as the fastest transforming business school in the country based on the formidable reputation that it has earned for its pragmatic approach to management education) initiated a management programme that incorporates design thinking into its mainstream management education.

# EVOLUTION OF DESIGN AND DESIGN EDUCATION IN INDIA

Five thousand years of Indian design stretch before us like a kaleidoscopic sequence of visual images. Historical, mythical, cinematographic – historically, design has been the way of life for the Indian people. Not confined to the craftsmen, design pervades all aspects of the Indian condition from the *rangoli* (colour pattern) on the doorstep to the *bindi* (traditional mark) on the forehead.

Design in India reached a high watermark in the ancient civilizations of Mohenjodaro and Harappa (3000–2000 BC). Here it was close to life, reflecting and borrowing from nature and men. Pottery, metal casting, and stone were employed to create objects of functional and cultural significance. Then came a shift in Indian history (after the thirteenth and fifteenth centuries), which led to an emphasis on metaphysical thought. This was marked by a period when rulers began to build huge empires by conquering and occupying neighbouring kingdoms. This led to the building of megastructures to showcase dominance.

Art schools began to rise, often due to extraneous influences such as the Gandharv School of sculpture, which was established under Greek influence. Some evolved from an agglomeration of craftsmen working at a single site over generations – for example at the sites of the great temples of Konarak, Dilwara, Khajuraho and Tanjore. Empire building reached its zenith with the advent of Mughals. They brought the change in design themes from the spiritual to the corporeal and courtroom. The Tajmahal, Miniature paintings captured patterns of flowing robes, glittering jewellery and ornate architecture. With the end of Mughal rule, the loss of royal patronage and the advent of British colonialism, India entered a period of economic impoverishment and cultural transfusion. As a nation, we learned that design cannot be imposed on a culture but that it must result from a natural coherence between the individual and his/her social milieu. After India gained independence from the British in 1947, the country entered into a phase of rapid industrialization brought on by the era of science and rationality.

Five-year plans were conceived and implemented from 1950 awards. Not surprisingly, the focus of the first five-year plan was on increasing agricultural production. The second plan was devoted to industrial development and concerned itself with infrastructural demands of the various industries. Many industries were set up in public sectors, including steel, energy and mining, often with help from industrially advanced countries. With the increase in demand for products and the availability of raw materials and government support, new industries in the private sector emerged to make products in collaboration with foreign companies, which provided technical assistance. Western multinationals began to set up production units both with and without local equity participation.

The Indian designer is an heir of two traditions. Tradition one is influenced by the German Bauhaus. Tradition two is a national and deeply rooted attitude of respect for older people. India was the first developing country to introduce design as a professional discipline – an experiment that began in the 1970s with the founding of the National Institute of Design (NID) at Ahemedabad and then the industrial design centre at the Indian Institute of Technology in Bombay.

The NID has been a pioneer in industrial design education, following the principles and methods of the Bauhaus and the Hochschule für Gestaltung Ulm (HfG Ulm) in Germany. It has had a key role in developing Indian design excellence. Among its other activities, the NID has introduced events around the theme 'Designed in India, Made for the World'. The NID's graduates have made a mark in key sectors of commerce, industry and social development by taking role of catalysts and through thought leadership.

Since the NID has a broad spectrum of design faculties and disciplines, a multidisciplinary approach is inherent in the scheme of things. Its strategic design management (SDM) programme draws upon several design disciplines available at NID, involving the faculties of industrial design, communication design, textile, apparel and merchandising design, and IT integrated (experiential) design. This two-year programme aims to enable students to develop design leadership and entrepreneurial skills and empower them to become wealth creators in the field.

## DESIGN, AN APT PARTNER FOR MANAGEMENT IN INDIA

According to Kanagasabapathi (2009) 'India is different than the rest of the world in terms of its economy, business practices and management. It is a society driven economy, its growth attributed to the inherent strengths of its people; the quality of putting ideas into practice, the initiative, energy and genius of the Indian people.' Another important unique factor about India is the dominance of its non-corporate sector in the Indian business system, which provides employment to a significant part of the population but also produces a large number of entrepreneurs, making India one of the largest entrepreneurial nations in the world. Even the corporate sector has different management needs in the Indian context, whether they be for the manual intensive production models, technology models for a large population, educated and uneducated, finance models for people who rely on their 'own' money rather than borrowed money or the marketing theories that take into account the fact that an Indian is not an individual but is a part of a family and community, which influence his buying decisions. Looking at these needs, only those management models will work in India that have taken into account the socio-cultural context and have built management paradigms around it. The management needs in India rely on being more intuitive rather than data driven. Management has to be holistic rather than in silos of functional specialization – marketing, finance, operations, systems and so forth.

The need for such an indigenous management model arose around 2006, when the economy was in boom. The need was felt even more during the economic slow-down in late 2008. It was evident during that time that Western models of management were not fully applicable in the Indian context. In fact, the conservative Indian belief in 'saving money' actually helped us during the financial crisis. Lessons had to be learned from within – economic, business and management models had to be 'designed' indigenously for future readiness. Management education itself needed fresh thought.

We found new concepts emerging at the intersection of divergent fields. Their combined strength could add a lot more value than they would individually. Interestingly, here design and management came out as the best partners, complementing each other in the lifecycle of a new emerging business. Design contributes a 'sensing/feeling' aspect that is not easily articulated in terms of market needs or well defined problem solving but more as tacit knowledge and pre-consciousness (Utterback et al. 2006). This aspect complements the management process, which often looks at people as customers and not human beings. It clearly identifies that managers could learn a lot more from the open-minded approach of designers and designers can gain by learning how managers manage resources and how they not only execute business but also help it grow. It was evident to us that a cross between management perspectives and a designer's way of thinking was the right blend of 'left-brain' and 'right-brain' thinking required by the intellectual capital of tomorrow.

As business and management become more complex in the face of competition, it is the design field that can 'hide' this complexity from the user/customer because, for them, complexity is not a necessarily a virtue. Whereas much management is based on number crunching and analytical methods of decision making, design brings the power of visualization and communication through prototypes that aid in the decision-making process. It can also be supplemented by the story-telling approach of design, to give a complete overview of the system (Utterback et al. 2006).

## DESIGN THINKING IN MANAGEMENT – RESONATING THOUGHTS FROM THE INDUSTRY AND ACADEMIA

A management school serves as a bridge between industry's human resource needs in the future and the aspirations of the youth who want to join the companies to serve in different managerial positions. It has to provide the best match between the two by grooming the latter in various functions of management, using the corporate network to give regular inputs on the type of skills the young people would require in their careers as future managers.

These thoughts resonate with some industry leaders as well. Kishore Biyani, the CEO of Future Group, a retail giant in India, spoke about bringing a tectonic shift in the very thought process and culture of the organization as the retail business is poised for a big leap. He wanted his managers to be able to think like designers, to be creative and to identify the pain points of the customers in his retail store from a very human perspective (Biyani 2007). Professor Dr Jagdeesh Sheth, Charles H. Kelistadt Professor of Marketing at Emory University, in one of his lectures, introduced a global perspective on design management and pointed out how the concept of 'design to us should mean creating a distinctly Indian solution to many of the problems which India is faced with.' Similar thoughts were echoed by other companies[2] from different sectors, forcing us to re-examine the entire MBA curriculum from the perspective of the designer.

Among those schools that now promote the concept of design thinking in management, Stanford D-school offers a module on design thinking to all MBA students in the inspiring atmosphere of their school. The Mälardalen[3] University in Sweden has a well established Idea-lab as a part of its Design, Engineering and Innovation Department, where students are encouraged to come up with new ideas and put them into practice under the guidance of professors from the Department of Innovation, who hold their hands and provide strong academic input on creativity and the process of innovation. Once the ideas become crystallized, the Munktell Science Park at the university provides an incubation hub for the budding entrepreneurs. The Dean of the Rotman School of Management in Canada, Dr Roger Martin, envisions business schools as places for 'Integrative Thinking'. The Rotman School has set up a consultancy under the name of 'Designworks' to provide design services to companies seeking business transformations. Similar approaches are under way at MIT and the University of California, Berkley in the US, the Stuttgart Media University in Germany and the Milan Institute in Italy. These act as comparators for the developments in India.

## NEW PLATFORM TO DISCUSS DESIGN THINKING IN MANAGEMENT EDUCATION IN INDIA

WeSchool[1] in Mumbai, India has chosen to create a new environment where 'design thinking' and business management are woven together to create a new 'integrated disciplinary approach to innovation and design thinking'. It provided a new platform to discuss Design thinking in Management by organizing a number of events to that effect, as shown in the sections below. Through these initiatives, it became possible to 'spread a buzz' around the industry about the merits of bringing design thinking into management, to benefit the companies operating in all industry sectors and also the design houses and consultancy firms as they would now find more buy-in for the value that design could create in the business.

WeSchool organizes events that promote design thinking in management as we are aware that any new thought has to be properly propagated to all stakeholders for its true value to be recognized. Some of the events have been as follows.

### 13 October 2006: 'Design: The Next Frontier'

A round table on the impact and use of design in various industry sectors like retail, banks, fast-moving consumer goods and capital markets. Interestingly all the eminent speakers spoke about how deeply 'design' affects their businesses. For some, 'design' was a road to 'Innovation' but, for others, it was the services that were 'designed' to achieve maximum customer satisfaction. The richness of the discussion that followed helped us to understand further how 'design' is to be redefined every time to gel with the context and how businesses can leverage on 'design thinking' in management within their organizations.

### 9–11 August 2007: 'Grow by Design'

A workshop by IDEO, USA. As WeSchool was looking to bring design thinking into the Indian business community, it brought IDEO, the leading innovation and design consultancy firm in the US, to conduct a focused, two-and-a-half day innovation workshop for a group of senior managers from a range of industries: retail, technology, agriculture and service. The workshop was part of an effort to bring creativity and innovative practices to this community. The goal of this workshop was to inspire and orient members of Indian business community to the IDEO approach to innovation.

### 29 October 2007: 'Three Approaches to Innovation'

A seminar by Professor Dr Sten Ekman and Dr Annalille Ekman, Mälardalen University, Sweden. The approaches discussed by the speakers in the seminar helped create a common knowledge base for the innovation imperative in a global context in interaction with design and leadership. The seminar was conducted in Mumbai as well as Bengaluru, and was very well attended by senior-level executives from Indian companies. The faculties and students of WeSchool also benefited from the seminar, seeing some of the possibilities that exist in the area of innovation, especially in the Indian environment.

### 27–29 August 2008: 'The Association of Indian Management Schools (AIMS) Conference'

The Association of Indian Management Schools – a body that represents around 3,000 business schools in India, including the Indian Institute of Management (IIM), Indian School of Business (ISB) and other leading business schools (Management Development Institute, Jamnalal Bajaj Institute of Management Studies, S. P. Jain Institute of Management and Research, etc.) – elected our Director, Professor Dr Uday

Salunkhe (co-author of this chapter) as their President in 2007. This gave us the op-portunity to inform other management education institutions about the concept of design thinking in management education. Dr Salunkhe included 'Grow by Design' as a key agenda topic for the AIMS conference in 2007, while the main theme at the August 2008 conference was 'Nurturing Thought Leadership through Management Education'. The sub-themes at the 2008 conference were:

1. India – The Knowledge Capital
2. Leadership Development at the Grass Root Level
3. The Innovation Imperative – Expectations from B-Schools
4. B-Labs – Incubating the Enterprise of the Future
5. Ecology for Thought Leadership
6. Responsible Leadership
7. Voice of the Customer
8. New Tools in Management

The themes for speaker sessions were such that designers were brought to the fore-front as speakers in a management forum, which greatly assisted the dissemination of design thinking in management education through out the country.

### 20 June 2009: 'Innovation Roundtable'

The innovation roundtable was organized by We School for top-level executives from industry to explain to them the benefits of design thinking in management and to promote innovative thinking in future managers.

### Ongoing Endeavours

Spreading design thinking through executive education – the business design team at WeSchool's Management Development Centre also conducts 'innovation leaders' workshops for company executives as part of their executive education.

To enable the new thought to percolate into the very culture of the organization and transmit its perceived image to the outside world, WeSchool transformed itself from being a business school (Welingkar Institute of Management) to a WeSchool – a new identity that signifies an openness to learn from other disciplines where doors and minds are always open to ideas and innovation (see Figure 8.1).

## THE FIRST BUSINESS DESIGN PROGRAMME IN INDIA

Sensing the opportunity for a new paradigm in management education in the emerg-ing context in India, WeSchool launched the first management programme with

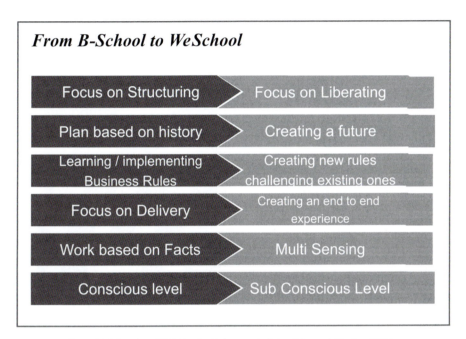

FIGURE 8.1    From B School to WeSchool. © Agarwal, Salunkhe and Vanka, 2011.

design thinking and innovation as its backbone in July 2006. The program was called the Post Graduate Diploma in Management (Business Design).

## Programme Objective

The objective of the programme is 'To foster future leaders with a unique blend of Management competencies, Design thinking, Innovation and entrepreneurial spirit to develop a holistic vision to manage business, environment and the social responsibility towards the stake holders.'

## Programme Structure

To weave design thinking into mainstream management it is necessary to have a multidisciplinary approach to management education. Social sciences provide an insight into what human beings want, their cognitive behaviour, the psychological aspects, but at the same time technology needs to be leveraged to take certain ideas forward. The pure design courses on form, colour and material give a new vocabulary for the effective communication of ideas among global citizens, cutting across their professions.

The programme is designed to keep human beings and their psychological, social and societal needs in the centre, using the 'design-thinking' approach to identify their unarticulated needs and generate plausible ideas for solutions. It includes subjects that

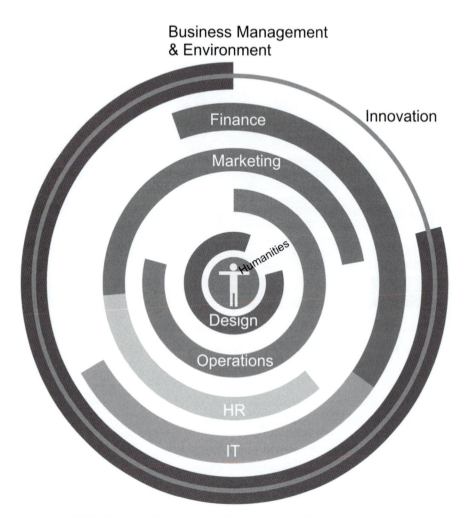

FIGURE 8.2    PGDM Business Design, 2011. © Agarwal, Salunkhe and Vanka, 2011.

will help to design a business around this approach and that teach the various functions of management required to run businesses in the most effective way. The entire course intends to bring in the process of 'innovation' so that the student leaves the programme with an understanding of the entire lifecycle of an innovative idea, knowing how to build a full-scale business around it and subsequently manage it in a sustainable way.

## DESIGN SCHOOL CULTURE VERSUS THE BUSINESS SCHOOL CULTURE

Compared to the design schools, management education is imparted in a quasi-corporate formal environment. The focus is to give the student a good insight into the way companies function and accomplish goals. The students are exposed to the various functions of management – planning, organizing, leading, coordinating, controlling, staffing and motivating through an analytical and case-study-based approach.

## New Pedagogy

Acknowledging that design and management education emanates from very different cultures, it is important to adopt new pedagogical methods for the subjects that are a blend of business and design – methods that encourage learning a new approach to management.

In Richard Klimonsky's (2007) words, 'What comes through clearly is that learning-by-doing is not only common to the human condition, but also that we as management educators would benefit greatly if we better understood the processes involved and adapted them to our teaching efforts.' Klimonsky further adds, 'If our students were to develop the capacity to efficiently learn from experience and also were to find ways to internalize the powerful benefits of seeking out experiences from which to learn, we would have produced individuals with the potential to be life-long learners.' 'Thus practice-based experience, if transformed using a practice epistemology into new knowledge, could be used to complement what we teach in our classrooms.'

It is generally agreed that most blended subjects should be introduced to students in this format. Any learning that happens by asking questions that come from live experiences is real knowledge creation, otherwise it is mere information dissemination from teacher to student. When students encounter issues and try to find solutions, they understand the practical implications of theoretical models and constructs. In the programme, therefore, conventional management subjects are given a new format so that they are taught in a team-teaching mode, with professors coming from different management functions. The presence of a design professor in the team brings a radical perspective to the topic being discussed and enthuses the session with more energy and a different line of thought.

## Integrated Projects

For the business design program, integrated project-based learning is used to give a holistic view of business. The biggest advantage of project-based learning is the development of the skill of defining the problem first and then thinking of 'breakthrough' solutions. This paves the way for innovation – whether in products, processes, services or environment as a whole.

The projects carried out by the students in a business management programme usually involve a lot of secondary research and analysis of data and are done in a particular field of specialization – for instance, finance or marketing. After sensitizing the students to design and the value that it brings to thoughts, execution and communication, they are given projects that involve inputs from different fields of study in design, social sciences, technology and management. This gives the students a chance to understand the lifecycle of any product, process, service or idea from inception to execution stage.

### Differentiating Business Design from Other Management Programmes

We find, in India as elsewhere, that business design students find diverse solutions to management problems. This happens because the business design students are groomed to first identify and formulate the problem after doing 'design research' as the primary research method. They are asked to use ethnographic tools to capture insights from the context and 'people-on-the-street'. The designer's skills of empathetic observation and holistic understanding of the context are new and valuable to these students as they enable them to develop varied and innovative proposals. Conventional market research is used to test out some proposed solutions. The use of experiential learning for real-life problem identification and the application of a 'design-thinking' approach to find innovative solutions to the identified problems differentiated these students from regular management students.

The conventional case-study method of analysing a problem is one of the time-tested ways for students in management schools to absorb situations and propose solutions. We found that the additional designer's toolkit, taught to them in the design studio and design insights, helped the students to add a lot more value to their understanding and their experience of tricky management situations.

## ROLES OF 'BUSINESS DESIGNERS'

The skills inculcated in the students going through the business design programme helps them to create new product/services/processes/customer experiences, manage business transformations and lead innovation initiatives, which may ultimately spawn independent businesses. Risk taking is an important design attitude, so these students are more prepared to form or join start-up innovative ventures. The students graduating from this programme have found their careers as 'design thinkers' in conventional management roles apart from carrying out the role of innovation evangelists in their organizations. They have found their way into mainstream management where they have displayed their abilities to think like designers while acting like managers – and this has brought a fresh perspective in their job profiles. These students have been absorbed by top companies across various industry sectors such as Retail, banking and finance, information technology and telecommunications, FMCG and consumer durables, healthcare and pharmaceuticals, travel and tourism, management and design consultancies. The students have chosen to move into companies that offer them job profiles like: new product development, innovation, consumer division, user interface, branch managers

FIGURE 8.3   InnoWe – Centre for Innovation and Memetics at WeSchool, 2011. © Agarwal, Salunkhe and Vanka, 2011.

in banks and management consultancy. These are the roles where the students can effectively apply their 'design thinking' approach to management to the maximum.

### Innovation Lab – An Ideal Place for Cultivating Design Thinking

Viewed from an innovator's perspective, an idea converted into a practical and innovative solution, then a business plan and then a venture provides the school and its students with a tremendous opportunity to create, incubate and scale-up new businesses.

Having an innovation centre is a step towards realizing the above objectives. The innovation lab at WeSchool, Mumbai campus, serves a larger purpose of spreading a culture of using 'design thinking' as a process that leads to 'innovation' in the entire organization. The innovation lab is a space where thinking is stimulated, ideas germinate and are nurtured to become paradigm-defining businesses. It is a prototype lab; students are encouraged to voice their ideas with optimism in a risk-free environment where they will be allowed to germinate, cross-pollinate and become fertilized. It creates a platform for people to develop their innovative ideas using a contest-like format. The innovation lab encourages the formation of companies on campus.

The innovation lab takes research and consulting assignments for corporations in India, especially those that require the generation of new ideas and the identification of future trends. Some of the projects carried out in the past have been with a retail giant and through this our students have been involved in research in the field of memetics, which is an emerging science pertaining to trend forecasting. Our students, along with the faculty, work on generating solutions for this. Such assignments have been highly successful and are appreciated by our clients.

## CHALLENGE ON OUR WAY

### *Industry's Apprehension*

No initiative is exciting if it is not met with obstacles and challenges on the way. In this instance we faced a number of hurdles both internally within the organization and externally from the eco-system in which we operate.

The synergy between design and management is not very obvious upfront. In India, as elsewhere, the two disciplines have been seen as two parallel streams, intersecting at specific points but again going back to their own paths and not really flowing as one wider, larger river. The world looks at the disciplines very differently. Business management is seen as a more of a money-management and a resource-oriented stream, with little consideration for people's emotions; design is viewed as a more humane, empathetic discipline with a mission of creating a comfortable, beautiful and convenient life for all.

Although, initially, industry was very excited about the new concept, it did not see how we would combine the two disciplines, so we created presentations about the concepts. Slowly, industry began to appreciate our approach and even saw value in building up a culture of innovation within organizations. However, often businesses were not sure where they would find a place for employees who came to them with the new skills of combining design and management together. Most companies did not have a separate department of 'innovation' as such or a defined role and responsibility that these 'business designers' could take on. We believe it will be three to four years [from 2006] before the companies are able to understand how to fit them into their organization structure. The culture of innovation in management is new in India therefore it is difficult for them to appreciate the innovation initiatives at the lower management level. Often companies feel that such strategic initiatives should be at the higher level of company hierarchy.

It is still the case, as in most of the world, that a student typically either chooses to become a 'manager' pursuing his MBA or a 'designer' pursuing a full-time course on 'design'. Institutions have started offering design management courses to the designers to help them understand the nuances of management. At WeSchool, the mission is different. Being a management institute dedicated to the cause of business and management education the aim is to ensure that management education is imbued with thinking about the future and is as contemporary as possible. That is where learning from fields like design can influence the programme and the students.

*Prospective Students' Dilemma*

Externally the disconnection between management and design at first appeared too great. Prospective students of management initially find it too risky to embark on a course like this. Students whose aim was to enter into the corporate world as management trainees found the course curriculum and the objectives of the course very exciting but they were in a constant fear of an identity crisis. They have never heard of a 'business designer' in industry. Since the business design programme at WeSchool is a general management course with design thinking and innovation as the specialized area, this was also too new for prospective students as they had only heard of specialization in marketing, finance, human resources and so forth. 'What kind of job would we land into?' was one of the foremost concerns in the minds of the students. This is deeply embedded in the education system in India because we are so structured and rigorous in following the syllabus at the school level that students, by the time they graduate, have mostly lost their sense of creativity – or so they think and feel. Being creative and innovative at the postgraduate management studies level was usually perceived as 'unthinkable' by most Indian students.

*Internal Faculty's Perspective*

Bringing design thinking into management courses was also a new way of approaching management education. The faculty found it challenging to change their tried-and-tested successful ways of imparting management education. The course was new, the pedagogy had to be different, the examples and case studies had to be built up all over again. It was a while before our own faculty understood what had to be done and how to do justice to the concept.

## COVERING MILESTONES

Although it is too early to define our success, we can share how we are supported to work towards integrating the disciplines. Our Postgraduate Diploma in Management (Business Design),[4] which is currently running its second cohort of students, is gaining popularity among students all over India, as indicated by the fifteenfold increase in the number of applications that we receive from prospective students opting for this course. In addition, the course has attracted students with a good profile – most of them have work experience of three years in industry, there is a mix of engineers, commerce graduates and science graduates with very high scores in the common entrance tests conducted nationally and at the state level for MBA aspirants. Our first cohort, which passed out in the year 2008, is working with reputed companies in various industry sectors like retail, information technology, banking and finance and design houses. Many of them have been inducted as a part of innovation teams in their respective companies. They are well connected and return to share their experiences and express how the different thought process has been beneficial to them in their work environment. From the educational standpoint, the business design students have become more sensitive to the importance of clarity of thought, aesthetics, visual vocabulary for effective communication, use of ethnographic studies for research, the development of soft skills like empathy – all this for better business development and management.

Our initiative, spreading design thinking to achieve innovation in the companies through the executive education programme conducted at WeSchool, has received encouraging feedback from the participants. The business design team is invited to speak at various forums at the national level on design thinking and management and, more specifically, on how this concept is being integrated with the other streams of management and how will it affect the business and management style in future.

## LEARNING ALL THE WAY

As with success, it is also too early to identify failure at this juncture. Our initiatives in Indian management education – the business design programme, the innovation lab and other related activities – are continuously being improved as we mature. We are constantly guided by our advisory board and other eminent consultants in the field of both management and design.

A vital point of learning for us during this journey has been the realization that the advantage of holistic and integrative thinking is to bring different people together for a common goal. We recognize that synergy between the seemingly far-flung disciplines has been beneficial; we will not be surprised if we could find more and more examples, in India, where bringing different disciplines together will create new avenues and opportunities.

The way design can influence the decision process is only too evident. During our meetings with the various companies from different industry sectors in India, we have been emphasizing the fact the design can make a strategic difference to business and management. The students of the business design programme who have been groomed in management skills with design thinking as their philosophical backbone are the ambassadors for spreading design thinking and bringing the design industry to the forefront of management in India.

## NOTES

1. Ranked eighth in the private school category, sixteenth in the top fifty B-schools category and ranked tenth by the 'Aspiring MBAs' according to the latest surveys (*Outlook Magazine* 2008; *India Today* 2008), WeSchool has the privilege of acquiring accreditations and recognition from the prestigious (South Asian Quality Systems (SAQS), National Board of

Accreditation (NBA), All India Council for Technical Education (AICTE) and Mumbai University, India.

2. See Mahindra & Mahindra (manufacturing), Phillips (consumer durables), IDIOM (design house), Tata Motors (automobiles), Kansai Nerolac (paints).

3. Mälardalen University, Eskilstuna, Sweden, www.mdh.se.

4. In June 2008, We School launched a two-year full-time Masters level programme in business design called the Post Graduate Diploma in Management (Business Design) – backed by the recommendations of the Advisory Board, and with our own research in the area in tandem with the wishes expressed by industry, and approved by the governing body – the All India Council for Technical Education (AICTE).

## REFERENCES

Biyani, K. (2007) *It Happened in India*. New Delhi: Rupa & Co.

Johansson, F. (2004) *The Medici Effect*. Boston, MA: Harvard Business Review Press.

Kanagasabapathi, P. (2009) *Indian Models of Economy, Business and Management*, 2nd edn. New Delhi: Prentice Hall of India.

Klimonsky, R. (2007) 'Introduction: Promoting the "Practice" of Learning from Practice', *Academy of Management Education & Learning*, 6(4): 493–4.

Ray, S. and Sinha, A. (2005) 'Management Education in India – Let a Thousand Flowers Bloom Amidst a Hundred Questions.' *Decision*, 32(2): 1–18.

Rowe, A. J. (2004) *Creative Intelligence*. Singapore: Pearson Education.

Utterback, J., Vedin, B.-A., Alvarez, E., Ekman, S., Sanderson, S.W., Tether, B. and Verganti, R. (2006) *Design Inspired Innovation*. Singapore: World Scientific Publishing.

# 9

# User-centred Design and Design-centred Business Schools

## DAVID DUNNE

### INTRODUCTION

In a long tradition of self-criticism and in recent commentaries, management education is painted as a technocracy, governed by a rigid, self-serving rationality that is ill-equipped to deal with rapidly changing conditions and with messy social problems. Business schools are seen as the incubators of this style of thinking and of disconnection from human issues. Ashkenazy (2007), in a special section of the journal *Academy of Management Learning and Education*, identified two unifying themes: change, and the need to humanize management education.

I argue that management educators can learn from the field of design. Confronted with the need to understand and engage users, the design profession developed user-centred design (UCD), which focuses on developing a rich understanding of human experience as the basis for the design of products, services, and experiences.

As an example of critiques of management education in the literature, Bennis and O'Toole (2005) argue that business school professors are beholden to a scientific model of management that has little relevance to real-world management practice. Mintzberg (2004) asserts that business schools teach MBAs the science of management while ignoring its craft. Ghoshal (2005) argues that management educators have failed to impart an appropriate set of ethical values to their students, instead promoting a set of 'ideologically-inspired amoral theories' (Ghoshal 2005: 76), which allow students to divorce their actions from their impact on humans.

A design perspective can shed light on these issues. Designers have a characteristic approach to problem solving that has relevance for managers (for example, Boland and Collopy 2004) and, moreover, could transform business education (Dunne and

Martin 2006). More particularly, user-centred designers develop products, services and experiences that solve problems faced by users. To this end, user-centred designers develop a deep understanding of users, coupled with an attitude of empathy that appreciates the challenges they face. Beyond this, user-centred designers take trouble to understand the context within which products and services are used, and the changing business environment.

I begin by tracing the development of UCD and its implications for design practice. This is followed by a discussion of the relevance of this way of thinking for managers. I conclude with a discussion of the implications of this perspective for the curriculum and teaching methods in business schools.

## THE EMERGENCE OF USER-CENTRED DESIGN

The term 'user-centred design' refers to the design of products, services, and experiences to meet the needs and capabilities of those who will be using them. While it may appear self-evident that design should take into account users' needs, there has been disagreement over the extent to which users should dictate the design process. Donald Norman (1988), in his classic work *The Design of Everyday Things*, argued that many designers had lost touch with users, with the result that common devices were often difficult, inconvenient and even dangerous to use. Norman advocated the introduction of UCD, which should make use of the natural properties of people and of the world, exploiting natural properties and constraints. As much as possible, design should operate without instructions or labels.

UCD resulted from a number of converging movements in the design field in the late twentieth century. The field of human–computer interaction (HCI) is concerned with the design of computer interfaces that are natural and easy to use (Gerlach and Kuo 1991). HCI arose from a realization that interfaces were difficult to use, resulting in user frustration and decreased productivity due to steep learning curves and underutilization. This resulted in a shift in research emphasis from the technical aspects of computing to modelling human behaviour in relation to computer systems (for example, Card et al. 1983; Norman 1986; Olson and Olson 1990). A related field is computer-supported cooperative work (CSCW), in which computer environments are designed to support collaborative work practices and researchers explore the role of computers in facilitating social interactions (Wellman 2001). According to CSCW, technological systems must relate to existing orders of social practice and remain adaptable to the emergent needs of user groups (Ackerman 2000).

Joint application design (JAD) was developed by Chuck Morris and Tony Crawford of IBM in 1977 as a process for involving users in system design (Asaro 2000). It grew out of existing methods of design and engaged users, designers, and external experts together in design; its focus was essentially pragmatic, on achieving the most efficient process of user-oriented system design. By contrast, participatory design (PD) originated in Norwegian and British movements to develop more democratic workplace technology. Its socialist underpinnings provoked reflection within the design research community on the political and ethical implications of workplace technology.

Despite the differences between these approaches, their core idea was the same: the need to develop an intimate understanding of users as an essential component of the design process. This gave rise to the adoption of ethnographic research methods to develop close connections with users and understand their interaction with designed objects. Because users often had difficulty in articulating their needs in terms that designers could understand, these methods allowed designers to observe their interactions and draw insights from them. Norman (1986) showed that the principles of usability lie in users' mental models, their understandings of how things operate based on their experience, learning, or the usage situation. To develop some insight into these mental models, designers needed to understand not only users' responses to proposed designs but also the context of use and users' personal perspectives.

Users also became involved in design not only through ethnography and the more traditional focus groups, but through engagement in the design process itself. The methods for accomplishing this included collaborative workshops in which users and designers worked together to develop and test designs. This was the basis of both the JAD and PD approaches, with differing emphases: in JAD, the user was a necessary but often subordinate party to the design process, while under the 'techno-populist' PD (Asaro 2000) users were seen as equal partners in the design process.

For design to be truly user centred, however, it needed to go further than merely adapting products to users' physical and cognitive needs and capabilities. Norman (2004) shows that users' *emotional states* also influence how they respond to design and that this response in turn affects the functioning of the design itself: objects that feel better actually work better. Hence designers need to develop a rich, deep understanding of the emotional context that users bring to designed objects.

To go further still, designers need to pay attention to the users' overall experience, which is not confined to their reaction to the physical attributes of the product. Experiences comprise not just the product itself, but services, interactions, processes and environment. However, each user's experience is essentially a subjective and only partially observable event: each user creates his or her own experience through the usage of the product in its particular context of use and in conjunction with the user's own physical, cognitive, and emotional perspectives. Hence the traditional, instrumentalist view of users among businesses as 'consumers' of design is called into question: if each user is essentially a creator of his or her own experience, he or she is on an equal plane with the designer, as proponents of PD would advocate for more political reasons.

These ideas have echoes in the management literature. Leonard and Rayport (1997) show how users' unarticulated needs can be observed through a process of empathic design and there is an extensive literature on the use of ethnographic methods in market research (for example, Arnould and Wallendorf 1994; Underhill 2000; Mari-ampolsky 2006). Prahalad and Ramaswamy (2003, 2004) take up the question of user engagement in the design process, arguing that consumers today are more connected, informed and active than ever before. They contend that, as a result, the traditional firm-centric view of value will give way to 'co-creation' of value between firms and consumers in which firms develop capabilities to respond flexibly and quickly to customer needs. For Prahalad and Ramaswamy, the task of innovation is one of developing experiences in partnership with customers, rather than products targeted to customers.

It is one thing to say that designers are engaging users in the creation of value. UCD, however, also implies a different way of *thinking*, which emphasizes interrelationships within systems and empathy with users. I explore this way of thinking and its implications for management in the next section.

## HOW MANAGERS COULD THINK LIKE USER-CENTRED DESIGNERS

Problems in management are increasingly characterized by complexity and instability (Boland and Collopy 2004). In this environment, managers need to develop an ability to understand 'wicked' problems: complex, dynamic problems involving multiple stakeholders that defy easy resolution (Churchman 1967). Failure to appreciate the full complexity of such problems can lead to disastrous results (Hackett 2007). Since Herbert Simon (1969), in *The Sciences of the Artificial*, called for a new management curriculum based on design, several authors have argued that managers can learn a great deal from the approach taken by designers (for example, Senge 1994; Boland and Collopy 2004; Dunne and Martin 2006).

Because designers are traditionally engaged for their creativity, it is with this quality that they are most closely associated (Kelley and Littman 2001). However, a great deal of research and reflection is required to develop ideas. Designers frequently need to reinterpret a brief to identify the underlying problem, to visualize abstract solutions and to integrate information from multiple sources. Conley (2004) argues that design competencies, such as the ability to frame problems in a meaningful way and integrate the components of a solution, can be applied to managerial problems; as an example, Kumar and Whitney (2003) show how data from ethnographic research conducted in a wide variety of contexts can be integrated through the use of thought tools for analysis and synthesis. Boland and Collopy (2004) go beyond skills and argue that a design 'attitude' views managerial problems as opportunities for invention and development of elegant solutions. Schön (1983) represents design as a 'reflective conversation with the situation' in which the designer attempts a solution, reframes the problem and tries a new approach.

A representation of the design process, adapted from the processes used at the Institute of Design at the Illinois Institute of Technology in Chicago, is shown in Figure 9.1. This is by no means the only way of approaching design problems, but

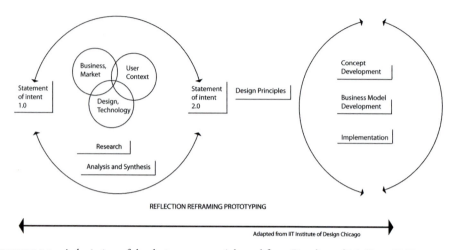

FIGURE 9.1    A depiction of the design process. Adapted from Paradis and McGaw, 2007: courtesy of IIT Institute of Design, Chicago.

it offers a reasonable representation of the process applied by many user-centred designers. The emphasis is on developing a deep understanding of the problem before attempting to develop solutions.

A notable feature of this process is that problem definition is provisional and iterative: the design team begins with 'Statement of Intent 1.0' and modifies this according to the findings of its research into users and their context, business and market issues and design and technological constraints. Several statements of intent may be developed before a definition of the problem is agreed upon and design principles developed. Throughout the process of problem definition, the team experiments with tentative solutions and explores aspects of the design problem through research and prototyping.

With an agreed set of design principles, the design team proceeds to use creative techniques to develop solution concepts and business models and to implementation of the design. This part of the process is also iterative as the team delves into its research on users and business issues to refine and tailor its solutions.

With UCD in particular, the critical element is the impact of the design on human beings. It follows that designers need to understand users' needs intimately and integrate this information with information from other sources. As noted earlier, ethnographic research provides a deep understanding of users' physical, cognitive, and emotional perspectives. To interpret this information, user-centred designers approach it with an attitude of empathy and employ systems thinking. *Empathy* is defined by Rogers (1959) as:

To perceive the internal frame of reference of another with accuracy and with the emotional components and meanings which pertain thereto as if one were the person, but without ever losing the 'as if' condition. Thus, it means to sense the hurt or the pleasure of another as he senses it and to perceive the causes thereof as he perceives them, but without ever losing the recognition that it is as if I were hurt or pleased and so forth.

When Leonard and Rayport (1997) refer to 'empathic design', therefore, they are not merely discussing how customers can be treated as instrumental objects of study, but of engaging in an intimate process of feeling and sensing with other human beings. Observational methods in ethnography include 'participant' observation, in which the researcher interacts with the subject: in the traditional anthropological/sociological approach which is the root of ethnographic methods in design, the researcher becomes an 'insider' by engaging directly with subjects and assuming a role in the family or community over an extended period of time (Mariampolsky 2006). The effect of such exposure is to develop a close identification with subjects.

'User-centred management' can have a broader interpretation than UCD, where management is concerned not only with creating value for customers, but also developing tools or methods for the administration of the business. Hence the 'user' of a balance sheet may be a financial analyst, or the 'users' of an organization design may be employees. Managers need to be just as engaged with these users as user-centred designers are with users of their designs: as human beings who bring a personal context to their engagement with the initiative.

Designers need to develop *systems thinking* – an ability to think broadly about the design problem – for two reasons. True empathy with users is only possible if one understands the user's context of use: not merely the usage situation, but the user's personal perspective based cultural, linguistic, and emotional factors (Mariampolski 2006). The second reason is that effective design is not limited to products alone but provides value to users from the integration of resources (Carr 1999). Thus effective design requires the designer to understand both the user's context and that of the business.

In understanding the context, user-centred designers consider the entire system of use: for example, the design of an office chair needs to take into account other elements in the office, such as the desk and table; the user's own physical needs and the type of work that will be conducted. At a broader level, the designer needs to consider the user's relationships with other users and his or her level of physical activity. Because the design of a chair affects, and is affected by, its context, the designer needs to think about the relationships between the elements of a system.

Russell Ackoff (1999) argued that humanity was leaving the 'Machine Age', which had been characterized by analytical, reductionist, and deterministic thinking: the idea that there exist clear, independent cause–effect relationships between phenomena that can be identified by breaking down the relationship into small pieces and testing their effects. In the 'Systems Age', by contrast, problems are more complex: interdependent variables together combine to form a system, defined as a set of interrelated elements that loses some of its essential properties when it is taken apart.

Systems thinking has two components: analysis and synthesis (Allio 2003). Analysis takes a system apart to reveal how it works but, because the system loses its essential nature when taken apart, analysis alone is not capable of understanding the system as a whole. Synthesis reveals why the system works as it does. Systems thinking integrates analysis and synthesis to understand the system.

In systems thinking, the problem solver attempts to understand the nature of the system and frame the problem accordingly. Since designers (and managers) are boundedly rational (Simon 1996), the problem frame chosen will restrict the set of solutions available. Using Donald Schön's classic example, Boland and Collopy (2004: 9) express this as follows:

> If we think of an urban neighborhood as a blight, it evokes a particular problem space where certain types of design intervention are seen as most appropriate (cutting out the blight, curing the sick, bringing in a fresh form of life) . . . if we look at the same situation as a folk community, we may instead not see it as a problem but as an opportunity, and develop plans to support its fragile infrastructure.

Designers have long been aware of the importance of systems: the field of CSCW evolved because of a growing consciousness among information systems designers that technology could not be separated from the social system in which it operates (Ackerman 2000). Yet reductionist analytical techniques that narrow the problem space are inappropriate for understanding the interconnectedness of systems (Ackoff 1999). To appreciate the relationships that form the system, designers both use traditional analysis and develop a synthesis of the system as a whole. The role of the systems perspective in the design process was captured by Gharajedaghi (1999: 23) as follows:

> Designers seek to choose rather than predict the future. They try to understand rational, emotional, and cultural dimensions of choice and to produce a design that satisfies a multitude of functions. The design methodology requires that designers learn how to use what they already know, learn how to realize what they do not know, and learn how to learn what they need to know. Finally, producing a design requires an awareness of how activities of one part of a system affect and are affected by other parts. This awareness requires understanding the nature of interactions among the parts.

Because management problems are often shifting, difficult situations characterized by complex interrelationships and multiple stakeholders, they also defy easy solutions. Systems thinking has therefore been of interest to management scholars for some time. Jackson (2000) traces the history of systems thinking and argues that its growth has been limited by a perception among managers that it is too theoretical for practical problems; nevertheless, the popularity of Senge's (1994) book *The Fifth Discipline* attests to the desire for a fresh approach to complex problems. In user-centred management, as in UCD, the key considerations are the relationships between consumers, employees, shareholders, managers and other stakeholders.

## A MODEL OF MANAGEMENT EDUCATION BASED ON UCD

While the basic idea of putting users at the centre of management thinking is simple, it has profound implications. As noted earlier, the term 'user' needs to be defined broadly: the 'user' of a financial statement may be investors, financial analysts or creditors. What is important is an explicit appreciation of the full range of users and an intimate understanding of them.

In management education based on the principles of UCD, students would develop skills in systems thinking and an attitude of empathy. To accomplish this, they would be required to solve 'wicked' problems by framing the problem, understanding users intimately, thinking abductively about possible solutions, using analysis and synthesis to develop an understanding of systems and their component parts, and collaborating in diverse teams (Dunne and Martin 2006). This does not, however, mean that design courses would supplant those courses currently being taught in business schools: what is needed is mostly a shift in attitude and focus rather than a large new body of material.

Nevertheless, these principles have important implications both for curriculum and teaching methods. These implications are discussed below.

## CURRICULUM

Master of Business Administration students learn a wide variety of techniques for analysing business problems but typically apply them to well-defined problems. Problem sets, exams and cases, for example, often spell out the alternatives available for comparison. Missing from the education of a typical business student is a discussion of how to identify the correct problem to work on, and how to think about new, untried alternatives. While business students learn some models designed to help them look at the bigger picture (Porter's 1979 model of competitive forces would be an example), there is scope to go much further and consider the implications of problems for users, markets, and societies.

Essential to a user-centred approach is an attitude of empathy: that users are not just 'consumers' to be targeted, but real human beings with thoughts, feelings, and needs; that employees are not merely factors of production; and that collaboration with others means understanding how the world appears from their perspective.

In a user-centred MBA curriculum, students would learn the following topic areas. In principle, all of these could be woven into existing courses. However, foundational courses in these topics would encourage students to approach all their courses with a different frame of mind.

*Problem framing.* To solve problems rather than merely treating their symptoms, students must learn to identify the underlying problem. With the considerable research in framing since Kahneman and Tversky's (1979) articulation of prospect theory, students would learn that one's perception of a problem and therefore one's readiness to accept solutions depends on how it is framed. From fields such as root cause analysis (Wilson et al. 1993), students would learn practical methods for understanding the dimensions of a problem. The important issue here is perhaps an attitudinal one: the understanding that real-life problems are rarely what they appear to be and that extensive research and reflection are required before solution development can begin.

*Ethnographic research.* As noted earlier, ethnographic research methods are used extensively by designers and are becoming popular in business (McFarland 2001; Mariampolsky 2006). Qualitative methods, including user observation, are currently included in many market research courses. One difficulty with this, however, is that the epistemologies and assumptions underlying ethnographic methods differ fundamentally from those associated with quantitative methods. Interpretivist approaches, for example, assume that truth is a set of socially constructed realities and the researcher's task is to look for meanings held by participants; other traditions, tracing their roots to Foucault and Marx, emphasize the interrelationships between power and knowledge (Harlos et al. 2003).

In a user-centred MBA, there is room for both quantitative and qualitative approaches. Both would be used in concert to develop a subjective, intimate, understanding of the user, in contrast to the distancing and dehumanizing effect of regarding consumers as statistics. Because of the differences in underlying epistemologies it makes sense to offer separate courses but framed by a common philosophy of user intimacy.

*Abductive reasoning.* Abductive reasoning, in contrast to deductive reasoning and inductive reasoning, is thinking about what might be possible. In Aristotelian logic, inductive reasoning is generalization from specific instances, while deductive reasoning involves inference from logical premises. Charles Peirce (1903) described abductive logic as 'the process of forming an explanatory hypothesis. It is the only logical operation which introduces any new idea.' Abductive reasoning proceeds by the observation of a surprising phenomenon that confronts pre-existing beliefs, reflection on the assumptions that led to the surprise and revision of these assumptions (Quilici Gonzalez and Haselager 2005); it includes creativity, which Boden (2004) defines as transformation of the conceptual space. To learn about abductive reasoning, students would learn to identify their own implicit beliefs and assumptions and to confront these by generating alternative solutions to problems through creative processes (for example, Czikzentmihalyi 1990; DeBono 2000). They would additionally learn how organizations can be managed to encourage abductive reasoning (Amabile 1998).

*Synthesis.* As noted earlier, the components of systems thinking are analysis and synthesis. Students would learn to integrate analytical and synthetic methods to arrive at an appreciation of the larger context for business problems. This does not mean abandoning a reductionist approach, but rather means learning that the relationships between components of a problem are just as important as the components themselves. The approach has already been widely applied in operations research: for example, Daellenbachand and Petty's (2000) application of the MENTOR system originated by Belton et al. (1997) has three stages: problem formulation, modelling, and implementation. Synthetic and analytical methods are used throughout iterations of the process, in which identification of the problem and the system are emphasized.

*Collaboration.* The prevailing approach in business schools to working with other students is a confrontational one in which ideas compete for acceptance (Dunne and Martin 2006). This works against the need to confront one's own assumptions in framing and solving problems, and the element of surprise when these assumptions are confronted. In a user-centred business school, students would learn to work collaboratively, rather than confrontationally, in groups. However, groups of relatively homogeneous business students are unlikely to go very far in shaking each other's assumptions and there is a role for external intervention, either through facilitation that pushes students to reflect or through group organization to maximize diversity (Kelley and Littman 2001). There is also an opportunity to increase group diversity through alliances with other institutions.

## TEACHING METHODS

Contemporary management problems are characterized by instability, unpredictability, and conflicting interests among multiple stakeholders – in other words, 'wicked' problems (Rittell and Weber 1973), a 'class of social system problems which are ill formulated, where the information is confusing, where there are many clients and decision makers with conflicting values, and where the ramifications in the whole system are thoroughly confusing' (Churchman 1967). A significant portion of students' efforts in a user-centred MBA would be devoted to learning to deal with wicked problems.

Standard teaching methods in business schools – lectures and cases – are capable of providing students with some of the concepts and tools of user-centred management. However, since these methods typically present problems as well-defined and indeed often provide students with alternatives to compare, they will not be successful in imparting some of the concepts and skills of user-centred management. The role of lectures and cases would be to help students understand the concepts in simplified form; students would then apply these skills in real-world projects that defy easy definition and require them to generate their own alternatives based on their understanding of users.

## EXAMPLE: A DESIGN-BASED BUSINESS COURSE

To provide a more concrete perspective on how UCD can be incorporated into a business school, the following example may be useful. In May 2007, a course in 'strategic innovation' was offered at Johannes Kepler Universität in Linz, Austria. Twelve students took the course: all undergraduate business students in their final year, although they came from a variety of national backgrounds. The course consisted of a project based on the design process shown in Figure 9.1. The course description in the syllabus was as follows:

> This course is a reflective practicum in the process of innovation. Students will be introduced to a user-centred design process and, in teams, will apply this process to develop either a new product/service idea or a business strategy for an existing product/service. Lectures and discussions will highlight the key stages in the process, and students will reflect regularly on their own individual approach to strategic innovation. By the end of the course, students will have developed a generalized, individual framework for strategic innovations they will face in the future.

There were five plenary sessions over the course of a month, interspersed with work in small groups. Following an initial session introducing them to the process, students proceeded to develop an initial Statement of Intent (an overview of what they were trying to accomplish), conduct user research, develop insights and design principles, develop a reframed Statement of Intent and present initial concepts. The focus throughout was on designing a customer experience; the students were divided into three groups, each of which selected a project: a zoo, a financial service and a retail art store.

Students were assessed in groups according to the quality of their final presentation and individually on a reflective journal that discussed their learning about the innovation process. Data on the course were collected from student journals and from depth interviews conducted by a graduate student who participated in the course.

In the initial phase, students identified problems to work on and developed a preliminary Statement of Intent, encapsulating their overall goals for the process and what they hoped to design. The financial services group, for example, was interested in providing low-cost loans to students.

Students collected secondary data and observed and interviewed potential users for their experience. This led them to rethink their original problem statement: the finance group, for example, saw that there was little demand for student loans but a great deal of interest in making sense of financial information, and revised their State-ment of Intent accordingly.

For students accustomed to a linear 'formulate-then-solve' approach, reframing the problem in this way was a new experience. Several students commented on this in interviews, for example 'We actually uncovered a totally different problem from what we actually believed.'

The discovery that they had misidentified the problem came quite late in the course for some students, after several phases of research. At the time, it appeared as a setback but students eventually came to appreciate the nonlinearity of the process. One stu-dent wrote as follows:

> The redefinition of our research question helped us to focus the problem. So I learned the right question can help me to [understand] the problem and to focus on [users'] pain points . . . each redefinition made us get two steps ahead.'

The students were also introduced to brainstorming and developed conceptual prototypes of their proposals, in the form of collages, as a means of refining and communicating the idea (see the photographs in Figure 9.2). The idea of develop-ing physical representations of ideas as part of the problem-solving process was a new and intriguing technique for the student groups. Physical representation brought out ideas that could not be expressed verbally and helped the team members understand each other:

> It is important to be able to express the thoughts and emotions not only with words, but even with pictures and other creative techniques because not every feeling can be expressed only with words. Only if there is something . . . your col-leagues can touch, can they start to understand the meaning of the idea.

While the overall experience of the course was positive for the students, there were nevertheless limitations. The limited time available for the course restricted the stu-dents' ability to absorb and fully implement the new process: more time would have allowed for richer, deeper user research and more thoroughly elaborated implementa-tion plans. In addition, students had difficulty with the ambiguity of the final assign-ment and sought clearer direction.

In spite of the limitations, several students found the approach transformative, as exemplified by the following quote:

> Having a look on my everyday life, I often find myself looking out for possible solu-tions when I recognize a problem. I cannot stop that, it is crazy! But I like it. Next week, I start an internship in the international human resources management at an Austrian production concern. In my mind, I already have so many approaches

FIGURE 9.2    Photographs from Strategic Innovation Course in Linz, Austria. (Top) Using Collages to Prototype the Experience. Photograph by David Dunne. (Bottom) Group Work on Problem Framing. Photograph by David Dunne.

to innovate the existing tools and habits that I would like to go there and change everything.

## CONCLUSION

Management education can learn from the user-centred movement in design. This involves building innovative solutions to problems based on an intimate understanding of users and the context of use, in addition to the circumstances faced by the business.

The change required of business schools is both attitudinal and epistemological. Business students need to learn to put users, broadly defined, at the centre of their efforts as managers and to collaborate to provide them with value. Attitudinally, this entails true identification with users and the circumstances in which they find themselves, along with a high degree of openness and curiosity. On an epistemological level, students can be provided with research methods and thought tools to focus their attention on users. To achieve these changes, business schools need to develop a curriculum that pushes students to experience user's issues face-to-face by working on challenging, dynamic, ill-structured problems.

## REFERENCES

Ackerman, M. (2000) 'The Intellectual Challenge of CSCW: The Gap between Social Requirements and Technical Feasibility', *Human-Computer Interaction* 15(2): 179–203.

Ackoff, R. (1999) 'Our Changing Concept of the World', in R. L. Ackoff (ed.) *Ackoff's Best: His Classic Writings on Management*. New York: John Wiley & Sons, Ltd.

Allio, R. (2003) 'Russell L. Ackoff, Iconoclastic Management Authority, Advocates a "Systemic" Approach to Innovation', *Strategy and Leadership* 31(3): 19–26.

Amabile, T. (1998) 'How to Kill Creativity', *Harvard Business Review*, September–October: 77–87.

Arnould, E. and Wallendorf, M. (1994) 'Market-oriented Ethnography: Interpretation Building and Marketing Strategy Formulation', *Journal of Marketing Research* 31(4): 484–504.

Asaro, P. (2000) 'Transforming Society by Transforming Technology: The Science and Politics of Participatory Design', *Accounting, Management and Information Technologies* 10: 257–90.

Ashkenazy, N. (2007) 'Critiques of Management Education and Scholarship and Suggestions for Change', *Academy of Management Learning and Education* 6(1): 102–3.

Belton, V., Elder M. and Thornbury H. (1997) 'Early Experiences of MENTORing: Design and Use of Multimedia Materials for Teaching OR/MS', *Omega* 25: 659–76.

Bennis, W. and O'Toole, J. (2005) 'How Business Schools Lost Their Way', Harvard Business Review (May).

Boden, M. (2004) *The Creative Mind: Myths and Mechanisms*. London: Routledge.

Boland, R. and Collopy, F. (2004) 'Design Matters for Management', *Managing as Designing*. Stanford, CA: Stanford University Press.

Card, S., Moran, T. and Newell, A. (1983) *The Psychology of Human-computer Interaction*. Hillsdale, NJ: Lawrence Erlbaum Associates.

Carr, N. (1999) 'Visualizing innovation.' *Harvard Business Review*, September–October: 16.

Churchman, C. (1967) 'Wicked Problems', *Management Science* 4(14): 141–2.

Conley, C. (2004) 'Leveraging Design's Core Competencies', *Design Management Review* 15(3): 45–51.

Cooperrider, D. and Whitney, D. (1999) 'Appreciative Inquiry: A Positive Revolution in Change', in P. Holman and T. Devane (eds) *The Change Handbook: Group Methods for Shaping the Future*. San Francisco, CA: Berrett-Koehler Publishers, pp. 245–61.

Cross, N. (2003) 'The Expertise of Exceptional Designers', in N. Cross and E. Edmonds (eds), *Expertise in Design*. Sydney: Creativity and Cognition Press.

Csikszentmihalyi, M. (1990) *Flow: The Psychology of Optimal Experience*. New York: HarperCollins.

Daellenbachand, H. and Petty, N. (2000) 'Using MENTOR to Teach Systems Thinking and OR Methodology to First-year Students in New Zealand', *The Journal of the Operational Research Society* 51: 1359–66.

DeBono, E. (2000) *Six Thinking Hats*. New York: Little, Brown & Co.

Dunne, D. and Martin, R. (2006) 'Design Thinking and How it will Change Management Education: An Interview and Discussion', *Academy of Management Learning and Education* 5(4): 512–23.

Gerlach, J. H. and Kuo F. (1991) 'Understanding Human-Computer Interaction for Information Systems Design', *MIS Quarterly* 15(4): 527–49.

Gharajedaghi, J. (1999) *Systems Thinking: Managing Chaos and Complexity: A Platform for Designing Business Architecture*. Burlington, MA: Elsevier Publishing.

Ghoshal, S. (2005) 'Bad Management Theories Are Destroying Good Management Practices', *Academy of Management Learning and Education* 4(1): 75–91.

Hackett, J. (2007) 'Preparing for the Perfect Product Launch', *Harvard Business Review* 85(4): 45–50.

Harlos, K.P., Mallon, M. and Jones, C. (2003) 'Teaching Qualitative Methods in Management Classrooms', *Journal of Management Education* 27(3): 304–22.

Jackson, M. (2000) *Systems Approaches to Management*. New York: Kluwer Academic/Plenum Publishers.

Kahneman, D. and Tversky, A. (1979) 'Prospect Theory: An Analysis of Decision under Risk', *Econometrica*, 47: 263–91.

Kelley, T. with Littman, J. (2001) *The Art of Innovation: Lessons in Creativity from Ideo, America's Leading Design Firm*. New York: Doubleday.

Leonard, D. and Rayport, J. (1997) 'Spark Innovation through Empathic Design', *Harvard Business Review*, November–December, pp. 102–13.

Mariampolski, H. (2006) *Ethnography for Marketers: A Guide to Consumer Immersion*. Thousand Oaks, CA: Sage Publications.

McFarland, J. (2001) 'Margaret Mead Meets Consumer Fieldwork', *Harvard Management Update* 6(8): 5–6.

Mintzberg, H. (2004) *Managers not MBAs: A Hard Look at the Soft Practice of Management Development*. San Francisco, CA: Berrett-Koehler.

Norman, D. (1986) 'Cognitive Engineering', in D. Norman and S. Draper (eds) *User Centred System Design*. Hillsdale, NJ: Lawrence Erlbaum Associates.

Norman, D. (1988) *The Design of Everyday Things*. New York: Basic Books.

Norman, D. (2004) *Emotional Design: Why We Love (or Hate) Everyday Things*. New York: Basic Books.

Olson, J. and Olson, G. (1990) 'The Growth of Cognitive Modeling in Human Computer Interaction since GOMS', *Human-Computer Interaction* 5 (2–3): 221–66.

Paradis, Z. and McGaw, D. (2007) *Naked Innovation*. Chicago: IIT Institute of Design.

Peirce, C. (1903) 'Pragmatism as a Principle and Method of Right Thinking', in A. Turrisi (ed.) *The 1903 Harvard Lectures on Pragmatism*. Albany, NY: State University of New York Press.

Porter, M. (1979) *How Competitive Forces Shape Strategy. Harvard Business Review*, March–April.

Prahalad, C. and Ramaswamy, V. (2003) 'The New Frontier of Experience Innovation', *Sloan Management Review* 44(4): 11–18.

Prahalad, C. and Ramaswamy, V. (2004) *The Future of Competition: Co-creating Unique Value with Customers.* Cambridge, MA: Harvard Business School Press.

Quilici Gonzalez, M. and Haselager, W. (2005) 'Creativity: Surprise and Abductive Reasoning', *Semiotica* 153(1): 325–41.

Rittel, H. and Webber, M. (1973) 'Dilemmas in a General Theory of Planning', *Policy Sciences* 4: 155–69. Reprinted in Cross, N. (ed.) (1984) *Developments in Design Methodology.* Chichester: John Wiley & Sons, Ltd.

Rogers, C. (1959) 'A Theory of Therapy, Personality and Interpersonal Relationships, as Developed in the Client-centred Framework', in S. Koch (ed.) *Psychology: A Study of Science.* New York: McGraw-Hill.

Senge, P. (1990) *The Fifth Discipline: The Art and Practice of the Learning Organization.* Sydney: Currency.

Schön, D. (1983) *The Reflective Practitioner: How Professionals Think in Action.* London: Temple Smith.

Senge, P. (1994) *The Fifth Discipline: The Art and Practice of the Learning Organization.* Sydney: Currency.

Simon, H. (1996) *The Sciences of the Artificial*, 3rd edn. Cambridge, MA: The MIT Press.

Underhill, P. (2000) *Why We Buy: The Science of Shopping.* New York: Simon & Schuster.

Wellman, B. (2001) 'Computer Networks as Social Networks', *Science*, New Series 293(5537): 2031–4.

Whitney, P. and Kumar, V. (2003) 'Faster, Cheaper, Deeper User Research', *Design Management Journal* 14(2): 50–7.

Wilson, P., Dell, L. and Anderson, G. (1993) *Root Cause Analysis: A Tool for Total Quality Management.* Milwaukee: ASQC Quality Press.

# 10

# Mutual Inspiration and Learning between Management and Design

RALPH BRUDER

The mutual inspiration between business leaders and creative people is not a recent phenomenon. There are many examples of close relationships between top managers and architects or designers. The German designer Otl Aicher can be taken as an example of a creative person working closely with top management (for example, ERCO, Bulthaup, FSB). Aicher was known for having a strong influence on product design and the overall visual appearance of a company.

In many of the cases where creative people had an inspirational effect on the development of companies it was because of their personal relationships with top managers. The managers feel themselves inspired by the personality of artists, architects and designers. But the responsibilities are clearly assigned, which means that it is the task of the management to define the strategy of a company and that the creative counterpart (or sometimes alter ego) should help making this strategy visible and tangible.

This clear assignment of roles between management and design is based on the assumption that the experts in the respective domains have the knowledge and the skills to solve the open questions in their area. And this assumption might be true if the problems to be solved are well known and the methods to solve those problems are well established.

For those well known problems in management, traditional economic theories can be applied assuming that every management task can be summed up in a problem that can be clearly defined and for which an optimal solution can be found. Not least as a result of the increasing global competition on deregulated markets, this theory, which works on stable, growing markets, has long reached its limitations.

Today's markets, however, are increasingly unstable and unpredictable. They evolve in unforeseeable ways with unforeseeable consequences. Confronted with this kind of radical uncertainty, managers can never know precisely what they are trying to achieve or how best to achieve it. They can't even define the problem, much less engineer a solution. (Lester et al. 1998)

As a result of these uncertainties, two different kinds of behaviour can be observed in management. One approach is to concentrate on those fields that can be controlled. A typical example is concentration on quality aspects and trying to avoid mistakes (for example, Total Quality Management, Six-Sigma Approach). In this context the introduction of key performance indicators for productivity should also be mentioned. The aim of these measures is mainly the reduction of costs. Reduced costs can build a buffer for flexibilities on market demands and they are a way to react to customer wishes for lower product prices.

But, besides the fixation on reducing costs, a second behaviour pattern can be observed: intensively looking for innovation and new products/services. The equation in this case is very simple: the better products you have, the more customers you will get. The second part of this equation is: if your product is really good and innovative, then your customers will stay with you for a long time. The role model for this behaviour is Apple, with its iPod, iPhone, iPad series. Many companies want their products to be as 'sexy' as the Apple products. Even very typical industrial products like the control unit of capital equipment (Hofmann and Holzkaemper 2009) should be desirable to use.

The supposed magic power of design to attract customers and users makes the discipline highly attractive to management in an economic phase where most managers feel that they no longer understand their customers and where the variety of products is dramatically increasing. In this context design is mainly reduced to the innovation of new products and services.

Looking at the success story of Apple, many top executives realized that it is not only the excellent products that can explain the attraction of iPod, and so forth, but also excellent brand management executed by excellent creative leaders. So the question arises of what management can learn from those creative leaders. Lester et al. (1998) pointed out that managers should learn from design to deal successfully with unstable and unpredictable markets. They studied the research and design activities in different cases (like Levi's) and introduced interpretive management as a typical way of a design-oriented problem solving strategy.

It is not really a new thing that the design disciplines are expected to contribute to the solution of problems in the economy and society. Looking, for example, at the publications from the Ulm School of Design, this holistic problem-solving approach can clearly be detected (for a comprehensive overview of the history of Ulm School of Design, see Spitz 2002). What is new, after more than ten years, is that it is not only the designers, architects and artists themselves that are claiming to have solutions for existing problems in economy and society but that other disciplines are asking design for support. A protagonist in that context is Roger Martin, the Dean of the Joseph L. Rotman School of Management at the University of Toronto (Martin 2004, 2010).

But with this growing interest, design as a discipline is facing enormous challenges. Design cannot react to these new challenges if it is reduced to the role of a mere provider of aesthetic solutions for formal problems. If a design approach is taken seriously an answer must be found to the question of how design can contribute to strategic development of companies (and even countries and societies).

## CROSS-RELATIONS BETWEEN DESIGN AND MANAGEMENT

In Table 10.1 a matrix is presented to categorize the links between design and management. The contribution of design to different levels of strategic management is visualized within this matrix. The framework represented by the matrix is based on the assumption that there is a strong relation between design and strategic management (see, for example, Mintzberg 1990, Liedtka 2004, Liedtka and Mintzberg 2007). Strategic management refers to different hierarchical levels: corporate level, business level and operational level. According to Porter (1987) corporate strategy concerns two different questions: what businesses the corporation should be in and how the corporate office should manage the array of business units. Strategy at business level is concerned with how to create competitive advantage in each of the businesses in which a company competes (Porter 1987, 1996). The focus on operational level (see Drucker 2006) is on day-to-day practice and refers to the functional responsibility of departments.

Despite the multiple definitions of design there seems to be one common understanding: 'Design must be understood as a word that describes both a process and an outcome' (Dumas 2000). Typically those outcomes are products (3-D or 2-D), services and communication processes. Design as a process describes the procedure that

**TABLE 10.1: Contributions from design to different levels of strategic management**

| Strategic Management Design | | Corporate | Business | Operational |
|---|---|---|---|---|
| **WHY** | Motivation/ Attitude/ Perspective | Creative leadership | Design of business | Design thinking |
| **HOW** | Tools/Methods/ Process | Brand design | Design thinking | Design processes standards |
| **WHAT** | Products/ Services/ Communication | Brand experience | Business of design | Multidisciplinary design teams |

is needed to come from a first idea to the finalized outcome. In many companies the design process itself is standardized to an extent. Within the design process several methods and tools are applied. These methods are part of what is known as 'design thinking' (cf. Brown 2008, 2009).

In addition to the *what* and *how* of design, the personality behind the designed products and the design process is increasingly becoming the focus of interest. The wish of many creative people to change imperfect scenarios by actively interfering through design is highly appreciated. Such an attitude seems to be very much needed when there are many options and it is difficult to decide which way to follow. Bruce Mau (a Canadian designer) asks: 'Now that we can do anything, what will we do?' (Mau and The Institute Without Borders 2004).

This emphasis on the importance of the personality of designers is also raised by Peter Gorb (the founder of the Design Management Unit at London Business School) when he claims: 'You're out there not to design, but to run the world' (Gorb 2003).

The nine elements of the matrix presented above correspond with the seven themes identified by Lockwood (2004) how corporations can use design as a business resource:

> One: Organizational structure
> (see Design Thinking)
> Two: Design Management processes
> (see Design Processes Standards)
> Three: Design to enable change
> (see Design of Business)
> Four: Design-facilitated corporate strategy
> (see Brand Design)
> Five: Design to realize innovation
> (see Design Thinking)
> Six: Design teams and collective purpose
> (see Multi-disciplinary design teams)
> Seven: Design-minded corporate culture
> (see Creative leadership)

## DESIGN AND CORPORATE STRATEGY

### *Creative Leadership*

If design should become an essential part of a corporate strategy then it seems to be helpful (if not necessary) to have creative people sitting on the board of directors. 'If design will be the byword of the 21st century, designers will have to take their places as its leaders' (Farson 2008). This close relation between the strategic orientation of a company and the disciplinary background of its leaders can be seen in many cases. For example, in the tradition of the German automotive industry it was, for a long time, an unspoken matter of course that members of the board of directors should 'have fuel in their blood', which means that they should have a strong background in engineering. If you are dealing with needs, wants and wishes of your customers it seems appropriate to ask for a strong background in creative disciplines. Steve Jobs of Apple is a good example for the strength of creative leadership.

## Brand Design

If the decision is made that a company should be known for its friendly face to the customer and for its orientation towards customers needs, wants and wishes, the question arises how to achieve this goal. And that is not only a question of which products to produce and sell or which business model to choose but it concerns the appearance of the company in its totality. This total appearance of the company to others like customers, stakeholders, employees is known as brand identity.

The application of knowledge from human sciences, technology, materials experiences, as well as from aesthetics and communication sciences, can be used by design to be part of the strategic process to formulate a brand identity for companies. Such an approach is described by Bevolo and Brand (2003) for the brand design process at Philips Design: 'In summary, Philips's brand design process offers a unique design management approach to delivering a strategic brand direction.'

In this context it should not be forgotten that brand design is not only about products, services or communication strategies but also about the integration of the employees into the brand identity (de Chernatony 1999).

## Brand Experience

A key lesson for many companies was that it is simply not enough to have excellent brand management regarding the marketing strategy, the brand name or the engagement of your employees. In the long run it is also necessary to fulfil or even surpass your customers' expectations with respect to the products or services offered to them. The importance of product and service experiences in the liaison between customers and companies was introduced to a broader public by Pine II and Gilmore (1999), Jordan (2000) and Jensen (2001).

The essential contribution of creative people to make the brand a tangible experience for customers was pointed out by Neumeier (2003). A successful brand experience is only possible if there is no gap between the strategy definition and its implementation. Engaging designers already in the phase of brand definition (see brand design) is necessary for a coherent brand appearance but it is also necessary to have excellent designers for the transformation of brand strategies into desirable products.

In this context the effect of the physical appearance of companies on the brand experience should also be mentioned. It is the field of what is called 'corporate architecture' (Messedat 2005). Lockwood (2004) presents the case of British Airways, which used design not only for its business-class customer traffic but also for the company's new headquarters. It is a typical example for a company that is committed to design throughout the organization.

## DESIGN AND BUSINESS STRATEGY

### Design of Business

A major concern for every company must be the design of a valuable business model. Existing methods of business management must be developed in order to establish new rules of competition, draft new business models and create new markets. And that is what might be called a design task. Design is all about finding solutions for abstract, vague problems. Design is also about making decisions and not just analysing situations (see Raney and Jacoby 2010). Because of the similarity between designing and managing it is not surprising that Roger Martin is arguing: 'Business people don't need to understand designers better: they need to be designers' (Martin 2004).

How design might matter for management is also part of the 'managing as designing' approach established at the Weatherhead School of Management by Richard J. Boland and his colleagues (Boland and Collopy 2004). The case of Frank O. Gehry and his projects had been very helpful for the experts in management at the Weatherhead School of Management to see the design process at work and to learn about the importance of the designer's personality for the success of (in this case architectural) projects.

### Design Thinking

'Thinking like a designer can transform the way you develop products, services, processes – and even strategy' (Brown 2008). This citation from Tim Brown (CEO of IDEO) shows the close linkage between design and strategic management. The success and acceptance of IDEO is proof that 'design thinking' has entered the business world. The basis of design thinking can also be found in Herbert Simon's understanding of design as 'science of the artificial'. This science of the artificial is, above all, driven by 'reflective practitioners' because they bring with them the awareness of the problems involved and the experience needed to formulate the relevant questions. To formulate the right questions is a key to innovation in a confusing situation like today with complex markets. It is important to realize 'that design thinking alone is not sufficient, but when mixed with solid business thinking, it can produce a combustible mixture' (Merholz, 2009).

*Business of Design*

There are many studies showing the positive effect of integrating design into the business strategy for the economic success of companies (Design Council 2008) but also for national economies (Moultrie and Livesey 2009).

Exemplary findings of the British Design Council for the Design in Britain 2008 are:

- 23 per cent of businesses thought design had become a lot more important over the last three years in enabling the firm to achieve its business objectives;
- 30 per cent said it was a little more important and only 5 per cent considered it less important;
- more than half (52 per cent) of firms agreed or agreed strongly that design is integral to the country's future economic performance – 16 per cent disagreed or disagreed strongly;
- 59 per cent of firms agreed or agreed strongly that there was a clearly positive link between investment in design and profitability – 13 per cent disagreed or disagreed strongly;
- over a twelve-month period 34 per cent of firms had seen design expenditure grow moderately or rapidly whereas only 5 per cent had seen it decrease;
- over half (54 per cent) of firms believe design will help them stay competitive during the economic downturn.

See Design Council (2008) for more details.

A recent study in Germany conducted by the German Design Council, the German Brands Association and the corporate communications agency Scholz&Friends came to similar results with respect to the importance of design for corporate success in Germany (Scholz&Friends 2010).

## DESIGN AND OPERATIONAL STRATEGY

*Design Thinking*

If a company wants to have a full integration of design within its strategy it is not enough just to formulate a business strategy based on design thinking. This strategy should be implemented by as many staff members as possible. That is what makes a design-driven enterprise. So besides the benefit of design thinking for the formulation of business strategies, the perspectives of designers should also be part of the operational strategy. In the understanding of Tim Brown and his colleagues from IDEO, design thinking is not only a collection of tools and methods from design-related disciplines nor is it a pure description of a design process (for example, from inspiration to ideation to implementation and back and forth). Design thinking might be also part of a personality's profile. According to Brown (2008) some of the characteristics of design thinkers are:

- empathy;
- integrative thinking;
- optimism;
- experimentalism;
- collaboration.

## Design Processes Standards

So when design addresses the shaping of processes, the significance of the design process itself increases. It is not only the result of the design process (for example, the formal design of products) that must be assessed. It should also be asked how effective the cooperation between the players in the design process is, how suitable the capabilities of these players are for the process, where it is possible to rationalize the design process and where resources are purposefully being saved in individual cases in order to optimize the process as a whole.

## Multi-disciplinary Design Teams

Successful product (or service) development is never done by one discipline alone. A product like Apple's iPhone is based on the innovative ideas from engineers concerning a new touch-screen technology. Porsche cars are not only known for their extraordinary technology but also (and especially) for their outstanding design. Nevertheless there is a tendency in companies to have a kind of a hierarchy between different departments that are involved in the product-development process. Sometimes product development is dominated by engineering departments, sometimes the marketing department is most influential and in some cases it is also the design department that is the driver for the design process.

It was shown by Lockwood (2004) that those companies having established a collaborative team approach in product development are successful when it comes to employee creativity (StorageTek was used as an example for this relation). It might be an interesting and challenging role for creative people to become the coordinators or facilitators for such multidisciplinary design teams. Creative people are used to referring to different sources of knowledge within their problem-solving process and they mostly have a holistic approach to problem solving. Recently many creative people have been acting within web-based social networks, which is a good way of learning how to get on with different perspectives and contributions. Nevertheless it will be important for designers and architects to learn, in their courses at colleges, art schools or universities, that it is not only their individual genius that is important for successful product development in practice. They have to learn that it is really important to have a strong design attitude based on ethical principles and to integrate this design attitude fruitfully into a collaborative purpose.

## TEACHING MANAGEMENT AND DESIGN

### Overview

A good overview of existing programmes in the context of design management can be found on the website of the Design Management Institute (www.dmi.org/dmi/html/education/grad_s.htm). Table 10.2 contains some suggestions for programmes to support the integration of design to strategic management.

The concept of an MBA in business design will be explained in more detail in the following section.

### A Concept for Teaching Business Design – A Case Study

*Background*   In 1999 the idea was born that a new school was needed at a former coal mining area in the western part of Germany (Zeche Zollverein in Essen), which could work to stimulate new job creation in the so-called creative businesses. Finally, in December 2003, the Zollverein School of Management and Design was founded as a private institution for teaching and doing research in the field of business and design. The purpose of the Zollverein School is to create a platform for mutual exchange between the often separated fields of business and design. At the Zollverein School, managers become familiar with the views and ways of thinking of designers or architects and vice versa. Both sides move away from their traditional viewpoints and link their activities to create innovative and sustainable strategies for future businesses. What makes the Zollverein School very special and unique is that it is neither a business school nor a design school but an institution where those different disciplines define a space of mutual respect.

**TABLE 10.2: Examples for teaching management and design**

| Management Design | | Corporate | Business | Operational |
|---|---|---|---|---|
| **WHY** | Motivation/ Attitude/ Perspective | Executive programmes | MBA business design | BA/MA design management |
| **HOW** | Tools/Methods/ Process | Design modules within MBA courses | | |
| **WHAT** | Products/Services/ Communication | MFA 'business design' | | |

Another specialty of the Zollverein School concept was the opportunity to design a new building just for the purpose of the new school. The architects of this building are Kazuyo Sejima and Ryue Nishizawa (SANAA) from Tokyo, the Pritzker Architecture Prize Laureates of 2010 (see Figure 10.1). For more information about the basic idea and the architectural concept of the Zollverein School, please refer to Bruder (2006).

*MBA in Business Design*   The first academic degree established at the Zollverein School of Management and Design was an MBA in business design. The idea behind

FIGURE 10.1   Zollverein School of Management and Design, Essen, Germany. Photograph by Thomas Mayer, courtesy of Thomas Mayer.

this MBA was to increase the acceptance of people with a background in design-oriented disciplines taking up leadership positions and the MBA degree is still valuable for such leadership positions. But the MBA degree is not only intended to support the careers of designers and architects. The integration of design thinking and design methods into the traditional MBA programme is a step towards the often-demanded reformation of existing MBA programmes (see Martin 2004, 2010). Nevertheless there are also the core courses in finance as a basic part of the MBA degree, which should be taken seriously in a business context. The Zollverein School offered only postgraduate and doctoral programmes. That means that students at Zollverein School have already had some years of professional experience. They should also have a first academic degree even though the course of study for this degree is of less importance. The Zollverein School looks for diversity in the disciplinary backgrounds and actual professions of the students.

*Course Concept in General*    The interpenetration of management and design has to be integrated in different types of courses. Within the lecture-type courses there are joint courses taught by teachers with different backgrounds. So, for example, the professors of strategic marketing and culture/society offer some lectures together and discuss the relationship between actual cultural studies and strategy formulation. In seminar-type courses (like the one on innovation) the students have to apply theoretical findings to real case studies. Such case studies would involve looking at the innovation level of the companies they are working for (or running themselves) and to come up with some ideas how to improve this innovation level. Project studies are another integrative part of the MBA programme. Those projects last for three months and are related to practical problems. The aim of the project is the implementation and validation of skills, concepts and methods that the students have learned in different courses. The learning and working happens within interdisciplinary teams.

The projects are guided by different MBA tutors and, again with their various disciplinary backgrounds, they deliver different perspectives to the project. To handle those different perspectives is not very easy for the students but it could be shown that the sometimes controversial discussions were the breeding ground for innovative and creative thinking. Finally, when looking for a topic for their master's thesis the students are encouraged to make use of the new skills and tools they learned at the Zollverein School. Therefore several sessions are arranged to discuss with all students about the topics of their master thesis and what these topics have to do with 'business design'.

*'Design Studies' Module*    The intention of the 'design studies' module is to give an overview of different fields of application for design and the design processes that are typical for those fields of application. The lecturers within this module are experts from the design professions – for example, product design and architecture – who reflect

on their practical activities. Using case studies, they discuss the processes underlying certain successful design projects. Nevertheless it is not intended to teach 'practical' design skills in the context of, for example, product design or graphic design.

Experts in design theory also teach in the module 'design studies'. In their courses, they present different approaches to and distinct applications of design methodology, which are then explored in regard to their applicability to different contexts. Finally, the teachers from 'design studies' are integrated in many of the other program modules. Thus, design perspectives have a great influence on the student projects mentioned above.

*Cross-disciplinary Team Work*   Referring to GK VanPatter (Bruder and VanPatter 2006) it is extremely important to teach innovation process mastery and cross-disciplinary team dynamics mastery. The Zollverein School demonstrates that it is indeed challenging to ask professionals with cross-disciplinary backgrounds to work together in teams, especially in a school setting. While the professionals are used to working in cross-disciplinary teams in their everyday work life, they are also used to clear role assignments in such professional teamwork. For example, one is either the client or the contractor. Further, in their professional lives, they are used to being part of a hierarchical system that can affect the efficiency and the dynamics of teamwork substantially. In many companies, people with a business (often financial) background still dominate the higher positions. This explains why it is still quite rare to find professional teams with mutual consideration of perspectives from business and design. This means that, even (or especially) for students with professional backgrounds, it is difficult to use teamwork as a source for innovative problem-solving strategies. It is naïve to think that it is enough to assign students with different background to teams and just wait for them to master cross-disciplinary team dynamics by doing teamwork. Because of the importance of team dynamics mastery the cross-disciplinary teamwork topic forms part of different modules of the MBA business design programme (for example, in the 'organization and leadership' or 'innovation' modules).

It is not enough to look at the dynamics of cross-disciplinary teamwork from an outside perspective. The projects of the MBA students (which are executed in cross-disciplinary teams) are accompanied by a module called 'leadership skills'. The aim of this module is to support personal development and the leading skills of our students by introducing self-reflective elements into their team building and teamwork.

Some of the topics within this module are:

self-organization;
management of complexity and difference;
structures of power within organizations;
leadership and organization;
participation.

*Personal Profiles of Business Designers*    The composition of the first group of students, who started their Executive MBA programme in February 2005 gives an idea of this multidisciplinary and multiprofessional approach. Within this group of eighteen students there had been designers (with a focus on products, communication or media), architects, economists, and engineers. They had been working, for example, as heads of marketing department at a supplier for the automotive industry, at consultancies with focus on design strategies, or running their own business. In the first year (2005) of its existence, the MBA in business design was attracting professionals from the so-called creative industries. In the following years, the Zollverein School became increasingly attractive for professionals who have no basic education in any design-related disciplines but who feel that understanding design processes and adapting design methods might be helpful for their future professional careers. Besides the diversity in disciplines and professional activities it was also a definite aim of the Zollverein School to attract students with different cultural backgrounds. Bringing management and design together is a topic that is not only of interest for certain local economies but has a global relevance.

An experiment called 'Defuzzing WHO' (Bruder and Van Patter 2006) was conducted with the eighteen first year students at Zollverein School. The experiment was done in two steps. In the first step the students had to decide in which of the following phases of a project process they saw their strengths:

1. Formulate problem/opportunity.
2. Formulate solution.
3. Implement solution.

Nine students thought their strengths were in formulating problem/opportunity, six more students thought their strengths were in formulating solutions and another three thought their strengths were in implementing solutions.

Then we continued with the second part of our research experiment by using an interactive innovation profile tool. This tool asks students eighteen times to describe the way problems are solved. With each question they had four possible answers that they had to arrange in an appropriate order according to their own problem-solving style (for example: 'How do you solve problems? Alert – ready – poised – eager'). After answering those eighteen questions each student's own special innovation profile is presented.

The innovation profile tool distinguishes four possible innovation types (generator, conceptualizer, optimizer and implementer) and the student gets a ranking according to which of those basic types fits best and which fits least to his or her own innovation profile.

Nine innovation profiles for the first-year MBA students at Zollverein School of Management and Design were analysed. Among the nine profiles we had one generator, five conceptualizers, one optimizer and two implementers. The results of the innovation profile were very much in line with the ratings of the self-estimated strengths according to the first part of our little experiment. So the generator and most of the conceptualizers were rated to have strengths in formulating problems/opportunities. The optimizer and one of the conceptualizers were rated to have strengths in formulating solutions. Finally, the two implementers were rated to have strengths in finding solutions.

We discussed the results of this experiment with out second class of MBA students. Again, most of them felt that they had strengths in formulating problems/opportunities and solutions.

It is interestingly to see that there is a disproportionately large number of conceptualizers and a small number of generators and optimizers in the group of creative people interested in business design. It will be left to the mastery of creative leaders in companies to engage the right combination of experts with different innovation profiles.

*Lessons Learned from the Zollverein School*    In 2009 the Zollverein School of Management and Design became part of the Ruhr Campus Academy (the institution that is responsible for advanced education at the public University of Duisburg-Essen) and was no longer an independent organization. With this integration into a public university the concept of the Zollverein School of Management and Design will only be implemented partially. There are several lessons from the development of the Zollverein School that are relevant to design management in general:

- It is an advantage to start a new school in the area of management and design right from scratch. In this case there is no dominance from specialists from business sciences or from creative disciplines, as might happen when starting a new domain at an existing school.
- Hybrid institutions between management and design are very innovative and they are raising curiosity to a high level but at the same time they are extremely difficult to communicate. In the best case they are seen as something really new taking the best out of both worlds (management and design). In the worst case, they are not accepted by either community. For the Zollverein School it was difficult to be seen as an institution with close links to the business world and not just as another design school.
- To start a new school means that you have to raise awareness of this so-far unknown institution. The Zollverein School received worldwide recognition after only a short period of its existence (for example, 'The new Zollverein School of Management & Design, which opened its doors to executive students in 2005 and full-time students in 2006, is the buzz of the Continent' – Hempel and McConnon 2006). But it takes more than

three years to establish a new programme on the market. One of the reasons for this foundation period is the absence of alumni within the first years. It is much easier to convince applicants (and investors) about a new school if you can show successful alumni to them. This phase of foundation and stabilization requires a business model that is not dependent on student fees.

- For a new school in management and design it is important to cooperate closely with industry. This cooperation might involve companies sending promising staff members to school for courses or companies giving donations to the school. Again, it takes some years to demonstrate to industrial partners that they might benefit from a partnership with this new school. In case of the Zollverein School, there had been many discussions with representatives from companies and in many of the discussions the outcome was that those representatives first wanted to be sure about the success of the school before giving money to it. This attitude is understandable and very much to be respected but it makes it difficult to start something new. To come back to the beginning of this chapter, it is very helpful to form personal contacts with top managers who are enthusiastic about the concept of a new school (see for example the engagement of Hasso Plattner, one of the founders of SAP, for the d-school at Stanford University).

## CONCLUSION

It is evident that close collaboration between design and management will be helpful for them to survive in increasingly complex markets. Design might be the decisive input to make the difference. But it is not enough to see design just as a creator of new products. Design has to become an integral part of business strategies. That means there must be a mutual respect between experts from the creative disciplines and from business. Design can contribute to business strategy on different management levels, from corporate to business to operational. It seems important to examine more in detail what the contribution at the strategic management level might look like. Furthermore it is important to establish more teaching programmes at colleges, art schools and universities to bring this new way of thinking about design and management to many creative enablers.

However it is still difficult to generate confidence in this combination of disciplines and understanding from business and the student community. It takes leadership, entrepreneurialism and political robustness to make it work, either in existing organizations or as new independent ventures.

# REFERENCES

Bevolo, M. and Brand, R. (2003) 'Brand Design for the Long Term', *Design Management Journal* (Winter): 33–9.

Boland, R. J. and Collopy, F. (2004) *Managing as Designing*. Palo Alto: Stanford University Press.

Brown, T. (2008) 'Design Thinking', *Harvard Business Review* (June): 84–92.

Brown, T. (2009) *Change by Design: How Design Thinking can Transform Organizations and Inspire Innovation*. New York: HarperCollins.

Bruder, R. (2006) 'The Zollverein School of Management and Design: Using Innovative Design Skills on Business Models', in K. Feireiss (ed.) *The Zollverein School of Management and Design Essen, Germany. Kazuyo Sejima + Ryue Nishizawa SANAA*. Munich: Prestel.

Bruder, R. and VanPatter, G. (2006) 'Business Design Academy. Understanding the New Zollverein School', *NextD Journal* 25: 1–16.

de Chernatony, L. (1999) 'Brand Management Through Narrowing the Gap Between Brand Identity and Brand Reputation', *Journal of Marketing Management* 15: 157–79.

Design Council (2008) *Design in Britain 2008,* www.designcouncil.org.uk/our-work/In sight/Research/How-businesses-use-design/Design-in-Britain-2008/ (accessed 7 April 2010).

DMI (2010) 'Seminars/Education', www.dmi.org/dmi/html/education/grad_s.htm.

Drucker, P. (2006) *The Practice of Management*. New York, HarperCollins.

Dumas, A. (2000) *Theory and Practice of Industrial Design,* Report for the EC funded project INNOREGIO: dissemination of innovation and knowledge management techniques, www.adi.pt/docs/innoregio_theor_design.pdf (accessed 30 March 2011).

Farson, R. (2008) 'Leadership is THE Strategic Issue', www.di.net/articles/archive/lead ership_strategic_issue/ (accessed 5 April 2010).

Gorb, P. (2003) 'The Design Management Interface', *The Association of Registered Graphic Designers of Ontario,* www.designthinkers.com/pdf/pgorb.pdf (accessed 7 April 2010).

Hempel, J. and McConnon, A. (2006) 'The Talent Hunt', *BusinessWeek* (9 October), www.businessweek.com/magazine/content/06_41/b4004401.htm/ (accessed 25 April 2010).

Hofmann, T. and Holzkaemper, P. (2009) 'Intercultural Accessibility in Capital Equipment', in *Proceedings of 17th World Congress on Ergonomics* (IEA 2009), Peking, China, 10–14 August 2009.

Jensen, R. (2001) *The Dream Society: How the Coming Shift from Information to Imagination will Transform Your Business*. New York: McGraw-Hill.

Jordan, P. (2000) *Designing Pleasurable Products*. London: Taylor & Francis.

Lester, R., Piore, M. and Malek, K. (1998) 'Interpretive Management: What General Managers Can Learn from Design', *Harvard Business Review*, March–April: 86–96.

Liedtka, J. (2004) 'Strategy as Design', *Rotman Management* (Winter): 12–15.

Liedtka, J. and Mintzberg, H. (2007) 'Time for Design', *Rotman Magazine* (Winter): 24–9.

Lockwood, T. (2004) 'Integrating Design into Organizational Culture.' *Design Management Review* Spring 2004: 32–9.

Martin, R. (2004) 'The Design of Business', *Rotman Management* (Winter): 7–11.

Martin, R. (2010) 'Management by Imagination', http://blogs.hbr.org/cs/2010/01/manage ment_by_imagination.html (accessed 2 April 2010).

Mau, B. and The Institute Without Borders. 2004. *Massive Change*. London: Phaidon Press.

Merholz, P. (2009) 'Why Design Thinking Won't Save You', http://blogs.hbr.org/merholz/ 2009/10/why-design-thinking-wont-save.html (accessed 1 April 2010).

Messedat, J. (2005) *Corporate Architecture: Development, Concepts, Strategies*. Ludwigsburg: AV Edition.

Mintzberg, H. (1990) 'The Design School: Reconsidering The Basic Premises of Strategic Management', *IEEE Engineering Management Review* 19(3): 85–112.

Moultrie, J. and Livesey, F. (2009) *International Design Scoreboard: Initial Indicators of International Design Capabilities*. Cambridge: University of Cambridge, Institute of Manufacturing.

Neumeier, M. (2003) *The Brand Gap. How to Bridge the Distance between Business Strategy and Design*. Berkeley, CA: New Riders.

Pine II, B.J. and Gilmore, J. (1999) *The Experience Economy*. Boston, MA: Harvard Business School Press.

Porter, M. (1987) 'From Competitive Advantage to Corporate Strategy.' *Harvard Business Review* (May–June): 43–59.

Porter, M. (1996) 'What is Strategy?' *Harvard Business Review* (November–December): 61–78.

Raney, C. and Jacoby, R. (2010) 'Decisions by Design: Stop Deciding, Start Designing.' *Rotman Magazine* (Winter): 35–9.

Scholz&Friends (2010) *The Beauty of Added Value*. Berlin: Scholz&Friends.

Spitz, R. (2002) *hfg ulm: The View behind the Foreground*. Stuttgart: Edition Axel Menges.

# 11

# Manifesto for the M(B)A in Designing Better Futures

## LUCY KIMBELL

In a world of conflict, crisis, and challenges ranging from the ecological to questions to equity, what is the role of design management? This chapter offers a manifesto for a new postgraduate course, and possibly a new type of education. Messily interdisciplinary, it draws on design and management but also traditions in the social sciences, the humanities and the arts that are attentive to situated knowledge, experiences and aesthetics. As a manifesto, it is not rooted in any one educational institution's legacy of what counts as a discipline, what it teaches or how to teach it. Nor does it have to offer an argument that bends to any one institution's strategic goals or that juggles with limited resources. Like other manifestos, it aims to be imaginative and provocative. It proposes what should be, without being overly shaped by what is.

This manifesto is not about trying to improve management education (for example, Boland and Collopy 2004; Adler 2006; Dunne and Martin 2006; Gosling and Mintzberg 2006; Starkey and Tiratsoo 2007; CCA 2009; Starkey and Tempest 2009; Datar et al. 2010), but rather about creating better futures. Attentive readers will quickly note that the curriculum discussed here does not organize itself along the established lines of the conventional MBA, such as finance, economics, accounting, organizational behaviour, operations management, marketing or strategy, although aspects of these disciplines are present. Nor does it rely solely on established ways of teaching design in art schools. Neither an MBA, but concerned with organizing for the future, nor an MA, but concerned with the imaginative practices of art and design, this fictional course in some ways resembles programmes that already exist but with important differences. The first is the approach to knowledge and its limits. The second is concerned with the boundaries between disciplines.

The course is underpinned by the idea that, in contexts in which uncertainty and ambiguity are high, then knowledge is of relatively little value. Concepts, theories, methods and tools derived from the past and its certainties are not necessarily help-ful for the future. Funtowicz and Ravetz (1993) distinguish between 'post-normal' and 'normal' science. For Funtowicz and Ravetz, when systems uncertainties and decision stakes are high, knowledge and expertise are contested, making it difficult for policy-makers and managers to make decisions since the knowledge which shapes them is open to challenge from other disciplines and from diverse stakeholders (see Figure 11.1) Existing problem-solving strategies rest on normal science in which human values are not acknowledged, high degrees of certainty are required, and qual-ity assurance is managed relatively informally by peer review. Funtowicz and Ravetz argue that when there are high levels of uncertainty around ethics and epistemology (what counts as valid and reliable knowledge) and when decision stakes reflect con-flicting purposes among stakeholders, the methodologies of normal science are inef-fective. In such contexts, decision-making processes should acknowledge uncertainty and include in dialogue all those with a stake in the issue. 'New methods must be

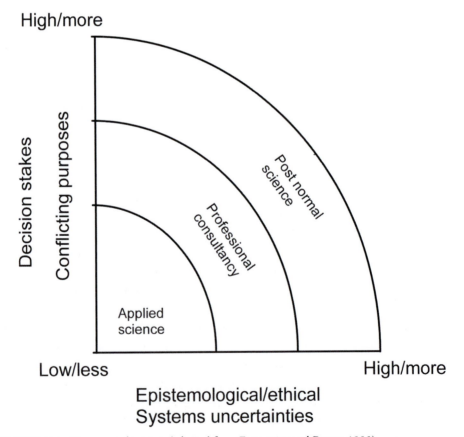

FIGURE 11.1   Post-normal science (adapted from Funtowicz and Ravetz 1993).

made to make our ignorance usable' (Funtowicz and Ravetz 1993: 743) and new ways to determine legitimacy and competence will extend peer communities to broader social and cultural institutions. Post-normal science requires uncomfortable knowledge and clumsy solutions (Rayner 2006) rather than selecting between well-defined alternatives based on knowledge, the certainty and quality of which have been assured. What this means for this M(B)A is that it rests on a *lack* of knowledge that stimulates conversation, humility and question-making about purposes and values, a space in which design theory and practice play an important role in moving from research to action.

A second foundational idea is the attention paid by scholars within science and technology studies (STS) to the ways scientific knowledge and technologies are produced, legitimized and institutionalized (cf. Foucault 1972; Bijker et al. 1987; Callon 1987; Law 1987; Latour and Woolgar 1986; Latour 1987; Latour 2005; Sismondo 2004). Space does not allow even a modest attempt to characterize what is a large and interdisciplinary field but two ideas are drawn out relevant to the project in this manifesto. The first is the attention within STS to understanding science and technology as thoroughly social processes, in which scientists, engineers, managers, designers and others are engaged in struggles for resources and in which conflicts are not to be ignored but rather are sites that illuminate what is of concern. The second idea is the favoured methodology of scholars working in this tradition. Through detailed, local ethnographies of scientists and engineers and their working practices, this research has presented evidence of the messy, contingent, worlds in which knowledge is constructed and action takes place. Science, technology and the knowledge involved in constituting them are not neutral, nor are they unproblematically human centred. What this means for this manifesto is that truth claims about design or management are not just accepted or presented without qualification.

Thus far we have considered the limitations on the knowledge base for this imaginary M(B)A. We now turn to the second important difference between this M(B)A and existing programmes, which is the approach to disciplines and their boundaries.

Claims about interdisciplinarity and new kinds of knowledge fused between disciplines are rooted in expectations of accountability and the relevance of research to stakeholders (Nowotny et al. 2001). But the abstract notion of interdisciplinarity promoted by policy makers and funders can turn out to be more complicated in practice, which raises questions for the project to combine design and management. In an empirical study of several projects crossing design and IT, social science and design, and art and science, Barry et al. (2008) found that an idealized interdisciplinarity was more complicated than its advocates suggest. Barry et al. identified three modes of interdisciplinarity. One, the *service mode*, involves one discipline being in service to another – for example design being in the service of management, an idea central to earlier work in design management (cf. Borja de Mozota 2003), or management in the service of design. The second, the *integrative-synthesis mode,* involves disciplines integrating – for example efforts to fuse design and management into a new holistic educational programme in which design is a kind of management and management a kind of design. A key example here is the project that conceives of managing as designing and aims to complement the well-established decision attitude in management education and practice with a 'design attitude' (Boland and Collopy 2004). The third, the *agonistic-antagonistic mode*, is forged as those advocating change question disciplinary commitments to ideas of what constitutes reality and what constitutes knowledge. In this mode, the encounter between design and management may be destabilizing, but may also be illuminating and productive.

Barry et al.'s findings raise questions about what might happen at the boundaries of design fields and management disciplines. The service mode and the integrative-synthesis mode may be the aspiration for project or organizational teams, but the agonistic-antagonistic mode may also produce important new questions and new knowledge. What this means for our imaginary M(B)A is that it will not rely on simple attempts to fuse design and management, or consider the management of design, or the (re)design of management. This M(B)A will privilege design as the activity of conceiving of, creating and organizing for better futures realized in the shifts between knowledge and action, but it will also acknowledge the agonistic-antagonistic mode that will, at times, produce irreconcilable differences between disciplines.

To summarize, in contrast to conventional presentations of new postgraduate programmes, this M(B)A manifesto attends to both knowledge and its limits. Established bodies of knowledge are not sufficient when stakes are high and purposes are contested and disciplines, while they can serve one another and integrate fruitfully, can also present one another with important and impossible differences. The resulting uncomfortable fluidity and hybridity means this M(B)A may not ever be able to come into existence because, as conceived of here, it is unlikely to meet the requirements for awarding degrees at either a design school or a management school because of the ways disciplines are governed. But let us move on nonetheless to explore it further. This is, after all, a manifesto.

## ENQUIRIES INTO FUTURE PRACTICES

Better futures require new ideas and new ways of doing things but novelty cannot be easily assessed based on current knowledge. This is why this manifesto conceives of working towards better futures as a kind of design, which is centrally concerned with exploring, proposing and testing new kinds of arrangement. However while it continues to change and adapt, design education is rooted in creating determinate objects within craft traditions shaped by industrialization and underlying narratives of economic production and consumption (Fry 1999). Contemporary design has seen an important shift away from designing objects towards interactions, experiences, services and changed behaviours but theory and practice have so far failed to engage deeply with theories of the social (cf. Ingram et al. 2007; Julier 2007; Shove et al. 2007; Fry 2009). There are some postgraduate programmes in design schools specifically concerned with futures, placing a particular emphasis on sustainability and environmental concerns (e.g., Goldsmiths 2009; Fry 2011), but they are not well connected to management theory and practice. In contrast, this M(B)A is concerned with designing better *futures*, rather than any specific kind of object, instantiated through practices which are understood as arrangements of bodies, minds, things, structures, processes, knowledge and agency (Schatzki 2001; Reckwitz 2002).[1]

Theories of practice are ways of analysing action in the world that work not at the level of individuals, explained by a person's individual purposes, intentions and interests, nor at the level of collective order, explained by social norms and values. Practice theories 'highlight the significance of shared or collective symbolic structures of knowledge in order to grasp both action and social order' (Reckwitz 2002: 246). They describe how practices are carried by individuals in their routinized or mundane ways of understanding and moving through the world, knowing how to do things, the objects they desire and do things with, how it feels, and the structures that are (re) produced in day-to-day action. The important idea here is not simply that graduates of this M(B)A will go out into the world to design and undertake activities that shape futures, but rather that this involves changing the symbolic ordering of things, minds and bodies. Arguably, design practitioners are already adept at creating new kinds of symbolic ordering (e.g., Ravasi and Rindova 2008), but this M(B)A goes further than contemporary design education and most management education and argues that designers and managers are designing (future) practices.

Three problematic concepts underpin this M(B)A and are explicit in its title.

## Designing

Aimed at many kinds of professional working in different kinds of organization, this course draws on Simon's (1996) argument that design is the core activity of the professions: 'Everyone designs who devises courses of action aimed at changing existing situations into preferred ones' (Simon 1996: 111). Simon made a distinction between the sciences, concerned with studying what is, and design, concerned with what could be. Although it rests on a simplistic definition of the sciences that ignores how are they are produced by a variety of social actors, Simon's work has been influential among scholars working in management and organization studies. His work can be read as a reduction of design to problem solving (Hatchuel 2001), which is often contrasted with a constructivist approach advocated by Schön (1983). Building on Simon but departing from his bounded rationality, Hatchuel and Weil (2009) argue that new concepts and new knowledge result from a process of expansion during design. During a design process, concepts are generated that cannot yet be evaluated as true or not. This expansion of concepts and knowledge is for Hatchuel and Weil – and for this M(B)A – why design is important for innovation by generating new practices creating value in new sets of relations.

The decision to use the verb form 'designing' in the title of the M(B)A course echoes Weick's (1979) exhortation to attend to organizing (the verb) rather than to organizations (the noun). What matters in this curriculum is helping students under-stand and learn how to do design. An important influence is the notion that design-ers (should) design for change over time (Tonkinwise 2003) rather than designing objects that are then seen as complete once designing is over, but which nonetheless persist when they break or are discarded. Further, it is not only designers who are involved in designing. From the perspective of anthropology and sociology, many people play important roles in (re)constituting the meanings and effects of designs through their practices (for example, Suchman 1987; Julier 2007; Shove et al. 2007; Ravasi and Rindova 2008). Designs – the final form of products and services, for example – remain incomplete (Garud et al. 2008). Thus designing never ends and many people are involved in doing it.

*Better*

The second term, *better*, is just as problematic. The aim to make things better in the future is fundamental to design. 'Science raises the question "is this proposition valid or true?", while design asks "will it work better?"' (Jelinek et al. 2008: 317–18). But design fields have been slow to take up the challenges and insights from sociology, an-thropology, cultural studies and other fields that have emphasized how difficult it is to determine what 'better' might mean, for who and why. If we think of a service, for example, to determine what is meant by designing a 'better' service could involve the organization delivering the service describing its goals and resources; identifying the market it is operating within, and any legislation or cultural and social practices that might constrain the service and the organization's resources; articulating what con-stitutes the service; identifying its customers and end users; and agreeing a point in time at which to gather data in order to determine whether there are improvements. Questions to be considered might include determining the boundaries of the service and of the organization. Should the organization only think about 'better' from the point of view of its existing customers – the people who write the cheque (who for a business-to-business service might be colleagues in finance, rather than the team who use a service)? What about potential or past customers? Or employees who co-create the service encounter? For a business-to-consumer service, to what extent should the organization consider what 'better' means for members of the customer's family who might also have knowledge and practices that shape the way the customer co-creates a service? Or communities whose habitats and wellbeing are implicated in the resources involved in constituting and experiencing the service? If we take an organization that is committed to reducing its carbon footprint, how far does the organization go to create a better service if this increases the work that end user has to invest in it? Thus 'better' may be an aspiration – but whose aspiration is it? For whom? Judged by what criteria? Determined by whom? The tension between a design approach – the desire to change existing situations into preferred ones – and approaches in the humanities and social sciences, that query assumptions drawing on extensive studies of power re-lations, difference, identity, governance and accountability are evident here. Ways of making judgments about what is, and what could be, will rely, whether explicitly or not, on ethics and aesthetics, the study of which must be part of the curriculum. This M(B)A sets out to encourage students to design better futures but it will remove any certainty about what that means.

## Futures

The third term, *futures*, is also not so simple. The future is understood here as un-known and unknowable. But as Fry (1999) points out, 'The future is never empty, never a blank space to be filled with the output of human activity. It is already colo-nised by what the past and present have sent to it' (Fry 1999: 11–12). We cannot know what will happen although we may speculate, imagine, model, propose, or hope and fear. Ideas about the future are necessarily connected to individual desires, which in turn relate to group, organization or community hopes and fears. Ways of think-ing about the future range from theories rooted in religion (overcoming uncertainty through belief), philosophical traditions (limiting uncertainty through reasoning), science (dominating uncertainty by modelling what is known), economics (bound-ing uncertainty through calculating probability) and the arts (mocking uncertainty by imagining alternatives).

The management field that sees its role as thinking about the future from an orga-nization's point of view is strategy. As in other management fields, theories about what constitutes a good strategy and how to create one have shifted over the years. This M(B)A will enable students to explore some of the important contributions but it will also learn from the practitioner field known as futures, in particular the approach that involves creating scenarios. The contribution of this methodology – developed and tested in organizations that plan in long timeframes – is its assumption of irreducible indeterminacy and ambiguity (van de Heijden 2005). Instead of forecasting and simu-lations, which involve projecting the past into the future, the scenarios approach asks strategists to construct futures based in different underlying structures. Conversations then take place around these scenarios and their structures, turning planning into an ongoing learning proposition. In the approaches used by contemporary futures prac-titioners, however, the role that design and art practices can play in generating fu-tures is relatively unexplored. This M(B)A will also attend to explorations of futures in the wider sphere such as films and literature (for example, Atwood 2003; Ballard

2006; McCarthy 2006) and in critical design (for example, Dunne 1999) in which designers speculate about the future through making prototypes, sculpture, films and other provocations. Creating fictions in a material or literary form is one way that designs for better futures have been proposed, from which students can learn and which they will be asked to construct during the course. Finally, following Fry (1999) and Tonkinwise (2000), this M(B)A will foreground questions of what should be sustained, rather than making overly simple claims about designing for sustainability. Thinking about what Tonkinwise calls 'sustainments' enables designers and managers to reframe their practices as designing organizations, systems and environments with the ability to sustain.

### Designing Better Futures

Described in this way, this imaginary M(B)A programme may not look promising. So far the course is underpinned by concepts that are troubling and promise to generate yet more problems – not obviously the best way to create a new educational programme. But this is exactly what this manifesto claims. The course pushes students (and in their teaching, teachers too) to embrace uncertainty and ambiguity. The goal is to think and act in ways that acknowledge and engage with the grave challenges facing the world and its human and non-human communities and habitats while avoiding simplistic accounts of what individual actors can achieve. The course draws in students with a bold, even ridiculous, claim: 'We will help you learn how to learn how to design better futures!' but at the very same time it says, 'Designing never ends, it's practically impossible to agree what constitutes better, and the future is, by definition, unknown.' And yet it is essential to try.

## A CURRICULUM FOR DESIGNING BETTER FUTURES

In terms of pedagogy, the M(B)A looks rather more like a course in a design school than one in a management school, even when it is teaching organization and management theory. The tradition of case-based teaching at leading business schools gives students a sense that with the right concepts and tools they can go out and refashion the world, or at least bits of it, although their conclusions are never tested during their education. Cases offer a constrained view of the world in which much of the messiness is taken away. Further, they privilege the student as knowledgeable actor. Students are asked to consider 'What would I do in this situation?' reinforcing an identity of the manager as rational being, somewhat at odds with theoretical traditions in organization studies that emphasize deterministic ecological and institutional factors (for example, Astley and Van de Ven 1983) and without committing students to action or seeing the consequences of their decisions.

In contrast, influenced by calls for changes to management education to enable students to try out what they are learning away from the lecture theatre (for example, Gosling and Mintzberg 2006; Datar et al. 2010), this course will force students over two years to research, develop, test and iterate ideas by designing interventions into projects and organizations at different scales, from international policy to cities to the home, resulting in designs for future practices in the form of new policies, processes, organizations or other kinds of arrangement. Combining studio work and desk-based learning with research and practice outside of the academy, the course will create opportunities for students to learn by engaging with people, artefacts and structures in formal and informal projects and organizations. Conceiving of this as a design practice, students will come to think of their activities as enquiries into futures that serve to shape them, emphasizing social, cultural, political and ethical matters played out in the world. Depending on the learning goal of a particular part of the course, students will produce presentations, written and illustrated reports, sketches or prototypes and supporting research materials, but also fiction, performance, film or animation and web resources. Their work will be assessed by stakeholders of the organizations with whom they engage as well as their peers and teachers. Part of the assessment will be deferred because the impact of the students' activities may not be clear until some time after their designs have been introduced, adopted, adapted or ignored.

This course combines elements that are not typically found together. While there exist MBA courses that teach art and design approaches, and design schools that teach aspects of management, and a few that try to fuse the two, this course draws in addition on theories and practices in the social sciences, arts and humanities.

## Design Theory and Practice

A foundation in the course is introducing students to important theories and concepts of design in particular Simon's (1996) problem solving, Rittel and Webber's (1973) wicked problems, Buchanan's (1992) design thinking, Fry's (1999, 2009) defuturing, Krippendorff's (2006) human-centred design and Hatchuel and Weil's (2009) analysis of the expansion of concepts and knowledge through design. Students will develop skills, knowledge and understanding of the practices of both design professionals, managers and artists. They will develop their own tacit and explicit knowledge as they engage in analysis and synthesis through iterative design processes applying design methods, techniques and tools. The curriculum will include researching experiences (for example, Bate and Robert 2007); synthesizing and interpreting data to generate design concepts (for example, Squires and Byrne 2002); visualizing, modelling and prototyping (for example, Buxton 2008); developing principles (for example, Lawson 2006), design rules and modularity (for example, Baldwin and Clark 1999); and using participatory and co-design approaches (for example, Kensing and Blomberg 1998).

However this curriculum will not overly privilege the doings of designers. Students will also attend to the practices of stakeholders who co-constitute the meanings and outcomes of design as they engage with designed artefacts and learn, from cultural anthropology and other social sciences, how to research and analyse. What happens with objects once they are in homes, work places, schools, public spaces and other contexts may serve to augment or undermine the intentions of designers: designers' plans become displaced through stakeholders' situated, embodied practices. This way of conceiving of design moves it away from the territory just of professional designers, or professional managers, towards a situated, emergent set of practices (Schatzki 2001; Reckwitz 2002).

## Institutions and Markets

Much design activity happens in the context of what we currently call markets, knowledge about which has until recently been dominated by the field of economics and underpinned by positivism. Accounts of management and organizational life rooted in social and cultural approaches over the past forty years have reframed how scholars see markets and organizations (cf. Astley and Van de Ven 1983). While organizations are centrally concerned with value and value creation, how one should go about assessing what value means remains an area of scholarly enquiry researching concepts such as commodities, gifts, costs and prices and the boundaries of the firm. This M(B)A course will be informed by the shift in contemporary organization and management studies away from thinking about exchange value towards value-in-use (Ramírez 1999; Vargo and Lusch 2004, 2008) and from value chains (Porter 1985) to systems (Ackoff 1973), value constellations (Norman and Ramírez 1993), networks and ecologies (Aldrich 1999) and hybrid arrangements (Ong and Collier 2005). In addition, the M(B)A will draw on socio-cultural approaches to understanding how markets work and how consumers consume (for example, Schor and Holt 2000) and how institutions form and change in response to innovation (for example, Hargadon and Douglas 2001). The implication for designers and managers concerned with designing better futures is to think not (only) at the level of products, services or experiences, or customers, end users and stakeholders but to consider how to act at the level of markets, systems and institutions and attend to their underlying symbolic structures.

## Aesthetic play

Management and organization studies founded on a desire to create a scientific body of knowledge have neglected the role of aesthetics in understanding what goes on in organizing (Strati 1999; Barry and Rerup 2006). In contrast, art and design practices hold a privileged place for discussions of aesthetics. But within art and design education, aesthetics is built into the iterative, studio-based mode of teaching in which students learn through handling and experiencing materials relevant to their discipline or field. The studio enables a more sensuous, embodied set of learning practices than the lecture theatre. Like other postgraduate courses linking design and management, this M(B)A will give students opportunities to develop their practices of material thinking (Carter 2004) by attending to both visual and non-visual aesthetic practices such as performance and choreography. Bringing to professional education opportunities for exploring the poetic and the beautiful, this course will be unusual in making a serious commitment to aesthetics in organizational life. Embracing the importance of play and its potential (Kane 2004; Guillet de Monthoux 2004; Guillet de Monthoux and Statler 2006), this M(B)A will help managers and entrepreneurs see the connections between ethics and aesthetics and learn how to create new playful practices in their teams, projects, organizations and policies.

## Publics and Engagement

Managers and designers have to find ways to persuade people of their ideas and attend to how they know and how they might act. They must create or engage with regimes of accountability and governance that shape what is possible and how action is understood. Whether this is conceived of as rhetoric (for example, Buchanan 1995) or enrolment (for example, Callon 1987), the important challenge for managers and entrepreneurs designing better futures is to find ways to represent and to involve stakeholders in conversations and design activities. The course explores different ways in which matters of concern are made public, drawing in part on work by artists and others whose projects assemble or draw together in novel ways (for example, Latour and Weibel 2005). What in design is conceived of as co-design or participatory design, is also present in contemporary arts practices in which artists convene new sets of relations between people and objects as part of their work. Bourriaud's (2002) concept of 'relational aesthetics' foregrounds art production and the aesthetic experience as located in new ways of producing and consuming art, although it raises questions about the politics of participation (Bishop 2006). Work by artists and designers who create new kinds of temporary institutional arrangements within which publics engage will serve as resources, including artists Nina Pope and Karen Guthrie (Somewhere 2009), architects ZeroZero (ZeroZero 2009) and artist-activists The Yes Men (The Yes Men 2009).

## Designing Socio-material Arrangements

Within management and organization research, one important intellectual tradition on which this M(B)A draws sees organizing and organizations as constituted by both social and material arrangements (for example, March 2007; Orlikowski and Scott 2008). The course will introduce students to work by leading researchers across management fields such as accounting, marketing, strategy, organization design, orga-nizational behaviour and operations. Instead of seeing organizations as static or stable, the course will attend to ideas of improvisation and change (for example, Orlikowski 1996; Weick 1998), practices (for example, Whittington 2006), narrative (for ex-ample, Czarniawska 1997), communication as language games (Boland and Tenkasi 1995), boundary objects (Carlile 2002) and sensemaking (Weick 1995). Students will be encouraged to see their professional work as the designing of new kinds of socio-material arrangements in which new practices will emerge.

## Scaling and Scale

People, teams and organizations have to find where they want to operate but, as geographers well know, scale is not a simple matter. Within the social sciences, scholars from opposing traditions still question the validity and worth of doing research at different scales, resulting in a serious division within scholarly com-munities between those who want to take a bird's eye view and those who attend to local, situated specific activities (Saïd Business School 2009). As Yaneva (2005) showed in her ethnography of architects, practitioners design buildings through scaling up and scaling down, but not in a linear or orderly way. Within manage-ment, how organizations scale has been a consistent concern from Taylor (1911) to lean production (Womack et al. 1990). For designers and managers working at project, team, departmental, organizational, city or policy levels, conceptualizing at different scales and being adept at moving between them is part of the work. On this M(B)A students will learn to inhabit both the bird's eye view – the supposed place of strategic thinkers – and the detailed, local perspective of ethnographers, designers and operational doers.

## New Arrangements and their Impacts

Where might the graduating students end up? They will work at different scales, rang-ing from international policy addressing global challenges such as climate change and conflict resolution, to national and regional discussions about equity and equality, through to local projects in communities or in their own personal networks. They will bring to these contexts knowledge and skills that reconceive of these activities as enquiries into futures by imagining, creating and testing new kinds of practice and ar-rangement. They will lead a shift away from instrumentalist discussions of policy and technologies towards a new understanding of how things come to be and how they can be changed, in which minds, bodies, things, structures, processes, knowledge and agency are entangled.

They will bring to the well established structures and practices of public and commercial institutions a willingness not just to 'take risks', as in the discourse of contemporary management, but to create new ways of exploring and representing the public good through aesthetic play. Several will become entrepreneurs creating ventures working across social and public sectors. Some will test the limits of organizational boundaries, and find that they have to invent new kinds of institution. Scaling up and down as they research, imagine, play and create new kinds of entity, these students will find that their ability to be comfortable with being uncomfortable will make them valued members and leaders of teams and projects.

They will use knowledge from art, design, the social sciences and management but they will also encounter the limits of knowledge and engage with design-based practices to work through these. They will be willing and able to discuss the limits of defining what constitutes 'better'. They will be able to tolerate and work through challenging encounters with and as stakeholders and other kinds of professional who have different stakes in the past, present and future. They will design new institutions, practices and arrangements and, along the way, create new methods and tools. The graduates from this M(B)A will try to design better futures. Some of them will succeed. This will be something that matters.

## ACKNOWLEDGEMENTS

Thanks to these people whose comments improved this chapter: Andrew Barry, Fred Collopy, Philip Hill, Javier Lezaun, Noortje Marres, Steve New, Rafael Ramírez, Sue Tempest, Laurene Vaughan and Nina Wakeford.

## NOTE

1. Space does not allow a full discussion of practice theories here. Reckwitz (2002) presents an ideal type of practice theory in his discussion of Bourdieu (1977), late Foucault, Giddens (1979), Latour, Schatzki and others, showing how their ideas of practices contrast with other accounts of the social which emphasize symbolic structures of meaning.

## REFERENCES

Ackoff, R. (1973) 'Science in the Systems Age: Beyond IE, OR, and MS', *Operations Research* 21(3): 661–71.

Adler, N. (2006) 'The Arts and Leadership: Now That We Can Do Anything, What Will We Do?' *Academy of Management Learning and Education* 5(4): 486–99.

Aldrich, H. (1999) *Organizations Evolving*. London: Sage.

Astley W. and Van de Ven, A. (1983) 'Central Perspectives and Debates in Organization Theory', *Administrative Science Quarterly* 28 (2): 245–73.

Atwood, M. (2003) *Oryx and Crake*. London: Virago.

Baldwin, C. and Clark, K. (1999) *Design Rules: The Power of Modularity, Volume 1*. Cambridge, MA: MIT Press.

Ballard, J. (2006) *The Drowned World*. London: Harper Perennial.

Barry, A., Born, G. and Weszkalnys, G. (2008) 'Logics of Interdisciplinarity', *Economy and Society* 37(1): 20–49.

Barry, D. and Rerup, C. (2006) 'Going mobile: Aesthetic design considerations from Calder and the Constructivists', *Organization Science* 17(2): 262–76.

Bate, P. and Robert, G. (2007) *Bringing User Experience to Healthcare Improvement: The Concepts, Methods and Practices of Experience-based Design*. Oxford: Radcliffe Press.

Bijker, W., Hughes, T. and Pinch, T. (eds) (1987) *The Social Construction of Technological Systems: New Directions in the Sociology and History of Technology*. Cambridge, MA: MIT Press.

Bishop, C. (ed.) (2006) *Participation*. Cambridge: MIT Press.

Boland, R.J. and Collopy, F. (2004) 'Design Matters for Management', in R. Boland and F. Collopy (eds) *Managing as Designing*. Stanford, CA: Stanford University Press.

Boland, R. and Tenkasi, R. (1995) 'Perspective Making and Perspective Taking in Communities of Knowing', *Organization Science* 6 (4): 350–72.

Bourdieu, P. (1977) *Outline of a Theory of Practice*. Cambridge: Cambridge University Press.

Bourriaud, N. (2002) *Relational Aesthetics*. Paris: Les Presses Du Reel.

Borja de Mozota, B. (2003) *Design Management: Using Design to Build Brand Value and Corporate Innovation*. New York: Allworth Press.

Buchanan, R. (1992) 'Wicked Problems in Design Thinking', *Design Issues* 8(2): 5–21.

Buchanan, R. (1995) 'Rhetoric, humanism and design', in R. Buchanan and V. Margolin (eds) *Discovering Design: Explorations in Design Studies*. Chicago: Chicago University Press.

Buchanan, R. (2001) 'Design Research and the New Learning', *Design Issues* 17(4): 3–23.

Buxton. B. (2007) *Sketching User Experiences: Getting the Design Right and the Right Design*. San Francisco, CA: Morgan Kaufmann.

Callon, M. (1987) 'Society in the Making: The Study of Technology as a Tool for Sociological Analysis', in W. Bijker, T. Hughes and T. Pinch (eds) *The Social Construction of Technological Systems: New Directions in the Sociology and History of Technology*. Cambridge, MA: MIT Press.

CCA (2009) *MBA in Design Strategy*, www.cca.edu/academics/graduate/design-mba (accessed 28 November 2009).

Carlile, P. (2002) 'A Pragmatic View of Knowledge and Boundaries: Boundary Objects in New Product Development', *Organization Science* 13(4): 442–55.

Carter. P. (2004) *Material Thinking*. Melbourne: Melbourne University Press.

Czarniawska, B. (1997) *Narrating the Organization: Dramas of Institutional Identity*. Chicago: University of Chicago Press.

Datar, S., Garvin, D.A. and Cullen, P. (2010) *Rethinking the MBA: Business Education at a Crossroads*. Cambridge, MA: Harvard University Press.

Dunne, A. (1999) *Hertzian Tales: Electronic Products, Aesthetic Experience and Critical Design*. London: RCA, CRD Research Publications.

Dunne, D. and Martin, R. (2006) 'Design Thinking and How it Will Change Management Education: An Interview and Discussion', *Academy of Management Learning and Education* 5(4): 512–23.

Foucault, M. (1972) *The Archaeology of Knowledge*. London: Tavistock.

Funtowicz, S. and Ravetz, J. (1993) 'Science for the Post-normal Age', *Futures* 25(7): 739–55.

Fry, T. (1999) *A New Design Philosophy. An Introduction to Defuturing.* Sydney: UNSW Press.

Fry, T. (2009). *Design Futuring: Sustainability, Ethics and New Practice.* Oxford: Berg.

Fry, T. (2011) *Master of Design Futures Preview.* http://www.griffith.edu.au/__data/as sets/pdf_file/0011/184790/master-of-design-futures-preview.pdf (accessed 9 March 2011).

Garud, R., Jain, S. and Tuertscher, P. (2008) 'Incomplete by Design and Designing for Incompleteness', *Organization Studies* 29(3): 351–71.

Giddens, A. (1979) *Central Problems in Social Theory: Action, Structure and Contradiction in Social Analysis.* London: Macmillan.

Goldsmiths, (2009) *MA Design Futures,* www.gold.ac.uk/pg/ma-design-futures/ (accessed 20 October 2009).

Gosling, J. and Mintzberg, H. (2006) 'Management Education as if Both Matter', *Management Learning* 37(4): 419–28.

Guillet de Monthoux, P. (2004) *The Art Firm: Aesthetic Management and Metaphysical Marketing.* Stanford: Stanford Business Books.

Guillet de Monthoux, P. and Statler, M. (2008) 'Aesthetic Play As An Organizing Principle', in D. Barry, and H. Hansen (eds) *The Sage Handbook of New Approaches in Management and Organization.* London: Sage.

Hargadon, A. and Douglas, Y. (2001) 'When Innovations meet Institutions: Edison and the Design of the Electric Light', *Administrative Science Quarterly* 46(3): 476–501.

Hatchuel, A. (2001) 'Towards Design Theory and Expandable Rationality: The Unfinished Programme of Herbert Simon', *Journal of Management and Governance* 5(3–4): 260–73.

Hatchuel, A. and Weil, B. (2009) 'C-K design theory: An Advanced Formulation', *Research in Engineering Design* 19: 181–92.

Ingram, J., Shove, E. and Watson, M. (2007) 'Products and Practices: Selected Concepts from Science and Technology Studies and from Social Theories of Consumption and Practice', *Design Issues* 23(2): 3–16.

Jelinek, M., Romme, G. and Boland, R. (2008) 'Introduction to the Special Issue: Organization Studies as a Science for Design: Creating Collaborative Artifacts and Research', *Organization Studies* 29(3): 317–29.

Julier, G. (2007) 'Design Practice within a Theory of Practice', *Design Principles and Practices: An International Journal* 1(2): 43–50.

Kane. P. (2004) *The Play Ethic: A Manifesto for a Different Way of Living.* London: Macmillan.

Kensing, F., and Blomberg, J. (1998) 'Participatory Design: Issues and Concerns', *Computer Supported Cooperative Work* 7: 167–85.

Krippendorff, K. (2006) *The Semantic Turn: A New Foundation for Design.* New York: Taylor & Francis CRC.

Latour, B. (1987) *Science in Action: How to Follow Scientists and Engineers through Society.* Cambridge, MA: Harvard University Press.

Latour, B. (2005) *Reassembling the Social: An Introduction to Actor-Network Theory.* Oxford: Oxford University Press.

Latour, B. and Weibel, P. (eds) (2005) *Making Things Public: Atmospheres of Democracy*. Cambridge, MA: MIT Press.

Latour, B. and Woolgar, S. (1986) *Laboratory Life: The Construction of Scientific Facts*. Princeton, NJ: Princeton University Press.

Law, J. (1987) 'Technology and Heterogeneous Engineering: The Case of Portuguese Expansion', in W. Bijker, T. Hughes, and T. Pinch (eds) *The Social Construction of Technological Systems: New Directions in the Sociology and History of Technology*. Cambridge, MA: MIT Press.

Lawson, B. (2006) *How Designers Think: The Design Process Demystified*. 4 edn. Oxford: Architectural Press.

March, J. (2007) 'The Study of Organizations and Organizing since 1945', *Organization Studies* 28(1): 9–19.

McCarthy, C. (2006) *The Road*. London: Random House.

Norman, R. and Ramírez, R. (1993) 'Designing Interactive Strategy: From Value Chain to Value Constellation', *Harvard Business Review* 71(4): 65–77.

Nowotny, H., Scott, P. and Gibbons, M. (2001) *Rethinking Science, Knowledge and the Public in an Age of Uncertainty*. Cambridge: Polity Press.

Ong, A. and Collier, S. (2005) *Global Assemblages: Technology, Politics, and Ethics as Anthropological Problems*. Malden, MA: Blackwell.

Orlikowski, W. (1996) 'Improvising Organizational Transformation Over Time: A Situated Change Perspective', *Information Systems Research* 7(1): 63–92.

Orlikowski, W. and Scott. S. (2008) 'Chapter 10: Sociomateriality: Challenging the Separation of Technology, Work and Organization', *The Academy of Management Annals* 2: 433–74.

Porter, M. (1985) *Competitive Advantage*. New York: Free Press.

Ramírez, R. (1999) 'Value co-production: Intellectual Origins and Implications for Practice and Research', *Strategic Management Journal* 20: 49–65.

Ravasi, D. and Rindova, V. (2008) 'Symbolic Value Creation', in Barry, D. and Hansen, H. (eds) *The Sage Handbook of New Approaches in Management and Organization*. London: Sage.

Rayner. S. (2006) 'Wicked problems: Clumsy solutions – Diagnoses and Prescriptions for Environmental Ills', Jack Beale Memorial Lecture on Global Environment, www.sbs.ox.ac.uk/research/Documents/Steve%20Rayner/Steve%20Rayner,%20Jack%20Beale%20Lecture%20Wicked%20Problems.pdf (accessed 11 November 2009).

Reckwitz, A. (2002) 'Towards a Theory of Social Practices: A Development in Culturalist Theorizing', *European Journal of Social Theory* 5(2): 243–63.

Rittel. H. and Webber, M. (1973) 'Dilemmas in a General Theory of Planning', *Policy Sciences* 4: 155–69.

Saïd Business School. (2009) 'From Scale to Scalography: A Provocation Piece: For an International Workshop on Scalography, 8th July 2009', www.sbs.ox.ac.uk/centres/insis/Documents/Scalography_provocation_22.pdf (accessed 29 November 2009).

Schatzki, T. (2001) 'Practice theory', in Schatzki, T., Cetina, K.K. and von Savigny, E. (eds) *The Practice Turn in Contemporary Theory*. London: Routledge.

Schön, D. (1983) *The Reflective Practitioner: How Professionals Think in Action*. New York: Basic Books.

Schor, J. and Holt, D. (eds) (2000) *The Consumer Society Reader*. New York: New Press.

Shove, E., Watson, M., Hand, M. and Ingram, J. (2007) *The Design of Everyday Life*. Oxford: Berg.

Simon, H.A. (1996) *The Sciences of the Artificial*. 3rd edn. Cambridge, MA: MIT Press.

Sismondo, S. (2004) *An Introduction to Science and Technology Studies*. Oxford: Blackwell.

Somewhere (2009) *Somewhere*, www.somewhere.org.uk/ (accessed 22 October 2009).

Squires, S. and Byrne, B. (eds) (2002) *Creating Breakthrough Ideas: The Collaboration of Anthropologists and Designers in the Product Development Industry*. Westport, CT: Bergin & Garvey.

Starkey, K. and Tempest, S. (2009) 'The Winter of Our Discontent: The Design Challenge for Business Schools', *Academy of Management Learning and Education* 8(4): 576–86.

Starkey, K. and Tiratsoo, N. (2007) *The Business School and the Bottom Line*. Cambridge: Cambridge University Press.

Strati, A. (1999) *Organizations and Aesthetics*. London: Sage.

Suchman, L. (1987) *Plans and Situated Actions*. New York: Cambridge University Press.

Taylor, F. (1911) *The Principles of Scientific Management*. New York: W. W. Norton.

Tonkinwise, C. (2000) 'Sustainments: Environments with the Ability to Sustain', Paper presented at the Shaping the Sustainable Millennium Conference, Queensland University of Technology, July 2000, www.changedesign.org/Resources/EDFPublications/Articles/ArticlesMenuMain.htm (accessed 22 October 2009).

Tonkinwise, C. (2003) 'Interminable design: Techne and Time in the Design of Sustainable Service Systems', paper presented at the European Academy of Design Conference, Barcelona. Available online: www.ub.edu/5ead/PDF/8/Tonkinwise.pdf (accessed 22 October 2009).

van de Heijden, K. (2005) *Scenarios: The Art of Strategic Conversation*. 2nd edn. Chichester: John Wiley & Sons, Ltd.

Vargo, S. and Lusch, R. (2004) 'Evolving to a New Dominant Logic in Marketing', *Journal of Marketing* 68(1): 1–17.

Vargo, S. and Lusch, R. (2008) 'Service-dominant Logic: Continuing the Evolution', *Journal of the Academy of Marketing Science* 36(1): 1–10.

Weick, K. (1979) *The Social Psychology of Organizing*. 2nd edn. Reading, MA: Addison Wesley.

Weick, K. (1995) *Sensemaking in Organizations*. London: Sage.

Weick, K. (1998) 'Introductory Essay: Improvisation as a Mindset for Organizational Analysis', *Organization Science* 9(5): 543–55.

Whittington, R. (2006) 'Completing the Practice Turn in Strategy Research', *Organization Studies* 27(5): 613–34.

Womack, J., Jones, D. and Roos, D. (1990) *The Machine that Changed the World*. New York, NY: Rawson Associates.

Yaneva, A. (2005) 'Scaling Up and Down: Extraction Trials in Architectural Design', *Social Studies of Science* 35(6): 867–94.

*Yes Men* (2009) Available online. http://www.theyesmen.org/ (accessed 22 October 2009).

ZeroZero, (2009) *00:/*, www.architecture00.net/ (accessed 22 October 2009).

# PART III

# Design, Management and the Organization

# Editorial Introduction

SABINE JUNGINGER AND RACHEL COOPER

Matters of implementation dominated studies of design management from its origin. This operational focus, pursuing questions of 'how', remains today as this section aptly demonstrates. Our aim is to provide a sense of the topics and issues that interest designers and managers as they continue to explore problems of design in the organization. We have therefore chosen to include a wide range of practices and methods, from different organizational and industrial contexts, presenting the breadth of questions that occupy the field at the moment. We have grouped these chapters around four topics: embedding design in the organization; strategy in design management; measuring and evaluating design and, finally, design leadership. The following paragraphs explain this grouping in more detail and provide a synopsis of the chapters in each of these subsections.

## EMBEDDING DESIGN IN THE ORGANIZATION

One of the key questions we face today is where to place design. Since Mark Oakley's chapter on 'Organizing Design Activities' (Part I), a rising number of studies have expanded our understanding of how design practices, design attitudes and design skills can become embedded in traditional organizations. There are a number of economic sectors where design has not played a conspicuous role so far. However, as Peter Gorb and Angela Dumas made clear in Part I, design activities do take place and have been taking place in many organizations even though we may fail to notice them. In this subsection, five chapters approach design management in five different 'organizational' environments. Each of these environments has different implications and poses distinct challenges for design management.

In the first chapter, 'Embedding Design Practice within Organizations,' **Angela Meyer** reflects on the challenges and opportunities of embedding designing in the organizational context. She argues that embedded design practice can aid an organization in fulfilling its mission as one of seven intrinsic benefits of embedding design practice within organizations. Meyer points out the characteristics of design blockers and design enablers, people within an organization who either seek to advance or to deter the intertwining of design activities with organizational processes. Meyer's contribution can be understood as advancing some of the issues of 'managing across organizational boundaries' (Gunz 1990) from a design perspective.

The next chapter by **Paola Tombesi** and **Jennifer Whyte** presents a study where the 'embedding' environment constitutes a particular industry sector. Titled 'Challenges for Design Management in the Construction Sector,' it provides interesting insights into the potential for design management in an area that has not attracted much attention in the past. With that, their chapter highlights the possible roles of, and the changing demands on, design management practices in areas beyond products, services and branding.

Services are not merely a problem for the service industry but are as important to manufacturing organizations, the public sector and charitable organizations. **Bill Hollins**' chapter on 'A Prospective of Service Design Management' shows how new and emerging design practices and design professions, too, provide new contexts for design management research and practice. He points out how service design was initially understood to be a variation of engineering design and looked at as 'just another type of product'. He explains how this idea has changed and discusses the consequences for the role of design in organizations.

**Davide Ravasi** and **Ileana Stigliani** situate design in Small and Medium-sized Enterprises (SMEs). Their chapter 'Successful Design Management in Small and Medium-sized Businesses' explores the factors that lead to successful design projects in SMEs. Ravasi and Stigliani highlight the problems that emerge when design consultancies engage with managers who are often unfamiliar with professional design activities and processes. Their findings provide useful insights into issues of design education for managers, which we address in detail in Part II. However, their chapter offers the perspective of design consultants and what they wish managers in organizations would know about design. With that, they offer a fresh perspective on a classic theme, which has been memorably put in words by David Walker in 'Designers and Managers: Two Tribes at War?' (Walker 1990).

**Thomas Lockwood**'s chapter 'A Study on the Value and Applications of Integrated Design Management,' closes this section with another take on embedding design in the organization. He uses the findings from his study into visual branding to draw insights into integrated design management. For Lockwood, the role of integrated design management is one of enabling innovations in distinct design disciplines present in an organization while ensuring an overall coherent design experience. Despite this obviously significant role, he finds that research and knowledge about integrated design management are lacking both in the academic community and in the practice community.

## STRATEGY IN DESIGN MANAGEMENT

Ever since Kotler and Rath highlighted design as a strategic tool (see Part I), the role of design in strategic matters has been a topic of great interest in design. The opportunities for design to tackle strategic questions has been advanced by two key developments in the field of strategy. For one, the concept of 'a strategy' has changed. No longer is strategy considered a finished product that needs to be executed. Instead, a strategy is interpreted to be an action and to be in action. This shift in interpretation has brought the concepts of product development closer to those of strategy making. The second key development concerns a shift in the nature of competition. Here, blue ocean strategy by Chan and Mauborgne (2005) has had an impact. A blue ocean strategy envisions the competitive landscape to be a wide ecological environment where everyone can carve out a living based on unique adaptive, collaborative and inventive skills. This is a departure from the red ocean strategy, which represents an ocean where the survival mode is 'eat or be eaten'. The chapters in Part III therefore focus on different aspects of the relationships between design and strategy.

Fittingly, the opening chapter by **Kyung-won Chung** and **Yu-Jin Kim** concentrates on the 'Changes in the Role of Designers in Strategy.' Chung and Kim offer a brief taxonomy of previous research on this topic. This research, they argue, points to two roles designers have in the development of product and service strategies: The first can be called 'interactional'. The role of designers here is to initiate projects, generate participation and navigate matters of subordination. In the 'functional' role, designers' involvement in strategic matters is limited to that of form giver, solution provider, concept generator and service initiator. Case studies from the manufacturing and Internet service industries illustrate these two roles for designers within businesses and explores the implications of these two roles for in-house design.

A different perspective is provided by **Brigitte Borja de Mozota**. In 'Design's Strategic Value Revisited: A Dynamic Theory for Design as Organizational Function', she traces the shifting foci in research on design and value in management. She argues that a first 'research loop' centred on design as value. A second round of research looked into design value in the value chain. This, Mozota states, was followed by research into the design value chain in the value management. Today, she says, the focus is on design strategic value as core competency. This last shift, in her view, is significant because it lines up with resource-based theories and thereby values design thinking as a competitive advantage in strategy that adds unique value to an organization.

The rise of branding has presented both problems and opportunities for design management. **Lisbeth Svengren** reminds us that the core ideas of brand management derive from legal and territorial concepts as much as those of ownership. Svengren sees a renewed opportunity for design management to come to the fore as an important area of research and practice. She illustrates her point by tracing key developments in strategic management and pointing out all the opportunities where design, particularly a design process, could and would have made contributions. In 'Design Management as Integrative Strategy', Svengren argues that the future role of design in strategic management is one of integrating functional, visual and conceptual strategies. The practice and profession of strategic design is a recent development and therefore poses new challenges and opportunities for design management.

'Designers, by their personal inclination, education and skill set, tend to have a set of preferences, values, tools and behaviours that closely match requirements of innovation', says **Bettina von Stamm**. And yet, she finds that the role of design in innovation continues to be hampered in today's businesses because of a lack of shared understanding, the fact that many people in businesses remain unaware about their design roles, communication issues and educational differences. In her exploration of 'The Role of Design in Innovation: A Status Report', she presents lessons from five case studies that illustrate these problems and that also show how the situation could be improved. To unfold its role in innovation, she concludes, design has to become part of the organizational fabric.

The strategic role of design in marketing is closely linked with the business decisions of an organization, observes **Margaret Bruce** in 'Connecting Marketing and Design'. Decisions to enter a new market, to increase market share, or to merge with another company represent strategic changes that initiate investment in design. However, Bruce finds that for design to have a strategic function within the company can require significant organizational changes. In her view, marketing and design have a key role here because they are close to consumers. This means that marketing and design professionals are at the forefront of change and can so contribute to sustainable and competitive businesses.

## MEASURING AND EVALUATING DESIGN

Like fine art and sometimes architecture, design suffers from being in that category of activities which everybody thinks he can do. This, of course, is not so. But even if it were, design costs money because, whether it is easy or not, it takes time and it needs space. And time and space mean money. (Hughes-Stanton 1967)

Researchers still seek answers to what the cost of design is and what it contributes to the bottom line. As a result, different ways of measuring and evaluating are being proposed. We include two chapters that approach this problem from two different perspectives.

**Tore Kristensen** and **Gorm Gabrielsen** seek to deliver evidence for the old claim that 'good design' also translates into 'good business'. In 'Design Economy – Value Creation and Profitability', they present the findings of a quantitative study that they conducted with a random sample of twenty-five large Danish companies. The study served to test three hypotheses. Hypothesis one states that the quality of product design is positively covaried with financial performance. Hypothesis two claims that logo design is associated with financial performance. Hypothesis three finally argues that a product's Web design positively covaried with the financial performance of a product. Their findings indicate that both product and logo design improve business. Interestingly, they did not find the same correlation for Web design.

Another way to measure the value of design to a business is put forth by **David Hands**. His chapter, titled 'Design Transformations: Measuring the Value of Design', turns to design audits as a tool to understand the contributions of design in the context of a specific organization. With that, Hands emphasizes the activities of auditing over the activities of measuring, the latter, in his view, lending themselves better to subjects that can be captured quantitatively. Design audits, in contrast, may serve an organization to provide it with 'a unique insight into its true value to obtain new and strategic horizons for continual growth and sustainability in uncertain times'.

## DESIGN LEADERSHIP

Early on, the role of the designer in the organization encountered questions of leadership. It would be difficult to argue that Peter Behrens, Charles and Ray Eames or Terence Conran lacked leadership skills. Yet, leadership was often treated implicitly and therefore secondary to the actual tasks of the design manager, which were described by Michael Farr as 'problem-solving, planning, briefing, communications and co-ordination' (see Farr, Part I). Today, the activities, principles and practices of designing are increasingly considered key tools for future leaders – and not only in the realm of management. The following two chapters explore issues of design leadership in design practice.

Leadership is a relatively new research topic in design management and **Alan Topalian** counts as one of the early contributors in this area. In 'Challenges for Design Leaders' he reports on a study that he conducted to find out about the skills and knowledge designers need to move into leadership positions. He derives a list of recommendations based on findings from interviews conducted with UK design leaders over a four-year period. Though this research is informed by the situation of design leaders in the UK, the recommendations have relevance for designers who seek to move into design leadership positions in other countries as well.

In contrast, **Frans Joziasse** argues in the following chapter, 'Design Leadership: Current Limits and Future Opportunities', that effective design leadership depends on four key factors. These include: (a) the psychological disposition of the designer; (b) the economic factors related to a project and an organization; (c) cultural context and (d) the role of sustainability. In his view, designers have yet to address design leadership in SMEs. Joziasse also tries to clarify the boundaries between design managers and design leadership. He presents some EU level initiatives that begin to address some of these limits and that indicate that design leadership is gaining in importance and salience, even if many of these initiatives do not – yet – involve actual design researchers or design practitioners.

## REFERENCES

Chan, K.W. and Mauborgne, R. (2005) *Blue Ocean Strategy*. Boston, MA: Harvard Business School Press.

Gunz, H. (1990) 'Corporate Approaches to Design Management', in M. Oakley (ed.) *Handbook of Design Management*. Oxford: Blackwell.

Hughes-Stanton, C. (1967) 'The Cost of Industrial Design', *Design Journal* 227: 43–7, www.vads.ac.uk (accessed 10 October 2010).

Walker, D. (1990) 'Designers and Managers: Two Tribes at War', in M. Oakley (ed.) *Handbook of Design Management*. Oxford: Blackwell.

## RECOMMENDED READING

Farr, M. (1965) 'Design Management – Why is it needed now?' *Design* 200: 38–9.

Gorb, P. and Dumas, A. (1987) 'Silent Design', *Design Studies* 8(3): 150–6.

Kotler, P. and Rath, A. (1983) 'Design: A Powerful but Neglected Strategic Tool', *Journal of Business Strategy* 5(2): 16–21.

Oakley, M. (1984) 'Organising Design Activities', in M. Oakley, *Managing Product Design*. London: Weidenfeld & Nicolson.

# 12

# Embedding Design Practice within Organizations

## ANGELA MEYER

When I began practising design, I found the studio to be a special place that harboured powerful tools, practices, and a magical way of working together with a group of intelligent and creative people. Somehow our whole was far, far more than the sum of our parts, and, to me, that was (and still is) one of the most astounding aspects of the experience of working as a designer. No matter what problem a client presented us with, we could manage to make the problem interesting, challenging, meaningful, and of course, solve-able. I used to think that this dynamic was unique to the design studio, a special world unto itself.

But as I got involved with larger, more strategic projects, my work started to take me increasingly out of the safe harbour of the studio and more and more into my clients' environments. I learned that instead of the old model of clients bringing discrete problems to the studio, designers must learn how to take the studio into the messy organizational context of the problem. Eventually I began working as a consultant on projects where the explicit goal was to take design thinking and practice into large organizations. Through my experience working inside both public and private organizations, I have seen that design practice can be every bit as powerful to business analysts, marketers, product and channel managers, vice presidents, and CEOs as it is to professional designers.

### HYPOTHESIS: DESIGN IS GOOD FOR ORGANIZATIONS

While design as an organization-wide function or practice is a relatively new concept, it is one that is gaining traction as companies and public-sector organizations look for ways to become more competitive and productive. Once they have exhausted the panoply of traditional business and management tools, whose purpose is to tighten operational focus, increase efficiencies and stabilize core capabilities, organizations often find themselves locked into a mechanistic operational model and undifferentiated in the marketplace. They may be very good at what they do, but they are unprepared for changes in the market or meeting the needs of new markets. These organizations are often unsure about how to identify and act upon growth opportunities, and may have even set up systems and structures that actively work against new value creation. A small number of companies and organizations are beginning to accept design (design thinking, design management, design methods and practice) as a way to address the gaps in the marketplace and, more interestingly, to address functional challenges inside the organization. And although there is no question that design helps organizations create value externally in the marketplace, there are some interesting possibilities to explore about the way that design can add value to organizations internally.

In their critique of contemporary management education, David Newkirk and Edward Freeman (2008) posit that business is a fundamentally human institution where we create value for each other by cooperating, trading, and specializing our labour. They assert, however, that 'the scientific view of management makes it difficult to incorporate the human spirit. It leaves little room for ethics, or for other fundamentally human considerations such as creativity, trust, initiative, and will' (Newkirk and Freeman 2008: 141). This perspective on the shortcomings of management science illustrates design's potential for offering organizations an improved method for value creation. Design's essentially humanistic agenda is precisely what makes it such a powerful dialectical tool for enriching organizational life.

Design is fundamentally about value creation. In the business world, the design of products, services, processes and systems can unlock new markets, drive new revenue and keep an organization running efficiently. Design can launch a new company, or it can renew and sustain an existing one. The aspect of design that has been far less explored and tested is what I will call the intrinsic value of design. Intrinsic value includes the many influences and outcomes that result from engaging customers, employees, and organizational dynamics in the design process. In other words, it is value inherent in the means, rather than the ends. Such values include social cohesion, ideological coherence, and strategic alignment that emerge from participation in design activities. Designers are not unfamiliar with these intrinsic benefits, but rarely are they the primary driver for design engagements with client organizations. Furthermore, value generated intrinsically is far more difficult to track and measure, and so it is often hidden or overlooked. Key to capturing the intrinsic side of the equation, however, is to understand design as a set of activities: methods, approaches, and techniques that provide its practitioners with a way of working together in a highly productive way. Furthermore, we must see design not as one specific set of activities, but as a malleable set of practices that include hallmarks such as collaboration, visualization, experimentation, ideation, and iteration.

## VALUING DESIGN'S INTRINSIC BENEFITS

Regardless of what a company makes as its product, there is a value in taking design approaches inside the organization. In an intrinsic context, the design outcomes are defined by successful working relationships, stakeholder support and agreement, more robust problem solving, more effective outcomes, and more humanistic values espoused by the organization. These may have either a direct or an indirect influence on an organization's products or outputs, but they have deep influence either way. Because design approaches encourage a quest for quality and dignity in both process and outcomes, this benefits both employees and customers. Design encourages pride in not only what we make but in *how* we make it. Design approaches insist that we engage with a problem holistically and honestly and provide the impetus for more disciplined thinking and decision-making – important tools for good strategy and management.

Specific qualities characterize the intrinsic benefits of design. Because these are humanistic qualities, they have both an organizational and an individual or personal dimension that should not necessarily be seen as distinct. It is worth noting that these same qualities are often those cited in corporate mission statements, but very often ring hollow for employees who know that they are abstractions that are often at odds with organizational systems and culture. The following list may not be exhaustive, but it captures the strongest and most typical intrinsic outcomes of participation in design.

### Engagement

Engagement is the active participation and connection among staff, customers, and management with the organization and its purpose. It fosters energy, motivation, and buy-in to organizational challenges.

### Ethos

Ethos is the collective ownership of specific values driven by habitual adherence to certain goals and practices. Ethos embodies knowing what's important and what we care about as an organization.

### Collaboration

Collaboration is people thinking, working, and solving problems together, in a way that they actively create new knowledge together. It is often confused with other types of consultative teamwork that do not require true conversation and sharing.

## Vision

Vision is both literal and figurative, in that visualization techniques can allow people to 'see' and develop an aspirational future state. Also, by visualizing the future, people develop a shared and unified concept of where their goals will lead.

## Coherence and Alignment

Coherence and alignment reflect the clarity and consistency of organizational strategy and action. They require that we all think and act strategically in concert. Design raises persistent questions about purpose, which forces ongoing alignment of intent and action.

## Accountability

Accountability is the linkage between people and results; it reinforces a sense of discipline and encourages processes that ensure that problems, especially customer problems, are identified and addressed.

## Learning/Reflection

Learning and reflection are attitudes as much as activities. Learning is the basis for change and renewal; reflection allows thoughtful growth. People who learn and reflect tend to deepen their investment in and dedication to their work.

All of these benefits are responsible for making people care about their work and the mission of the organization. Through these experiences, people can realize more meaningful professional lives and the organization can become more productive and sustainable.[1] Qualitatively, it is easy to see the value that is intrinsic to design practice. But the intrinsic value is difficult to measure in quantitative terms. How do these benefits result in return on investment (ROI) for the organizations that commit to realizing them?

Return on investment measures around design are often difficult and contentious for two reasons. Firstly, positive impacts do not always have direct measures in dollars. While many extrinsic benefits are easily measured, such as increased sales and market share, customer loyalty, and brand strength, many intrinsic benefits are harder to measure. A new product design that allows us to expand our customer base or product range is economically measurable. When employees feel happier with their work because of increased engagement or accountability, for example, that may translate into both measurable and non-measurable benefits. Productivity and employee retention may increase, which we can easily measure, but it might be harder to gauge the value of employee satisfaction or shared organizational vision as a quantifiable value.

Secondly, many intrinsic benefits are realized slowly, over time, making them harder to identify and measure in a landscape that is constantly shifting. Where we might be able to measure profits easily over a six- to twelve-month time period, many of the intrinsic benefits can take several years to bear recognizable fruit. It is the fact that these benefits are often a longer term investment in company culture that makes them a harder sell. If the average tenure of a CEO is six years,[2] what incentive is there to commit to a programme that builds longer-term value for the organization? And to what degree is culture valued in organizations focused on short-term returns?

Those of us who work in design full time are quick to point out the immense personal rewards and satisfaction to be found in design practice, and we are quick to scorn 'corporate types' who choose what we see as less interesting or meaningful types of jobs. But there is no inherent reason why working for a large organization necessarily has to be a mind-numbing and soul-destroying experience. That we accept this tradeoff shows a poverty of thinking around the potential for organizational life. The operational systems that are required to keep a large organization running efficiently have a tendency to institutionalize and dehumanize people's daily activities, but design practices can help to return organizations to a human-level scale. While these intrinsic benefits are merely by-products of the design process, the effects are significant enough to become the ends in themselves and not just the means.

## DESIGN/DESIGNING ORGANIZATIONS

There are many organizations that have very explicitly and deliberately incorporated design and design approaches to core of their mission and activities. Many of these are companies are known for excellence in delivering consumer products such as Apple, Samsung, BMW, Herman Miller or Dyson. Others, like Morningstar, Zappos, or the Mayo Clinic, deliver exceptional service-based products. Part of the brand these companies trade on is the intense focus on design practice as a core competency. But, to varying degrees, these companies may or may not be what we would call 'design organizations', where the incorporation of design thinking and processes infiltrate every organizational nook and cranny. There is a spectrum of design activities that can play out in any organization that designs, from running a centralized and deeply specialized design group, to separate and unique product development and design departments in each division, to ensuring that there is broad participation in a shared methodology and adherence to organization-wide design vision, practices, and culture.

The Australian Tax Office (ATO) is an organization that began a journey in the late 1990s to transform itself from the policy level on up using design. Rather than simply redesigning tax forms or using superficial design tactics to try to improve communications about an already overly complex system, they decided to begin redesigning the tax system from its very foundations, using clarity and ease of use as their core design principles. They started by reconceptualizing the form and structure of the underlying code from which the taxation system is administered. In this way, the tax system as a whole was designed, with explicit attention to how the intent behind the law that forms policy should be captured and expressed and how tax policy as a whole should be structured and maintained. From this, a system of creating, communicating, and administering federal tax was created, with a specific emphasis on user-centred design. The ATO established a department for this purpose, called Integrated Tax Design.

In order to complete this organizational, policy and content-level transformation, the ATO created a design hub for specialist services, the ongoing development of design competencies with people at all levels of the organization, and the establishment of organizational structures, processes, and tools to support design activities throughout the organization, based on a user-centred design methodology that was customized specifically for the needs of the ATO. The goal was to create an organization that has as its sole mission the idea of improving the tax experience for all taxpayers by making encounters with the Tax Office easier, cheaper, and more personalized. The organization rejected the idea that tax had to be an innately unpleasant experience, and set about creating a system of positive interactions based on trust, understanding and fairness. They challenged themselves to make tax a good experience by encouraging people to feel engaged as citizens. In this organization, every project is required to utilize a user-centred design approach. In the first five years of the transformation process, the ATO was supported by the Sydney-based strategy and design consulting firm 2nd Road and several US-based design experts. Once they had systems up and running with a critical mass of trained staff, they left the consultants behind and continued transforming themselves from within. The project is expected to be a ten-year commitment. This may be the largest and most complete design transformation project to be documented to date.[3]

Another example is the Australian company, Suncorp, a top-25 Australian company that provides financial services (also a client of 2nd Road). In 2006 this organization began working to install design capability with the intent of transforming itself into a world-class customer-centred organization. But it has taken a more organic approach than the ATO. By planting design efforts around the organization at the level of everyday projects and supporting these through a small, centralized leadership hub with targeted training, communications and support, they hope gradually to win the organization over to design from the grassroots level. While slow, it has made steady progress. To date more than 900 employees on the operations side have received design training and over 1,000 employees have elected to join the design community of practice (6 per cent of the total workforce). This was achieved without any executive-level mandates and relies on voluntary participation and individual initiative. The enthusiasm that has been awakened for design in this organization is remarkable.

The design organization, as such, is an emergent species. Design organizations are those for which design is the culturally dominant way of working and thinking but many of these are still experimental or works-in-progress. Full transformation to a design organization is not always the goal, nor does it need to be, though for any organizational design initiative to thrive there must be a tolerant, if not supportive, environment or culture and so some aspect of organizational change will always have to accompany the creation of an internal design capability.

## WHO ARE THE DESIGNERS?

In their position paper on transformation, the UK Design Council notes that the design community is 'experiencing two important shifts: firstly, in *where* design skills are being applied, and secondly, in *who* is actually doing the designing' (Burns et al. 2006: 10). When we introduce design into an organizational context, we are inviting many people to the table as 'designers'. Some members of the design community regard this as somewhat dangerous. The argument is that it is better to leave design to the experts, and outsource design activities to consultancies that specialize in these competencies and excel in delivering high-quality results. And in truth, there are significant barriers to introducing design to organizations not historically aligned to design thinking and practice. There is the risk that too little knowledge can be dangerous, or that if the level of commitment is not serious enough, misappropriation of design approaches can lead to disastrous outcomes and a backlash. That design is its own profession with deep specialty areas is an important counterpoint to the assumption that 'anyone can design'. Certainly there are many companies out there who prefer to outsource their design projects, even when they have an organizational culture that is supportive of design and a degree of internal design competency.

There are two problems with this argument, however. One is that, at some level, design thinking is innate to all of us as human beings trying to change outcomes in our world. Clearly there are important specializations that professional designers bring to the table but that shouldn't mean that non-professionals cannot participate at any level or that design skills cannot be taught and applied in new contexts. The second problem with leaving design to the 'experts' is that as the subject matter of design expands beyond information and objects, into areas like service, policy, and organizations, it is not clear whose expertise is most appropriate in the domain of a given problem space. Most professional designers are educated by schools that provide skills in graphic, interface, and industrial design. But as we move into a twenty-first-century economy, professional designers are now not the only ones with subject matter knowledge relevant to making new things. More and more, the things that companies have to sell are less about the concrete value of objects and more about the abstract value of concepts – proxies, relationships, and systems. People learning to design within organizations do not replace the need for trained professional designers; in fact they may amplify the need for them as organizational design efforts create new kinds of design opportunities.

## THE CHALLENGE OF EMBEDDING DESIGN

Bringing design into an organization is not about simply introducing killer marketing or product concepts that might define a company superficially as a 'design' company. It is about introducing ways of thinking and working into an organization's people. This requires the adoption of design methods and practices, roles, structures and processes, and environments. Furthermore, it is critical to identify human-centred design as the key methodology in the organizational context. It is critical not only because of its central focus on the customer or end user but because it is also an inclusive humanistic approach that takes into account the needs of all stakeholders, which include internal users as much as customers of an organization.

Design can be introduced in a bottom-up or top-down fashion, and that might mean different things in different organizational cultures. For example, one client was convinced that design thinking and practice could not be successfully 'mandated' and therefore looked for a way of introducing design from the bottom up, pushing from the project team level to general management and then to senior leadership. This is the viral approach mentioned earlier. Another client, conversely, came to the conclusion that design culture must be actively driven top-down from the senior levels of the organization, with accountability to this goal from each subsequent level. There is no single correct approach. The success of onboarding any kind of cultural change relies on a solution specific and appropriate to that organization.

Tony Golsby-Smith, who founded 2nd Road, has identified thirteen key organizational systems that can be positively influenced by design and that will together make a design transformation possible. Many of these systems function as distinct departmental functions in large organizations, such as marketing, human resources, information technology and project management; other systems include cross-business functions such as planning, rewards, and risk management. Golsby-Smith argues that these systems must all be addressed in order to allow the uptake of design practice to be successful. What this means is that embedding design in organizations requires a broad, holistic and systemic approach. If design is to be sustainable and productive, it cannot be penned in by traditional silos or limited to lower levels of organizational hierarchy.

Furthermore, design seems to be at its most effective in an organizational context where it has focused equally on process and outcomes, where design becomes the focus of improving customer experience and providing a strong methodology for problem solving. The balancing of process and outcome is critical. With too much emphasis on process we can lose sight of keeping design agile, scalable and focused. With too much emphasis on outcomes, we quickly revert to efficiency-driven behaviours that undermine meaningful, effective solutions and value creation. This balancing act becomes a critical tool for organizations to prioritize and evaluate their efforts and keep actively moving toward growth opportunities. In this way, design can help organizations get better at evaluating and adjusting strategies in real time.

One of the organizations 2nd Road worked with to install internal design capability is Suncorp (mentioned above). Suncorp's Executive Manager of Design, Peter Vozvoteca, made the following observations about the organizational benefits they began to realize several years into the programme:

> Apart from the obvious skills gained, I have found that people are beginning to think differently. Their approach to projects is different. People now have confidence to push back and ask the tough questions such as, 'What is the real problem we are solving?' 'Are we are jumping to solution-mode too soon?' or 'But what does the customer say?' This means we don't blindly go down the path of doing activity for activity's sake. We stop and think things through and apply some judgment. This will mean a better capital allocation process. We are trying to position Design as a great de-risking methodology in this current environment.

The notion of de-risking is another important intrinsic benefit. Design is often viewed as an approach that requires more time and resources, and therefore is seen as riskier than traditional business problem-solving approaches. It has the potential, but no guarantee, for big rewards down the line. But by engaging design in a variety of projects around the organization, Suncorp has discovered that whether or not a great new product or service emerges at the end of a design project, the learning acquired along the way always has value. Vozvoteca cites more than one project that was abandoned or drastically rescoped after design research and prototype testing activities demonstrated that the proposed project was misaligned with customer need. As much as new products drive new revenue, getting more intelligent about what initiatives not to invest in can equally impact the bottom line. Design can actually help avoid risk in the market.

A big part of Vozvoteca's job was the ongoing challenge of persuading people to take a chance on design's potential. Bringing about change in a large organization takes tenacity, patience and clever positioning of the benefits. But with a foot in the door, it often doesn't take a lot to convince people once they begin to see evidence around them. At Suncorp, as at other organizations that have attempted to embed design as a core organizational competency, design emerged slowly. Design began to happen when teams applied design approaches to everyday problems and projects. Most of the day-to-day work at any organization is dedicated to sustaining and making improvements to existing products and systems, and this provides a steady stream of opportunity for design interventions. Design does not have to focus exclusively on new products; instead, it can uncover breakthrough opportunities in the things organizations are already doing.

Because design projects tend to shift organizational focus away from internally driven dynamics and towards customer experience, it is not difficult to trace dramatically different results from projects that adopt a design approach. For example, one such project involved a review of an existing cross-selling initiative coordinated by two different divisions of the organization. A successful pilot programme was underway in a limited geographical region, and yet management was unsure about precisely what elements were making the initiative a success. Before launching the programme nation-wide, management wanted to understand more specifically what made the pilot initiative successful and why, so that an effective training and support programme could be created. The project team nominated themselves for a design approach, and the team members, including the team's manager, enrolled themselves in Suncorp's design training programme.

Armed with a new toolkit, the team began the design process by revisiting the strategic intent of the project and launching a three-week exploration in the field. They interviewed all the staff involved in the pilot initiative, visiting them on site and observing the initiative in action, and then they met with customers, both those who had been successfully pitched for the initiative and those who had not. Their careful dissection of the different players involved (from two different organizational silos) and their shared customers, revealed far more nuanced and detailed information than was trickling up through the normal chain of command. The project team learned what was really going on, and they realized that they had mischaracterized their customers' needs. Although their pilot initiative had been a success, by their initial projections, they realized that they were still leaving a lot on the table. The result was that they were able to then design a much more effective nationwide marketing programme with the added benefit that they were using their staff much more efficiently to deploy the programme. With this new design, the project grew from its originally projected AUD23 million to an AUD116 million pipeline. Not only did the conversion rate more than double, but the value of converted sales increased by more than 800 per cent. The team later insisted that they never could have realized this result without using a design approach.

Once a handful of projects like this one began to sprout around the organization, more people began to hear about design and, more importantly, they began to see the direct impact that it made on projects similar to their own. This kind of spontaneous generation of grassroots design teams and design projects began to drive curiosity and persuade people around the organization to take notice and give design a try. However, organizations can be incredibly hostile to design, even when strong champions for design strategy and practice are present. Golsby-Smith of 2nd Road likes to say that organizations send out antibodies to fight and reject new approaches, and this is what design must contend with when it enters an organization. It must be positively viral in order to embed itself successfully.

This viral success includes some key ingredients: a combination of training, coaching and a supportive learning environment; demonstration projects with broad organizational participation and visibility; a design centre or hub to maintain the integrity of the methodology (this can be the ATO style of directly servicing the organization or the Suncorp style of coordinating distributed capabilities); leadership commitment to supporting and sponsoring design initiatives; measures and metrics that appropriately gauge design success; HR support of design-specific capabilities as job requirements and rewards for design achievement and last but certainly not least, commitment to maintaining a customer/user research capability that services the entire organization.

## BLAZING NEW TRAILS FOR DESIGN

Richard Buchanan has noted that there is a taxonomy of subject matters within organizations – classes of a new type of 'product' that exists within the organizational environment, which are ripe for design thinking and approaches. 'Better understanding of the potential of traditional forms of design practice and the articulation of new methods for integrating design as a strategic tool in organizational life should be one of the goals of management education' (Buchanan 2004: 55). The UK Design Council has also elaborated on the challenge for traditional management approaches:

> Traditionally, organizations have been designed for a complicated rather than a complex world. Hierarchical and silo structures are perfectly designed to break problems down into more manageable fragments. They are not, however, so effective handling high levels of complexity. For this reason, many of our most long standing institutions are now struggling to adapt to a more complex world. (Burns et al. 2006: 8)

These emergent problem spaces also expand the list of reasons why design can benefit organizations, because many of the organizational problems we may assign to design are not strictly product or service related. They relate to the formation and perpetuation of the organization itself. These problems might include as their subject matter strategy and planning,[4] business models, internal processes, or other organizational infrastructure such as roles and governance. And these are some of the subject matters that have the most to gain from a design approach. But to succeed in utilizing this kind of design, the understanding and practice of design must be pervasive and not just constrained to one specialist team.

One tool that allows us to leverage these new and exciting problem spaces is the recruitment of internal expertise. Installing 'design managers' became one of the key elements of the transformation programmes I helped to support at 2nd Road. These design managers were not created – rather they were recruited from among the existing ranks of organizational management. What we learned was that every organization has a few of these individuals who may not instinctively self-identify as designers or design thinkers but who display an immediately recognizable set of behaviours that tag them as design minded. These design managers had as much to teach us about the business as we had to teach them about design. Our collaboration was one of the most exciting places to see design unfold.

A counterpart to the design manager's role is the notion of people who are growth catalysts inside their organization. In their research on the managers who function as growth leaders from within organizations, Jeanne Liedtka, Robert Rosen and Robert Wiltbank note that growth leadership is inherently a learning activity. This learning attitude includes both the inclination to seek out new knowledge and experiences and the willingness to experiment. I would draw the parallel that design features these two behaviours and can also be defined as a learning activity. The catalysts that Liedtka et al. studied are the equivalent of our design managers, but practising in non-design organizations. The catalyst individuals engage in what the authors term a 'virtuous cycle' of mindset and behaviours that allows them to identify and pursue growth opportunities effectively for their businesses (Liedtka et al. 2009: 37–43). These are the same behaviours encouraged by design practice. As 'designers', operating managers can utilize a different toolkit that allows them to use resources more creatively and effectively to facilitate organic growth in their organizations. More importantly, Liedtka et al. examined whether such thinking and behaviour was innate or whether it could be taught, and they concluded that these were learnable strategies.

## TRANSFORMATION IN PROGRESS: DESIGN BLOCKERS AND ENABLERS

Some examples follow of the mechanisms and dynamics that allow design to grow and thrive in the organization, as well as some things that prevent design from being successfully engaged and embedded. Despite best efforts to introduce a comprehensive plan of attack, design transformation programmes inevitably find early victories in some areas while running up against roadblocks in others. Organizational blockers and enablers can be either systemic or cultural.

*Enablers include:*

- Adaptation of an organization-specific design methodology (as seen at the Australian Tax Office or Suncorp, two large organizations who have undertaken design transformation).
- Establishment of learning communities that bring together people from different parts of the organization.
- Support for cross-silo initiatives that bring together different disciplines.
- Recognition and rewards for learning-driven, not outcome-driven, behaviours.
- Demonstration projects that provide visibility and a forum for learning.
- High visibility for customers and robust research into customer experience.
- Management made accountable for design outcomes.
- Championship for the design cause at the executive level.
- Willingness to frame problems or opportunities in more open-ended ways.

While design generally needs to be introduced in a bottom-up, rather than a top-down fashion, it will never get its foot in the door without full endorsement from senior leadership. Most organizations familiar with culture change efforts will attest to the difficulty of mandating change. But when it is allowed to grow organically from within, using the example of early adopters and recognition and support from above, design culture can flourish.

*Blockers include:*

- Lack of design leadership or a coherent programme for on-boarding design practices.
- Old values are not challenged, or people receive mixed messages about what kind of behaviours will be rewarded.
- Projects undertaken are not ambitious enough or are seen as remedial; they resemble solutions-in-search-of-problems or are simply too restrictive in scope.
- The business case process does not accommodate design.
- Falling back on consultative, not collaborative, ways of working.
- Higher-ups do not participate directly in design and expect to delegate their participation.
- Lack of funding and time allocation for adequate exploration and research.
- No support for knowledge creation beyond individual teams; no shared organizational learnings.

A common mistake made in trying to install design practice is when design methods are translated into arbitrary stage-gate or TQM/Six Sigma-type processes. The biggest cultural challenge is to increase people's comfort level with ambiguity around problem solving. A good encapsulation of the type of cultural clash that organizations experience when traditional management practice meets design is related through Roger Martin's conception of reliability versus validity (Martin 2005). By analogy, business thinkers favour consistent outcomes (reliability), where design thinkers look for appropriateness or correctness of outcomes (validity). It is easy to see how each might label the other approach wrong.

By taking into account the blockers and enablers of a successfully integrated organizational design capability, a picture of intrinsic value begins to emerge. These are the things that play a key role in making the organization a more humanistic enterprise.

## CONCLUSION

Brand strategist Marty Neumeier has asked, 'if design is such a powerful tool, why aren't there more practitioners working in corporations?' (Neumeier 2009: 26). It is clear that both design and management fields have a strong interest in the ways that design thinking and practice can be adapted to the cultural context of the organization, but the success of embedding design has as much to do with organizations embracing design as it has to do with design embracing organizational challenges. Introducing design into an organizational context is no simple undertaking and much work remains in determining how this can best be done. Design has a great deal to offer as a counterpoint to management science but it is not a silver bullet and bringing design into organizations will require designers and managers alike to grapple with the limits and the appropriate uses of design thinking and practice. For designers this is a huge opportunity; it will vastly increase the relevance and value of the work that we do; it is a chance to move from twentieth- to twenty-first century design.

In this chapter we have taken a closer look at the intrinsic benefits design has to bestow on organizational life and why design approaches offer critical value not only to the end products that organizations discharge into the world but, more essentially, to the success of organizational culture itself. The recent global economic crisis has shown us the deficit of moral and ethical practice in many organizations, and it seems clear that a human-centred approach to value creation is now more appropriate than ever. Organizations that embrace the pursuit of becoming design organizations stand to benefit richly from a commitment to design thinking and practice, whether for the sake of developing a more innovative, productive culture or simply creating the type of environment where people feel valued and are enabled in creating value. Design is a worthwhile organizational endeavour if for no other reason that it allows people inside organizations recapture their interest, energy, motivation, and engagement through the humanizing influence of design. And while humanizing organizations would seem a noble purpose, it is also an eminently pragmatic one.

## NOTES

1. In his book *Drive*, Daniel Pink cited two recent studies where financial incentives actually correlated with reduced performance (Pink 2009: 41) and that, furthermore, extrinsic rewards were shown to be effective for algorithmic (for example, formulaic, left-brained) tasks, but ineffective for right-brained tasks like problem solving, invention, or concept generation (Pink 2009: 46). Pink argues that it is intrinsic rewards – which he characterizes as autonomy, mastery, and purpose – that motivate people most effectively in job performance.
2. CEO tenure data available online from multiple sources, including Forbes.com and weber-shandwick.com (accessed 28 April 2009).
3. John Body, who served as Assistant Commissioner of Integrated Tax Design, has written at length about this project in *Design Issues* (2008: 55–67). It is worth noting that because of the ATO's success with this design transformation, the New Zealand Department of Inland Revenue has recently embarked on a similar programme of design transformation.
4. Jeanne Liedtka, in her essay, 'In Defense of Strategy as Design' (2000: 8–30) proposes that design is a powerful analogy for strategy, cataloguing the key ingredients that they share.

Strategic thinking is (like design) synthetic, adductive, hypothesis-driven, opportunistic, dialectical and value-driven. From this perspective, strategy can be positioned as one of the purest forms of design thinking.

## REFERENCES

Body, J. (2008) 'Design in the Australian Taxation Office', *Design Issues* 24(1): 55–67.

Buchanan, R. (2004) 'Management and Design: Interaction Pathways in Organizational Life', in R. Boland and F. Collopy (eds) *Managing as Designing*. Stanford, CA: Stanford University Press, Chapter 4.

Burns, C., Cottam, H., Vanstone, C. and Winhall, J. (2006) 'Red Paper 02: Transformation Design', UK Design Council, www.designcouncil.info/mt/RED/transforma tiondesign/ (accessed 28 April 2009).

Liedtka, J. (2000) 'In Defense of Strategy as Design', *California Management Review* 42(3): 8–30.

Liedtka, J., Rosen, R., and Wiltbank, R. (2009) *The Catalyst*. New York: Crown Publishing Group.

Martin, R. (2005) 'Validity vs Reliability: Implications for Management', *Rotman Magazine* (Winter): 4–8.

Neumeier, M. (2009) *The Designful Company: How to Build a Culture of Nonstop Innovation*. Berkeley, CA: New Riders.

Newkirk, D. and Freeman, R. (2008) 'Business as a Human Enterprise: Implications for Education', in S. Gregg and J. Stoner Jr. (eds) *Rethinking Business Management: Examining the Foundations of Business Education*. Princeton, NJ: Witherspoon Institute, Chapter 10.

Pink, D. (2009) *Drive*. New York: Riverhead Books.

# 13

# Challenges of Design Management in Construction

PAOLO TOMBESI AND JENNIFER WHYTE

To focus on the challenges of design management in construction is to immediately raise a number of questions. What is design? What is management? What is construction? Generally accepted answers to such questions stress the linearity of production processes. Design becomes associated with the work of professional actors, such as architects and engineers. Management is seen as a project management responsibility, the role of managers to monitor designers' progress and outputs. The term 'construction' is narrowly applied to a process of on-site assembly. In descriptions such as the conventional Royal Institute of British Architects (RIBA) stages, work in design is passed into a construction stage after the main design work has been completed.

Based on such definitions, design management in construction has been explored in two main ways. First, it has been described as the management of the specialist professional expertise within consultant architecture and engineering firms at a pre-construction stage. Here researchers have discussed the commercial management of design firms (Winch and Schneider 1993; Emmitt 2007) and the differentiation and integration of design tasks into the conceptual design, interior layouts, façade, mechanical and electrical, structural and other subsystems. Second, it has been described as the management of vertical connections between phases of the design and construction process. Here researchers have articulated processes and protocols for the flow of materials, information and value for customers across the project lifecycle (Ballard and Koskela 1998; Winch 1998; Austin et al. 2001; Gray and Hughes 2001; Winch 2003; Cooper et al. 2005).

Our approach is more aligned with the latter tradition, but in this chapter we go further to argue that the conventional definitions, descriptions and understanding of design do not adequately address the new challenges of design management that are emerging in practice. From our perspective, the challenges in construction are different from those within consumer products industries because the construction sector involves the production of socially and technically complex products and hence requires close interactions between users and producers throughout the process (Hobday 1996; Gann 2000). There is a need for innovation in design of the built environment to engender careful use of resources in production; to shape sustainable lifestyles and to mitigate and respond to a changing climate. At the same time, designers and managers operate in complex organizational settings, increasingly involving public-private partnerships and the distribution of design activities across global networks of manufacturing and use.

In developing our argument, we use the broader meaning of the term 'construction' to invoke the whole sector of the economy with a number of associated industries involved in the production and maintenance of the built environment. We draw on work on industrial and social organization to see design not as a professional role or stage of construction but as a set of synchronic and diachronic activities that involve clients and end users, fabricators, component suppliers and tradesmen as well as professionals. Management of design involves differentiating and allocating activities, motivating people to work on them and coordinating the results of their labour: it also involves managing expectations, identifying institutional gaps and ensuring consistency across multiple strands of ongoing activity. After discussing the 'professional system of design' we introduce a number of case studies from our own research to illustrate good and poor design management from this perspective.

## DESIGN, PROFESSIONS AND THE BOUNDARIES OF EXPERTISE

The architectural and engineering design professions have dominated discussion of design in the construction sector. From the beginning of architectural practice as a modern profession, architectural institutions have been active in the advocacy of a particular image of the building process that legitimates the social necessity of the architect's role (Kostoff 1974; Saint 2007). Since trades, contractors, and developers gained a strong operative and decisional presence, this image had to account for the actual spread of responsibilities without attaching it to a release of expertise. Authority 'on the subject' of architecture was connected with authority 'over the operations' of the building process (Tombesi 2009). This could only happen by promoting, as John Soane famously did in the late eighteenth century, a principle of non-compatibility between the production of advice (i.e., design) and the production of goods (i.e., construction), and by concentrating the 'design' function with architects insofar as they were professional.

Much writing in this vein refers to a triangle of practice, with trust relationships between a client in a commissioning role, design as a technical agency and construction management as a delivery function. Within this triangle of practice it was the ability to instruct and monitor the work of others that gave architects their professional status.

This implied a technologically exhaustive function, unfolding prior to construction and organized progressively, which was to be reflected in the technical dimension of the architect's work: define and organize the information necessary to envision the overall idea for the building, produce its various physical components, and determine the processes required for the implementation of the plan.

During the twentieth century this design function became increasingly distributed as processes of design and construction grew in complexity. In the twenty-first century relationships are now undergoing rapid change, due to new building types and forms of contracting, such as design and build, new tools and new approaches to delivery such as public-private partnerships and the increasing importance of a digital infrastructure for the delivery of project work (Whyte and Levitt 2011). As implied in the right-hand diagram of Figure 13.1, contracts are now available that facilitate official design assistance from several other parties, and procurement methods have been developed that parcel out the design effort in recognition of the actual distribution of technical expertise across the industry and overlay it with construction (Bennett and Ferry 1990; Pietroforte 1997).

Hence, current relationships can be seen as involving a polygon of practice, in which there are an increasing number of design roles involving subcontractors engaged in product engineering, testing and management. Yet, in many construction contexts, the theoretical faculty to provide overarching instructions, originally ascribed to the architect as main agent of the principal, has left an indelible mark on the

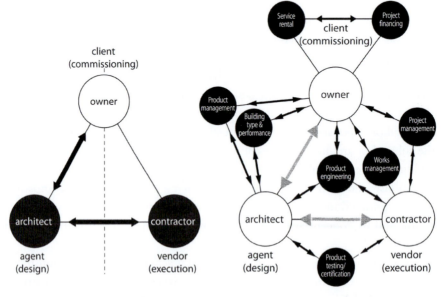

The traditional triangle of practice                    The current polygon of practice

FIGURE 13.1   Relationships in construction practice. © Tombesi and Whyte, 2011.

way design is conceptualized – a largely autonomous activity, closely associated with the work (and the services) of the architect plus fellow engineering professionals, and culturally separated from delivery.

## EXAMPLES OF DESIGN COORDINATION CHALLENGES IN CONSTRUCTION

The legacy of this view of design and construction is the multitude of projects featuring construction shortcomings of different types, mostly ascribable to insufficient coordination between design dimensions. These include interface coordination; workmanship control and maintenance planning; relationships between parts; clashes between functional programme and the assembly or between procurement design and the expected performance; institutional vacuums and analytic gaps. Table 13.1 gives examples.

Considering the examples in Table 13.1 (based on Tombesi 2004, 2009; Tombesi et al. 2008), the Ara Pacis Museum is still considered to be a successful building by many of the actors involved. This is because there is a perception that the defective solutions highlighted do not belong to the life of the project, and are rather part of the unavoidable contingent risk ever present in the industry. The building facades at Federation Square are indeed the indirect by-product of a combination that never contemplated building users' needs, partly because the sequence of design operations would not allow for it and partly because the specific brief was not there. The Turin Polytechnic campus and the Athletes' Village in Turin could be seen as victims of the procurement methods chosen for their development, and the consequent fragmentation of design monitoring responsibilities. Even the Scottish Parliament building at Holyrood, Edinburgh, has been discussed as the result of an uneasy relationship between different parts of procurement design (Bain 2005). Here, however, together with the examples of Simmons Hall and the Kursaal Centre, it speaks of the distance between building design and assembly design. The bridge in Venice and the Disney Concert Hall show that both designers and project managers often operate with significant uncertainties, which are sometimes created by the loose socio-technical circumstances.

What this set of examples shows is that the management of design in construction goes beyond the detailed allocation of resources, the evaluation of alternatives, the monitoring of decisions, or the effectiveness of communications within the project team. Managing the project from a design perspective also means ensuring consistency between the multiple streams of intellectual activity carried out across construction's supply chains, which ultimately have a bearing over the quality of the building produced.

Construction can be seen as a process characterized by 'interdependence and uncertainty', in which important factors can reach beyond the project's immediate contractual environment, and thus the network of transactions enabled within it. As Ruskin

**TABLE 13.1: Examples of design coordination challenges from across Europe, USA and Australia**

| Example | Problems Encountered and Results |
| --- | --- |
| Ara Pacis Museum, Rome, Italy, 2006 | Here there were problems combining construction components with different degrees of technological sophistication, seemingly without appropriate interface coordination, workmanship control and maintenance planning. The result is a conspicuous collection of as-built defects that will be hard to redress, and a series of nondesigned postoccupancy solutions that may go against its performance in the long run. |
| Australian Federation Square, Melbourne, Australia, 2002 | Here there were problems with the relationship between the external boundary design and office design. In this case, the technical composition of the envelope and its conflict with interior space is not only a result of the architects' decisions but also a reflection of how their services were procured and divided, how tenancy was selected, and how the brief was defined (Wong and Teoh 2008). Workspaces behind the highly intricate triangle-based outer façade of the main building do not have enough protection from the sound emanating from the square, variations in external temperature, or the incidence of daylight on the internal environment. |
| Polytechnic of Turin, Turin, Italy, 2006 | Here the functional programme clashes with the development and assembly of its manufactured component systems, particularly in a situation where construction is being procured while budget-related space planning is yet to be finalized. At the Polytechnic, changes in the building's section of parts of the complex have combined with completion pressures and excessive underbidding under a particular form of design-build integrated contract to generate under par subcontracted built work |
| Winter Olympics Athletes' Village, Turin, Italy, 2006 | Major inconsistencies in the reinforced concrete structure of some of its buildings, largely due to the fact that different design stages were assigned to different agencies under different contracts, and that these had found it difficult (or not convenient) to coordinate decisions at a time when handover dates would allow technical details to be overlooked. |
| MIT's Simmons Students' Hall, in Cambridge, MA, USA, 2002 | The decision of the architect and the structural engineer to conceive the envelope as a giant Vierendeel truss, made of precast waffle-like panels (so that the skeleton of the building would conform to the architectural idea), turned the university dorm into a nearly impossible construction undertaking, and eventually the most expensive student structure in America per bed. In the end, the solution for the panels was arrived at by the general contractor, through the construction of elaborate mockups that were used to design and reproduce the trade sequences required (Gardiner 2008). |

| | |
|---|---|
| Southern Cross Station, Melbourne, Australia, 2005 | Problems, however, can occur also when the overall development process goes out of phase, creating friction between building procurement and expected performance. The passive control design solutions developed for the building were placed under scrutiny shortly after the opening of the facility to the public, when serious internal air pollution problems emerged. Eventually, it became clear that the problem did not reside in how the fumes exhaust system had been designed but in the fact that delays in the construction programme had deferred the erection of a wall needed for the section of the terminal to work, environmentally, as planned (Drake 2008). |
| Scottish Parliament, Edinburgh, UK, 2004 | A critical relationship exists between the detailed design of the building and decisions related to its constructability or the monitoring of its construction. In Edinburgh, the Debating Chamber of the parliamentary building had to close in March 2006, after a strut from the roof structure came loose, later to discover that the failure was due not to the design of the component but to the workmanship of the subcontractor (The Holyrood Inquiry 2004). |
| Kursaal Centre, San Sebastián, Spain, 2000 | A similar situation occurred in the foyer of the Kursaal Theatre in 1998, when the main concrete stairs collapsed, following the structural failure of their anchorings to the steel facade, after they had been redesigned during construction to simplify assembly procedures (Serna et al. 2000). |
| New Bridge, Venice, Italy, 2008 | Even successfully completed artefacts can reflect coordination shortcomings (or institutional vacuums) in their development process. Although this was hailed as a structural *tour de force*, it is now waiting the addition of a steel-frame, egg-shaped cabin for mobility-impaired passengers because, at the time of the planning permit, it did not (or was not made to) respect the inclusive design legislation that applies to public works in Italy. |
| Disney Concert Hall, Los Angeles, USA, 2003 | In another example of such an institutional vacuum, here the external volumes had been defined in accordance with the city's guidelines for vertical surfaces. The decision to substitute the original stone cladding with steel panels for value engineering purposes created light refraction and heat gain problems for surrounding buildings, which caused, in the end, the partial covering of the roof with grey tarpaulin. |

(1853) had already intimated, the division of design responsibilities has socio-cultural/ technical connotations that are institutionalized and predate (or precede) project-specific agreements but can influence the quality of the work performed under them by laying out opportunities for integration that are either conducive to effective collaborations or mindful of potential barriers/hurdles.

If one accepts the disciplinary openness of this proposition, the boundaries of 'design management' become blurred with those of 'project design'. Since the procurement of any building project requires a number of autonomous design dimensions to be concurrently active within the construction field, the design management function requires mutual relationships between different design processes. Determining such relationships on the basis of an understanding and judgment of the project's aims, the resources available, the socio-technical context, the timeframe and the rate of change internal to each design dimension is itself an act of design – that is, of project design.

## DESIGNING AND MANAGING: KEEPING IT REAL

To take the notion of design in construction beyond the cultural boundaries defined under the 'professional system' requires attention to the characterizing traits of the related activities. Design has variously been discussed as a problem-defining, problem-solving, information-structure activity that, on the basis of ideas and (partial) knowledge, defines a specific course of action (Schön 1983; Hatchuel 2001). In these terms, design activity enters all of the dimensions of the building procurement process: building production, building erection, building use and maintenance, and project definition and control. This reconceptualizes a building process as a 'system of design production' independent of the profession – a cycle, that is, within which all of the information necessary for the implementation of the building is conceived and either produced or assembled (Gray et al. 1994). It focuses attention on interactions with users, the trades and the hidden design work that goes on throughout the supply-chain – in components; in temporary works and in assembly sequences.

Design involves a process of representation, using drawings and models to develop and consider possible future states. This is the great skill of the professional designers but the great danger is that the drawing rather than the building becomes seen as the end-product of their work. Design quality indicators were developed in the UK as part of an attempt to engage a range of stakeholders in design processes (Gann and Whyte 2003). Their strength is in providing a structured process for engagement and discussion of design with the various stakeholders and end-users at different stages, within the wider context of performance measurement initiatives. This wider process of design management through interaction between professionals and suppliers requires a constant validation of models with reality.

While the production of the built environment is often institutionalized in ways that do not enable interactions across this cycle, there are examples of the building process conceived as a system of design production. First, Buntrock (2002) describes a context in which architecture is embedded within a building process, with design details negotiated on site by contractors, manufacturers and architects and great attention paid to maintaining collaborative relationships. Her study was conducted in Japan, where the construction sector is comparatively consolidated, with large contractors offering a broad range of construction and professional services, doing in-house research and development, financing some projects during construction and holding the economic power to obtain support from subcontractors and suppliers. She argues that top architects, working in construction firms and in independent practice, are able to benefit from these resources.

The implications of her study are quite different to those from the study of the Japanese car industry, which have been used to promulgate a culture of monitoring and measurement with information delivered ahead of action, instead implying a design process that continues as work is delivered on site, with little information delivered in advance. As design continues on site and there are no binding drawings, the reduction of change orders is not a goal. Construction is perceived as large-scale assembly and the plans and elevations of a building are reworked and further refined long after the structure is up.

Decisions about detailing and material sizes are based on *in situ* conditions and made by people in daily contact with the building while it is under construction. For example, measurements for a door will be taken after the opening is roughed out. This means that there is no need to use tolerances to account for the possible variations in the initial stages of construction, and it accounts for the neatly finished appearances of the related buildings.

Great architects and designers sustain an engagement with the materials beyond the drawing. The success of the team that designed and constructed the roof of Heathrow Terminal 5 depended on close working relationships between the architects, steel erectors, steel casting foundry, engineers, temporary works designer, the construction contractor and a host of other members of the supply chain. It was organized around a set of hybrid practices (Harty and Whyte 2010), using a range of physical and digital models to explore aspects of design and progress understanding, with the architects spending time working with fabricators in the steel foundry, as well as temporary works and construction managers. Hence in successfully accomplished design work, the process of making a design real involves a dialogue between professional designers with a focus on the overall architecture, with those concerned with subsystem design, components, fabrication and assembly. As the architect for the roof described:

> The most important thing is that the people around the table, who speak for their
> own firms, understand what you're talking about, and get it, and can go back and

do their work, and we are all literally again in one room. We were constantly building large-scale models of the roof, and then we started to integrate through model
building, how the roof actually negotiates the substructure.

## DESIGN MANAGEMENT IN CONSTRUCTION AS JUDGMENT AND MEASUREMENT

This chapter suggests new directions for research design management research and
practice in construction. The idea that 'project design' and 'design management' are
two segments of the same function was contained in Caudill's famous 1970s writings
about design, where he stated that every building project should be led jointly by a
'troika' of experts, each sponsoring a discipline of practice: management, design and
technology (Caudill 1971). In Caudill's world, however, project design and building
design were considered on the same level, almost interchangeably, and were likely to
be developed by the same professional agency. The twenty-first century construction
sector is organizationally complex, with design activities distributed widely. It faces
substantial and new challenges.

   In construction, there is a need for project management and control, but a culture of monitoring and measuring can drive out the iterative experimentation that is
required in a design inquiry. Writers have articulated the difficulty faced by designers
in a performance measurement culture (Allinson 1993) and have drawn attention to
the different judgment- and measurement-based approaches to design quality (Gann
and Whyte 2003). The judgment-based approach to design is prospective, concerned
with the future; whereas the measurement based approach is retrospective, measuring the past and checking progress. It is this kind of design activity that has inspired
management scholars to draw parallels between design and management seeing both
as forms of reflective practice or inquiry (Boland and Collopy 2004). For those from a
judgment-based approach, managing occurs through the activity rather than separately
from it – those engaged in design activities act as design managers, rather than design
management emerging as a separate profession.

   Hence Caudill's formation could be recomposed at project level by way of technical coalitions strategically 'designed' to respond to the reflective management tasks at
hand. In the case of the roof subproject team, the design management role was passed
around the team as the design developed, starting with the architect and then passing
to the steel fabricator, with the construction manager keeping the team on track in
terms of deadlines and compliance. For all of the parties involved this was not business as usual but an experience of what construction could be like in a collaborative
culture.

   Computers offer great opportunities and also great threats to this collaborative culture. New digital tools for design, coordination and governance of large construction
projects include centralized project archives that can be used to track design progress

by the various parties; shared digital models and prototypes; automated search facilities; simulation programmes and communications technologies. These technologies are having a dramatic impact on the way design work is organized, for example enabling design work to be distributed globally over multiple offices to speed up its delivery. New practices that combine digital and physical ways of working are emerging (Harty and Whyte 2010) and their implications for design quality and for practices of design management need to be better understood. These technologies also herald a host of new intermediary roles, as document controllers, 3D CAD technicians and others become more salient in the coordination of professional design work.

Hence the aim of this chapter is to open up debate, highlight the broad challenges faced in design management in construction and provoke the reader to consider the alternative approaches and futures rather than to provide definitive answers. By opening up our definitions of design management to understand the production of the built environment as iteratively achieved and involving a wide range of players, we inevitably have to face its complexity. The focus shifts from the traditional triangle of practice to a wider set of actors – a polygon of practice, as shown in black within Figure 13.1.

What strategies can be used for design management in construction in this context? Our work suggests that there are a number of areas in which design management can be improved generally through work that negotiates common objectives between different production processes and achievable internal parameters, defines contingency plans to be activated in the event of design processes' relative failure, monitors the procurement of products with different lead times and degrees of uncertainty built into them, builds robustness into the design system by working to understand how elements can be removed from the critical path of the project, and plans interfaces between design products of different scales and intensity.

The management of any production process involving socially varied investments and returns is intrinsically political. Design management in construction is no exception: different approaches to it have different production and product implications. Hence, the role of the design manager cannot be detached from the relational environment in which the position exists. Design management may indeed involve shaping and being shaped by the context within which design work happens – for example through interactions with clients, regulators and funders, as well as directing the detailed work of designers.

## REFERENCES

Allinson, K. (1993) *The Wild Card of Design: A Perspective on Architecture in a Project Management Environment*. Oxford: Butterworth-Architecture.

Austin, S., Baldwin, A., Hammond, J., Murray, M., Root, D., Thomson, D. and Thorpe, A. (2001) *Design Chains: a Handbook for Integrated Collaborative Design*. Tonbridge: Thomas Telford.

Bain, S. (2005) *Holyrood: the Inside Story*. Edinburgh: Edinburgh University Press.

Ballard, G. and Koskela, L. (1998) 'On the Agenda of Design Management Research.' Sixth Annual Conference of the International Group for Lean Construction. Guaruja, Sao Paulo, Brazil, 13–15 August 1998.

Bennett, J. and Ferry, D. (1990) 'Specialist Contractors: A Review of Issues Raised by their New Roles in Building', *Construction Management and Economics* 8: 259–83.

Boggs, C. (1993) *Intellectuals and the Crisis of Modernity*. Albany, NY: State University of New York Press.

Boland, R.J. and Collopy, F. (2004) *Managing as Designing*. Stanford, CA: Stanford University Press.

Buntrock, D. (2002) *Japanese Architecture as a Collaborative Practice: Opportunities in a Flexible Construction Culture*. London and New York: Spon Press.

Caudill, W. (1971) *Architecture by Team*. New York: Van Nostrand Reinhold.

Cooper, R., Aouad, G., Lee, A., Wu, S., Fleming, A. and Kagioglou, M. (2005) *Process Management in Design and Construction*. Oxford: Blackwell Publishing.

Drake, S. (2008) 'Falso Allarme', *Il Giornale dell'Architettura* 7(65): 6.

Emmitt, S. (2007) *Design Management for Architects*. Oxford: Blackwell.

Gann, D. (2000) *Building Innovation: Complex Constructs in a Changing World*. London: Thomas Telford.

Gann, D. and Whyte, J. (2003) 'Editorial: Design Quality, its Measurement and Management in the Built Environment', *Building Research and Information* 31(5): 314–17.

Gray, C. and Hughes, W. (2001) *Building Design Management*. Oxford: Butterworth Heinemen.

Gray, C., Hughes, W. and Bennett, J. (1994) *The Successful Management of Design: a Handbook of Building Design Management*. Reading: Centre for Strategic Studies of Construction.

Harty, C. and Whyte, J. (2010) 'Emerging Hybrid Practices in Construction Design Work: the Role of Mixed Media', *Journal of Construction Engineering and Management* 136(4): 468–76.

Hatchuel, A. (2001) 'Towards Design Theory and Expandable Rationality: The Unfinished Program of Herbert Simon.' *Journal of Management and Governance* 5(3/4): 260–73.

Hobday, M. (1996) *Complex Systems vs Mass Production Industries: A New Research Agenda*. Brighton: Complex Product Systems Research Centre.

Kostoff, S. (1974) *The Architect: Chapters in the History of a Profession*. Oxford and New York: Oxford University Press.

Pietroforte, R. (1997) 'Communication and Governance in the Building Process', *Construction Management and Economics* 15(1): 71–82.

Ruskin, J. (1853) *The Nature of Gothic. The Stones of Venice II: The Sea Stories*. London: Kelmscott Press.

Saint, A. (2007) *Architect and Engineer: A Study in Sibling Rivalry*. New Haven, CT: Yale University Press.

Schön, D. (1983) *The Reflective Practitioner: How Professionals Think in Action*. New York: Basic Books.

Tombesi, P. (2004) 'Sloppily Built or Precisely Loose? The Technology of the Curtain and the Ideology of Disney Hall', *Architectural Research Quarterly* 8(3–4): 247–59.

Tombesi, P. (2009) 'On the Cultural Separation of Design Labor', in P. Deamer and P. Bernstein. *Building (in) the Future: Recasting Labor in Architecture*. Princeton, NJ: Princeton University Press.

Tombesi, P., Gibello, L., Milan, L. and Chiorino, C. (2008) 'Se le Archistar sbagliano', special report. *Il Giornale dell'Architettura* 7(65): 1, 4–9.

Whyte, J. and Levitt, R. (2011) 'Information Management and the Management of Projects, in P. Morris, J. Pinto and J. Söderlund. *Oxford Handbook on the Management of Projects*. Oxford, Oxford University Press.

Winch, G. (1998) 'Towards Total Project Quality: a Gap Analysis Approach', *Construction Management and Economics* 16(2): 193–207.

Winch, G. (2003) 'Models of Manufacturing and the Construction Process: the Genesis of Re-engineering Construction', *Building Research and Information* 31(2): 107–18.

Winch, G. and Schneider, E. (1993) 'The Strategic Management of Architectural Practice', *Construction Management and Engineering Economics* 11(6): 467–73.

# 14

# A Prospective of Service Design Management: Past, Present and Future

W. J. HOLLINS

In the UK, in the 1980s, there was a rapid decline in the manufacturing sector. Britain moved away from manufacturing, partly fuelled by poor management, lack of investment, poor industrial relations but also by government policy. Over the same period, the financial sector grew fast. 'Yuppies' and 'Barrow Boys' emerged in London, those with sharp minds but few academic qualifications made fast money, not by providing artifacts but by buying and selling stocks and commodities. Those employed in manufacturing became fewer and poorer, whereas those employed in the service sector – especially the newer financial or IT sectors – became much more wealthy.

This aspect of manufacturing decline was encouraged by the government of Margaret Thatcher, who stated that she wanted Britain to become an 'all service economy'. The counterargument was given by her own Minister for Trade and Industry, John Butcher who stated that to have a balance of payment surplus in an all service economy, the UK would need 55 per cent of the world market for services. This would be quite an impossible aim; however, as will be shown later, the claim was not necessarily accurate. Therefore, in the 1980s, in the UK, there was a rapid decline in manufacturing and an increase in employment in the service sector. This has been a worldwide phenomenon and, even in Japan, employment in the service sector now exceeds that in manufacturing (Sakao and Shimomura 2004).

## DESIGN UNDERSTANDING

What constitutes the development of new products has 'grown' over the past forty years. Design as undertaken by art school designers was once considered to be an end in itself, which resulted in products that were for an 'educated elite' (Walker 1989) but were invariably difficult to make, expensive to buy and had limited appeal. This type of design was a throwback to the Arts and Crafts movement, pioneered by William Morris at the end of the nineteenth century, but fails to fulfil the needs of those operating in mass markets. In the mid-1970s, as supply began to approach and then exceed demand in many product areas, companies considered design to start at the concept stage and also included consideration of ease of manufacture (for example, Anderson 1975).

Through the 1980s, there were significant advances in our understanding of design with the realization that designers were not 'artists' that worked in isolation but that their success depended on their understanding of the markets, being able to make the product and being able to sell it. Stuart Pugh (1982) proposed 'Total Design' in which the market was given prior consideration and a thorough specification was developed all before the concept stage of design. The broad stages of the process were:

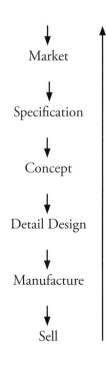

A written design process is often called a 'design model'. This still widely accepted model ends at the selling stage but the scope of 'total design' has been extended to include everything up to and including, 'disposal' such as servicing, marketing, service recovery, redesigns and recycling (British Standards Institution 1989; Andreasen 1994; Hollins and Pugh 1989) and much more to do with the customer experience.

The understanding of design has continued to grow, with designers now considering strategic objectives of the organization. Consideration of how design teams can be more effective has also emerged with the realization that the days of the lone designer are over. The management of people is a major part of any process. Another change that has taken place since the early 1980s is that there has been a greater use of IT through all aspects of the design process, well beyond the use of only CAD and CAM.

## THE BRIC COUNTRIES

Another major change in the world has been the emergence of the BRIC countries, Brazil, India, Russia and China. These vast countries with well-educated but low-cost labour seem likely to corner manufacturing worldwide. Governments adopt a laissez-faire approach of non-intervention and very limited support to manufacturers in the UK. Currently, due to investment in those countries by Western organizations, Western nations can still design products that are manufactured in, mainly, China for sale around the world. In some aspects service design companies are acting as a service provider but are gaining a financial contribution from these overseas manufacturers – thus the fear expressed of the need to get 55 per cent of the world market for services was wrong as such potential income was not considered. Having said this, the trade deficit between the EU and China has grown to monumental proportions and (excluding the very large swings of business cycles) shows no sign of reaching a plateau.

The future may not be so rosy for Western engineering designers as the educated workforce in these BRIC countries will eventually be designing products for themselves. It has already been speculated that there are too many designers currently being turned out by Western universities (Tether 2008). Certainly, many of these designers are being trained in declining areas of design. Only a few are currently being educated in service design although there is clear evidence that the emphasis in university education in this area is now growing. There is also evidence that British universities are reducing courses in design management, which may reduce the future effectiveness of leadership in design.

## SERVICE DESIGN

It seems surprising that huge efforts are aimed at improving design for the declining manufacturing sector while almost no effort targeted the emerging and fast-growing service sector. The first book that looked at the management of service design (Hollins and Hollins 1991) was written with the realization that there was little manufacturing in and around London and most people were employed in the service sector. It would seem obvious that this showed a need for research and understanding of this growing area yet few scholars became involved. In 1994 the British Standards Institution published the world's first guide to managing service design (British Standards Institution 1994) but sales of this British standard were disappointing.

In 1995 the first professor of service design management, Birgit Mager was appointed in Cologne International School of Design. She has taken a practical approach to the subject and produced successful service design projects since (cf. Mager 2003). One of her ex-students, Stefan Moritz (2005) has written an excellent report that outlines the history of service design and the tools that can be used. This also includes some useful examples of the application of service-design techniques. Joe Heapy and Sophia Parker (2006) also produced an excellent report. Apart from these efforts, the apparent lack of interest in service design and its management continued in the UK and in most countries over the subsequent years from 1990 to the point today where almost 80 per cent of people employed in the UK are now working in the service sector. From 1990, contrary to the anticipated and predicted decline in working hours, most people are now working harder and longer. So we now have a wealthy group of individuals who have little spare time but want to fully utilize that 'time for enjoyment' efficiently and effectively. This has resulted in the growth of leisure services such as health clubs and short overseas breaks – all of which would benefit from effective design.

In the UK early in this century came the wider realization that, not only could services be designed but that companies would be prepared to pay for it. Direction Consultants had been involved in service design since 1989 but they were joined by Live/Work in 2003, PLOT in 2004 and Radar Station in 2006. Engine started as a service design consultancy in 2000 and the Design Council introduced a knowledge base website on service design in 2004 and have undertaken a lot of good work in this direction since. IDEO have also become involved in service design.

Most recently there have been some new additions that work in the area of service-design consultancy. These include EMC, Seren, Snook, Service-Design Gbr, Spirit of Creation, Think Public and Continuum. Some universities both in the UK and abroad are also now offering service design to clients.

At the end of the first decade of this century the UK led the world in the understanding and management of service design. This is probably due to its rapid decline in manufacturing. Scandinavia has also taken a greater interest in the management of service design, especially Denmark and Finland, and there have been advances from Italy. Germany is also quite effective and this has also been demonstrated by the fast growing Service Design Network which, although based in Cologne in Germany, has many organization and individual members worldwide. The US seems to lag behind and it would appear that the US view of service design is that it is an extension of branding. However, it is much more than that. Service design consultancies will continue to increase in number and will spread across the whole world. This will mainly be fuelled by product design consultancies that are forced to refocus as their traditional customers continue to decline. This will go hand in hand with the adoption of design management techniques by service companies. Much of this will depend on the realization by management in service companies that service design will improve their competitiveness and, therefore, profitability.

In an interesting piece of research, Voss and Zomerdijk (2007) described how this is already happening by looking at innovation in experiential services. This report describes a series of case studies where service design has been used and this includes examples where it is applied to drive the strategy of some service organizations.

Leadership is the key. Perhaps, this leadership will come from individuals previously employed in manufacturing. Perhaps, this leadership will result from better training and learning by existing managers. For this to occur, design management must be taught in management programs such as the MBA and there is evidence that the baton has been taken up by universities especially in Scandinavia and other parts of Europe.

## WHAT IS A SERVICE?

The fact that a service is different from a manufactured product greatly affects how it is designed and how that design is managed. Service design can be both tangible and intangible. It can involve artifacts and other things including communication, environment and behaviours. Most services differ from manufactured products in up to five ways:

- *Tangibility*. One can physically touch a manufactured product but most services are intangible. One cannot touch legal advice or a journey, although one can often see the results.
- *Transportability*. Most services cannot be transported and therefore, exported, (although the means of producing these services often can) but the ability to export services is fast changing due to new technology.
- *Storability*. Because services tend to be intangible, it is usually impossible to store them. For example, a car in a showroom if not sold today can be sold tomorrow but an empty seat on an aeroplane is lost once the plane has left.
- *Quality*. In manufacturing, quality tends to be measured against drawings, etc. The measures tend to be quantitative. Due to the intangible nature of services and as production and consumption often occur at the same time, it is difficult to ensure consistency and quality. The quality of a service is often down to the person giving it. As a result, the measures of quality in a service tend to be qualitative and there are few quantitative measures. The effect of this is a wider variability in services and it is more difficult to control the quality of a service.

- *Customer contact.* Generally, in manufacturing the customer may be unaware of how the product came about. In services, production and consumption tend to occur at the same time (simultaneity).

In the design of services, there are more 'customers' (stakeholders) to be considered than in manufacturing. For example, consider the customers/stakeholders in primary school education (pupils, parents, governors, local government, central government and even taxpayers) or even in a hospital (patients, general practitioners, government and taxpayers). This makes the design of successful services more difficult as it is necessary to understand and provide the needs and the relative importance of each of these stakeholders to succeed. Another difficulty is that the relative importance of each of these stakeholders can change at different times throughout the life of the service. Most services cannot generally be patented and therefore intellectual property in services is more difficult to protect and copying of competing services easier. This is a good reason to keep applying serial innovation (British Standards Institution 2008a) to retain that competitive edge.

## FEATURES OF A SERVICE

As services are the growth area, well designed services can be very profitable. The opportunities for innovation through technology, marketing and throughout the life of a service are currently changing the whole way that customers are contacted, served and retained. Service design can be applied at all these stages where customers interface with the organization to improve their satisfaction and company profits. By putting customer convenience and satisfaction at the forefront of Total Design, designers are forced to think (and then design) the customer experience.

As production and consumption occur at the same time in a service (Kelley et al. 1990), customers cannot fail to notice if the service has been poorly designed (Edvardsson and Olsson 1996). Of course, this relates to the physical surroundings but, increasingly, users are looking to the 'totality' of the service. What is offered must, at least, meet their perceived expectations. These customer expectations are continuing to rise. Service that was acceptable in a shop, hospital outpatients or railway station just a few years ago is now considered unacceptable. Many of these necessary and ongoing improvements (*Kaizen* – improvement is a journey, not a destination and one should endeavour to aim for 'continuous improvement') can be brought into the service through the application of good design. Quality starts with design and quality needs to be built into the design of the service provision rather than being added later. The application of tools such as SERVQUAL (cf. Parasuraman et al. 1988, 1994; Mills 1990; Mattsson 1994) is an attempt to match (or exceed) service provision with customer expectations. Companies were called on to go beyond just satisfying customers and instead seek to 'delight' them (Deming 1986) by going beyond what customers expect if they wish to retain them in increasingly competitive markets.

### Importance to Public Services

Often, public services are serving large numbers of people and must operate within tight financial constraints and budgets. Although it may not be possible to increase the finance available, especially in recessionary times, through service design it is often possible to make the finance available stretch further. Increasingly, local authorities are applying service design principles and employing service design consultancies to improve their planning especially at the early stages of projects. Such projects include the more effective provision of care for the elderly and in planning urban developments.

The Design Council in London has a programme called Public Services by Design (2008), championed by the Chairman, Sir Michael Bichard. He believes there are three key ways in which design can make public services better.

- It can redesign the way we deliver services allowing us to 'build or reshape our services around citizens, around clients, around customers'.
- It can help the development of better policy 'ensuring that ideas are tested before having scarce resources invested in them on a national basis'.
- 'Design can help us in the public services to be more innovative. We need to be conscious that today's problems are just not going to be addressed by yesterday's ideas and yesterday's solutions . . . we need a whole new approach to policy over the 10 years.'

### Importance to Charities

Most charities both raise and spend money but receive money from and dispense money to quite distinctive groups. The needs of both of these are likely to be very different. When designing charities it is necessary to balance the funds raised from one group with the commitment (spending) to the other (Hollins and Hollins 1991). In practice this requires a service-design process for raising the finance and a design process for spending the finance. These two processes run in parallel (concurrency) and are highly iterative and dependent on each other.

### Importance to Manufacturing Organizations

Twenty-five per cent of people employed in manufacturing organizations are actually involved in the service side. As companies involved in manufacturing look at the whole life costs and benefits of their offering they are now applying design management techniques to the post-manufacturing stages of their products where service aspects are more prevalent and profitable.

Furthermore, companies that once manufactured now do (effectively) 'badge engineering' where they tie up with other manufacturing organizations and then market this output through their marketing channels. This is common with electronic and car companies where the service side then becomes the major part of their offering. Although profitable and relatively easy in the short term, it can be a very risky strategy for the longer period when the actual manufacturers begin to develop their own brand and eliminate the 'middle man'. It is difficult to protect a set of service techniques from being copied by the competition.

## THE APPROACH

In the early days of understanding service design, the process was viewed as a variation of engineering design, as if services are just another type of product. This approach could be justified in that the well developed and understood approaches used when designing artefacts were applicable, especially in the early stages of the design process. In services, as in manufacturing, it is necessary to understand the market – the main reason for product failure is 'not understanding the customer requirements' (Cooper 1988 and still true). Evidence shows that this main reason for product failure is the same in services as it is with manufactured products and the other main reasons for failure are similar to each – technical failure, the service doesn't work, and financial failure the service cannot be developed within the required budget.

It is certainly necessary to write a full specification that bounds the subsequent stages of the process (although the emphasis in the various elements may be different). The next main stage is the concept stage, where various options are delivered, for example, the idea could be that one may want to deliver parcels. In the concept stage one considers all the ways that this can be done – for example, by helicopter, catapult cannon.

Like engineering design, the next stage is to identify the detailed design so that it can be produced. Generally, the subsequent stages in service design and product design differ. Then services tend to be implemented, whereas artifacts need to be manufactured and sold.

So it is quite acceptable and understandable that engineering designers could divert their skills towards designing services as so many of their skills will apply. It is also acceptable as the areas where they do apply are in the early front end of design, when 80 per cent of the management decisions are taken and 80 per cent of the funds committed, but only 15 per cent of the actual expenditure is made (Figure 14.1) (Berliner and Brimstone 1988).

**Relatively early in the design activity the decisions taken will commit
the operation to costs which will be incurred later**

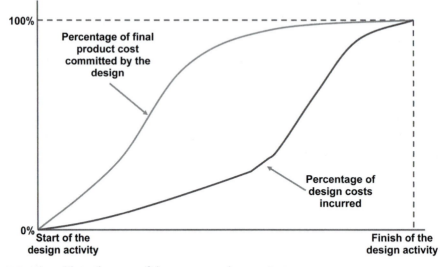

FIGURE 14.1   The early stages of the process are the most important.


## RESEARCH INTO THE UNDERSTANDING OF SERVICE DESIGN
## BY PRACTITIONERS

Research was undertaken in 2003 (Hollins et al. 2003) who investigated how managers developing new services used good practice of service design management. The
purpose of the research was to identify whether service design was, by then, effective
in organizations and which aspects of service design management were most used.
Most (68 per cent) stated that they were actively involved in the development of new
services for their organizations.


### Research Findings

The overwhelming finding was that service design was still not widely managed in an
organized manner. As a result, such companies are not in control of their design function. British Standards on design management (the 7000 series) were not being used.
Only one-third of the companies questioned had a product strategy document and
only one-fifth had a written process for the delivery of new services. If the management
of new services is the organization of the process for developing these services – then
why no process? How do the managers given this responsibility know how to start?

Forty-eight per cent did no research for new services prior to their development.
This research would typically be done near the start – at the low-cost end of the

process. It is here that the most obvious failures should be detected and eliminated and better ideas more fully thought out whilst still 'on paper'. This will result in a more efficient use of the resources available within tight constraints. If no market research is undertaken then the designers are working in the dark when it comes to satisfying the potential customer's wants and needs. Sadly, it is not until after these services have had all the development costs pumped into them that they will be shown up as failures when in the full view of the market and competitors.

When seeking new ideas several stated that they look at their competition or the market leaders. This 'me-too' attitude is generally accepted as being an unsuccessful route to new product success. One cannot overtake the competition by just copying what it does. Quite a few companies seek ideas only from inside their organizations, such as ideas from directors, senior managers and even suggestion boxes. This would be all right as long as it is backed up by some market research to show that there are customers out there who want the benefits that these new ideas may provide. 'Customer complaints' was quoted as a source of new ideas. It is wondered if those customers were still be around for these company's products when they have sorted out these complaints. Design specifications (the controlling documents) tended not to be written. Forty-eight per cent of the respondents had not seen a specification for the development of a new service in the past seven years (remembering that most were actually involved in the development of new services). A guide for specifications in the service sector is currently available – BS 7373 part 3 2004. One of the main reasons for new products failing is that the full set of requirements is not considered at the early stage of the process (Hollins and Pugh 1989). It is here, right near the start, that all the compromises (new products are all about compromises) need to be resolved. For example, can the company actually produce the new service? Can they market it? Can they afford to develop it? Can it be made to work (Hollins and Hollins 1999)?

All of these decisions need to be confronted in the early stages. It is this low-cost, front end of the process where most product and service failures are rooted and yet this research shows that it is here that the service companies are most inadequate.

What the research did show was that the few companies that did appear to be effective (about 16 per cent) were very good and that most of these generated greater than 30 per cent of their turnover from services developed in the previous three years. This suggests that the few companies that actively adopt good design practice are very likely to gain quick and reasonable benefits over their more ignorant competitors. Those companies that used more of the tools and processes in their design management were faster growing than their competitors. A follow up study a couple of years later showed only a slight improvement (Hollins and Shinkins 2006).

In 2005, the role of design in manufacturing and service firms was compared in a survey of 1,500 firms for the Design Council (Tether 2008). In this study,

manufacturers, finance and business services and consumer services were asked about
the role of design in their organizations. Questions sought to answer, for example:

- whether it was integral to the firms operation;
- if it had a significant role to play;
- a limited role or no role at all.

The results were not encouraging and confirmed the lack of design awareness found
in the earlier study. This is shown on Figure 14.2.

In the same study (Tether 2008) the same cohort were asked how design was used
in firm's new product and service development. Manufacturers, finance and business
services and consumer services were asked the role of design leadership. They were
asked:

- if the design manager/design team leads and guides whole process.
- if designers are used in all stages.
- if designers are used in some specific stages.
- if designers are not included in the process.

Figure 14.3 shows the rather depressing result. It is clear that design management
has a long way to go before it is used regularly in the service sector.

This study showed that the lack of design understanding was partially based on
the kind of leadership and involvement of designers in services when compared with
manufacturing.

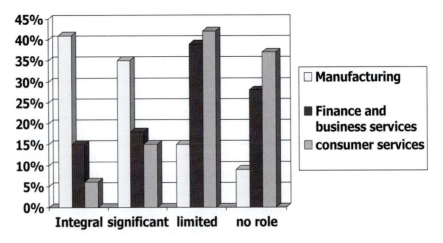

FIGURE 14.2   The role of design. © Tether, 2008.

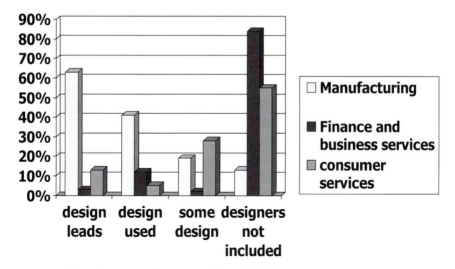

FIGURE 14.3   How design is used in new products. © Tether, 2008.

### Terminology

In another piece of research undertaken by the author (Hollins and Shinkins 2006), various managers working in the service sector were asked if they understood the meaning of various words used by designers. The list is shown in Table 14.1. The results show that the words in the table were not understood by a sizable number of the research cohort.

   If business people cannot understand the words then how can they understand the designer or the process? The words that we designers use may give us a certain degree the power of an expert to confuse the 'uninitiated' but more likely will drive a wedge between ourselves as designers and other business people. Part of this problem may be addressed by the British standard on terminology used in design management, which defines the words that we use (British Standards Institution 2008b). It is hoped that this glossary will be widely adopted.

### Blueprinting

One aspect of design that seems to be far more successful in service design is blueprinting. Blueprinting, as a method for designing services, was first proposed by Shostack, (1984) and later developed by De Brentani (1991), Randall (1993) and others but not widely adopted at the time. Recently, and increasingly, blueprints have been used much more widely in design and quality improvement in services (for example, see Meyer and De Tore 2001). The technique seemed to lie dormant until used by service designers as a fundamental building block by companies such as Live/Work; who used it to develop service prototypes. It helps to identify the 'Touchpoints' and 'bottlenecks', which can then be improved through applying service design techniques.

**TABLE 14.1: Words used by designers not understood by many business people**

Blueprinting. Brand Identity. Brand Architecture. Brand Value. Buy-in. Concept. Core Service. Corporate Identity. Critical Drivers. Data Mining. Design. Design Management. Design Process. Experience prototype. Innovation. Interdisciplinary Team. Internal Customers. Iteration. Launch Champions. Moments of Truth. Pilot Test. Points of Integration. Product Champion. Project Configuration. Robust Design. Scenario. Service. Stage Gateway. Stakeholder. Tangible Evidence. Touch-points. Trigger. Value System.

Writers and researchers have not fully realized that design models – the process by which product and service design can be managed – can be linked with the use of service blueprints. This allows certain existing process design tools and techniques to be used to greatly advance the design, development and improvement of services. It will now be shown that blueprinting fits neatly into the stages of the service design process.

### The Link between Design Models and Blueprints

Design models are normally presented as a vertical sequence whereas the service blueprint is normally presented horizontally so where does the blueprint cross the design model? In a simple blueprint that covers just the progression of the customer, and parallel activities that occur as the customer passes through the process, the blueprint can almost be considered as a production process. This being the case, much of the production process design techniques can be applied to improve the blueprint. After all, it was the linking of product and process design in manufactured products in Japan that was one of the keys to improving quality and lowering costs that first enabled them to win world markets.

From this it can be said that a service blueprint is really a demonstration of service process design and this fixes its position firmly in the total design process from the detail stage through to implementation and subsequent product and process improvement (although it may help to clarify things at the specification stage of design).

## INNOVATION

Innovation, being an important subset of the design process, is poorly applied in the service sector (Hollins et al. 2003). Innovation can occur in all stages of the whole life of a product, especially (and increasingly) at the service end when customers are more likely to be directly involved with the delivery of the service. Innovation is generally easier with services as there is less of an existing infrastructure to be replaced by the new. As such, customers more readily accept changes brought about through innovation. This can lead to 'Serial Innovation' (BS 7001 2008) as observed by Alan Topalian where innovation in, say, the product can subsequently lead to innovations in the service side, which can then result in other innovations along the value chain. For an example, which may not be realistic but demonstrates the principle, pizza would be less successful if it were not delivered – the first serial innovation. Payment can now be made by phone – another serial innovation. So there is no reason for the person delivering the pizza to knock on the door if the pizza would fit through the letter box. So why not make pizza rectangular so it would fit through the letter box?

## HOW SERVICE DESIGN CAN BE IMPROVED

A blueprint shows the stages that customers pass through when they use the service and the parallel activities that occur at the same time. If sufficiently well constructed, it will also identify bottlenecks that may slow the process, or reduce its quality, which can lead to redesigns and greater efficiencies and improved performance. It will also show who will be involved in the delivery of the service (and therefore potential skill shortages). From this can be identified the timescale for the service delivery and the potential costs. If other activities occur in parallel it will also show the critical path upon which any delay will result in a delay in the total process.

Blueprints should always be presented with a base of time and this is essential for determining the parallel stages, the concurrencies, the total time and, therefore, cost. Blueprinting can be used to apply more of the tools of TQM to further enhance the quality of the process such as benchmarking, production planning techniques and process layout. It can be used to identify critical paths where an improvement on that critical path can improve the whole service activity. Blueprints are also used where 'mystery shoppers' can identify service delivery problems. Other production tools that can be applied in this new context will include Value Analysis, and Line of Balance, queuing techniques and JIT.

As a more recent development, blueprints can be used to plan, predict and eliminate problems in service recovery. It has been shown that customers have more empathy with an organization if, when something goes wrong with the service, the organization solves these problems quickly and efficiently more than if the problem didn't occur in the first place. This is what practitioners of service recovery endeavour to do.

## AND WHAT IS NEXT?

The increase in the power and availability of information technology and ease of communication and other technology will make transactions faster (especially across borders) and more repeatable making it easier to control and increase the quality of the service. But the standardization brought about by the application of technology could reduce the personal interaction and thus the 'individual' nature of services. On the other hand, further 'discrete' applications of advanced technology can allow the benefits of apparent 'individual' and culturally focused services combined with the benefits that can be achieved with repeatability and 'selective' standardization. It could also allow the service providers to spend more time with customers.

A serious prospect is that as the BRIC countries become richer, they will design their own products and then their own services. They will then be better placed to take a stake in our Western business. This should ensure employment for service designers into the future but perhaps for Chinese employers.

There is currently an explosion in research and publications relating to service design that hopefully will lead to improvements in practice within organizations although evidence from the recent past has shown a slow uptake of such new techniques.

The most recent developments are that more specialists are coming together to combine their knowledge and skills to make service design much more inter and multidisciplinary. These include interaction designers, branding experts, psychologists, ethnographers, experience designers and interface designers amongst others. This has resulted in new service design tools and methods that are used (mainly by service design consultancies) to increase the accuracy resulting from the service design process. Many new user research techniques will be used to observe, probe and explore and understand the nature of potential customers and stakeholders through, for example, scenarios and personas and hence inform the designs needed to satisfy these.

Then during the later stages of the service development, potential users are increasingly engaged to test out the idea. This is research for validation of the design thinking. This kind of research requires prototypes of different kinds including acting out and fine tuning parts of the blueprint. Increasingly, this prelaunch research reduces the failure rate of new services.

Another recent focus has been more systematic post launch research, which leads to early insights, changes in design specifications and further advances in the service redesign.

It is anticipated that these new methods and tools will eventually filter down to practitioners outside of consultancy companies to improve general business design practice. And would it be too much to hope that we will hone in on one common design language to describe what we are doing to each other and also to those in business? Perhaps British Standards Institution (2008b) would be a good place to start.

## REFERENCES

Anderson, N. (1975) *From Concept to Production: A Management Approach*. London: Taylor & Francis.

Andreasen, M. (1994) *Design Model*. WDK Workshop on Evaluation and Decision in Design. Technical University of Denmark, Lingby, May 2-3 1994.

Berliner, C. and Brimstone, A. (1988) *Cost Management for Today's Advanced Manufacturing: The CAM – 1 Conceptual Design*. Boston, MA: The Harvard Business Press.

British Standards Institution (1989) *BS 7000 Guide to Managing Product Design*. London: British Standards Institution.

British Standards Institution (1994) *BS 7000-3 Guide to Managing Service Design*. London: British Standards Institution.

British Standards Institution (2004) *BS 7373-3 Guide to Specifications in the Service Sector*. London: British Standards Institution.

British Standards Institution (2008a) *BS 7000-1 Guide to Managing Innovation*. London: British Standards Institution.

British Standards Institution (2008b) *BS 7000-10 Design Management Terminology*. London: British Standards Institution.

Cooper, R. (1988) *Winning at New Products*. London: Kogan Page.

De Brentani, U. (1991) 'Success Factors in Developing New Business Services', *European Journal of Marketing* 2(2): 33–59.

Deming, W. (1986) *Out of the Crisis*. 2 edn. Cambridge, MA: MIT Press.

Heapy, J. and Parker, S. (2006) *The Journey to the Interface*. London: Demos.

Edvardsson, B. and Olsson, J. (1996) 'Key Concepts for New Service Development', *Service Industries Journal* 16(2): 140–64.

Hollins, B. and Hollins, G. (1999) *Over the Horizon. Planning Products Today for Success Tomorrow*. Oxford: John Wiley & Sons, Ltd.

Hollins, W. and Hollins, G. (1991) *Total Design. Managing the Design Process in the Service Sector*. London: Pitman.

Hollins, W. and Pugh, S. (1989) *Successful Product Design*. Oxford: Butterworth.

Hollins, W., Blackman, C. and Shinkins, S. (2003) 'Design and its Management in the Service Sector – Updating the Standard', Fifth European Academy of Design Conference, Barcelona, 28–30 April.

Hollins, W. and Shinkins, S. (2006) *Managing Service Operations: Design and Implementation*. London: Sage.

Kelley, S., Donnelly, Jr., J. and Skinner, S. (1990) 'Customer Participation in Service Production and Delivery', *Journal of Retailing* 66(3): 315–25.

Mager, B. (2003) *Service Design, a Review*. Cologne: Service Design Network.

Mattsson, J. (1994) 'Improving Service Quality in Person-to-Person Encounters: Integrating Findings from a Multidisciplinary Review', *The Service Industries Journal* 14(1): 45–61.

Meyer, M. and De Tore, A. (2001) 'Creating a Platform-based Approach for Developing New Services', *The Journal of Product Innovation Management* 18: 188–204.

Mills, P. (1990) 'On the Quality of Services in Encounters: an Agency Perspective', *Journal of Business Research* 20: 31–41.

Moritz, S. 2005. *Service Design. A Practical Access to an Evolving Field*. London: KSID.

Parasuraman, A., Zeithaml, V. and Berry, L. (1988) 'SERVQUAL: A Multi-Item Scale Measuring Consumer Perceptions of Service Quality', *Journal of Retailing* 64: 12–37.

Parasuraman, A., Zeithaml, V. and Berry, L. (1994) 'Reassessment of Expectations as a Comparison Standard on Measuring Service Quality: Implications for Further Research', *Journal of Marketing* 58(1): 111–24.

Pugh, S. (1982) The Design Activity Model. Loughborough: Loughborough University.

Randall, L. (1993) 'Customer Service Problems: Their Detection and Prevention', Service Superiority Conference, Warwick Business School, 25–26 May 1993.

Sakao, T. and Shimomura Y. (2004) 'A Method and a Computerized Tool for Service Design',
    Design 2004 Conference, Cavtat, 17–20 May, 2004: 497–502.
Shostack, G. (1984) 'Designing Services that Deliver', *Harvard Business News* (January–
    February): 133–9.
Tether, B. (2008) 'Service Design: Time to Bring in the Professionals?' In L. Kimbell and V. P.
    Seidel, *Designing for Services – Multidisciplinary Perspectives: Proceedings from the Explor-
    atory Project on Designing for Services in Science and Technology-based Enterprises. Said Busi-
    ness School.* Oxford: University of Oxford.
Voss, C. and Zomerdijk, L. (2007) 'Innovation in Experiential Services – An Empirical View',
    in DTI (ed). *Innovation in Services.* London: DTI, pp. 97–134.
Walker, D. (1989) *Design or Decline.* Video. Design Management Series, Open University.

# 15

# Successful Design Management in Small and Medium-sized Businesses

## DAVIDE RAVASI AND ILEANA STIGLIANI

### HOW DESIGNERS CAN IMPROVE THE PERFORMANCE OF SMALL AND MEDIUM-SIZED ENTERPRISES

Past studies have documented that the effective use of design and designers can positively influence the business performance and competitiveness of small and medium-sized enterprises (SMEs) (e.g. Black and Baker 1987; Walsh et al. 1992; Bruce et al. 1995; Hertenstein et al. 2005).

Although it is not unheard of, only a few entrepreneurs possess a design background. James Dyson and Giulio Cappellini, founders of their namesake companies, are rare exceptions. Internal design centres, where a team of designers works on company projects on a permanent basis, are equally rare in SMEs, as the small size of the business and, often, the relatively low rate of introduction of new products make it difficult to cover the associated fixed costs adequately. Therefore, should they want to entrust the industrial design of their products to a professional rather than leaving them in the care of company technicians or engineers, SMEs most often tend to turn to the services of external design consultancies.

Systematic collaboration with design consultancies helps SMEs overcome some limits associated to their small size, narrow scope and limited available competencies (Bruce et al. 1999). Typically, entrepreneurs and small business managers possess a good knowledge of their market and their customers and often a certain degree of technical competence in their field. The development of new entrepreneurial ideas, however, requires contributions from a range of actors, whose knowledge and skills are complementary to those of the entrepreneurs and small business managers (to simplify, we will refer from now on to both as entrepreneurs) and must be obtained from industrial, commercial and research partners, consultants and designers (Birley 1985). Research has shown, however, that, for an SME, the benefits of turning to the services of a design consultant may go beyond the mere enhancement of product aesthetics or user interface. In fact, we find that design consultants can act in one of the following three capacities. As 'knowledge brokers', Designers can help SMEs access useful technologies available in unrelated industries. As 'Brokers of languages', designers can connect SMEs with evolving artistic and socio-cultural trends. Finally, by bringing an outsider's view, they may encourage them to challenge current industry conventions.

## Design Consultants as Knowledge Brokers

As knowledge brokers, design consultants may help SMEs envision and develop new products based on the creative combination of technologies available in different industrial domains. By working with many clients in various sectors, design consultancies come in contact with different ideas and technological solutions that can be fruitfully transferred and applied across industrial boundaries to introduce product innovation. In doing so, design consultancies act as 'knowledge brokers' (Hargadon and Sutton 2000), helping small businesses to compensate their narrow scope and to tap into pools of ideas and technologies that would be otherwise unreachable.

## Design Consultants as Language Brokers

Research on Italian design-based companies (Verganti 2003) suggests that more aesthetically oriented designers may act as 'language brokers' to help SMEs embed their products within the broader design discourse and cultural context. The notion of a knowledge broker refers to the more technically oriented, functional aspects of industrial design whereas the notion of a language broker points at the semantic implications of product form. Consumers value products not only for what they do, but also for what they mean and in particular for how these products can be used to express appealing social and personal identities reflecting current socio-cultural models (see Ravasi and Rindova 2008). Decoding changing cultural values and symbolic needs, however, is not a common expertise and SMEs typically lack the resources required to set up internal centres dedicated to systematic research or to support large-scale studies on socio-cultural trends. Italian producers such as *Kartell* (plastic furniture), *Alessi* (kitchenware) and *Zanotta* (home furnishing) owe their competitive advantage to their capacity to effectively partner with inspired designers who assist these companies in endowing their products with conceptual and stylistic features that are simultaneously original and in touch with current socio-cultural trends.

*Design Consultants as Challengers of Industry Conventions*

Finally, design consultants may help entrepreneurs reconsider the tacit assumptions underpinning their product policies and business models. Research shows that companies' product and market strategies tend to reflect widespread conventions in their industry about issues such as how to design products and how to serve customers (Porac et al. 1989). Scholars refer to these conventional beliefs as 'industry recipes' (Spender 1989) and observe how difficult it is for competitors to even envision new strategies that depart from these conventions. Design consultancies, however, often enjoy the privilege of approaching a project as an outsider. Without preconceived notions about the product or its context, they are capable of taking a fresh look at the market, the product, the customers and other relevant elements to develop a new and possibly different understanding. By refusing to take the assumptions the client embodied in the project brief as a given and by bringing in a fresh perspective, therefore, design consultancies may encourage small businesses to challenge entrenched beliefs about the appropriate way of designing, producing and communicating products, opening the way for the introduction of substantial innovations.

Despite the considerable potential benefits, however, many small and medium enterprises still seem reluctant to turn to the services of design consultancies – and when they do, the outcome of the collaboration is often unsuccessful (Bruce et al. 1999). Perhaps for this reason, national and international agencies and institutions, such as the British Design Council or the European Commission, have recently engaged in activities aimed at promoting the use of design among small and medium business enterprises. One of the aims is to educate entrepreneurs about how to make the most of the collaboration.

In this chapter, building on previous research and on our own expertise, we address some of the key issues affecting the collaboration between design and business. We adopt the perspective of designers and we propose some guidelines for entrepreneurs and small business managers for how to improve the outcome of joint design projects.

## THE DESIGNER'S PERSPECTIVE: WHAT CAN GO WRONG WHEN DESIGNERS WORK WITH SMALL AND MEDIUM-SIZED ENTERPRISES

Past research has pointed out a number of problems that causes the under-exploitation of the opportunities offered by design. Four of these seem to be particularly true for SMEs (Kotler and Rath 1984; Walker 1990; Bruce et al. 1995): (a) low design literacy; (b) design decisions driven by cost concerns; (c) behaviour and business patterns anchored in old traditions; and (d) high risk aversion.

### Design Illiteracy of Entrepreneurs/Small Business Managers

While the number of managers who have a background in design or the arts is growing, in SMEs there are still only a handful of managers who can combine their entrepreneurial skills with design literacy. Like most of us, managers in small businesses, too, have aesthetic tastes and preferences as well as opinions about functional and style-related issues. A lack of understanding of design and design-related fields however, often hampers their ability properly to appreciate the quality and the potential of the proposals they receive from design consultancies. Most small business managers and entrepreneurs tend to have either a commercial or a technical background. Their evaluations are often based on general (and personal) aesthetic assessments – in terms of like/dislike – and they tend to underestimate the potential of new forms and languages that break with established conventions.

### Cost Concerns

Limited budgets are characteristic of SMEs. The net income of small business owners is directly and negatively affected by cost increases. Many small business managers and entrepreneurs, therefore, seem to refrain from using the services of a design consultancy under the assumption that design (both the fees of the consultant and the costs of the implementation of the projects) is excessively expensive (Bruce et al. 1999). Large firms can easily afford the cost of periodic explorative projects and are only marginally affected by the occasional failure of a project. The same may not be true, however, for the limited funds available to SMEs, increasing their reluctance to squander their resources in projects of unclear return.

### Behaviour Anchored in Traditions

Innovation and creativity lie at the heart of design. When engaging in a new project, industrial designers are generally inclined to develop proposals that challenge product concepts, functions and languages. However, while some degree of rule breaking and innovation is at the heart of entrepreneurship, as we have mentioned earlier, it is common for managers of small businesses to become embedded in a web of taken-for-granted conceptions of the product and the market and conventions about the 'right' way of competing. Long tenure and experience at the helm of their companies is likely to increase both their confidence and their inability to question these entrenched beliefs in the face of designers' proposals that radically depart from industry conventions.

### Risk Aversion

Finally, designers' proposals frequently involve a certain degree of risk, to the extent that they suggest the departure from consolidated – but relatively safe – ways of operating, they explore uncharted segments of the market or they require additional specific investments for their realization. Common wisdom points at small businesses as more flexible and less risk-averse than their large counterparts. Common wisdom, however, may be overemphasizing the representativeness of young entrepreneurial firms in hi-tech sectors, while neglecting the many more small enterprises in less dynamic and more mature businesses.

In an effort to explore what prevents SMEs from utilizing design to their full advantage, we began by looking into the experiences of design consultancies with their SME clients. We surveyed more than one hundred industrial design consultancies and inquired into the problems they encountered when they worked with different types of clients. The clients of the consultancies we surveyed included large marketing-oriented multinationals, SMEs, design-oriented enterprises, high-tech enterprises, individual entrepreneurs and inventors (Ravasi et al. 2008).[1] Of all these company types, small businesses were perceived to be the most problematic: on average, our respondents reported that, in the last three years, almost one-third of their projects with this type of client had failed.

Our results, summarized in Figure 15.1, suggest that, consistent with past research, designers indicated risk aversion and lack of design literacy among the main causes of failure of collaborations between design consultancies and small business. However, these responses should be taken with a pinch of salt, as the designers who completed the survey might have overstated – in good faith – the deficiencies and responsibilities of small business managers. It is interesting, however, that the design consultants did not perceive these two aspects to be the most critical. Also notable: while small businesses, on average, did seem to be more cost conscious than other types of clients, this preoccupation was not ranked among the most frequent causes of failure. In fact, our respondents reported that the decisive factors for success, in their experience, were their client's skills for managing the collaboration process throughout the various project stages and, in particular, their capacity to produce an effective brief and to secure internal support for the project. In other words, many projects failed due to their client's lack of design management skills (see also Bruce et al. 1995).

## Lack of Clarity of the Brief

The majority of the design consultants we interviewed reported that the most frequent problems undermining the success of their collaboration with SMEs was related to fundamental ambiguities in the initial instructions they had received, frequently, but not always, in the form of a brief. As a knowledgeable informant reported, most small business managers tend to have a look at their portfolio, identify an apparent gap and ask a designer to 'do a nice chair'. Lack of clarity about the fundamental goals and boundaries of the project ('what did the client really want?') and absence of adequate commercial information about the client (i.e. current and desired positioning, price levels, distribution, brand values) often result in proposals that are considered

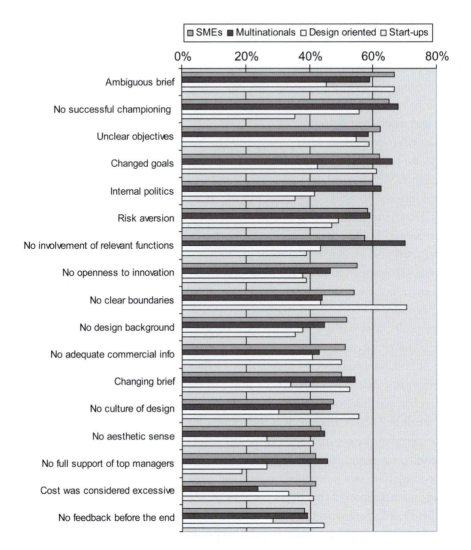

FIGURE 15.1   Reasons for the failure of projects in small and medium businesses. © Ravasi and Stigliani, 2011.

incoherent with the – poorly articulated – expectations of their counterparts. While, to some extent, these problems seem to affect all types of clients, they were reported more frequently with respect to SMEs. This possibly points to a general lack of experience, familiarity and skills in the management of design–business collaborations.

### Lack of Power and Internal Support

Finally, many of our respondents noted that their counterparts seemed incapable of securing the needed approval of top management for the continuation and realization of the project. In a small and medium enterprise, especially if privately owned, power tends to be highly concentrated at the top. Yet, only rarely are design-business collaborations handled directly by the top – a feature that characterizes, for instance, the few, widely popular Italian 'design factories'. In companies like Kartell, Alessi or Artemide, top managers personally engage in design management. In most small businesses, design consultants interact with marketing managers or engineers, who will eventually need the support of more senior managers or of the small business owners themselves to make decisions. The inability to successfully champion the project with the top seemed to prove fatal for many projects.

## TOWARDS SUCCESSFUL DESIGN–BUSINESS COLLABORATIONS

The results outlined in the previous paragraphs point to some areas of improvement in the relationships between design consultants and SMEs. We will use the remainder of this chapter to propose a set of guidelines that are meant to improve design management skills in two key areas: the briefing process and the management of the relationship. We build our recommendations on previous research as well as our own experiences. We will begin with the briefing process and then turn to suggestions for how to manage relationships with a small business's top management.

### The Briefing Process

A design brief is a written document aimed at thoroughly explaining the problem to be addressed by designers. Ideally, the design brief should be prepared after a design consultancy has been selected for the job, possibly – but not exclusively – on the basis of a less detailed 'Request For Proposal'. The purpose of a succinct Request For Proposal is essentially to invite preliminary ideas (although at this stage designers are reluctant to disclose their best ideas) and screen potential consultants. Instead, a design brief should be the outcome of a jointly held reflection and discussion, carried out by representatives from the business client and the design consultant and resulting in a comprehensive description of the goals, scope, constraints and timeline of a project (see Phillips 2004). In reality, however, briefs are often prepared unilaterally by the client and handed over to designers. At other times, projects are launched even in the absence of a formal brief and based only on verbal agreements between the parties. Even when a brief exists, it is often ambiguous or incomplete. Indeed, our research indicates that an ambiguous brief (57 per cent), unclear objectives (56 per cent), lack of adequate commercial information (44 per cent) and changes in the mandate embodied in the brief itself (46 per cent) are among the most frequent reasons that lead to the failure of a project. In fact, almost half of the surveyed consultancies declared how they often received no brief at all. Some designers we interviewed told us how, after a preliminary meeting, they often have to prepare some sort of brief themselves to be submitted to their clients for approval, to try to compensate for the uncertainty and ambiguity surrounding the mandate.

On average, designers working with SMEs seemed to encounter more frequent problems associated with the briefing phase than for other types of clients. These findings seem to reflect a relative lack of familiarity of small business managers with good briefing practices – which, in turn, has a negative impact on collaboration, at least according to designers. Therefore, learning to prepare an adequate brief, containing all the relevant information for the project, seems crucial to make effective use of the services of design consultants. While an extensive discussion of good briefing practices falls outside the scope of this chapter, what follows are some general guidelines for small business managers (see also Phillips 2004).

First, a good design brief should be focused and concrete. Lack of familiarity may bring entrepreneurs to view design as pure 'art' or 'magic', rather than a goal-directed problem-solving activity. A designer we interviewed during our research, for instance, lamented how often entrepreneurs would not articulate clear goals, but vague 'dreams' – placing the burden on its consultants to interpret these dream and make them come true. Unfortunately, a vague mandate tends to reduce the chances that a designer's proposal will make her client happy. In our research, we have come across many examples of frustrated designers who saw the results of their creative efforts turned down by the client because 'it is not what we wanted'. Often, however, what the client wants is not really clear in the first place.

Entrepreneurs and small business managers are usually very busy people. However, when it comes to design, it is crucial that they will devote proper time and attention to the clarification of the goals of the project and the constraints to be considered (time, budget, available technologies, channels and so forth) – especially if the collaboration with the consultant is new. There are many services a design consultant can provide – from a simple restyling of product lines to the development of innovative product concepts – and many ways in which their services can improve business performance – from simple revamping of existing lines to spur new sales growth, to line extension or brand repositioning. Drafting a comprehensive brief will offer entrepreneurs an opportunity to reflect on the ultimate business outcome associated to the project, check whether these expectations are realistic and establish clear reference for both parties for the evaluation of the outcome.

The actual informational content of the design brief will then depend on the specificities of the project. In general its content should cover the following areas: current product strategies, target audience/users, brand values and positioning, main competitors, distribution channels and other elements of the marketing mix. Our research indicates that lack of adequacy of this type of information is rarely an issue when working with large companies, where designers' counterparts are professional marketers, operating within formalized policies and supported by advanced methods for collecting and elaborating market data. In SMEs, however, this is not always the case. Relatively informal and intuitive marketing policies may combine with lack of familiarity with design collaboration to increase the ambiguity and uncertainty surrounding the project. In this respect, engaging in substantial design collaborations may provide an additional benefit to SMEs by encouraging entrepreneurs to clarify their commercial strategy and to upgrade its marketing practices, by reflect explicitly about their brand positioning, target audience, etc.

In summary, paying close attention to the briefing process is likely to bring several advantages for both parties. First of all, it will facilitate the transfer of project-related information, eliminating ambiguity regarding expected outcomes, contextual constraints and so forth and reducing the likelihood of misunderstandings that may harm the effectiveness of the solutions later proposed by the designers. More relevant 'knowledge' is available within a firm than is often conveyed to the consultant in the early stages of a collaboration. Setting up a proper briefing process leading to the emergence of all the important information related to a project will help overcome this asymmetry and allow designers to have a more thorough understanding of the design problem that needs to be solved. In addition, working side-by-side with designers on the definition of the brief can facilitate the establishment and/or the reinforcement of mutual trust and understanding, laying down the ground for successful multi-project collaboration. In turn, a long-standing collaboration is likely to simplify the briefing process, as the designer has gained familiarity with the firm, its products, market and so on and developed a good working relation with a firm's staff.

Finally, but no less important, a well prepared brief will provide a solid reference for the assessment of designers' proposals. In our research, we observed how many small business managers who have little familiarity with design tend to evaluate proposals on the basis of personal tastes (and can be quite opinionated about it). Proposals may be dismissed simply because managers 'don't like them' – unless, of course, the proposal comes from a recognized *maestro*, in which case any product of the designers' superior talent is gratefully accepted without question. In neither case does the coherence of the proposal with a firm's goals, strategy and positioning seem to be considered. A certain degree of personal intuition, instinct and trust cannot – and probably should not – be removed from the evaluation of design proposals. However, clarifying the expected business outcomes and the relevant contextual information should facilitate a more 'objective' evaluation of the work of designer, whose proposals might be assessed based on their coherence with the agreed upon brief, rather than on the more ambiguous realm of personal tastes.

## Managing the Collaboration

Of course, a well prepared brief is not the only ingredient for a successful collaboration between design consultancies and SMEs. Even before sitting down to discuss the design brief, the careful selection of a consultant is required. Consistent with previous findings (Bruce et al. 1999), our research shows how designers tend to be selected by SMEs based on previous collaborations, the observation of past works and the general reputation of the consultant. Recent media coverage of the designers' work may be decisive, if the goal of the project is essentially to associate the firm with a prestigious name. When it comes to a designer's skills, lack of familiarity with design may bring entrepreneurs to assume that the main difference between designers is whether they are 'good' or 'bad'; whether their creations are more or less creative, artful or simply beautiful. In fact, designers tend to bring distinctive styles, philosophies and attitudes towards design. Some may be more ironic, provocative or colourful whereas others may be more sober, minimalist and essential. Some may emphasize ergonomics and user interaction, while others may pay more attention to environmental concerns. Some designers whom we interviewed remarked that, while as consultants they obviously make an effort to develop solutions that are appropriate to their client's context or the target audience, they tend to naturally design for 'people like themselves' with their own needs, problems, tastes and lifestyles. This is the audience they know the best and for whom they are likely to produce their most effective ideas. It is therefore important that, while evaluating a prospective consultant, entrepreneurs pay particular attention to assess, not only the invited proposals but more generally the whole portfolio of a designers, trying to assess whether a designers' style, skills and philosophy match the desired brand positioning, the desired level of innovation, the desired product language.

Second, entrepreneurs and small business managers, busy as they are, tend to envision design collaborations as a 'supply' relationship, in which design consultants disappear after having received a mandate, to reappear some time later with a final set of proposals. Our research suggests that when collaborations are structured in this way, the final output is more likely to disappoint the client. Design collaborations, instead, should be conceived as 'partnerships' (Phillips 2004), where both client and consultant take responsibility for the process and the final results. Our research shows that the most successful design consultants tend to interact frequently with their clients, periodically touching base with them and asking for feedback on emerging solutions, in order to verify the coherence of these proposals with the goals and the expectations of their clients. While this approach will obviously demand more time and attention from small business managers, frequent interaction with the client will help bring out information and assumptions that might not have been made explicit in the brief, reducing the chance that an unfortunate mismatch between the designer's proposal and the client's goals, preferences and constraints to be discovered only in a final stage.

Another important issue that should receive attention is who should be involved in the collaboration on the client's side. Obviously, the answer to this question depends on the type of project and on its relevance for a firm's strategy, positioning and growth. Ideally, strategic projects aimed at expanding or revamping product lines or at supporting or repositioning a brand, should see the direct involvement of the ultimate strategist of the organization – that is its owner-top manager. In fact, direct involvement of organizational leaders in design-related activities is a distinctive traits of design-oriented firms, from relatively large corporations like Apple or Bang & Olufsen (Ravasi and Lojacono 2005), to small and medium-sized Italian 'design factories' such as Alessi, Kartell or Artemide (Verganti 2006). The involvement of top managers in design activities clearly signals to the rest of the organization that 'design matters' and it does not merely have a decorative function and ensures direct coordination between design and strategy. In addition, direct involvement of top managers lends designers credibility and motivates them, by clearly signalling that their work is taken seriously inside the company.

Designers frequently mention how important it is that their counterparts 'understand design'. In fact, the substantial involvement of an opinionated small business manager, who overestimates his aesthetic judgment and design insight, may be more detrimental than beneficial to the process. Having a design background is not a necessary requirement for managing successful design–business collaboration but it certainly seems to facilitate the dialogue. In fact, direct engagement in intense collaboration with design consultants may provide an opportunity for small business leaders to enhance their understanding of design and improve their capacity to assess and discern good design. In the meantime, an experienced designer – hired as a consultant or on a permanent basis – may act as an intermediary between the company and the 'design world' and supervise and facilitate the management of design collaborations.

Finally, our research indicates that, whereas designers acknowledge the importance of a strong support from the top, they find it very important to be able to interact directly, right from the beginning of the project, with all the functions that will be involved in its implementation (marketing, engineering, operations and so forth). Having direct access to marketers, salespeople or production technicians will help designers gather rich, first-hand information about consumers' behaviour, sales practices, production constraints and so on, directly from the most knowledgeable source and in doing so develop a better understanding of the constraints of the project. Some designers we interviewed also observed how they find it particularly inspiring when it is possible to visit the firm's premises and production facilities and to understand the culture and the specificity of the organization. Offering prospective designers such visits in an early stage of the process and inviting them to familiarize themselves with one's technologies and production facilities may also facilitate collaboration by signalling support and openness to designers' contributions.

Involving all the interested parties right from the start will provide them with an opportunity to provide indications, clarify constraints, express concerns and point out objections at an early stage. In addition to the obvious informational value, early involvement of these parties represents an opportunity to get them 'on board' early on and secure their support throughout the process and during the implementation of the approved solutions. Our informants pointed out clearly how the success – or failure – of design collaboration depends often on internal political processes that need to be properly managed. In SMEs power tends to be concentrated on top – in the hands of the owner/senior manager, whose ultimate support is crucial for a successful collaboration. It is not infrequent, however, to see powerful senior marketers or production managers, whose trust and support needs to be garnered in order for the project to run smoothly and effectively.

## CONCLUSIONS

Design and designers can enhance the competitive advantage of small and medium enterprises in many ways, even if a company's ambition is not to replicate the design-focused strategies of the likes of Alessi, Hermann Miller, Vitra or Bang & Olufsen. Considering the potential benefits, however, SMEs still seem to be reluctant to use the services of design consultants or, if they do, they are not able to take full advantage of them. In this chapter, based on past studies as well as our own research, we have outlined a number of issues that usually stand in the way of successful collaboration between design consultants and their business clients. We have then proposed a few guidelines that are meant to suggest how entrepreneurs and small business managers may increase the chances of mutually satisfactory collaborations. These guidelines require that the top managers of a SME will dedicate more time and attention to design issues than is customary. Yet, deeper involvement in these activities may provide managers with a periodic occasion to reflect on the validity and coherence of their commercial strategy and brand positioning and on new opportunities for growth associated to a rethinking of the distinctive features of their product lines, languages and communication.

## NOTE

1. Our research relied on a questionnaire submitted to the senior designers of design consultancies located in Italy and offering industrial design services. Our sample was selected among the official members of ADI, the Italian Association of Industrial Designers. Moreover, we also scanned the list of recent winners of the 'Compasso d'Oro' award – the most prestigious industrial design award in Italy. This search produced a total of 225 design consultancies. After two rounds of recall, we had received 104 valid questionnaires, with a response rate of 46.2.

## REFERENCES

Birley, S. (1985) 'The Role of Networks in the Entrepreneurial Process', *Journal of Business Venturing* 1: 107–17.

Black, C. and Baker, M. (1987) 'Success Through Design', *Design Studies* 8(4): 207–16.

Bruce, M., Cooper, R. and Vazquez, D. (1999) 'Effective Design Management for Small Businesses', *Design Studies* 20: 297–315.

Bruce, M., Potter, S. and Roy, R. (1995) 'The Risks and Rewards of Design Investment', *Journal of Marketing Management* 11: 403–17.

Cooper, R. and Press, M. (1995) *The Design Agenda*. Chichester: John Wiley & Sons, Ltd.

Hargadon, A. and Sutton, R. (2000) 'Building an Innovation Factory', *Harvard Business Review* 78(3): 157–66.

Hertenstein, J., Platt, M. and Veryzer, R. (2005) 'The Impact of Industrial Design Effectiveness on Corporate Financial Performance', *Journal of Product Innovation Management* 22(1): 3–21.

Kotler, P. and Rath, A. (1984) 'Design: A Powerful But Neglected Strategic Tool', *Journal of Business Strategy* 5: 16–21.

Phillips, P. (2004) *Creating the Perfect Design Brief: How to Manage Design for Strategic Advantage*. New York: Allworth Press.

Porac, J., Thomas, H. and Baden-Fuller, C. (1989) 'Competitive Groups as Cognitive Communities: The Case of Scottish Knitwear Manufacturers', *Journal of Management Studies* 26(4): 397–416.

Press, M. and Cooper, R. (2003) *The Design Experience: The Role of Design and Designers in the Twenty-First Century*. Aldershot: Ashgate.

Ravasi, D. and Lojacono, G. (2005) 'Managing Design and Designers for Strategic Renewal', *Long Range Planning* 38(1): 51–77.

Ravasi, D., Marcotti, A. and Stigliani, I. (2008) 'Conditions of Success and Failure in Collaborations between Business Firms and Design Consultancies: The Designers' Perspective.' DIME Working Papers on Intellectual Property Rights, No 55, www.dime-eu.org/files/active/0/WP55-IPR.pdf.

Ravasi, D. and Rindova, V. (2008) 'Symbolic Value Creation', in D. Barry and H. Hansen (eds) *Handbook of New Approaches in Management and Organization*. London: Sage, pp. 270–84.

Spender, J. (1989) *Industry Recipes: The Nature and Sources of Managerial Judgment*. Cambridge, MA: Blackwell.

Verganti, R. (2003) 'Design as Brokering of Languages: Innovation Strategies in Italian Firms', *Design Management Journal* 13(3): 34–42.

Verganti, R. (2006) 'Innovating Through Design', *Harvard Business Review* 84(12): 114–22.

Walker, D. (1990) 'Managers and Designers: Two Tribes at War?', in Oakley, M. (ed.) *Design Management: A Handbook of Issues and Methods*. Oxford: Basil Blackwell, pp. 145–54.

Walsh, V., Roy, R., Bruce, M. and Potter, S. (1992) *Winning by Design: Technology, Product Design and International Competitiveness*. Oxford: Blackwell Business.

# 16

# A Study on the Value and Applications of Integrated Design Management

## THOMAS LOCKWOOD

As the benefits of design in business become increasingly evident, the importance of developing effective internal design organizations and models of design integration also grows. Truly, design has more potential to lead change, enable innovation, influence customer experience and add value to the triple bottom line than any other business function. In fact, every interaction between a company and its customers is influenced by design and every touch point is designed, yet there is very little published information about cross-functional or cross-discipline integrated design management to encourage both innovation and design coherency – the focus of this research.

This study is not about simply visual branding but rather how an organization can enable innovation per design discipline and yet at the same time present a coherent design experience. There seems to be a gap in knowledge about integrated design management in both the academic community and the practice community. This may be because, in both communities, the various design disciplines tend to be pursued in an independent and isolated way rather than as parts of the whole. For example, industrial designers study industrial design and then practise it when they design products; interface designers study visual design and interface, and then go to work designing interfaces; architects study and then design built environments; and so on, all of them doing so with little or no crossover with other design disciplines. This traditional single-discipline orientation of design education has ramifications for design practice and supports a functional or departmentalized design output. Businesses, too, are traditionally organized by function: product designers are typically in R&D, interface designers in IT, architects in facilities management and graphic designers in communications or branding. However, the responsibilities of the corporate design leader should cross all design functions, just as the customer experience encompasses all touch points. In fact, increasingly emphasis today is given to the cumulative effect of customer experience rather than simply customer satisfaction. Experience is a differentiator. Customer experiences are influenced in part by all of the interactions with a company or touch points and all touch points are designed – hence the importance of integrated design management.

This chapter outlines design management best practice techniques that integrate design functions within large multinational corporations in order to enable innovation in each design discipline, as well as design coherency and holistic customer experiences throughout the company. When I speak of 'design disciplines' I am referring to the visual aspects of design in five key areas: product design, interface design, communication design, environment design and identity design. The chapter is based on a multimethod, multicase inductive research study. I conducted the research over several years and it consisted of quantitative and qualitative research in two phases: a first phase including a survey ($n = 49$) and four group interviews ($n = 40$), followed by a second phase of individual in-depth interviews ($n = 53$). The study sample, which emerged from the first phase of research as great examples of design-aware companies, included Caterpillar, Kodak, Levi Strauss, Microsoft, Nike, Starbucks and (my former employer) Sun Microsystems. At each company, I interviewed the people responsible for design of each of the five design disciplines. Typical job titles included corporate design director, functional design manager (representing each of the main design disciplines) and marketing or brand manager. In addition I have expanded this research internationally to compare the findings on a global scale and have conducted similar interviews ($n = 9$) with design leaders at BMW, Braun, British Airways, Heineken and Samsung.

## THE INFLUENCE OF CORPORATE CULTURE ON DESIGN INTEGRATION

Every organization has a corporate culture that influences the behaviour of its employees. In broad terms, corporate culture refers to 'the way we do things around here'. It is the sum total of the values, norms, virtues and accepted behaviours of the organization. Culture is complex and difficult dimension to understand because it encompasses a wide range of behavioural areas (Bernick 2001; Kaplan and Norton 2004; Moneypenny 2004). It can be transmitted among individuals, as information that includes such things as ideas, values, beliefs and behavioural styles (Henrich 1999). Therefore, culture is not limited to company policy, processes, technologies or social structures; however, culturally transmitted preferences do influence employee behaviour and, in turn, the presence or lack of integrated design management.

Occasional learning, coupled with cultural transmission and a tendency to conform, can lead to the spread of group norms. This is evident in many ways, including how individuals perform, how teams perform and how design is utilized. The adherence to group norms must be recognized by design managers in order to better integrate design into the organization.

Throughout my interviews I found many examples of design being influenced by corporate culture. This culture is often based on the founding principles of the company – principles that are passed from employee to employee in the form of storytelling, as corporate norms and myths – and by the company founder as well. For example, at Nike, design management employees shared a sense that as an organization, they were committed to doing the right thing for the athlete and that design is a tool that helps enable this. According to one design manager: 'The whole foundation in Nike comes out of one guy, Bill Bowerman [the company's co-founder] . . . and all of design basically emanates from his perspective on it.'

The executive management team, both past and present, also has an influence on the value the organization places on design. There is also evidence that in design-aware companies, executive management teams, including presidents and vice presidents, recognize design's value for their companies. Of course, the influence of current executive management, past executive management and company history all play into the mix of corporate culture and influence the value and use of design, so one company's path to design awareness may not be the same as the next's. For example, at Nike, Starbucks and Sun, the commitment to design appears to stem from the individual founders. BMW, Braun, Levi's and Kodak appear to have come to an appreciation of design more as the result of overall company history and at British Airways, Caterpillar, Microsoft and Samsung, design awareness appears to be largely the result of business performance. However, in the end all of these are a function of corporate culture.

Many of the sample companies referred to their design history as a guide for design management. This is a form of organizational memory that is often spread by storytelling. According to a Levi Strauss design director: 'I think Levi's uses a lot of their history and also a filter of the values of the company in order to look towards future design . . . I think they have the ability to have a legacy, which is very different from a lot of other companies, particularly in apparel.' By telling stories, the design employees help keep the myths alive and this tends to lead to design adherence – the kind that is generally self-imposed, self-regulated and not a formal policy. This would imply that the ability to produce coherent design appears to live within corporate norms and culture, regardless of company policy. Of course, most companies have a fairly unassailable design policy regarding 'trade dress' (i.e., corporate logo and colours), which is generally driven by brand guidelines. However, an integrated customer experience based on design and integrated design management, goes far beyond simply trade dress.

## INTERNAL DESIGN STUDIOS SUPPORT
## DESIGN INTEGRATION

Although history, culture and executive management all have their effects on design within a company, it is also crucial that the company feels it 'owns' that particular design. This means that in order to have a uniquely visually branded company, one must take full responsibility for and control of and production of, the visual design output of the company. As a vice president of design at Starbucks said, 'To own your brand is to own your design.' And a design manager at Nike chimed in: 'We have the internal resources for everything.' For these companies, design is carried out primarily by internal design studios, with strong design leadership. Each one emerged in phase one research as demonstrating cohesive design innovation and visual design coherency, among other criteria and I do not think it is coincidental that each company also maintains comprehensive internal design capabilities with very capable internal design managers. As one Starbucks design manager concluded: 'We don't have to outsource.'

Indeed, none of these organizations needed to outsource design, unless they so desired for specific projects. However, most of them did rely on external design vendors for support: to explore new innovations, support growing workloads or just handle routine production design. These scenarios all require strong internal design leadership. The effectiveness of these internal design studios is due, in part, to the integrated nature of design into the business model, as well as the cross-functional capabilities of the design teams. And in most cases this was also due, in part, to the history of the organization and its commitment to design functions.

A general perception within the design community is that fresh design ideas come from outside designers, which is arguable, but not evident in this research. In most cases, the sample companies believed their internal design capabilities equalled or surpassed those provided by external design studios with regard to quality, service, cost and knowledge of the company, its products and its customers. One of Nike's design managers summarized the importance of a high-quality internal design department thus: 'If you're not on top of your (design) game, you're not going to be playing. You're not going to be making a good product. So this forces the entire team to step up and be demanding and we love the challenge. And that's what makes it better.'

All of the eleven companies in this sample have competent internal design studios that produce the vast majority of their design work, in each design discipline. This is a sourcing strategy and has been in place in each company for a number of years. This sourcing strategy allows for the use of external designers and design firms and external design resources are often critical to the success of a project but the externals are controlled and managed from the internal design management competencies. Consider, too, that in general the executive management teams in these companies understand the business value of design and their endorsement facilitates the effectiveness and integration of the overall design operations.

Size of the internal design functions varies at these companies, depending on overall design needs, such as company size, number of products, fast-moving versus capital goods, business versus consumer focus and so on. Samsung, for example, has close to 500 designers and Nike has between 350 and 400 designers at their headquarters in Beaverton, Oregon, alone. Because the companies in the sample have internal design management and design capabilities that include all of the design disciplines, they appear to be very able to manage innovation processes, design thinking and design coherency across all customer touch points.

## INTERNAL DESIGN FUNCTIONS SUPPORT INTEGRATION AND CREATIVITY

Another finding of this research is the level of collaboration in the design management functions. Design is like glue; it can connect corporate strategy to products and services delivered and connect the company to its customers. This requires a significant amount of cross-functional collaboration. The role of the design manager is in part a role of a collaborator. That said, it is difficult to measure levels of collaboration in corporations where design is concerned because measuring collaboration, as with many social sciences, is often an interpretative task. In my judgment, however, the level of design collaboration in this group of organizations is high compared to more traditional, 'function-based' design management. As one design manager stated: 'The way we are structured, it's almost by design . . . we have a connection in the interface with most high-level organizations, like human resources, public affairs and certainly with the marketing organization.'

Companies that use internal design studios and employ designers and design managers also find it bolsters their internal creativity. The process of designing involves creativity and creativity can be a tremendous business resource. There is much literature today stressing the intangible value of corporate creativity and the need for creativity as a business differentiator in the future (Florida 2002; Forbes and Domm 2004; Lockwood and Walton 2008). Creativity is part of a corporation's knowledge management system and is an intangible or 'soft' asset. Intangible assets add value to a company (Kaplan and Norton 2004; Lev 2004; Ulrich and Smallwood 2004; Low and Kalafut 2002) and staffing internal design employees builds companies' creative abilities. According to Lev (2004) intangible assets are what give today's companies competitive advantage. Directly related to this research regarding integrated design management is an argument by Ulrich and Smallwood (2004) that shared mindset, accountability and collaboration are also intangible assets. They cite the need for collaboration in working across boundaries to ensure both efficiency and leverage, stating that collaboration occurs when an organization as a whole gains efficiencies of operating through the pooling of services or technologies, through economies of scale or through the sharing of ideas and talent across boundaries, which is similar to the benefits of internal design capabilities and leadership.

### Design Insourcing Supports Integrated Design

There is much debate, especially during these difficult economic times, about the benefits of outsourcing versus insourcing. The question every CEO must make is whether design is core to the business or not. The fact is that everything made by human beings requires design and that in business, design requires management. And many companies are realizing competitive advantage by strategically managing design. What's more, many people argue that every company must find its new future, not simply reverting back to pre-recession ways and we must innovate our way forward. Design is what brings innovation to market. If this is the case, then it stands to reason that companies that use design as part of their business strategy are likely to reap a competitive advantage. Indeed, design is increasingly recognized as a competitive differentiator and with that realization has come a growing demand for competent design leaders and managers.

Outsourcing and offshoring are ways to achieve near-term savings in operations and are often based on transaction-cost economics; however this was not the evident in this study sample with regard to design services. Moreover, within these sample companies, it was clear that the internal design studios collaborated significantly with other departments in the organization. This type of interfirm coordination can work toward improving overall supply chain performance (Ho and Tang 2004; Parker and Russell 2004). This implies, if anything, a greater need for companies to develop advanced internal design management operations.

Companies bring activities in-house for one of two reasons: either they can execute those activities better than anyone else or, even if they can achieve superior performance by going outside, it still isn't worth the extra cost and bother of working with an outsider (Hagel and Brown 2005). From the viewpoint of organizational efficiency, this use of internal resources rather than sourced open-market resources goes much beyond the savings of transactional-cost economics to the benefits of interaction-cost economics (Hagel and Brown 2005). As noted, the internal design competencies within the companies in this study are influenced in large part by company history, culture and norms and the resulting commitment to controlling design as an internal resource. Indeed, the notion of interaction-cost economics in regards to in-house design studios is presaged in Ronald Coase's (1937) seminal essay, 'The Nature of the Firm'. Coase suggests that under certain circumstances firms provide a more efficient mechanism for accessing and using resources internally than through open-market transactions.

## EFFECTIVE DESIGN OPERATIONS REQUIRE EXECUTIVE-LEVEL DESIGN LEADERSHIP

In all of the sample companies, design was led at the executive level by a vice president or director. This not only helps keep design in a high position on the corporate agenda but also encourages collaborative work processes to include design. In turn, the design disciplines themselves become integrated. These senior-level design managers are not simply figureheads; they are involved with and informed about design and they make decisions about its management. This role is typically referred to as Chief

Design Officer, Design Director, Innovation Director or Design Manager, although sometimes the individual is called Creative Director or VP Brand; job titles for those responsible for individual design disciplines naturally varied. Many companies have very highly regarded and well recognized Chief Design Officers. Sometimes these positions report to the CEO, at other times to various vice presidents, such as VP Operations, Chief Marketing Officer, Chief Innovation Officer, VP of R&D or VP of Brand, etc. The trend I am seeing in companies like Apple, P&G, Target and Philips, among others, is for design to report directly to the CEO. It is, after all, ultimately the CEO who is responsible for the corporation's design output and more CEOs are getting actively involved in design today. My colleague Clive Roux, who is the executive director of the Industrial Design Society of America (IDSA), argues that every company should have a Chief Design Officer and I couldn't agree more.

Reporting to the chief of design (or whoever is in charge of design) is typically a host of design managers. Although these managers have direct responsibility for design functions, their responsibility or influence for the other functions of design is indirect. Often this is a dotted-line reporting structure, which is interesting in that it means the design responsibilities are only partially distributed. Therefore, there is not a simple explanation of design integration by having all design located in one studio and in one individual's organization.

### Design Organization Structures

The structure in all the companies in the sample included a senior design director with a staff of design managers, who oversee a staff of designers and other personnel. While this implies some design discipline overlap there is also some in different departments. In all cases the structure was a normal hierarchal organization structure, with a top person in charge and various levels below. For example, the structure at Levi's includes a vice president of design, with four design directors below, each with up to twelve designers. Kodak, for example, has a product design director with numerous design managers below and just one design manager, for example, is responsible for a team of up to twenty-five people. Every company has different policies about the grade levels and number of direct reports and so forth. The main point, regarding discipline overlap, is to maintain open lines of communications between the design managers and also amongst their teams.

Often companies split product design into two groups: a 'pre-commercialization' group and a 'commercialization' group (such groups are also often called 'advance design' and 'in-line'). The former focuses on future products, the latter on products that are current. In addition, some organizations split the teams, based on the need for speed to market. For example, Levi Strauss has design teams working on normal processes and schedules based on launch seasons, but they also have a 'quick release' process that allows them to get some products out to market very quickly in order to respond to fast-breaking fashion changes.

**TABLE 16.1:** Variations of centralized and decentralized design department structures

| Company | Centralized Design Studio | Satellite Design Studios | Business Unit or Sub-brand Studios |
|---|---|---|---|
| BMW | Yes | Yes | No |
| British Airways | Yes | No | Yes |
| Caterpillar | Yes | No | Yes |
| Heineken | Yes | No | Yes |
| Kodak | Yes | Yes | Yes |
| Levi's | Yes | No | Yes |
| Microsoft | No | No | Yes |
| Nike | No | Yes | Yes |
| Samsung | Yes | Yes | No |
| Starbucks | Yes | No | No |
| Sun | No | Yes | Yes |

Many companies also have design councils, which help to guide design integration and overall design strategy. I serve on several such councils as an outside expert. Nike, for example, has an internal 'design leadership team' that, on a seasonal basis, is responsible for providing design direction. The output of the design leadership team is funnel-like in that it provides broad input that informs future design initiatives for other design teams to work towards. Nike also maintains a sophisticated design library with a librarian and a trends researcher. The design library is cross-functional and collaborates with the design leadership team on the front end of the design process, as well as with the design teams at large on an ongoing basis.

Some of these companies – BMW, Braun, Caterpillar, Kodak, Levi Strauss and Starbucks, for example – are structured such that design occupies a centralized corporate role, with most of the designers in one organization. Others have a centralized design organization but decentralized departments; the design director or VP would thus play a centralized corporate role, with design directors and managers located in the business units. At Nike, the design council serves a centralized role and design functions take place in the business units. Table 16.1 illustrates centralized versus decentralized structure and the use of satellite design studios.

## DESIGN MANAGERS ARE 'HANDS ON' ABOUT DESIGN

My research showed the leadership of the internal design functions in these companies to be led by design managers with design backgrounds. The design manager is typically responsible for design strategy and execution, and must pull teams together and manage the process of designing. As with many management positions, the design manager must make business decisions as well as decisions pertaining strictly to design. Design managers are generally capable of and empowered to manage design work, approve design work and even occasionally do design work. This is not to suggest that the design work doesn't get approved again at a higher level but it does imply that these design managers are generally capable of doing design themselves as well as approving it. In this sample, the highest-ranking design manager at ten out of the eleven companies was a former designer. As one design manager at Nike said, 'I approve all the images'. This indicates the design manager is competent to approve from a design as well as a process point of view. Indeed, asked about who approves the design work in your company, a design director from Kodak, who manages a studio of nearly 100 designers, answered 'me'.

Hands-on design ability and responsibility for design management is a competency that can support the collaboration of design managers, design teams and individual designers, in single and multiple design disciplines. Looking back on my own experience as a practising design manager, I cannot imagine how I could have done my job well had I not been first a designer. It will be interesting to watch, with the greater recognition of the value of design in business and the rise of interest about design and design thinking by business educators, how business students may eventually cope with actual design management. Figure 16.1 summarizes the findings.

### Design Management Involvement based on Intimacy Theory

Intimacy theory offers a structure that may make it easier to understand collaboration issues – why, for instance people work well or don't work well together – and may also

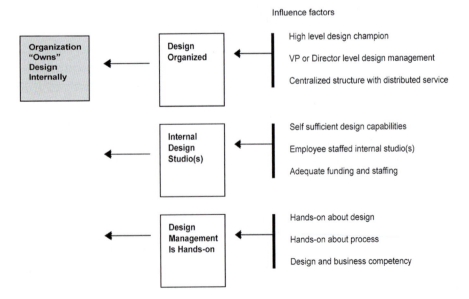

FIGURE 16.1   Summary of findings. © Lockwood, 2011.

be a good way of understanding why most competent design managers have backgrounds as designers and tend to be hands-on in nature. Intimacy theory (see Harmsen 2001) posits that the individual elements of a whole are dependent on each other to give the true picture of the whole. In this model, the elements may not be taken away and replaced without altering the perception of the whole. In this study, I found much evidence that integrated design management requires a participative management style, which explains its hands-on nature. Indeed, Kim (2002) presents an argument for participative management, positing that managers who share their decision-making power with employees will reap benefits of enhanced performance and work satisfaction. This links closely with the belief that involvement enhances learning – a theory that underlines the value of design leaders with design experience in their backgrounds.

## UNIQUE PROCESSES SUPPORT DESIGN AND IMPROVE INNOVATION

Now, here's a paradox: the companies in my research all demonstrated design innovation and design coherency yet all used various flexible, collaborative, informal processes rather than the typical well defined stage-gate process of design management. Much akin to design thinking, integrated design management requires integrated and collaborative processes. This involves collaborative exploration and active based research methods, collaborative design briefing process, which include the use of traditional briefs as well as visual briefs and briefs used to brief multiple design functions, collaborative design review process with much iteration of informal and formal review techniques, such as design previews, check-ins and reviews and finally, both informal and formal design approvals.

### An Integrated versus Traditional Stage-gate Process

At StorageTek and Sun Microsystems, we used the stage-gate process for new product development. But this was based on an engineering process to which the product design function was forced to adhere and frankly, it did not work all that well for innovation. In an innovation or design thinking mode, a more collaborative, flexible and more ongoing and continuous process is required. However, the design management process most often referenced and recommended in design management literature is the stage-gate process (Press and Cooper 1995; Bennet and Oakley 1998; Bruce and Bessant 2002; Borja de Mozota 2003), which harks back to Cooper's original stage-gate model. The stage-gate process indicates a linear, sequential workflow process: essentially a project moves through specific stages, passing through specific gates, but stopping at each point for approval. This process stems from the traditional line-production methods of Fordism and is a structured, linear, formal, clearly defined process intended to maximize productivity and efficiency as well as reduce rework. In a staged process, it is also typical that team members change throughout the different stages – the notion being that each defined stage requires different competencies in order to accomplish the defined task and proceed to the gate for approval. Members come in as appropriate and leave when their expert contribution requirement has been fulfilled. Each gate review is a formal process overseen by an individual or group of individuals with authority to approve. Figure 16.2 illustrates a visualization of a typical stage-gate process:

In sharp contrast, in my research and practice I find very little use of formal stage-gate processes outside of engineering and particularly not in the innovation or design thinking phases of work. The design management process that I have seen in my research is more integrated, flexible and collaborative. At no company did I find much evidence of the traditional stage-gate process; those I found were more concurrent and flexible. There were obvious similarities to the stage-gate model, in that there were discrete starts, briefs, reviews and approvals. But they were generally more flexible, ongoing and iterative, with design check-ins, previews, informal reviews and even informal approvals. The processes were based on the corporate culture, the needs of the project, the work styles of the employees and the high level of design competency of the internal design studios. These processes are able to adapt to new information, feedback from customer perceptions, design iterations and ideas from different designers and other people. More flexible processes allow the design teams to explore alternative design solutions (within acceptable ranges) and this permits continuous improvement as well as design innovations. A more flexible process also permits team participants to join and leave a project as needed throughout the lifecycle of the project, rather than only at pre-prescribed stages. There was no evidence of specific job roles required at specific stages – or even stages, for that matter – instead, employees contributed to the design process as needed, seemingly informally. This involved a rather continuous process of designing and refining and was always led by the design manager in charge of the project, indicating a proactive method of seeking collaboration from others as needed. I refer to this technique as using team 'players' rather than team 'members,' to indicate that they appear to come in and out of the process as needed, but are not formal, continuous team members in the traditional sense of a team or defined-stage participant. The collaboration within this process also helps the designers and design teams to better understand the project and also to understand the opinions of others, the company design style and the target customer.

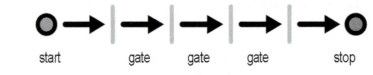

FIGURE 16.2   Typical stage-gate process. © Lockwood, 2011.

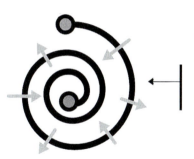

Players come and go as needed
Flexibility
Continuous improvement
Collaboration cross-discipline and cross-function
Ongoing design check-ins, previews
Design reviews

FIGURE 16.3    Integrated design management process. © Lockwood, 2011.

For lack of a better term, I describe these processes as 'roundabout design man-agement', and the reason why is evidenced in Figure 16.3. Granted, they do contain design reviews and design approval processes, which are similar to those found in stage-type processes. For example, a signage design project could not proceed to fabrication without approval and similarly, an web site should not be published without approval. In this regard, the roundabout process functions as an informal community of prac-tice, yet it is still a sequential process in regards to movement through time. That ac-knowledged, the integrated design management processes I have observed in the data, at least prior to final approval, are not fixed, linear stage-gate processes but collabora-tive, iterative, ongoing and flexible. I liken this to the way an artist makes a pot from clay: first it is thrown, then kneaded, then shaped and formed and then fine-tuned as part of a development process before it goes to the kiln. So, too, innovation design management involves discovery, shaping, forming and collaboration between multiple designers and design managers and other associated employees, in order to create the most appropriate design work per project. (I use the term 'appropriate' to mean appro-priate to the project, appropriate to the customer and appropriate to the company.)

Figure 16.3 is a simple visual representation of the integrated design management process, where the process is more like a roundabout intersection or an inward spiral, permitting a continuous process of discovery, shaping and forming and allowing team members or players, to enter and exit as needed as the design process iterates closer towards a final design solution in the centre. The grey arrows indicate different team players coming and going, collaborating as needed throughout the project.

This is not to suggest that a gate process with approval points is inappropriate be-cause ultimately all the design work is either approved or disapproved. But in these processes, the stages are not clearly separated or, in some cases, even defined. This re-search therefore suggests the stage-gate process to be a simplified explanation of what really is happening within the sample companies.

I suspect that one reason a formal stage-gate process is not commonly used in innovation and design is because it implies separation of involvement. This separa-tion attempts to structure the design creation process into discrete tasks and implies a sequential, linear process. In the stage-gate process, when a particular stage is finished those involved can pass the project off to the team for the next stage and so on, throughout the process. But in these companies it was evident that design is a process requiring different perspectives over time and is developed in an integrated and collaborative manner. All of the companies demonstrated processes for design reviews, some formal and some informal; these included design check-ins, design previews, design pass-arounds and design reviews. These processes had taken the place of formal stage-gate reviews, but they served the same purpose: to check the work often as it progresses. The data indicates not a sequential process but instead one that brings a set of perspectives together at the outset, refines these with visual briefs and builds an ongoing shared concept around which team players can contribute as needed.

## SOCIAL EXCHANGE THEORY AND EMPLOYEE INVOLVEMENT

To further understand this flexible and collaborative process and the structure of social networks, I turned to social exchange theory. The basic principle of social exchange theory is that social interaction involves a reciprocal exchange of material things or ideas, of emotions and of behaviour (Bothamley 2002). In my sample companies, the reciprocal exchange was one of ideas and knowledge; as one party gains information, the other benefits from improved design in the company and this is valuable because both parties value design and are committed to the company's success. Design managers share knowledge and research about customers, consumer trends and design in the service of developing good and appropriate design for the company. This is a form of organizational knowledge as social capital. Organizational knowledge is an important intangible resource, which can be a source of sustainable competitive advantage (Wiklund and Shepherd 2003). Knowledge is fundamental to organizational success (King and Zeithaml 2003; Chen 2005), and organizations gain knowledge by two methods: creating it or deriving it from external sources (Knott 2003; Spencer 2003). Throughout the sample companies, I observed knowledge creation and knowledge sharing, much like design thinking with regard to design research, design visualizations, fast and rough testing, user involvements and co-design. In addition the process of integrated design management should address usability and common experience and design to align with social cultural trends, as well as how to develop and maintain a corporate design style. Indeed, a corporate design style, as it evolves, is an example of innovation based on sustained knowledge creation. This is firm-specific knowledge, which can aid management in making such resource-deployment decisions as investment in R&D and product development within the organization (Kor and Mahoney 2005); it can also inform innovation strategies for knowledge sharing outside of the organization (Spencer 2003). Spencer suggests that, in certain circumstances, firms that share technological knowledge may achieve higher innovative performance than firms that do not share knowledge. While his research is focused on intra-firm knowledge sharing, it is relevant to this study, in that innovative performance goes well beyond patent acquisition to evaluate the benefits of knowledge sharing and knowledge acquisition. In this study, I found that one reason why internal design studios are effective at collaboration and knowledge sharing is because they have internal clients. The internal design studios are communities of practice, which create and share knowledge. Since the source of design is internal to the organizations in the sample, this alleviates a potential problem in sharing organizational knowledge because it is difficult for ad hoc supply chains to tap into organizational knowledge memory (Hult et al. 2004) and internal design studios thus reduce knowledge search and transfer costs (Haas and Hansen 2005). In the sample companies, transferring knowledge from one unit to another and from one design discipline to another was common practice.

Organizational knowledge is created and transferred within an organizational environment so it is tied to a social context. The social context is the internal community, yet organizations are knowledge communities embedded in a larger domain of industry communities (King and Zeithaml 2003). Knowledge-intensive firms need to help share knowledge by employees if they are to gain the most from their intellectual capital and sharing and integrating knowledge within the organization depends partly on building social capital (Swart and Kinnie 2003). Further organizational knowledge reflects an organization's ability to differentiate itself from competitors (King and Zeithaml 2003); consider, for example, the benefits of a unique design style. There are two broad categories of knowledge in my study regarding design style: explicit, such as the limited use of design standards, visualizations of briefs, mood boards and customer personas; and tacit, such as the shared knowledge about design style, informal design learning methods, knowledge of customers and intuition about design style. Explicit knowledge is knowledge that can be codified; tacit knowledge is difficult to articulate and difficult to transfer as well because it takes time to explain and learn (Martin et al. 2001; Pak and Park 2003; Hatch and Dyer 2004; Levin and Cross 2004) and is distributed and embedded within practice (Ethiraj et al. 2005; Swart and Kinnie 2003). This brief overview of the role of knowledge in the study sample is helpful as a foundation to discuss the influence of social learning as related to knowledge and to integrated design management.

In conclusion, as a practitioner and an analyst, I have studied design management in three broad categorizations: the corporate organization level, the process level and the project level. My objective in this brief chapter was simply to share some of my insights. There are many influences on behaviour in organizations, including the actions of the employees, the way in which management is carried out and organizational contexts such as company history, values and norms. And there are multiple facets of an organization, all of which may influence design. How this all comes together to empower innovation and great design, how design is leveraged, how design influences customers experience and how the results support the triple bottom line; this is the arena of integrated design management.

# REFERENCES

Bennet, D., Lewis, C. and Oakley, M. (1997) 'The Design of Products and Services', in M. Bruce and R. Cooper (eds) *Marketing and Design Management*. London: Thompson Business Press.

Bernick, C. (2001) 'When Your Culture Needs a Makeover', *Harvard Business Review* 79: 53–61.

Borja de Mozota, B. (2003) *Design Management*. New York: Allworth Press.

Bothamley, J. (2002) *Dictionary of Theories*. Canton, MI: Visible Ink Press.

Bruce, M. and Bessant, J. (2002) *Design in Business*. Harlow: Financial Times/Prentice Hall.

Coase, R. (1937) 'The Nature of the Firm', *Economica* 4(18): 386–405.

Chen, G. (2005) 'Management Practices and Tools for Enhancing Organizational Learning Capability', *Advanced Management Journal* 70(1): 4–21.

Ethiraj, S., Kale, P., Krishnan, M.S. and Singh, J. (2005) 'Where Do Capabilities Come from and How Do They Matter?' *Strategic Management Journal* 26(1): 25–45.

Florida, R. (2002) *The Rise of the Creative Class*. New York: Basic Books.

Forbes, J. and Domm, D. (2004) 'Creativity and Productivity: Resolving the Conflict', *Advanced Management Journal* 69(2): 4–15.

Haas, M. and Hansen, M. (2005) 'When Using Knowledge Can Hurt Performance: The Value of Organizational Capabilities in a Management Consulting Company', *Strategic Management Journal* 26(1): 1024.

Hagel, J. and Brown, J. (2005) 'Productive Friction', *Harvard Business Review*, February: 83–91.

Hatch, N. and Dyer, J. (2004) 'Human Capital and Learning as a Source of Sustainable Competitive Advantage', *Strategic Management Journal* 25(12): 1155–78.

Harmsen, K. (2001) 'Experimental Event Marketing', *Integrated Marketing Communication Research Journal* 7: 19–26.

Henrich, J. (1999) 'What Is the Role of Culture in Bounded Rationality', http://www2.psych.ubc.ca/~henrich/Website/Papers/boundedR.pdf, accessed 4 April 2011.

Ho, T. and Tang, C. (2004) 'Introduction to the Special Issue on Marketing and Operations Management: Interfaces and Coordination', *Management Science* 50(4): 429–30.

Hult, G., Ketchen, D. and Slater, S. (2004) 'Information Processing, Knowledge Development and Strategic Supply Chain Performance', *Academy of Management Journal* 47(2): 241–53.

Kaplan, R. and Norton, D. (2004), 'Measuring the Strategic Readiness of Intangible Assets.' *Harvard Business Review* 82(2): 52–63.

Kim, S. (2002) 'Participative Management and Job Satisfaction', *Public Administration Review* 62(2): 231.

King, A.W. and Zeithaml, C. (2003) 'Measuring Organizational Knowledge: A Conceptual and Methodological Framework', *Strategic Management Journal* 24(8): 763–72.

Knott, A. (2003) 'Persistent Heterogeneity and Sustainable Innovation', *Strategic Management Journal* 24(8): 687–705.

Kor, Y. and Mahoney, J. (2005) 'How Dynamics, Management and Governance of Resource Deployments Influence Firm-level Performance', *Strategic Management Journal*, 26(5): 489–96.

Lev, B. (2004) 'Sharpening the Intangibles Edge', *Harvard Business Review* 82(6): 109–16.

Levin, D. and Cross, R. (2004) 'The Strength of Weak Ties You Can Trust: The Mediating Role of Trust in Effective Knowledge Transfer', *Management Science* 50(11): 1477–90.

Lockwood, T. and Walton, T. (2008) *Corporate Creativity.* New York: Allworth Press.

Low, J. and Kalafut, P. (2002) 'The Invisible Advantage of Innovation', *Perspectives on Business Innovation* 8: 75–8.

Martin, G., Pate, J. and Beaumont, P. (2001) 'Company-Based Education Programs', *Human Resource Management Journal* 11(4): 55–73.

Moneypenny, N. (2004) 'Five Foundations for Developing a Corporate Culture', *The RMA Journal* 86(5): 22–5.

Pak, Y.S. and Park, Y.-R. (2004) 'A Framework of Knowledge Transfer in Cross-Border Joint Ventures: An Empirical Test of the Korean Context', *Management International Review* 44: 417–34.

Parker, D. and Russell, K. (2004) 'Outsourcing and Inter/Intra Supply Chain Dynamics: Strategic Management Issues', *Journal of Supply Chain Management* 40(4): 56–68.

Press, M. and Cooper, R. (2003) *The Design Experience.* Aldershot: Ashgate Publishing.

Spencer, J. (2003) 'Firms' Knowledge-Sharing Strategies in the Global Innovation System: Empirical Evidence from the Flat Panel Display Industry', *Strategic Management Journal* 24(3): 217–33.

Swart, J. and Kinnie, N. (2003) 'Sharing Knowledge in Knowledge-Intensive Firms', *Human Resource Management Journal* 13(2): 37–55.

Ulrich, D. and Smallwood, N. (2004) 'Capitalizing on Capabilities', *Harvard Business Review* 82(6): 119–27.

Wiklund, J. and Shepherd, D. (2003) 'Knowledge-Based Resources, Entrepreneurial Orientation and the Performance of Small and Medium-Sized Businesses', *Strategic Management Journal* 24: 1307–14.

# 17

# Changes in the Role of Designers in Strategy

## KYUNG-WON CHUNG AND YU-JIN KIM

It is widely recognized that the design profession has contributed significantly to corporate assets, including buildings, factories, showrooms, products and corporate identity. Cooper and Press (1995) classified design activities into three distinct sets: the development of corporate identity, the design of saleable products and the design of operating environments. Visual communication design (graphic design) has played a role in improving the quality of corporate communications through advertisements, publications and the Internet, whereas industrial design has been influential in enhancing the competitiveness of products and systems. Environmental design has a very strong influence on the value of corporate assets through developing interior and exterior physical environments, to meet day-to-day functional needs or create experiences. Environmental design is typically used as a term to describe environmentally concerned designs linked to issues of sustainability, conservation or the like (as in 'design for the environment'). However, the term also refers to a group of design disciplines such as interior, display, exhibition, architectural design and others for creating human-designed environment, especially corporate working environment. The use of environmental design in this context is slightly different from 'systems design' or the design of complete environment.

Along with the enlarged scope of managing design, there have been various attempts to indentify the actual contributions of design in practice. Through a content analysis of more than 700 articles published in DMI's periodicals over the past two decades, Kim and Chung (2007: 47) discovered that 'the role of design management has expanded from managing product development into leveraging strategic and competitive advantages, managing identity and brand as strategic assets and maintaining a cutting edge in the global and digital markets.' Based on their professional experiences, the Design Strategy Group (unpublished interview, 25 August 2005) at Continuum has found that design can connect companies and consumers in a meaningful way by presenting companies with the real voice of the consumers that they act upon, as well as creating competitive advantages by translating business strategies into actionable plans for products and services with superior emotional and experiential value in diverse industries.

In particular, as a result of the rapid growth of virtual business environments, the role of visual communication design in the development of Internet services is now a key study area (Kim and Chung 2007). Leading Internet-based companies such as Google and Yahoo! have located design at the centre of their service development to guide the differentiation of services as well as user engagement (Best 2006; also unpublished interview conducted with Benedict Davies at Google on 8 October 2008).

In this vein, this chapter discusses the actual role of designers in product and service strategies based on understanding expanded and diversified design activities in different industries. We suggest a taxonomy characterizing the designer's roles for actualizing product and service strategies in terms of their interactive and functional aspects. Through case studies in the manufacturing and Internet service industries, we also illustrate how designers play interactional and functional roles in new product and service strategy, respectively, as well as how these two roles are integrated for business success.

## A NEW TAXONOMY FOR THE ROLES OF DESIGNERS IN STRATEGY

The growing importance of design in the context of corporate strategies for new products, brands and communications has also led to a growing interest in the designer's role in strategy since the 1980s. According to Heskett (2005), strategies relating to product development are located along two axes: On the first axis, strategy runs from existing products to innovative, yet-to-be-created products. The second axis begins with product lines and ends with corporate strategies. These two axes intersect in the middle. In conjunction with the two axes, he classified four levels of design practices and defined designers' roles in each level: interpreting product specifications as interpreters, differentiating existing products as differentiators, creating systematic connections as system creators and suggesting new concepts and systems as planners. Perks et al. (2005) developed a taxonomy characterizing three roles of designers in the new product development process in UK manufacturing companies. Such roles include functional specialists, multi-functional team members and process leaders.

From a different viewpoint, several studies have been undertaken in different industries to identify the actual roles of designers in strategy formulation and implementation processes (Chung 1989; Seidel 2000; Valtonen 2007; Kim 2008). This chapter deals with two studies in greater depth, which were conducted in different industries: the manufacturing sector in the UK in the 1980s (Chung 1989) and the Internet service area in Korea since the mid-1990s (Kim 2008). From these studies, we can see that designers perform two distinct roles in relation to new product and service strategies: an interactional role and a functional role. The interactional role highlights that the designer has to interact with senior management and other experts in strategy formulation and implementation processes. As the term 'interactional' means mutual or reciprocal action or influence (Merriam-Webster Dictionary 2009), the two-way effect is essential in the interactional role. The functional role differs from the interactional role in that it points to the contributions designers make through their professional skills and capabilities, for example, when implementing a service strategy. Designers' professional contributions can range from the simple job of giving form to highly sophisticated work that initiates new services in line with a firm's strategic demands.

It is of particular interest and relevance to understand these two closely related roles of the designer in the strategic context from the aforementioned two studies. The details of each role, together with examples, will be discussed in the following section.

## Interactional Roles of a Designer

Designers interact with various levels of management and many different experts in new product development. Chung (1989: 77) identified three different strategic roles of designers in terms of developing new product strategy (initiative, participative and subordinate roles), along with the timing of their involvement in the strategy formulation and those with whom they interact (senior, middle and lower management), as shown in Figure 17.1.

- *Initiative role.* Designers can provide firms not only with the motives for new product development, such as new opportunities, new ideas or a new combination of existing ideas, but also a basic guideline for new product strategy for a certain product concept. Occasionally, a company can secure a license from design consultants for the commercialization of a new product. However, only a limited number of design agencies can play this kind of initiative role because various preconditions such as specific knowledge, technology and diverse experience of strategic issues are required. However, the majority of design agencies still focus on cultivating specialized design expertise and skills in a rather narrow scope.
- *Participative role:* A designer can work as a director who is in charge of corporate design activities and participate in the formulation of new product strategy. Sometimes a design consultant acts as an outside design director in the client company. Kenneth Grange of Pentagram, for example, served as design director for several companies, including Wilkinson Sword, Thorn Domestic

| Interactional Roles of Designers | | |
| --- | --- | --- |
| Types | Involved | Interact with |
| Initiative Role | Strategy Formulation | Senior Management |
| Participative Role | Strategy Formulation/ Implementation | Senior/Middle Management |
| Subordinate Role | Strategy Implementation | Middle/Lower Management |

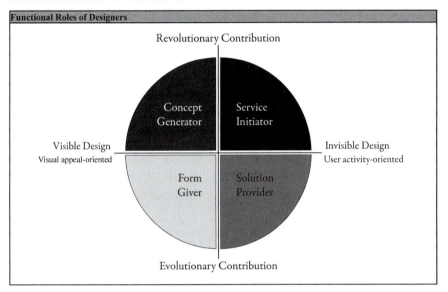

FIGURE 17.1   Two taxonomies of designer roles in strategy: Interactional and functional roles. ©
Chung and Kim, 2011.

Appliances and Kenwood. He influenced his client firm's new product strategy through his position on several marketing and product development
committees. He acted as a cohesive force between different functional specialists in the committee.

• *Subordinate role:* This can be referred to as the more generalized role of designers. Designers, whether internal or external, are usually involved in the
implementation of a firm's established strategy. In other words, a designer
usually contributes to the solution of problems, which are defined by the senior management of a company.

### *Functional Roles of Designers*

The roles of designers in developing Internet services can vary depending on corporate goals and demands. The activities of designers differ in terms of the types of design outcomes they are working on. Some designers work primarily on visible design
elements, which formulate the aesthetic and interactive visuals of Internet services
or brand images (Reynolds 1995; Shida 1996; Hanna 1997; Laar and Berg-Weitzel
2001). Other designers focus mainly on invisible design elements, which determine

the functional usability of Internet services or experiences of users (Swack 1997; Norton and Hansen 2000; Schmitt 2000; Nielsen and Loranger 2006). Meanwhile, these two types of design activities can contribute to Internet service innovation at a different level. Design practices including in-depth user research and brand identity building can be committed to radical business success revolutionarily, whereas designers focusing on more visual and aesthetic practices contribute to incremental innovation of Internet services evolutionarily (Waters 2003; Loveday and Niehaus 2007; Kim 2009).

In light of these perspectives, Kim (2008) discovered that designers play various roles in formulating and implementing service strategies in Internet businesses. These differing views on design activity highlight four different types of functional roles for designers in Internet businesses: designers can be form givers, solution providers, concept generators and service initiators. The bottom diagram in Figure 17.1 depicts the relative position of each type depending on whether its design outcomes are visible or invisible and whether its contribution is revolutionary or evolutionary.

- *Form giver.* Designers in this category decorate digital content to make it aesthetically pleasing to look at or dynamic to interact with. As a more traditional type of designer, they create the 'look and feel' of the site and evoke user emotion.
- *Solution provider.* In this role, designers materialize interactive Web services using visual script languages (such as Flash/Flex ActionScript) by connecting other designers' dynamic design concepts with the developer's Web programming. They therefore contribute to offering new service solutions (such as Rich Internet applications, RIAs),[1] developing innovative user interface designs and improving service usability using the latest technologies.
- *Concept generator.* Designers can build competitive brand images by creating unique brand concepts of Internet services and embodying them in emotional brand experiences using interactive digital story-telling techniques.
- *Service initiator.* Designers can create new user experiences by discovering the needs and demands of users, understanding business strategies, suggesting innovative services and then initiating a new Internet service development process or coordinating the whole process. In terms of the designers' initiative role in Internet service development, YouTube, the video sharing website, is a good illustration. Chad Hurley, who specializes in user interface design, took the initiative in realizing the tagging and video-sharing aspects that are YouTube's core service strategies. According to Wikipedia (2009), together with two computer programming experts, he contributed to the launching of an innovative service that enables anyone to use a computer to post a video that millions of people can watch within a few minutes.

## INTERACTIONAL ROLES OF DESIGNERS IN THE MANUFACTURING INDUSTRY

As manufacturing companies have faced urgent needs for improving their competitiveness in order to overcome a declining share of markets, the role of design in new product strategy becomes one of the most important corporate issues. Designers' strategic role can vary depending on the firm's intention and circumstances such as market movements and technological changes. In the case of more complex strategy (for example, the higher market movements, technological changes, collaboration of organizations), most of the companies hired professional design consultancies whereas their in-house designers participated in implementing the lower complexity of strategy.

### Roles of Design Consultants

Since the mid-1980s, there has been a change in the design consultancy profession in the UK in conjunction with the 'Enterprise Initiative' scheme of the Department of Trade and Industry, which aimed to encourage small and medium-sized companies to use consultancies to solve their business problems. To cope with these changes, some design consultants became members of large consulting groups and leading design consultants started to provide a comprehensive advisory service on new product strategies.

Despite the belief that design consultancies were potentially useful in new product development, relatively little was known about the specific contributions of a design consultancy to the formulation of a new product strategy (Moody 1980; Alexander 1985; Bruce 1985). In other words, the specific contribution of design consultants toward solving a problem already defined by senior management was well established. What was not established was the contribution of a design consultancy to integrate senior management activity, i.e. strategy formulation.

In order to identify the actual contributions of design consultants toward their client firm's product strategy in terms of the interactional role, Chung (1989) investigated six British products that were designed by design consultancies. It was discovered that new strategies for the products were successfully formulated and implemented by an effective collaboration between internal design-conscious product champions and external creative design gatekeepers. According to the role type of design consultancies in these relationships, two cases were found to fill an initiative role, three were identified to fill a participative role and one was found to fill a subordinate role.

Higher level roles involving design consultants (e.g., initiative and participative) in strategy formulation were noticeable in these cases. The senior management of firms who aim to discover new opportunities or to penetrate a niche market with new products usually contacts design consultants to obtain strategic support. In particular, the initiative role of design consultants was clearly demonstrated in the following collaboration between PSS and Frazer Design:

In order to cope with the ever-intensifying competition in the publishing sector across the world, an American publisher, PSS, decided to obtain a 'Questron technology' (a patented technology similar to barcode reading technology) license from London-based Frazer Design Consultants (FDC). PSS developed a new educational toy system, the 'Questron Electronic Wand' and an accompanying interactive book series in close collaboration with Frazer, an industrial designer and head of FDC. As the originator of the Questron concept, FDC's contributions toward PSS's product strategy for a synergy between books and toys has been significant since Frazer already had precise ideas on the target market and customer, price range, design features and production methods.

Frazer Design Consultants also provided PSS with strategic guidelines on market position, product name/trademark, styling and printed materials because the firm, as a publisher, had no experience or know-how in developing plastic toys. PSS subsequently formulated a deliberate strategy and developed the Questron wand product system on the basis of these guidelines. Frazer was also deeply involved in the formulation and implementation of the manufacturing strategy. The consultancy developed the final design of the Questron wand and suggested a production and assembly method for reducing the costs and maintaining the quality of the product system. Frazer Design Consultants has maintained a long-term relationship with PSS since the Questron project and PSS uses the consultancy as a research facility for new product development. In other words, FDC has played major roles in the development of PSS's other new products.

As company demands for strategic design services have increased along with ever intensifying competition, world leading design consultancies, such as IDEO, Continuum, Doblin Group and others, have been specializing in that area since the 1990s.

### The Rise of In-house Designers in Strategic Matters

Chung (1989) also studied three British products that were designed by in-house design teams in order to identify how those teams contributed toward new product strategy in terms of the interactional role. Interestingly, he found that the role of in-house designers tended to remain confined to the lower level of corporate management. From the three cases, two in-house teams filled a subordinate role (Quad Electroacoustics Quad 606, a power amplifier and JC Bamford Excavators Sitemaster, a backhoe loader), while the other in-house design team (Glasdon UK) played a participative role in upgrading their Topsy litter bins.

The main reason was that the position of the in-house design team in the company's organizational hierarchy was relatively lower than other business functions. When a company wanted to revamp existing product ranges, they preferred to use in-house designers in a subordinate role because its executives and managers already had a clear product strategy, as in the following Quad Electroacoustics case.

A high quality hi-fi manufacturer, Quad Electroacoustics, developed the 'Quad 606' power amplifier to be a market leader in the UK and the overseas hi-fi market, using its own current dumping system and electrostatic loudspeakers. Together with these innovative technologies, Quad uses design as part of a competitive strategy to maintain its reputation and to cultivate the competitiveness of its products. The senior management of Quad also clearly recognized design as a legitimate and effective tool for business. As a result the Quad product range has a certain corporate image, such that its products were instantly recognizable.

Quad has effectively used both in-house designers for creative product design and outside consultants for visual communication design. All of the Quad product range is designed by in-house designers without any assistance from outside since the designing of high quality hi-fi equipment requires a thorough understanding of highly specialized sound reproduction technology and specified manufacturing techniques.

In this design-conscious management environment, the 'Quad 606' power amplifier was designed by an in-house designer, Rodney Mead. The amplifier became the best selling single product in the Quad range. However, the product strategy was formulated by a new product committee headed by the design-minded Managing Director, Ross Walker. Although Walker had no design background, he had a deep understanding of how important design is in developing a competitive product, as well as how to develop a creative design. No designers served as regular members of the committee. The Quad in-house design team contributed significantly to the implementation of the product strategy rather than the firm's new product strategy formulation.

### In-house Designers Depend on Links with CEOs and Management

The hierarchical position of the in-house design team has risen since the 1990s in conjunction with the ever increasing interest in the strategic issues of design management. The role of design executives who led the in-house design function became higher profile because leading manufacturing companies, such as Apple, Sony, LG Electronics and others, operated a corporate design centre under the direct control of the CEO. Meanwhile, the role of design consultancies, which had no specialty for providing their clients with strategic contributions, was confined to the subordinate level for actualizing well thought-out strategies already formulated by the senior management of the company. Similarly, the role of in-house designers at the lower level of the corporate hierarchy was also limited to the implementation of existing strategies. In Korea, the role of in-house design teams within manufacturing companies has increased significantly since the First Korea Industrial Design Conference was held in the Blue House (the Korean presidential residence) in December 1999. The conference served not only to report the government's strategic plans for promoting the Korean design industry but also to reward companies that had used design as a strategic tool for coping with ever-intensifying competition throughout the world.

Samsung Electronics, the winner of the Grand Prize at the conference in 2000, revamped its in-house design group as the Corporate Design Center (CDC). This centralized design centre was put under the direct control of the Vice Chairman of Samsung Electronics and comprises five major parts: the Design Strategy Team and the Design Research Lab under the CDC and three design groups under three business units (Digital Media, Telecommunication Networks and Digital Appliances). The design centre was headed by the President of Digital Media Business (DM) and he acted as Chief Design Officer. Several senior design officers leading a group of designers dedicated to that business unit have been promoted to become Vice Presidents.

The interrelationship between each design group and business unit was more or less similar to that of a design consultancy and its client, even though all of them were part of the same Samsung group. The main reason for this was that each design group worked solely for its counterpart. Accordingly, designers could accumulate their own design expertise and know-how in a particular product area on a long-term basis, as they were able to collaborate and communicate with different levels of management within the business unit. The head of each design group reported directly to the President of each business unit. As a result, a design-initiated new product development process could be achieved in Samsung's corporate culture. Chung and Freeze (2009) identified an initiative role in Samsung designers for developing innovative products, even in a very saturated market. They also found that even young designers in a lower level of the organizational hierarchy were able to initiate new product strategies.

The DM business's urgent mission was to develop highly innovative LCD TV designs that could compete with foreign brands in the global market. At the end of 2004, Seung-ho Lee, a young designer in the digital media design group, developed a fairly new TV design concept that had a strong emotional association with the shape of a wine glass. A slim and delicate-looking design mock-up with a thickness of 83 mm (existing TVs were 130 mm) was made and reported directly to President Gee-sung Choi who liked it and asked for the creation of a prototype. However, the first prototype made by development engineers looked dull and was very thick (120 mm). Choi pointed out the problems of the prototype and asked the engineers to make a new one exactly the same as the design mock up. A taskforce team composed of a circuit engineer, structure engineer, marketer, designer and others was formed and a new prototype with a profile even slimmer than the design specification, at only 79.6 mm, was developed within twenty days. The new TV brand, named 'Bordeaux'[2] in reference to the original design concept, became the best-selling LCD TV in the world market since 2006. The findings from these case studies on the interactional roles of designers can be distilled in three key observations. Firstly, the contributions of design consultancies were significant both in the strategy formulation and implementation stages in conjunction with the DTI's 'Enterprise Initiative' scheme, while in-house designer contributions were somewhat limited to the strategy implementation stages in the UK in the 1980s. Secondly, the higher level roles of design directors in strategy have become more distinctive than ever as world leading manufacturers including Apple started to raise the hierarchical level of their corporate design centres from the 1990s. Jonathan Ive, the Senior Vice President of design at Apple, has initiated an innovative new product development process in close collaboration with CEO Steve Jobs. Some world-leading design consultancies have specialized in providing their client companies with strategic design services. Thirdly, even ordinary in-house designers, not just design directors, can play initiative roles in product strategy as evidenced by Samsung's unique corporate structure for enhancing mutual collaboration between the design groups and the business units since 2000.

## FUNCTIONAL ROLES OF DESIGNERS IN THE INTERNET SERVICE INDUSTRY

In the early days of Internet business, superiority in technology and price was important in achieving a competitive advantage. As the business environment matured with the increase in the number of Internet service sites and online customers, companies had to provide compelling and distinctive Internet services through new service ideas and attractive designs. In this sense, it is widely accepted that the strategic use of design contributes not only to creating innovative services, but also enhancing the brand value by meeting sophisticated user expectations for emotional and functional experiences (Young 2001; Waters 2003; Long 2004; Van Duyne, Landay and Hong 2007; Loveday and Niehaus 2007).

### Roles of Design Consultants

Along with the proliferation of Internet technologies and Internet service areas (for example, promotion, e-commerce, portals and communities) in Korea, design professionals, primarily in visual communication design, have founded design consultancies since the mid-1990s. These consultancies, specifically designated 'digital design agencies' have helped client companies develop distinguishable Internet services that are both aesthetically pleasing and functionally convenient.[3] They have also supplemented in-house design groups by accelerating the service development process and offering the latest digital design skills.

In recent years, a number of small agencies have been established rapidly since the Korean Government changed regulations to lower the barriers for entry into the digital design agency market. In order to provide more compelling design consulting services in the severely competitive digital design agency market, the current agencies have focused on expanding and specializing their design skills and capabilities.

In order to investigate diversified design activities of these design consultants in terms of the functional role, Kim (2008) analysed representative Internet service projects of three Korean digital design agencies. Each agency's dominant functional role in achieving the strategic goals of client companies was identified as follows:

- Design Fever participated in visually consolidating the robust corporate brand image of NEXON (an online game publisher) as a form giver.
- VINLY improved the usability of CGV (a multiplex movie theatre) online services using the latest Web programming technologies as a solution provider.
- Sugarcube contributed to building a distinguishable brand concept and image for Mple.com (an online open market) as a concept generator.

These findings demonstrate that the agencies' roles in Internet business have widened from the mere decoration of Web content to diverse strategic activities including improving user interface and interaction, building brand identities and providing new user experiences. Among the case studies, Sugarcube's project explains in detail how designers participated at a more strategic level of Internet service innovation.

Mple.com strategically collaborated with Sugarcube in order to propagate its brand and services through its pre-launch website, 'Mple Marble'.[4] Sugarcube created Mple.com's compelling brand concept, i.e. 'Interesting shopping friends' and wrote an imaginary scenario about the concept in a humorous and comical way as follows: 'Users meet nine popular shopping friends in characteristic mini rooms, see their hot selling items, attractive sitcoms and shopping know-how and then become fascinated with the items.'

Sugarcube designers also visualized these stories by integrating interactive videos into the website. They enabled users to navigate the website by spinning a roulette wheel and shifting walking human markers similar to playing a 'Monopoly' board game. By developing a distinguishable pre-launch promotion site based on creative ideas and interactive storytelling techniques, the designers played a crucial role in building a robust brand image and enhancing the brand recognition of Mple.com. They embodied Mple.com's unique brand experience in a more emotional and memorable way.

### Roles of In-house Designers

In the initial stage of Internet business design, designers in many Internet-based companies played a limited role in maintaining the Internet services that were developed by prominent digital design agencies. Along with the proliferation of Internet portal services, however, Internet-based companies, particularly leading Internet portal companies such as NHN Corp., DAUM Corp., Yahoo! Korea Corp. and so on, began to establish their own in-house design groups in 2003. These design organizations have tried hard to integrate design into their corporate culture by improving their design capabilities.

In this vein, Kim (2008) also conducted case studies on two in-house design centres and their major service projects, which can represent designer contributions to developing new service strategies in a newly emerging market as service initiators:

- SK Communications' UI Design Center and its Cyworld service and
- NHN's CMD (Creative Marketing and Design) division and its Naver Blog Season 2 service.

Among these services, the NHN's Naver Blog Season 2[5] case that follows suggests that in-house designers are committed to a more strategic level of Internet service innovation. The designers' noticeable achievements in strategic management activities forced the rest of the organization to recognize that design is a strategic issue, worth the attention of the highest levels.

Established in 1999, NHN, an Internet portal service company, operates Naver, Korea's top search portal and Hangame, the leading online game portal. NHN consolidated its design competitiveness by creating its own centralized design organization in 2003. NHN also inculcated design DNA into its corporate culture with the support of the CEO's design-centred outlook and built a design-driven Internet service development process. As designers have increased their influence on marketing, NHN took an unconventional step. It incorporated its marketing group into the design centr and then established a CMD division in 2007, appointing Sean Joh, a senior designer, as its Chief Creative Director. He facilitated mutual support between design specialists divided into four functions (designer, researcher, scripter and marketer) in order to create diverse user experiences in a consistent way, thus accomplishing the service brand goal and strategy.

The Naver Blog Season 2 service project shows how designers contributed to NHN's service innovation. This project was initiated by a UI designer in CMD and a new blog service strategy was established by in-house designers because its core innovative factor was the design approach under the theme of 'I'm a blog designer'. In order to enable users to design and manage blogs easily in their own style, like professional designers and developers, NHN's in-house designers developed three design tools: blog skin selection, layout selection and a built-in 'remote-con' webpage management tool. Recently, this blog service received high acclaim from industry experts and blog users due to NHN's efforts to employ new technologies such as Ajax and Flash as well as its innovative design-driven development process. Joh (unpublished interview, 15 February 2007) addressed the designers' contribution to the success of Naver Blog Season 2 as follows:

With the popularization of UCC, designers have two types of role. I want to explain these two roles using a metaphor comparing design activities to cooking. Recently, our designers have had to design food flavors as well as the dishes in which the food will be contained. In the case of Naver Blog Season 2, we designed the dishes and then provided a full set of ingredients for helping people to cook tasty food in a short time. In other words, we designed spaces, where information is displayed and modules, which enable users easily to design and manage the information as a professional designer and developer would.

NHN has been working ever since to launch the next theme for enhancing the competitiveness of its blog services and increasing its active user population. With a deep understanding of the user's desire to express and share online content in diverse ways, its in-house designers established renewal strategies for the blog and initiated the whole process of actualizing the strategies by putting together a new customized blog service experience.

The findings of case studies on the functional roles of designers can be summarized in two main points: first, designer roles in developing Internet services have become more diversified and strategic. Designers in the four functional role types have played complementary parts in achieving the different strategic goals and service demands of their companies. Together with concept generators, service initiators contribute to a more strategic level of Internet service innovation by suggesting new service strategies and concepts in the initial stage of the new service development process, while solution providers and form givers embody Internet services in more functional and emotional ways.

Second, in the early stage of Internet business design, digital design agencies were largely committed to the service development process by solving a variety of different service design problems for their clients while in-house designers were primarily involved in maintaining or modifying their existing services as form givers. Along with the increasing design value in Internet business success, companies established their own in-house design centres and established competitive design-driven development processes; thereby elevating the designers' hierarchical level in corporate management; for example, Sean Joh, CMD's director, now has one-quarter of the corporate decision-making power in NHN. Practically speaking, in-house designers have expanded their contributions into more strategic management activities with long-term perspectives for their own Internet service success, such as brand identity research and user experience research, as concept generators and service initiators. These designer roles in two design organizations have continuously evolved in parallel with the proliferation of Internet technologies and the diversification of Internet services.

## CONCLUSIONS

A growing number of companies have adopted design as a strategic tool and placed design at the top of the strategic agenda for successful innovation in both online and offline business environments. This chapter has discussed designers' work in many different roles ranging from operative tasks to strategic management through case studies.

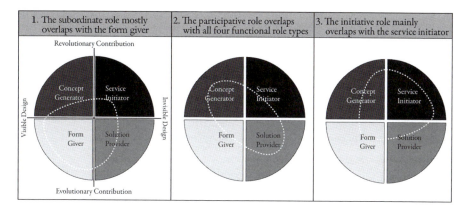

FIGURE 17.2 The relationships between interactional roles and functional roles. © Chung and Kim, 2011.

It can be seen that the product or service development cases chosen in this research are examples of what might be thought of as 'good practice' rather than a random or representative sample. However, it should be noted that studies of 'success' often provide a theoretical framework with some practical applications in the real world.

Although we described a new taxonomy that represents the evolving roles of designers in strategy in two different ways, (interactional roles in the manufacturing sector and functional roles in the Internet service industry), these two types of role can be closely associated with each other. For example, the initiative role is more or less similar to that of the service initiator, while the form giver is subordinate. The participative role can overlap with all four functional role types (Figure 17.2).

In both the manufacturing and Internet service industries, we also discovered distinguishable trends of design management in corporations. In the early stage of adopting design as a strategic tool, companies prefer to collaborate with skilful outside design consultants while their in-house designers are used at a rather subordinate level.

As well as spreading design value at a corporate-wide level with the support of the CEO, designers in in-house design teams have become increasingly committed to actualizing their corporate strategies by visualizing innovative solutions. These contributions to business success also provide designers with a greater opportunity to be promoted to executive positions and to participate in the decision-making processes at the highest corporate levels. Depending on the strategic demands of products or services, in-house designers are tending to work in closer cooperation with design consultants to conceptualize breakthrough solutions or execute their development strategies.

## NOTES

1. Rich Internet applications (RIAs) are Web applications that have some of the characteristics of desktop applications, typically delivered by way of proprietary web browser plug-ins or independently via sandboxes or virtual machines. *Source*: Wikipedia.

2. Bordeaux TV: www.samsung.com/us/consumer/tv-video/televisions/index.idx?pagetype= type
3. In this chapter, the term 'digital design agency' refers to a design consultancy that has a specialty in a wide range of visual communication design fields that are involved in envisioning Internet services on websites and mobile devices.
4. Mple Marble: www.sugarcube.tv/mple.
5. Naver Blog Season 2: (1) www.section.blog.naver.com/intro/UniqueBlog.nhn and (2) www.blogimgs.naver.com /imgs/event/0612/blogrenewal.html.

## REFERENCES

Alexander, M. (1985) 'Creative Marketing and Innovative Consumer Product Design – Some Case Studies', *Design Studies* 6(1): 41–50.

Best, K. (2006) *Design Management: Managing Design Strategy, Process and Implementation.* London: AVA Publishing.

Bruce, M. (1985) 'The Design Process and the "Crisis" in the UK Information Technology Industry', *Design Studies* 6(1): 34–40.

Chung, K. (1989) 'The Role of Industrial Design in New Product Strategy', PhD thesis, Manchester Metropolitan University.

Chung, K. and Freeze K. (2009) 'Samsung Electronics and the Pathway to Integrated Design Success', *Design Management Review* 20(3): 21–7.

Cooper, R. and Press, M. (1995) *The Design Agenda: A Guide to Successful Design Management.* Oxford: John Wiley & Sons.

Hanna, J. (1997) 'The Rise of Interactive Branding', *Design Management Journal* 8(1): 34–9.

Heskett, J. (2005) 'Design Is a Problem, Not a Solution', Better by Design Conference, Auckland, New Zealand, 31 March 2005.

Kim, Y.-J. (2008) 'A Study on the Strategic Role of Design in Internet Business: With Analyses Focusing on Design Management Research Trends and Korean Companies', PhD thesis, Korea Advanced Institute of Science and Technology.

Kim, Y.-J. (2009) 'Transforming the HyundaiCard Website 3.0 in Partnership with Strategic Designers', Design Management Institute Conference, Seoul, Korea, 4 December.

Kim, Y.-J. and Chung, K. (2007) 'Tracking the Major Trends in Design Management Studies', *Design Management Review* 18(3): 42–8.

Laar, G. and Berg-Weitzel, L. (2001) 'Brand Perception on the Internet', *Design Management Journal* 12(2): 55–60.

Long, K. (2004) 'Customer Loyalty and Experience Design in E-business', *Design Management Review* 15(2): 60–7.

Loveday, L. and Niehaus, S. (2007) *Web Design for ROI.* Berkeley, CA: New Riders.

Merriam-Webster Dictionary (2009) *Merriam-Webster,* www.merriam-webster.com (accessed 1 April 2009).

Moody, S. (1980) 'The Role of Industrial Design in Technological Innovation', *Design Studies* 1(6): 329–39.

Nielsen, J. and Loranger, H. (2006) *Prioritizing Web Usability.* Berkeley, CA: New Riders.

Norton, D.W. and Hansen, L. (2000) 'The E-Commerce Blueprint: Creating Online Brand Experiences', *Design Management Journal* 11(4): 25–35.

Perks, H., Cooper, R. and Jones, C. (2005) 'Characterizing the Role of Design in New Product Development: an Empirically Derived Taxonomy', *The Journal of Product Innovation Management* 22(2): 111–27.

Reynolds, C. (1995) 'Design for the Internet', *Design Management Journal* 6(4): 58–62.

Rudd, N. (1999) 'Going Direct: Design and Digital Commerce', *Design Management Journal* 10(4): 17–22.

Schenkman, B. and Jönsson, F. (2000) 'Aesthetics and Preferences of Web Pages', *Behavior and Information Technology* 19(5): 367–77.

Schmitt, B. (2000) 'Creating and Managing Brand Experiences on the Internet', *Design Management Journal* 11(4): 53–8.

Seidel, V. (2000) 'Moving from Design to Strategy: The 4 Roles of Design-Led Strategy Consulting', *Design Management Journal* 11(2): 35–40.

Shida, T. (1996) 'Corporate Identity and the World Wide Web', *Design Management Journal* 7(1): 56–60.

Swack, T. (1997) 'Web Design Analysis: Creating Intentional User Experiences', *Design Management Journal* 8(3): 71–6.

Van Duyne, D. K., Landay, J. A. and Hong, J. I. (2007) *The Design of Sites: Patterns for Creating Winning Web Sites*. New York: Prentice Hall.

Valtonen, A. (2007) 'Redefining Industrial Design: Changes on the Design Practice in Finland', PhD thesis, University of Art and Design Helsinki.

Waters, J. (2003) *The Real Business of Web Design*. New York: Allworth Press.

Wikipedia (2009) www.wikipedia.com (accessed 30 September 2009).

Young, S. (2001) 'What We've Learned So Far: Research Insights into Web Design', *Design Management Journal* 12(2): 43–8.

# 18

# Design Strategic Value Revisited: A Dynamic Theory for Design as Organizational Function

BRIGITTE BORJA DE MOZOTA

C'est ça, une théorie, c'est exactement comme une boite à outils il faut que ça serve, il faut que ça fonctionne. Et pas pour soi-même. S'il n'y a pas des gens pour s'en servir à commencer par le théoricien lui – même c'est qu'elle ne vaut rien ou que le moment n'est pas venu.

*Gilles Deleuze,* L'Arc, *May 1972*[1]

Theory matters and university professors as theorists are disciplinary subjects. The university is a disciplinary institution. Academics as professionals typically hold primary identities as specialists in a specific field. This chapter, therefore, is based on our expertise as an academic specialized in strategy. In the new field of design management, what are the relevant theories? The route towards a design management theory can start from design theories and/or from management theories. So, what is the role of design in the field of strategy in organizations in the light of strategy theory?

Most research in design management is rather practice-based and tends to describe design concepts in an organizational context through design theories: design project management, design strategy, managing a creative team. Design management is interdisciplinary and interdisciplinary work should not be seen as work that alters existing disciplines. Our vision of design management comes from management and from the science of organizations as a way to enrich design management. We follow the route of other researchers, such as Johansson and Woodilla (2008) towards a better partnership between design and management from the viewpoint of management theory. For clarity of organizational analysis, these authors are referring to Burrel and Morgan's analytical framework, with four quadrants using the two dimensions of objectivism versus subjectivism and regulation versus radical change.

This chapter offers an objectivist approach to design management research, which focuses on how the strategic design discourse is grounded in strategic theories. Starting in the functionalist paradigm of understanding the social world through 'sociology of regulation', we use the value chain concept and open a new route in radical structuralist change through the resource based view theory.

This longitudinal research on design strategic value emphasizes unity and cohesiveness between design value and management value. It attempts to explain through a broader view of value 'why' and 'how 'designers and managers may hold together. This 'Designence' model of design value is concerned with social order rather than conflicts, with integration and cohesion rather than contradiction.

It is indeed design management from the viewpoint of management theories and concepts. Yet, although this study could be perceived as an academic perception, it is not so. It is, instead, a result of mixing a scientific background of university studies in economics with ten years' professional experience of managing designers that has been theorized.

## FUNDAMENTALS

In my previous research based on management models (Borja de Mozota 1992) a convergent model of design management was developed based on two perspectives: the reactive or managerial approach and the proactive design approach.

The managerial approach involves enhancing design impact in organizations by accommodating management concepts. All management paradigms are examined in order to choose the ideas and methods that will make design more efficient. This can be achieved by linking design with concepts such as organization, reputation or strategy. This perspective applies the different theories of management – scientific, behavioural, decisional, systemic, situational – to design issues. The role of design management is to improve design with managerial knowledge.

In contrast, the design approach involves examining design as a new paradigm in order to arrive at ideas and methods that can be used to enhance the efficiency of management in general and design management in particular. Since a different vision of organizational reality emerges from the science of design, the management system is enriched by design. The objective of design management is, then, to improve management with design knowledge (Junginger 2006).

In this chapter, the longitudinal study of the theory of strategic value design in organizations refers to the managerial approach of design management and to the concept of strategy and competitive advantage.

The first research question is simple: 'what is the value of design for strategy in organizations?' So the theoretical background is both strategy theory and design theory. The main finding is that design is a function of organizations and, consequently, a management method. Therefore evidence is given for the second approach of design management based on a new theory of strategic management: the resource based view of strategy.

Having explained our research approach, why focus on value? Value is what motivates human activities. Value, according to Peter Drucker (1967) is the essence of what organizations are. Managing for value is not just about profit making; it is also about bringing about long-term changes such as shareholders' value. Our analysis is not about implementing simplistic economic metrics such as Stern Stewart's Economic Value Added (EVA™, New York) as a key measure of performance. Value is not only about change in the accounting system; it is about commitment to shareholders and to stakeholders' value. It is about facing resistance to change in order to implement the desired cultural transformation that will enable all the workers, including designers, to make value creating decisions.

The main contributions of value-based management (Haspeslagh et al. 2001) are: to understand where value is created or destroyed in companies, to make employees appreciate that capital has a cost, to make managers focus on the balance sheet, to allocate resources towards the most productive uses, to make sure that business units make a profit that cover the cost of capital, to make managers act like owners of the company, to improve communication between business units and the corporate centre, to improve stock prices and to make strategic planning less of a paper exercise.

Value is also the aim of the design activity. Every design project is targeted for value creation. Design research, publications and education are based on the certitude that the design community has high standards of ethics and shared values. Bringing value to society and to human beings is what design is about. Yet when it comes to assess these design values, designers tend to rely on either peer reviews – as in design awards for 'good design' – or on the numeric evidence of design – improving sales figures, brand market share and reputation. Hence, the design community is missing the broader point of the concept of value from the management community.

Indeed, it is an interesting paradox between the conviction of designers to bring value to organizations and society and their total ignorance of what organizations mean by – and how they create – value. This chapter explores the value paradox through the dynamics of theory building of value and design in two sections. The first section explains the foundations of the theory with its four loops. The second section discusses the professional and future developments of the theory of design management in strategy (see Table 18.1).

**TABLE 18.1: Design strategic value longitudinal retrospective**

| Research Loop | Period | Theory Background | Main Findings |
|---|---|---|---|
| 1. Design as value | 1980–90 | Peirce model Profit Performance | 1. Design is a generic activity across design disciplines<br>2. Design increases an organization's profit because it improves innovation management, the innovation itself and the marketing of innovation<br>3. Design value can be measured as perceived value, economical value *and* process value. |
| 2. Design value in the value chain | 1990–5 | Value chain | Organizing the variables describing the value of design into the model of the value chain. Design brings value to the principal, activities (differentiator) to the support activities (coordinator), and to the value chain system (anticipator). |
| 3. Design value chain in the value management | 1995–2002 | Michael Porter Strategy Competitive advantage | Demonstrating how design is positioned in the value chain gives a typology of Design management in three clusters:<br>– Design strategy as economic value<br>– Design strategy as managerial value<br>– Design strategy as resource value |
| 4. Design strategic value as core competency | 2002–9 | Resource-based view | Giving evidence of design as core competency.<br>Design as 'generic design' or 'design thinking' is an activity, a management skill, rare, inimitable and non-substitutable that can bring a sustained competitive advantage ('design you can't see' competitive edge). |

'Designence' value index history, B. Borja de Mozota.

# DESIGN STRATEGIC VALUE REVISITED: THE FOUNDATIONS OF THE THEORY

## Design Equals Value, 1980–90

How does one define design value in organizations? My PhD in business (1985–90) pioneered research on design as 'generic' design meaning and looked at design as an activity across different design projects. One of the main findings of this research showed that value by design was independent of the design disciplines involved. Eleven case studies could be described by the same semiotic model of Charles Peirce (Peirce 1931–58) – a sign as a *representamen*, an *object* (the sign's subject matter) and an *interpretant* (the sign's meaning). When asked how these various design projects created profit in their organizations, our respondents said that design increased profit because:

- Design improves innovation management through participative innovation, more creativity and market orientation of innovation.
- Design improves the performance of the company's portfolio through better use of technology, utility and image.
- The value of design was measured by its impact on profit, such as increased sales and portfolio margins. Then, the research gave evidence of the causal relationship between the substantial value of design and its measured impact on corporate image and reputation and, also, through process improvement in innovation and marketing. For example, there was evidence that design investment cuts marketing costs. Hence, a new value for design management for improving some organizational processes emerged.

## Design Strategic Value in the Value Chain, 1990–5

If design creates financial and substantial value, does this mean that design is creating a competitive advantage? Firms gain a competitive advantage from conceiving new ways to conduct activities, employing new procedures, new technologies or different inputs. Firms create value for their clients through performing these activities. To gain competitive advantage over its rivals, a firm must either provide comparable customer value but perform activities more efficiently than its competitors (at lower cost) or perform activities in a unique way that creates greater customer value and commands a premium price (as in differentiation).

The activities performed in competing in a particular industry can be grouped into categories that Porter (1985) calls the value chain. Activities can be divided broadly into those involved in the ongoing production, marketing, delivery and servicing of the product (the primary activities) and those providing procurement, research and development, human resources management or overall infrastructure functions to support the other activities (the support activities).

A company's value chain for competing in a particular industry is embedded within a larger system of activities, the value system of which includes suppliers and distributors. A company can create a competitive advantage by further optimizing or coordinating its links to the outside world. So where does this leave design in the value chain? Is design adding value to the primary activities only or also to the support activities? Can design help in transforming the value chain of an industry?

At this stage of theory building, the focus is to link design value with the concept of the value chain. A review of the literature points to a consensus among researchers that competitive advantage by design is covering the three activities of the value chain:

1. Design as differentiation or design creates value to the principal activities of the value chain. Design is strategic because it creates customer value through differentiation perceived by the market inducing customer behaviour. Developing semantic value. Improving market research methods. Creating brand value.
2. Design as coordination or design creates value to the support activities of the value chain. Design is strategic because it creates value through the coordination of functions by changing the NPD process: user oriented NPD. Visualization as coordination tool in innovation teams. Developing team management and concurrent engineering. Technological change.
3. Design as transformation or design creating value to the value chain system of the industry. Design is strategic because it adds value through the anticipation of changes in the firm's environment, whether internal or external: design is seen as a cognitive, psychological value in an 'interaction-driven' firm, which induces societal value, a new vision of the environment, develops cultural long term value for shareholders and partners. Inducing social value such as 'green design'.

Based on the consensus in the literature, can we find evidence of how managers were actually managing design in their value chain?

### Design Strategic Value: The Final Model, 1995–2002

If design creates value through its impact on the value chain, what are the drivers of design for creating unique competitive advantage? Are all organizations similar in their strategic vision of the role of design in the value chain? The European Design Prize, created with the aim of reinforcing the competitiveness of European firms, presented an ideal research context for investigating the managerial side of building design value (Borja de Mozota 2002).

One of the most interesting results was the importance of time and with that, the accumulation of years of experience in building competitive advantage through design. We can call this the learning curve of design. Experience in design was more important than company culture or company size. Yet, size was important in order to understand the impact of design on cost or for design to be viewed as an internal ability. The smallest companies were less convinced of the potential of design to reduce cost. The larger the company, the more design was perceived as a management asset with internal impact.

Geographic differences emerged in design management, which reinforced differences in European management styles: in northern Europe design was seen as a know-how that transforms processes, while in southern Europe design was limited to a useful tool in project innovation within a multidisciplinary team.

Twenty-one different variables characterizing design management were analysed and classified in order to isolate the variables with the highest scores in the data matrix (Table 18.2). Design, which creates a competitive advantage, arrived in the first position. The thirty-three European design-driven firms in the sample reported unanimously that design provided them with a competitive advantage.

Yet, there was no consensus in the way these companies connected design value with competitive advantage. A factorial analysis was conducted in order to group the twenty-one variables into significant clusters. The research proved that the discriminating factor between the economical versus the managerial vision of design value was: competitive advantage as market 'fit' versus core competency.

No correlation could be found between the variables describing economical value and the variables describing managerial value. Their independence was also reinforced by the fact that the variables describing the role that design plays in the innovation process were, on the contrary, highly correlated.

So, these companies might not be similar in the way that they see design and its impact on the value chain. All thirty-three European SMEs employed excellent product design strategies, but the research question was: Do they manage design similarly? How do the CEOs of these SMEs link design with value management? A perceptual map representing the matrix of all twenty-one variables' data along five significant axes was used to arrive at a typology of design management. It gave evidence that a variety of design management styles are being employed. Seven discriminating drivers emerged out of the twenty-one variables. These drivers are highlighted in Table 18.2 in bold type. The thirty-three SMEs, all excellent in design, were in four different classes according to design leadership.

This typology validates the research assumption of using the concept of the value chain to explain different design management values and gives further evidence on how the value management concepts apply to the field of Design Management. The findings are known under the 'Designence' model of design Management value.

The leaders of these thirty-three SMEs agree on three routes for strategic competitive advantage through design: core competency organizational process and economic value. I will explain in the next section what they imply.

**TABLE 18.2: Classification of the twenty-one variables of design management**

| Design Value Variables and Rankings* | Mean | Dispersion |
|---|---|---|
| 1. **Design creates a competitive advantage** | 5.39 | 0.55 |
| 2. **Design is a core competency** | 5.12 | 1.04 |
| 3. **Design contributes significantly to benefits perceived by consumers** | 5.00 | 0.97 |
| 4. Design changes the spirit of the firm that becomes more innovative | 4.94 | 0.86 |
| 5. Design develops exports | 4.88 | 1.15 |
| 6. Design increases market share | 4.75 | 0.94 |
| 7. **Design allows the company to sell at a higher price** | 4.69 | 1.16 |
| 8. **Design improves co-ordination between marketing and R&D functions** | 4.68 | 1.07 |
| 9. Design is a know-how that transforms the activity processes | 4.64 | 1.12 |
| 10. Design develops customer care in the innovation policy | 4.60 | 1.25 |
| 11. Design generates technology transfers | 4.22 | 1.47 |
| 12. Design gives access to a wide variety of markets | 4.19 | 1.55 |
| 13. **Design accelerates the launch of new products** | 4.07 | 1.28 |
| 14. **Design improves co-ordination between production and marketing** | 4.00 | 1.16 |
| 15. Design develops project management of innovation | 3.93 | 1.20 |
| 16. **Design creates a new market** | 3.90 | 1.72 |
| 17. Design improves the circulation of information in innovation | 3.80 | 1.34 |
| 18. Design means higher margins or costs reduction | 3.80 | 1.31 |
| 19. Design is difficult to imitate by competitors | 3.76 | 1.43 |
| 20. **Design changes relationships with suppliers** | 3.70 | 1.23 |
| 21. Design improves co-operation between agents | 3.64 | 1.18 |

*Rankings*: ***6** = fundamental; **5** = very important; **4** = important.

### Core Competency: Design as Resource Value

Six of the thirty-three firms show an orientation toward being 'market and client driven' when it comes to the strategy and a vision of 'external transaction cost' of design. Design management in these companies gives priority to the impact of design in terms of perspective and imagination and on continuous quality improvement. High scores are given to 'innovation driven by design' and design seen as a 'know-how that transforms the processes'.

## Organizational Process: Design as Managerial Value

The variables in this cluster share an 'innovation' vision of design management and a strategic orientation based on internal transactions costs. These sixteen firms, out of a total of thirty-three, perceive the competitive advantage of design by the value it creates on the management of the support activities and, in particular, on the role given to design as a source of ideas and innovative concepts. The variable 'design' changes the spirit of the personnel that becomes more innovative' is the one that has the highest score.

## Differentiation Value: Design as Economic Value

Five firms do not see the importance of design in innovation management. They attribute a lower score to each modality that tends to give a managerial value to design. These companies have an economic profit-oriented vision of design. The value created through design is judged by its impact on marketing-mix policies. Design management is operational and limited internally to product policy and externally to product performance.

Four firms are uncertain as to the value that design can create. Here, design integration seems conjectural. Design management shows no objective of creating a competitive advantage, only the willingness to innovate within the product portfolio.

## Design Strategic Value as Core Competency, 2002–9

This new evidence of 'design as resource' and 'core competency' raised the question if another theory in strategy, the resource based view (RBV) of strategy, is pertinent to design's strategic value. Could this RBV theory explain the value of design through 'design thinking'? The resource-based view of strategy highlights how the possession of valuable, rare, inimitable and non-substitutable resources may result in sustained superior performance. The resource-based view of a firm's competitive advantage emphasizes the importance of the invisible assets or the 'design you can't see' value (Figure 18.1).

Since the design profession is increasingly recognized as research and knowledge based, there is a paradigm shift in the design activity from project oriented to process oriented. Designers are more than just problem solvers; they are actors of the dynamics of knowledge building in organizations through research:

> The activity of design consists in the transformation of an input representation into an output representation. In an activity that functions by way of representations, knowledge plays a central role. Designing is a cognitive activity. (Visser 2006)

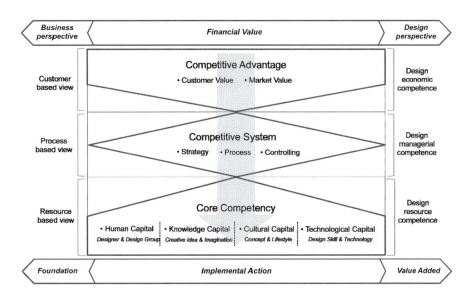

FIGURE 18.1   From design as fit to design as resource. © Borja de Mozota and Kim 2009.

Hamel and Pralahad (1994) argue that information-based invisible assets such as technology, customer trust, brand image, corporate culture and management skills are the real resources of competitive advantage, because they are difficult and time consuming to accumulate and can be used in multiple ways simultaneously. To design managers, it means assessing design value as a resource that is rare, inimitable and non-substitutable; it also means managing design with the long-term perspective of sustained competitive advantage rather than short-term view of project management.

In summary, when pleading for strategic design, design managers, designers and design educators should articulate and explain their definition of strategy, whether they refer to strategy as 'fit' or strategy as 'RBV'. Additionally, if they want to adopt a prospective and contemporary view of design strategy, they may turn to the Resource-Based-View theory and a vision of design as 'core competency'. In the following section we consider the tools and methods in order to implement design at a strategic level.

## THE DESIGN BALANCED SCORECARD: ASSESSING DESIGN'S STRATEGIC VALUE IN PRACTICE

One can only manage what can be measured. Therefore we need a model for design value to be assessed and measured. In this section, we introduce a model, 'Designence', which any designer can apply, whether for managing a design project or a design firm or a design department, as the scope of the design profession is changing

**TABLE 18.3: 'Designence' in practice: design value for company performance**

| | |
|---|---|
| How should we appear, through design, to our customers in order to achieve our vision? | To satisfy our stakeholders how can design help in the business processes we in which we excel? |
| **Design as difference** | **Design as performance** |
| Design management as perception and brand | Design management as innovation process |
| Market value. Customer value. | Innovation. Modular architecture. |
| Brand. Consumer research. Competitive advantage. | Time to market. TQM. R&D. Technology |
| **VISION** | |
| How will we sustain, through design, our ability to change and improve? | To succeed financially, how should design appear to our shareholders? |
| **Design as transformation** | **'Good design is good business.** |
| Design management as core competency | Societal value of design |
| Strategic value. Vision. Prospective | Financial and accounting value. ROI. |
| Change management. Empowerment | Value for the society. Stock market value. |
| Knowledge learning process. Imagination. | Socially responsible enterprise. |

with an 'activity-based' and 'research-based' view (Laurel 2003). The 'Designence' model is depicted in Table 18.3. Using this model turns design into an activity of the organization and a resource that improves its organizational, knowledge and information capital. The three design management clusters define the dimensions and the criteria that have to be considered when defining the design strategy.

Implementing design value is made easier by implementing the model through the management tool of the Balanced Scorecard or BSC (Kaplan and Norton 1996). The BSC model has many advantages for design managers. Widely known and used by auditing and strategy consultants, it has both a dynamic and a long-term vision that is coherent with design management. It provides a simple framework that can be applied to any design decision or design project. Finally, because most designers only see design value through the value of their artefacts and their creativity, it insists on other values such as organizational and financial value.

In order to achieve a strategic vision, the BSC model gives four different perspectives for design management: the market perspective, the performance perspective, the learning perspective and the financial perspective. Each perspective is equivalent to the design management clusters defined previously. Within each perspective, criteria for assessing design are selected (see the case study in Table 18.4).

**TABLE 18.4: Case study: Steelcase Co.**

## The Value of Workplace Design for Business Results

As businesses experience new dimensions of competition, more organizations see how workplace design affects bottom line results. Using the workplace as a leverage point, organizations can better facilitate structural realignment: implement new technology, redesign business processes and reinforce the organization's values, culture and image.

Measurements related to the workplace have typically focused on cost per workspace, space efficiency, reconfiguration costs, and energy use – the cost side of the cost/benefit equation. The workplace, however, significantly affects an organization's people, processes and technology. In the business results model shown below, the workplace is one of four key factors that drive business results. Efforts in all four areas must be integrated, balanced, and measured.

*Workplace 'Firm A': Improve Worker Interaction and Workplace Flexibility*

In a high-tech electronics firm 'A', the allocation of workspace was based on hierarchy, status and rank. As the firm re-engineered and moved to a more fluid, team-based work process, the design of the workplace impeded progress. Members of one team were located on multiple floors, conference rooms were unavailable at short notice, and moving one person took up to twelve weeks. When the firm redesigned the workplace, members of each team were co-located to encourage informal communication. Collaborative space was integrated into the teamwork setting to facilitate interaction. Freestanding furniture within panels cut the time required for personnel moves from twelve weeks to twelve hours. A modular network and lay-in cabling sharply reduced changes to network connections.

| | |
|---|---|
| **Value for the Market Position** | **Value for the Process** |
| Increase market share | Accelerate product development process |
| | Implement self-directed work teams |
| *Measure:* | *Measure:* |
| Market share contribution of new products | Time to market (before and after) |
| **Value for the Personnel and Knowledge Management** | **Financial Value** |
| Increase worker interaction within product development teams | Move people and equipment not furniture and cables |
| *Measure:* | *Measure:* |
| Workplace flexibility to support frequently changing work teams | Time and costs required for workplace moves, adds, changes reduced by 72 per cent |
| | ROI in five years. |

Design managers have to answer four questions:

- How should we appear, through design, to our customers in order to achieve our vision? Design knowledge applied to corporate difference-building and strategic market positioning.
- To satisfy our stakeholders how can design help in the business processes in which we excel? Design provides improvements in company performance and processes, these innovations in processes being totally invisible to outsiders.
- How will we sustain, through design, our ability to change and improve? Design-explicit knowledge is applied to strategic change, perspective, personnel empowerment and talent search.
- To succeed financially, how should design appear to our shareholders? Design explicit and measurable value of company reputation, stock market performance and societal responsibility.

Applying strategy theory and the BSC tool offers the design profession both a theory and a concept for communicating the value of the design activity within organizations. With that, design value can become part of the business jargon.

## DESIGN MANAGEMENT VALUE IN PRACTICE: TWO ROUTES FOR DESIGN INTO CORPORATE STRATEGY

It would be more interesting if models could be dynamic. It will make them more human.

*Prof. Larry Leifer, Director, Design Thinking Research Laboratory,*
*Stanford University*

Theory and models are relevant only if they are useful for practitioners. So what are the insights from strategy theory that may be useful for designing in the twenty-first century? Organizations' structures have to change: from modern or 'classical' machine view, to 'interpretive' image driven or 'post modern' organizations as a collage. The history of organization theory has followed a parallel route of design theory. Interdisciplinary design management is opening the silos. We can see a move away from organizations seen as classic or scientific 'machines' where design managers are rational actors with a focus on power based on engineering design methods. The trend is towards modern organizations and 'living systems', where design managers act more like an independent part of this living system applying systemic design tools. There is, further, a move towards organizations as interpretive or as 'symbols', where managers focus on artefacts or symbols of the organization and design managers focus on environmental interpretation and storytelling. There is further a shift to view current postmodern organizations as a 'collage', where design managers have to behave like an artist or theorist focusing on creativity, personal freedom and individual responsibility.

From a strategy perspective, there are two routes for strategic design in practice: the incremental strategy of 'meta design' and the radical strategy of 'core competency'.

### Design Management Incremental Route: The Meta Design Strategy

Our global context is redefining the existing design disciplines through changes such as sustainability, ethics and the digital economy. Traditional design disciplines have broadened to answer these changes; these include eco design, inclusive design, experience design, sensorial design and service design. Yet, even more important on a strategic decision level is the emergence of new meta disciplines that act as bridges between existing designs disciplines, in order to develop a coherent design strategy for the organization's value chain.

Tactical design management is a key issue. It is understood that design leadership has to be completed at a process decision level both for protecting design as a new function in the organization and for developing design coordination processes with all the other specialist functions such as R&D, brand, Customer Relationship Management.

An independent design function reports to the company board. Design strategic value is a research asset useful to understand the environment and to develop new vision. Design research methods are specific and differ from R&D and marketing yet share the same aim of providing the organizations with the knowledge and vision necessary to survive in the long term.

The 'artefacts' conceived by designers are 'embedded knowledge'. Design is a knowledge-based activity. Consequently, design managers can use their design-balanced scorecard to improve the integration of designers' implicit and explicit knowledge in all managerial decisions for value creation in value management.

### Design Management Radical Route: The Core Competency Strategy

The radical route explores how design knowledge might be exported and imported across the traditional borders of design. New design forms emergent roles for design that are value driven, such as designers as negotiators of value, as facilitators of thinking, as visualizers of the intangible, as navigators of complexity and as mediators of stakeholders (Inns 2007).

More CEOs are thinking of design not from the standpoint of the design outcomes, but from the standpoint of contemporary managers' challenges that can turn to design thinking for solutions and for inventing new ways of governance. Suddenly, design is 'a must for companies' performance'. Chief executives attend design seminars or send their managers to design studios for education in order to bring more creative thinking into their jobs. As an example, Procter & Gamble is changing its culture and integrating design in the company DNA.

This strategy is also broadening design skills to face the challenges of contemporary managers and society governors. The design activity is 'sense making' (Weick 1995) for organizations that are facing choices in their organizational model. So in practice, what are the current managers' challenges for design management?

- *Managing complexity*: applying 'design thinking', holistic view and 'information design' visualization skills to simplify complex environments, to build scenarios for system change.
- *Globalization and innovation*: companies have to be international: design teams are used to multicultural working environments and creation has no frontier. Designers can help changing to be both excellent in standardization and in personalization for 'glocalization'.
- *Process-oriented companies*: companies have to be more human centric, customer driven, process oriented; new information systems have to be invented for customer experience management. These managers' challenges are a perfect application for 'user-oriented design', provided that designers take a wider view of users' 'design for all' inclusive design: users as employees, shareholders, suppliers and so forth.
- *Socially responsible organizations*: while this business model is growing, methods are needed in order to implement it as a collective action, as 'sense building' (Weick 1995). Designers' input can go beyond 'advanced design' projects in eco design, towards inventing and implementing standard processes for change towards SRE enterprises.

By using the background of RBV strategy theory, it is possible to reverse the strategy map of the design ladder or of the necessity of years of experience in design management before reaching the strategy level. The design Balanced Scorecard shows another route that is dynamic, long term and actively 'human'.

Design can enter organizations from the intangible value of design skills – what we have named above the transformation value or the learning perspective in the BSC.

Additionally, designers have to avoid their myopia towards the financial value of their profession. Any design decision has both a financial and substantive value. This is particularly important when the new international accounting standards IRFS are evaluating the intangibles in the organization's capital, where design is involved through intellectual property or talented creative human capital or brand or customer relationships.

So, in practice, what is the route for this radical design management (Figure 18.1)? This alternative route starts with design skills as core competency. Design directors and CEOs in strategic committee contexts decide first which financial objective is preferred: productivity or growth. In order to reach this financial target, design managers have to follow four phases:

1. Insert design talent in the company capital and intangible assets – whether human capital, information capital or organization capital.
2. Align this intangible value with the corporate strategy through design strategy, job position, information technology and organizational charts.
3. Develop processes such as operational processes, customer management processes and innovation regulatory processes.
4. Finally, create three different types of customer value: product/service attributes, customer relationship, brand.

In the near future, design management will be driven by 'transformation design value' or by a 'design you can't see' attitude. Choosing design as core competency will transform both strategy and organization, then reconcile finance and culture: The Art Firm (Guillet de Monthoux 2004). The design value chain has changed priority. The design strategy map starts by the 'design you can't see' value. It is design as research and knowledge, reinforced by employees' training and empowerment, adequate organizational processes and technologies that in the end produce competitive market and brand value.

## CONCLUSION: GENERALIZING THE RESEARCH FRAMEWORK OF DESIGN SCIENCE IN ORGANIZATIONAL STUDIES (2010–)

During these postmodern times, interdisciplinarity is happening in a period in which the aging modern university still dominates the educational landscape. While disciplinarity survives intact, interdisciplinarity proliferates rather as counterdisciplines.

Design management, as a new interdiscipline, is like other interdisciplines: often housed in insecure research programmes and in poorly funded research departments. Rather than a counterdiscipline, it is 'a multidiscipline' built on the assumption that close cooperation among different disciplines is a wholesome thing. The postmodern version of design management interdisciplinarity seeks not to unify or totalize, but to respect the differences and not the dream of the end of the disciplines and their jargons.

What happens is always contamination. (Jacques Derrida, in Leitch 2003: 170)

The idea of Design management as contamination is not clear in terms of which connections might exist between the aims of this view of design management and today's market economy – with its need for rapid change renewal, its disregard of established ways and traditions and its preference for flexibility and temporary contracts.

Theory building is important for creating a mental image of the territory of design management. The example of the strategic theories, explored in detail over two decades, can be replicated with other management theories. With such interdisciplinary research, Design Management will be defined in organization studies.

Because the design activity is an activity that is integrated within society, institutions and organizations, analysing this activity from the viewpoint of organization studies is pertinent. Romme, in 2003, saw 'organization as design' and how scholars in organization studies can guide human beings in the process of designing and developing their organizations towards more humane, participative and productive futures.

So this viewpoint will provide insights for designers and design managers on how to adapt their roles to the different organizational contexts and also on understanding the reasons underlying the success of 'design thinking' in our postmodern economy.

Design strategic value is driven by the organizations' strategic platform in strategy definition, selection and implementation processes. When managing design value in organizations, it should not be forgotten that all organizations are different in the way in which they develop their strategies. Nonetheless, design managers can act as actors of change in the theory behind the corporate culture through the assessment of design strategic value.

## NOTE

1. *Author's translation*: 'This is what a theory is; it is exactly like a toolbox. It has to be useful. And not for yourself. If there are no people to use it, starting with the theorist himself, then the theory is worthless or its moment has not yet come.'

## REFERENCES

Borja de Mozota, B. (1985) 'Essai sur la fonction du design et son rôle dans la stratégie marketing de l'entreprise', PhD thesis, Université Paris I Panthéon Sorbonne.

Borja de Mozota, B. (1990) 'Design as a Strategic Management Tool', in M. Oakley (ed.) *Design Management: Handbook of Issues and Methods*. Oxford: Blackwell.

Borja de Mozota, B. (1992) 'A Theoretical Model for the Future of Design Education and Research', *Design Management Journal* 3(4).

Borja de Mozota, B. (1998) 'Structuring Strategic Design Management: Michael Porter's Value Chain', *Design Management Journal* 9(2): 26–31.

Borja de Mozota, B. (2002) 'Design and Competitive Edge: A Model for Design Management Excellence in European SMEs', *Academic Review Design Management Journal* 2: 88–103.

Borja de Mozota, B. (2003) *Design Management*. New York: Allworth Press/DMI.

Borja de Mozota, B. (2006) 'The Four Powers of Design: A Value Model in Design Management', *Design Management Review* (Spring): 44–53.

Borja de Mozota, B. and Kim, B. (2009) 'Managing Design as a Core Competency', *DMI Review* 20(2): 66–76.

Drucker, P. (1967) *The Effective Executive*. New York: HarperCollins.

Guillet de Monthoux, P. (2004) *The Art Firm: Aesthetic Management and Metaphysical Marketing*. Stanford, CA: Stanford Business Books.

Hamel G. and Prahalad, C. (1994) *Competing for the Future*. Cambridge, MA: Harvard Business School Press.

Haspeslagh, P., Noda, T. and Boulos F. (2001) 'Managing for Value: It's Not Just about Numbers', *Harvard Business Review*, July–August: 65–8.

Inns, T. (ed.). (2007) *Designing for the 21st Century: Interdisciplinary Questions and Insights.* Aldershot: Gower.

Johansson, U. and Woodilla, J. (2008) 'Towards a Better Paradigmatic Partnership between Design and Management', International DMI Education Conference: Design Thinking: New Challenges for Designers, Managers and Organisations, ESSEC Business School, Cergy Pontoise, France, April 14–15.

Junginger, S. (2006) 'Change in the Making: Organizational Change through Human Centred Product Development', PhD thesis, Carnegie Mellon University, Pittsburgh.

Kaplan, R and Norton, D. (1996) *The Balanced Scorecard: Translating Strategy into Action.* Cambridge, MA: Harvard Business School Press.

Laurel B. (ed.) (2003) *Design Research, Methods and Perspectives.* Cambridge, MA: MIT Press.

Leitch, V. (2003) *Theory Matters.* New York: Routledge.

Peirce, C. (1931–58). *Collected Papers of Charles Sanders Peirce (1931–1958).* Edited by C. Hartshorne and P. Weiss (volumes 1–6) and A. Burks (volumes 7–8). Cambridge, MA: Harvard University Press.

Porter, M. (1985) *Competitive Advantage: Creating and Sustaining Superior Performance.* New York: The Free Press.

Romme, A. (2003) 'Making a Difference: Organization as Design.' *Organization Science* 14(5): 558–73.

Visser, W. (2006) *The Cognitive Artifacts of Designing.* New York: Routledge.

Weick, K. (1995) *Sensemaking in Organizations.* Beverly Hills, CA: Sage Publications.

# 19

# Design Management as Integrative Strategy

## LISBETH SVENGREN HOLM

In a way, it is obvious that design has become accepted as an important competitive tool for business success. And yet, we still know surprisingly little about what constitutes design, especially if we refer to design strategy, or design as a strategic resource. As early as the 1980s, design was highlighted as a strategic factor in the increasingly intense global competitive market (Kotler and Rath 1983; Lorenz 1986; Gorb 1988). The focus was, however, primarily on design as a way to differentiate products and design as a communication tool in a world where technical performance and quality no longer were the main differentiators. I would argue that this approach represented a tactic by design management scholars and practitioners who felt that it served their interests if design management was viewed as a strategic issue (cf. Pilditch 1987; Gorb 1988; Olins 1989). At that time, design was also thought to involve only physical objects. Some researchers within design management in the 1980s and 1990s did, however, include issues of development and innovation (Roy 1986; Walsh et al. 1988; Walsh et al. 1992; Cooper and Press 1995; Svengren 1995) and argued for the design process as equally, or even more important if we wanted to understand not only the role of design for business success but also design for a wider scope of management (Johansson and Woodilla 2009; Junginger 2009).

Discussions within the field of design management often focused on the coordination of different design areas, particularly those of product design, communication design and environmental design. This categorization of design, argued Gorb (1988), reflected not only the different functions of a company but also the structure of the balance sheet. The goal of coordinating product design, communication design and environmental design was to create a clear and consistent visual identity, which eventually formed the concept of corporate identity (Gorb 1988; Olins 1989). Design management, in this discourse, was about developing a system of control to verify that all design work was done in accordance with a specific design policy, i.e., there was a need for a design manager with a mandate to act as controller and enforcer. In particular, this was a model for large companies, especially for multinationals with independently acting subsidiaries. This situation posed a problematic context not only for design and identity but for many different management subjects.

Differentiation and communication are important aspects of market positioning. The design of the product should communicate functional and emotional added value and at the same time symbolize the brand's distinctive values. Let us for a moment consider market perspective and a product's value. The conventional view in a marketing perspective is that value is something that is added to the product until there is an optimal balance between desires, costs and the price a customer is willing to pay. This view on how value is created is based on the fact that certain features of a product are standards. When adding something above standard there is an 'added' value. If this added value is unique, it is called a USP (unique selling proposition). Moreover, if it is of value to the market, it is a strategic competitive advantage. If this added value is difficult to imitate and will retain its value for a long time, it could then become the basis for a sustainable strategic competitive advantage (Barney 1991). This forms the basic reasoning in traditional marketing.

But this reasoning about design and added value is too narrow and involves too one-sided a focus on the result of the design process, on product design. The risk inherent in this reasoning is that it neglects the importance of the design process, the choice of design methods and design thinking in creative problem solving on the one hand and that it overlooks the link between strategic concept and innovation on the other. Integration of design as a strategic resource cannot have one without the other. The designed product, the design outcome, is a result of design capabilities throughout the whole process.

However, this discussion faded away when brand management evolved into an umbrella concept for the integration of all communication during the 1990s. Brand management hence became the hot topic at business schools and pushed design management into the background. The brand manager should have the mandate, not the design manager. The latter was mainly found in those design departments that worked with product development. The root of the brand discourse is quite different compared to that of design. If design is based on product development (regardless of which kind of products) with roots in art, craftsmanship and artistic practices, brand is rooted in law and related to ownership. It was, of course, important to communicate ownership and the brand – or more correctly: the logotype – became the most effective design element to do this.

What was neglected, however, when brand management took charge of the ideas of visual coordination was that the design process, based on the creative methods of industrial design in particular, as such also represented a new approach that, itself, could have constituted a new strategic resource if innovation had been more of an interest. The fact that innovation has become the cherished area within management since the late 1990s, not least with an interest in 'blue ocean strategy' (Kim and Mauborgne 2004), is an opportunity and strategic window for design management as a field of research and theoretical development. But this will require major theoretical development of design management related to both innovation and strategic management: at least more than what has been the case hitherto.

To contribute to design management theories from the strategic management perspective, and design as a strategic resource, we first need to define what we mean by strategic resource. This chapter will (a) discuss some of the issues that are relevant to the development of strategic management, (b) explain what creates strategic advantages through design and (c) show why strategic management is relevant for design management. I will present three different integration processes that together constitute design management at different levels within an organization. The aim of this chapter is to provide a framework for the concept of integrating design as a strategic resource. I will use case studies from my own research for illustration.

## STRATEGIES AND COMPETITIVE ADVANTAGES

Business is about how various resources that the company owns or has access to are managed to achieve certain goals. Financial resources and different capital assets have ceased to be the most crucial basis for a sustainable competitive advantage, especially since the market conditions changed due to significant influences like globalization and information technologies. In the early 1990s, researchers in strategic management tried to define what the conditions were for a resource that could be the basis for a competitive advantage. This research was referred to as a resource-based view (cf. Barney 1991). To constitute a resource for creating a sustainable competitive advantage the resource had to be (a) relevant for the target market, (b) distinctive and (c) difficult to imitate (Barney 1991; Amit and Schoemaker 1993). The research showed that good financial resources are important but not the basis per se for competitive advantages. Researchers (for example, Prahalad and Hamel 1990; Barney 1991) began to identify process-related issues as the basis for a resource difficult to imitate. This led to a series of questions centring on the possibility to develop superior skills that were distinct and difficult to imitate (Prahalad and Hamel 1990). This discussion parallels and links with the discussion of the knowledge-intensive firms (Starbuck 1992; Alvesson 2004) as well as with the discussion of intellectual capital as asset, a trend in management research that became popular in the 1990s (Edvinsson and Malone 1997). But this discussion had a firm focus on the human capital from an intellectual perspective and not from a skill or from an organizational perspective.

A company's business idea has to be viewed, according to Richard Normann (1975, 2001), not only from a market perspective, but also from an organizational perspective. The market is dynamic and changes can occur quickly. Organizational changes are slower and cause frictions. The development of a business idea is first shaped by the extent to which it requires change and second by the kind of change it involves. Some development can take place with existing resources and competencies. Some changes require new kinds of resources, competencies and skills and can therefore lead to deep changes that also affect the company's political processes and power positions. Dominating ideas and significant actors are within this perspective important to understand when we discuss opportunities and need for change and development. Although we know that dominating ideas and actors are crucial for how a company manages to develop a new course of direction, this perspective belongs rather to organizational development theories than to management models.

A market-oriented approach, which is a distinct basis for management models, represents more of a normative approach and states that companies need to be flexible to adjust to the market's changing demands. According to Prahalad and Hamel (1990) this could lead to a situation where companies are focused on the market alone, try to imitate leaders, and forget to find creative solutions based on their own core competencies. And, I would add, this could lead to a situation where companies do not work with visions informed, for example, by a wider understanding of their users and of trends in society. Prahalad and Hamel, however, called for a greater emphasis on visions in regards to strategic intent and that recommended companies utilize their existing competencies as leverage in developing future competitive advantages.

Design as a process could have been part of this discussion but it was not. Design was referred to, but mainly in relation to brands and as a tool for differentiation, i.e., as a marketing tool. In a resource-based view discussion, design related to products was of little value as product design can be quite easily imitated. In contrast, brands gained appreciation as a strategic resource: brands were market-oriented, distinctive and protected – hence difficult to imitate. Brands could even be measured financially in different ways. Design, especially when we discuss design as a process, is much more difficult to express in numbers.

Designing, as a specific skill for strategic development, was rarely on the agenda. When looking at how industrial designers work, one thing that is quite obvious is their capability to base new product concepts on a combination of the client's competence, skills, production capabilities and the brand image. One example for this is Bahco Tools.

Bahco Tools is a Swedish company founded in 1886 by businessman Berndt August Hjort. In 1892, it started to manufacture the adjustable wrench (in some countries called 'a bahco'). In 1991, a severe recession affected Bahco Tools and the company was acquired by Sandvik AB as part of their tools division. The tools division was then sold in 1999 to the American enterprise Snap-On, Inc.

At the end of the 1970s Bahco started to work with Ergonomidesign, a Swedish industrial design consultancy. Ergonomidesign had approached Bahco, the best tool maker in Sweden at the time, with a proposal for a new type of wrench. At the time when Ergonomidesign approached Bahco a general view at the company was that tools had reached their peak and the only way to improve them was through material development. The product development manager, however, was not very interested in the new wrench presented by Ergonomidesign. But he became interested in the method the designers had used to develop it. Ergonomidesign were commissioned to design a new screwdriver. The study of how users actually used screwdrivers resulted in a totally new screwdriver handle with room for two hands (as this was a common way to use a screwdriver for large screws), long thin handles for small screws (to enable fine adjusting) and always round handles with a special structure in the material to maximize the force of the hand without causing injuries. This led to the development of the Ergo concept for the whole range of tools within Bahco. Despite a change of owners in the 1990s, the collaboration between Bahco and Ergonomidesign continued.

The design process is a resource that can be the basis for a company to develop a superior and distinct competitive advantage. The design concept and the collaboration with Ergonomidesign remained a strategic advantage and strategic resource when Bahco was sold in 1992 and later on in 1999 to Snap-On. The collaboration with Ergonomidesign changed the perception at Bahco of what constitutes innovation for tools. The Ergo concept received a Red Dot award in 2002 as well as other international design awards.

In this way the skills contribute to a core competitive advantage (Prahalad and Hamel 1990). Discussions of strategic resources and of competitive advantages have, however, had an economic theoretical basis contingent upon the successful study of factors that could constitute a strategic resource. Design in this context has been ignored and reduced to giving form to the product. I do not think that the researchers rejected this aspect as less important. I believe that they did not see the process of design as anything special or different from what the company did within the context of innovation or product development. One of the reasons might be the difficulty of articulating the design process. We also know from many design management researchers that it remains a challenge to communicate the intrinsic value of design (Walker 1990; Thornquist 2005; Digerfelt-Månsson 2009).

While I believe that the design management literature has provided arguments as to why design can be a strategic instrument, design management research has ignored the need for integration of design skills with strategic skills. One reason for this could be that the models presented by strategic management are too narrow and economics oriented. But if design is to become something more than a tactical instrument for the purpose of differentiation, if design is to concern strategic intent, then design management competencies must take into account a company's traditions, its existing skills and values, and those aspects that are part of strategic management. These issues are discussed more in the management literature with an organizational perspective (for example, Mintzberg 1987, 1994; Normann 2001). Design management competence is the company's skill and ability to use the talents and knowledge to develop the company's resources, which simultaneously involves a complex interaction of resources

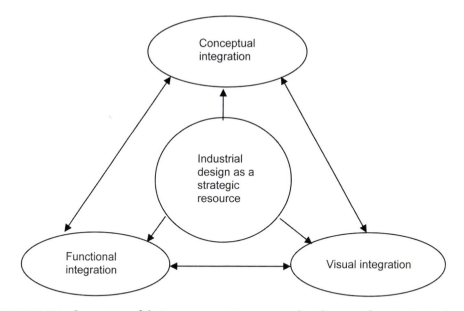

FIGURE 19.1   Integration of design as a strategic resource is based upon a functional, visual and conceptual integration. © Svengren, 1995.

(Leonard-Barton 1992; Bruce and Jevnaker 1998). I therefore believe that the market-oriented approach to the business concept and its view of this are limitations for the discussion of design as a strategic resource. Instead, I believe that an organizational perspective will contribute to a deeper understanding of design as a strategic resource and of the contributions of design to the company's strategic development.

Strategic management, marketing and brand management, in particular, have efficiency, differentiation, positioning and strategic change processes as key concepts. These concepts are also concerns for design management.

However, we still need to develop arguments and concepts for how design is contributing to these concepts as a strategic resource. Integration is a key concept for this discussion. Integration is not a new concept. However, we need to clarify what we mean by integration from a design management perspective. In this perspective it consists of three different categories of which design management can take place, namely functional, visual and conceptual. These three categories will form the structure of the remainder of this chapter.

## FUNCTIONAL INTEGRATION

Functional integration has to do with the designer being part of the product development team and process. A central argument in this context is that the designer should be involved at an early stage of product development projects to utilize the whole potential of the creative design process. Creative ways of thinking should be used to find

new solutions, then contribute to increased competitiveness and strategic development, for example through more efficient processes, innovations and better products.
Functional integration in this sense actually takes place on an operational level and
concerns the integration between different functions, disciplines and phases related to
product development and innovation. This discussion has been taking place since the
1960s, so it is definitely not a new subject. It is a fairly prescriptive and normative field
of research but it is still part of design management as a subject.

How companies organize product development has changed due to the outsourcing
of manufacturing, but also globalization and the development of information technologies. All these issues force companies to try new ways to organize their design. It also
requires a new approach for how to deal with the organization and the relationship
between different functions. The old fragmented organization, with walls between the
functions had severe communication and coordination problems, which project organization was supposed to solve.

Another way to deal with the new situation of the company is to see the company as
part of a network of alliances that needs to be orchestrated (Möller and Halinen 1999;
Visser 2009). Project-based and networked organizational models try to avoid the kinds
of communication and coordination problems that used to be common. These organizational forms enable and encourage multidisciplinary teams. Experiences show that
it is often easier to prescribe multidisciplinary team and networked organizations than
to make them work efficiently. People who enter the team bring with them established
attitudes and practices, which may lead to intransigent positions and misunderstandings in communication. We know from several design management studies that design
methods, design thinking and the physical environment with support for communication and collaboration (Kristensen 2004) can contribute to successful interactions
within multidisciplinary teams (Johansson and Svengren Holm 2008). Further evidence is provided by individual design consultancies that have been able to capitalize
on their expertise in project organization, multidisciplinary teams and flexible project
rooms, consulting management not only in design but also about how they can build
creative and innovative companies (Kelley 2001; Feldman and Boult 2005).

But let us put the organizational discussions aside and instead focus on another
type of reason for functional integration on an operational level – the integration of
the user in the product development and innovation processes. User driven and open
innovation is creating a lot of attention both in research and in praxis (von Hippel
2001a; Chesbrough 2003; Kristensson et al. 2004; Marzano 2005). Design methods
and design thinking focus on the relationship of human beings and things, which
means in particular that we take the dynamics of this intersection as a starting point
and frame of reference and not existing products and systems. As a result, the initial
creative design process strives for a release from existing structures with the purpose of
decoupling the discussion from existing problems. Design management, at this level,
is to develop a user-centred design process and methods of creating knowledge, and

to integrate this knowledge in a creative way in the innovation process where the dynamic relationship between the human being and object is at the centre. It is also a way to deal with the fact that the user, through the Internet and other media, has the power to influence companies' product development and communication to an extent that was unheard of only a few years ago.

Today's society suffers no shortage of good design and is almost experiencing what we might call a sort of aesthetic saturation. It is increasingly difficult for companies to stand out with the help of 'good design'. The picture is reversed when the interpretation of what 'good design' constitutes rests in the hand of the communicating users. Companies need to offer something more and we also see how the user, partly through the Internet, has been given a more active role in the design of products (von Hippel 2001b). This does not necessarily mean that we have better designs in general but we do see more personal designs that engage the user. Personal involvement is probably the key, but also the personal approach that has always characterized the designer's way of working in the development of new product concepts.

I have tried to find an explanation of what 'user orientation' or 'user's advocate' – which are key expressions in design – really mean to some companies. I understood them as meaning the involvement of the user in different ways in the development of products. It is also true with the way many designers and companies work. Some decades ago, when I interviewed Braun's legendary design director, Dieter Rams, he deemed it unnecessary to ask users about anything they might want. To conduct market surveys that inquire into users on various issues and concerns are activities still being criticized by the majority of design writers. Braun designers, to this day, claim that they are 'user oriented' like many other designers. But they reject market research 'because users cannot know what they want to have in the future'. What is at stake seems instead to be an ability or willingness to empathize with the user's situation, to intellectually identify with the user, to feel empathy with the user. We find that empathy for the user, and this interest in aesthetics, clearly formulated in Braun's basic conception of design, product development and communications.

What we can see in recent design research, though, is a willingness to systematize the way designers work, to develop methods to do user studies, often based on ethnographic methods and workshop methods (Westerlund 2009), developed together with other research disciplines. The cultural probe, for instance, is one of the methods that are recognized within different design fields (Gaver et al. 1999; Mattilmääki 2006). With this, designers avoid criticism of being subjective and relying too much on intuition, even if subjective, intuitive ability is central to this method and the interpretation of the results. With a new framework, design methods are developed with a newly acquired legitimacy and are considered valuable when the need for more creativity and innovation is discussed.

An empathic interpretation of the requirements of the user can naturally be seen as subjective and thus can be dismissed as unreliable and risky. Several project case studies show that this ability to find the 'right' solution is based on studies and obser-vations of active users, often with observations in the field, and decisions are based on a confidence in an intuitive ability to see what happens – both the details and from an holistic perspective.

While market-oriented interpretations of market research are described as defen-sive, the design-oriented interpretations are described as offensive. In my opinion, the interpretation and market-oriented processes are not necessarily opposites, but rather the ideal interpretation would be an integration of both the market and design orien-tations. There is a need for an offensive interpretation of users' needs and wants if the company wishes to act in an innovative manner, but there is also a need to acknowl-edge market forces and to be knowledgeable about what is happening in the market.

The challenges for the design function are growing. The critical elements in the innovation and development process used to be the first phase – the designer partici-pated and the creative methods were allowed to contribute to a better start. We now see that it is equally important that the designer is part of the whole process – to facilitate the communication between all functions, not least with the users. The role of the design function is to focus on the user and employ visual methods to communi-cate and interpret the various functional and emotional requirements from the prod-uct. This means that for design to be a strategic resource, companies must make use of design skills with a much broader approach than differentiation and communication throughout the development processes.

## VISUAL INTEGRATION

The design perspective on communication was based on the notion that every com-pany has a personality and a unique identity that develops over time. Visual coherence and coordination of all design elements have been the traditional and classical argu-ments for design management and the integration of design (cf. Gorb 1988; Olins 1989). An assumption was that the design should reflect the 'true' reality of the com-pany, or at least a credible visual statement of companies' strategies (Olins 1989). According to these authors many companies were ignorant about design, which cre-ated a gap between the true identity, 'who the company is', and what the design communicated – the image of the company. The company lost in credibility and image. A logical deduction from this is that if you analyse a company's design you can also trace its identity, personality and (real) strategies or the gap between these. Design management therefore had to be concerned with the identity and the personal expres-sion of the company.

Methods to investigate and visually analyse the identity of a company were offered by most design consultancies. In contrast to advertising agencies, design consultan-cies did not work with campaigns but rather with the logotype as a brand from a sys-temic perspective in relation to the whole organization and its structure. Olins (1989) categorized the identity structure as monolithic, endorsed, or product based. Olins argued for the need to develop the whole organization according to its identity in order to be credible. He therefore extended the concept of corporate identity to embrace the behaviour of the people in the organization. This extension, however logical, made the concept of design management and corporate identity very complex. It now involved the domains of organizational behaviour and human resources – large research fields in their own right (Albert and Whetten 1985; Johansson and Svengren Holm 2008). This line of argument came very close to brand management, as Olins (2001) noted in a book chapter that he titled 'How brands are taking over the corporation.'

What separates the discussion of the design management literature from a discussion of the market communication perspective is the role of design as a catalyst for strategic development and change. A comparison of different object design – product, information, and environment design – the company's business and strategic approach can give indications of the need for change – the business idea and its execution. There is reason to believe that a visual analysis can contribute to new perspectives and thus also to new measures. A visual analysis is not very meaningful unless it is simultaneously connected to the company's other conceptual discussions.

Within the design management literature, especially in the 1980s and early 1990s, the view on identity and visual integration was quite instrumental. Some argued that companies need design managers who are responsible for all design decisions so that all design always follows the visual manuals and design guidelines (for example, Blaich 1993). As this is a fairly complex process, the logo and its application to various items the company used in its operations (advertisements, letterheads, brochures and on the product itself) became the lowest common denominator.

Some companies addressed this problem by implementing strict guidelines for all graphic communications while trying to instil a certain design attitude and design knowledge in their product development. At Bang & Olufsen, for instance, a company with a strong visual product identity, all employees, including managers, were put through a series of educational seminars about the company's internal design policy. In addition, Bang & Olufsen endowed the person responsible for product design development with the authority to make design decisions. This happens to be an external design consultant, David Lewis. Lewis has worked for Bang & Olufsen for several decades and has acted as its design director. The case of Bang & Olufsen is, however, based on the fact that the external designer had a very long lasting relationship with the company and thus had an insight into the company's identity, strategies, competence and so forth, which takes time to acquire. Few consultants have that kind of long lasting relationship with their clients.

Let me go back to the discussion of design, visual identity and corporate identity. The definition of corporate identity includes not only the company's signature and artefacts, which can be given a certain look and form, but also the organization's behaviour. The concept of identity is usually found in the field of organization theory.

Within design management literature the concept of identity is rarely used to explain various organizational phenomena. Rather it is used to argue that the message design positively affects behaviour in the organization – if the message is quite clear. However, we know that establishing the business mission and getting organizational members to act in accordance with it constitutes complex identity and ideology-building processes. IKEA is a good example of how the identity has been maintained over the years and is a reflection of the ideology of the owner, Ingvar Kamprad. In 1976 Ingvar Kamprad wrote the little booklet *A Furniture Retailer's Testament* with nine points that came strongly to characterize IKEA's corporate culture and therefore any design.

This thinking in design management was also embraced by brand management as a kind of recognition that if the brand should be in accordance with the identity and the values of the company, building the brand required organizational development and included human resource management. *Living the Brand* (Ind 2001) and several brand management books (for example, Kapferer 1992; Aaker 1996 ) have included models specifying how to achieve this. Many design-oriented companies have instead linked their brand strategies with individual designers – often highly recognized personalities within the design community. *Alessi* is well known for this strategy. Another company that used this strategy to change its positioning in the market was ERCO Leuchten, a German lighting company (Svengren 1995).

ERCO Leuchten was founded in 1934 by Arnold Reininghaus, Paul Buschhaus and Karl Reeber – ERCO is an acronym for Reininghaus & Co. It developed and manufactured lamps and components for lamps. In the 1970, with a new managing director Klaus-Jürgen Maack, the business idea was defined as 'we sell light not lamps', which also led to a focus on the public and architectural market. To implement this new approach and business idea, ERCO commissioned a well known German graphic designer, Otl Aicher, to develop the new logotype for ERCO and a corporate design programme that guided product design and all communication. This new design strategy was also implemented in the development of the products for which Ettore Sotsass and other well known international designers were commissioned. The company did not use any famous architect for the architecture of the new company building, which was erected in the 1980s, . The reason for this was that these architects did not match the identity of ERCO. Instead a young architect, Uwe Kiessler, with ideas that corresponded to those of the company received the commission. All spaces and rooms in the new building reflected the mission and vision of the company as an advanced lightening company with the capability to innovate new lightening systems rather than products (Svengren 1995).

There are probably few short cuts when it comes to visual integration, creating a coherent visual integration as basis for corporate identity and, of course, for building the brand. One interesting aspect is that this is all very much a question of control, especially when we consider the large multinational enterprises with a need to integrate all subsidiaries and operations around the globe. However, a company will not be able to control the interpretation of the company's different elements. We can start with the example of a product aimed at a special target group in mind that is taken over by a different group – perhaps a group that the company did not even want to be associated with. But if this is a problem for some luxury brands this is probably not a huge problem for most companies. On the contrary, they might find it flattering that others see the potential or attraction of the product that the producers themselves did not see. It will probably increase the quality image of the product and stimulate new development and innovation. The problem is more of a desire to control what cannot be controlled. But one lesson to learn is how important the link is between functional and visual integration. There cannot be one without the other.

To serve as a tool for marketing communication, the visual identity and brand management should be flexible to respond to market changes. Flexibility is a key word frequently used in management literature. The development of 'the marketing concept' in the 1950s meant that companies should strive for flexibility so that they can respond to market changes. An increasing need to develop a unique and clear (communicable) identity or brand, in order to be able to position oneself in a market with ever more intensive information and the need to hold together an organization requires a degree of stability. Flexibility and stability can be seen as opposites, as Christensen (1992) has pointed out in his thesis on the ways in which companies organize their market communication based on these two needs. He concludes that the marketing department has a loose coupling to the rest of the organization and is able to adapt to the market by tailoring communication dimensions (in the widest sense) to the steady changes of the company's external conditions. By regularly listening to the market, the marketing department establishes a responsive communication system that signals both internally and externally that the company is market oriented.

Yet it is not clear that substantial changes in the company really happen. Lack of any substantial changes might, according to Christensen, lead to difficulties in establishing the credibility of what it communicates and the company's actual activities. In the same way as designers take a pragmatic approach to the task of shaping its design elements as unique and full of character, marketers take, according to Christensen, a pragmatic attitude towards the gap between the need for changes initiated by the market and the organization's actual operations.

What is the impact of this interpretation on the process of integrating design? Although designs are affected by the need for change, the design function, in product development or in market communication can hardly be loosely linked to the organization if the design should be a reflection of the company's business concept, rather the opposite. Several studies suggest that successful companies have a close and deep collaboration, even when they are working with design consultants (cf. Dumas and Mintzberg 1991; Cooper and Press 1995; Svengren 1995; Borja de Mozota 2003). It is hence not a question of whether the company has a design department, employing designers or not. Visual integration is hence not a question of a design department within the organization but the mandate of the designers working within or with the company, which in turn is a question of the conceptual integration of design.

## CONCEPTUAL INTEGRATION

Conceptual integration in a design perspective means that the company's mission, vision and business idea is visually interpreted in the company's different strategic concepts and dialogues. This obvious, simple definition is based on a series of complex integration and development processes: in product development and innovation, in marketing processes as well as the overall visual interpretation processes of the company's business mission and vision. The complexity is often not recognized even within design management literature or in design management research. One reason for this is probably that design management is still quite a young discipline that has tried to get into the management curriculum. To become accepted many design management researchers have linked their findings into mainstream management, which in some ways reduces the value of design (Johansson and Woodilla 2009). Today, we denote this approach of emphasis on the design process as design thinking, and this term has generated a lot of attention even in management research (Austin and Devin 2003; Brown 2008; Rylander 2009). Design thinking is discussed in contrast to analytical thinking, which dominates management thinking (Martin 2007; Junginger 2009; Rylander 2009).

In the late 1980s Dumas and Mintzberg (1989) discussed ways to integrate design in what is referred to as an 'infusion' of the design approach, meaning that design thinking and design methods are used for developing the relationship between a company and its surroundings, one part being the core offers – the products. They note that this is the hardest way but when it succeeds the most effective way. Even in their follow-up article, they discuss various organizational solutions for the integration of design in different situations and needs (Dumas and Mintzberg 1991). Dumas and Mintzberg meant that a common aspect of organizational solutions is that manager's job is to create consistency (fit) between the products and functions.

Managers are designers in this regard because of their concern with the fit of product with context, the fit across different forms of design expertise, and the fit of these with the sister functions of production and marketing (Dumas and Mintzberg 1991).

This would make the manager a kind of 'silent designer' (Gorb and Dumas 1987). This is partly built on Simon's reasoning in *The Sciences of the Artificial* (Simon 1969/1981), although a risk with referring to Simon's definition and approach to design in this context is that design is too general and synonymous with 'management'. It could for example mean that the design process, methods and approaches, and what we refer to as unique with design thinking, are not allowed to affect the organization's management and approach to innovation and communication. Silent design exists in every organization but it becomes more important to clarify the role of conscious design in a strategic perspective.

When we discuss conceptual integration of design in strategic development it does not only mean that design methods and design thinking are actively used by top management but also that the values found in the relationship between the company's design objects and its stakeholders are on the agenda. It also means that within a company's vision and mission design is seen as a field of knowledge for the development of the company's products and communication, including its brand-building communications. This also means that top management does have to support and have knowledge about design and what design stands for in the context of its business.

The study I conducted with Ericsson Mobile Communication in the beginning of the 1990s testifies that without the active support of top management of design issues, conditions for conceptual integration of the design are missing (Svengren 1995, 1998). Without clear guidance from management on the role of design in times of pressure for both the marketing and the product development side, there is also no opportunity to change practices and methods. The statement that design would be the number one means of competition became more an expression of a wishful thinking than actual behaviour.

Factors like a technically oriented corporate culture and a technologically advanced product play a role. My study shows that Ericsson Mobile Communication with frequent CEO-changes and a geographically dispersed management makes it difficult to provide continuity and guidance for action. This brings us back to the discussion of management's role as a significant player in the development of a company's business concept. The dominant ideas held by top management are, for various reasons, more powerful than the other parts of the company idea system and support the company's power structure. These dominant ideas seem, in each company, to be reflected in a small group of actors, characterized by those who have the real and usually – though not always – also formally the highest positions of power in the company. Within Ericsson Mobile Communication the dominant ideas were strongly related to those of the corporate Ericsson – a business-to-business, technology-based company with no experience of a consumer-driven market. This changed when Ericsson merged its mobile phone division with Sony and formed Sony Ericsson. Through that merger the design department received a new role and grew in its influence. This probably led to survival of Ericsson within the mobile communication market but we can still conclude that Sony Ericsson has yet not been able to achieve any market-leading position within the global mobile business. Within the mobile system market, Ericsson is still one of the world's leading companies.

Whether the role of a dominant idea system is negative or not has been discussed by, among others, Björkegren (1989). He discusses how the existence of dominant ideas within a company has been considered to have negative effects on the organization's learning capability because it limits the employee perspectives. Only what is considered relevant from a narrow perspective is part of learning and development. Björkegren finds, however, in his research, that if the dominant ideas are allowed to change and develop to a high degree, the organization's learning capability benefits and allows for changes. This means that if the dominant ideas within a company are open to adopt design thinking and if the company sees the value design methods, the organization can also develop its design management capabilities and relevant knowledge structures for design. This in turn makes the conceptual integration of design more likely.

The risk in a company with a strong technical culture is that professional development is about technological development only. Other areas, such as, for instance, market communication, must not have the same degree of development and therefore can operate in parallel without being integrated with design in product development. In companies like Bahco Tools, and also Braun, where for instance ergonomic design provides highly relevant knowledge for their products, this can facilitate the integration of conceptual design. This was obvious in my study (Svengren 1995), and is also supported by Björkegren's study, which indicates that an open knowledge production and relevant knowledge structures facilitate learning and reduce the need for revolutionary advancement of knowledge.

It is likely that if the knowledge structure for integrating conceptual design is perceived as irrelevant to an organization's knowledge production, design is not seen as a strategic resource – design skills cannot contribute to the company's progress. If this is true it would mean that it is only when design is perceived as relevant knowledge that conceptual integration of design – design as a strategic resource – is possible.

People often state that they have become more creative in their own discipline through working together with designers (Johansson and Svengren Holm 2008). Designers themselves rarely change the way they act, they are more acting as catalysts. To change institutionalized behaviour requires that the organization perceives the new methods and thinking as important to existing patterns of behaviour (Svengren 1995). If the integration of design as a new approach requires changing behaviour, the integration of design as a strategic resource, therefore, is a question of learning at the organizational level.

## THE CASE OF ELECTROLUX

When the board of directors at Electrolux decided to change product positioning from medium to premium segments it would have consequences for, among other things, the design activities in the company. It would not be enough to change the communication of the brand if the company wanted to have a new corporate identity. The company had to look through the whole product range in order to match the products to the new strategy. The design department was given a strategic role but also had to coordinate its activities with many other departments and it was decided to go the whole way to transform the company. Electrolux management did its homework and created a new department: the consumer insight lab. This lab was put in the centre with the design department, the market communication and the R&D department as major stakeholders. It required a learning process for the whole organization. Electrolux also employed ethnographic researchers in the consumer insight lab, participating in the market research as support to the transformation process. With such a large global company like Electrolux, which also owns several other brands, the change process will take several years to implement. It will require reorganization, new skills and training of the employees. Major investments are needed to make the whole remake trustworthy both internally and externally. But management has decided that this is so important that they have to put the whole organization on the school bench with a focus on learning about consumer trends, new product development methods where design methods and design thinking has a major role and how to build the new Electrolux brand.

The example of Electrolux shows that various strategic decisions have implications for the design department, but also require the design department to coordinate with other departments in a new way. Above all, the example shows that repositioning of a brand is not just a question of giving the brand new value words. It is about integrating design thinking and design methods in several processes, where product development and marketing are not the only ones.

One conclusion of the discussion in this chapter is that conceptual integration is a prerequisite for design to have a role as a strategic resource and is a prerequisite for the design process – its practice, methods and thinking – to be used for strategic business development. It is obvious that traditional effectiveness and profitability arguments are not enough to motivate the integration of design in a strategic sense. Top management needs to 'discover' the potential of design through practical experience rather than theoretical readings. Even this does not seem to be enough to create an organization that use design as a strategic resource. As we can see from the Electrolux case a long process that includes a new organization for those functions that are related to the product and a new physical environment with facilities for encounters and a series of learning seminars are needed. This requires investments and resources.

## INDUSTRIAL DESIGN AND STRATEGIC DEVELOPMENT – SOME CONCLUDING REMARKS

The development of design as a strategic resource is dependent on at least three issues. One concerns the status of design in the organization and how top management supports and deals with design. This is probably the most common argument in the design management literature and it is confirmed by several studies (for example, Cooper and Press 1995; Bruce and Jevnaker 1998; Freeze 1998; Jevnaker 2000; Borja de Mozota 2003). Top management defines what is important and does so both symbolically through its actions and decisions and operationally through its diffusion of resources. A related issue is the mutual understanding and respect of each knowledge area and its contribution to innovation and development.

The second issue has to do with the frequency and continuity of communication between the design function and other functions within the company. Frequent and continuous communication can contribute to a mutual understanding of goals and conditions of each project. Integration of design in product development largely focuses on how the communication processes between the various functions work but also on how the company physically organizes the conditions for the communications to take place. We know from some case studies that a studio environment that allows more physical discussions is positive for communication and understanding in development projects (Kristensen 2004; Johansson and Svengren Holm 2008).

In companies where top management emphasizes that the business idea must be visible in the design of all its objects, it is likely that a certain design style emerges that is coherent and creates a strong visual identity. A risk with this strong emphasis is that this style may become so identified with the organization that changes become more difficult. The interpretation of the business idea, ways to define and visualize this, is something that requires awareness of the importance of design at various levels and the design decisions in strategic decision making. My thesis, in line with the design management literature's arguments for a management dedicated to design, is that it is precisely these processes that require the active involvement of top management that give the people the mandate to work with visual and conceptual integration of the company's properties and business. In the same way as a business grows with the development of specific company competencies, you cannot untie the visual and conceptual integration from the organization's historical development and therefore the skills the organization has developed (cf. Normann 1975; Prahalad and Hamel 1990). This suggests the need for a more organizational theoretical perspective on design management and design as a strategic resource rather than a strict communication perspective.

In this sense design management has the same argument as brand management. A company's values and visions should be reflected in the physical objects and behaviour of the company. The difference is that within the design management context, design methods and design thinking are the tools for achieving this. There is also a claim in the design management literature that design is a catalyst for questioning and advancing existing design concepts. For this to happen, design must be part of the strategic dialogue within the company – at top level as well as throughout the organization. Hence it is necessary to have conceptual integration in order to achieve both a functional and a visual integration. Design is certainly not the only issue for strategic development but it is an important one. To become design oriented and to use design as a strategic resource is, for some companies, a conscious choice. A fundamental argument for the integration of design appears to be about a specific approach to design within the organization. This approach is based on a concern for things or 'passion for artefacts', a strong emotional and aesthetic sense of the integrity of things, but also courage to make decisions about the integrity of the products and its users. In theses concluding remarks I will go through what I mean.

By aesthetics, I mean a sense of those emotional and cognitive messages that the product's design expresses. Several case studies show that it is not just about some key actor's passion for the aesthetics of things but also about commitment to things related to the user based on a vision of how things can make our lives easier, better, funnier and so forth. Empathy is the concept that I consider most appropriate to describe this here. Empathy emphasizes the ability to become emotionally and intellectually familiar with the user and empathy is the capacity for the intellectual identification with another person. In the design context, this means empathy in how users perceive things. Here we also find the difference between the approach of the designers, who study the user and try to create new product concepts based on the user's needs and wants, and the approach of marketing people. Design-inspired ethnographic methods, the use of visual tools, or personas have already caught the attention of marketers and management. This is one reason why some business magazines like, for instance, *Business Week,* ask if the D-schools are the new B-schools.

A passion for artefacts is also a passion for detail – a passion for details that affect the aesthetic value of a product. What characterizes the designer's way of looking at a product can be described as a constant commuting between the product in its entirety and the small details of the product. This could be called a zooming process. These are rarely observed by the customer who sees the product, but they are crucial to the overall impression, unless the customer/user is an expert and knows to appreciate these details. Several cases demonstrate the importance of these details that the designers sometimes have to fight for, especially if the detail costs extra to produce.

Other elements that I have observed in several case studies, in addition to passion and empathy, which influence whether a company integrates design as a strategic resource is courage. There may be many people with a passion and aesthetic sense of things but it also requires courage to embrace an aesthetic approach and to act upon what is the vision for the user – even if it requires investment beyond the short-term bottom line. To work in a larger organization means that there are many requirements to consider – managers, board and the financial side, with the stock market evaluation every day. That is when you need to have courage to withstand demands that would violate the integrity of the product and the vision for satisfying the user and the customers. When these elements interact – vision, empathy, aesthetics, and courage to pursue them – you can tell if there is a strategic design philosophy at hand.

Highlighting design, aesthetics and semantics as the driving forces for development is not obvious in an economic context. There is therefore a need for an organization where members, especially managers, have the courage to prioritize criteria that are not always justified economically or technically. Sometimes it is also about resisting occasional fashion trends. It would be wrong to claim that companies I have studied always prioritize design requirements rather than economic or technical requirements if they happen to be contradictory. But several case studies testify how key actors have resisted demands that would have encroached upon the basic design concept or the integrity of the design quality. It is, as one Bahco Product development manager commented, important to be able to compromise the right way but not with the basic concept of integrity. The problem that remains is that it is impossible to judge in advance the cost as well as the income. Therefore courage is needed to have faith in the choices made. It is also a reflection of short- and long-term goals, where design goals can be important in the long term.

## REFERENCES

Aaker, D.A. (1996) *Building Strong Brands.* New York: Free Press.

Albert, S. and Whetten, D. (1985) 'Organizational Identity', *Research in Organizational Behavior* 7: 263–95.

Alvesson, M. (2004) *Knowledge Work and Knowledge-Intensive Firms.* New York: Oxford University Press.

Amit, R. and Schoemaker, P. (1993) 'Strategic Assets and Organizational Rent', *Strategic Management Journal* 14: 33–46.

Austin, R. and Devin, L. (2003) *Artful Making. What Managers Need to Know about how Artists Work.* Upper Saddle River, NJ: Financial Times Prentice Hall.

Barney, J. (1991) 'Firm Resources and Sustained Competitive Advantage', *Journal of Management* 17(1): 99–120.

Björkegren, D. (1989) *Hur organisationer lär* [*How Organizations Learn*]. Lund: Studentlitteratur.

Blaich, R. with Blaich, J. (1993) *Product Design and Corporate Strategy. Managing the Connection for Competitive Advantage.* New York: McGraw-Hill.

Borja de Mozota, B. (2003) *Design Management. Using Design to Build Brand Value and Corporate Innovation.* New York: Allworth Press and Boston, MA: Design Management Institute.

Brown, T. (2008) 'Design Thinking', *Harvard Business Review* 86(6): 84–95.

Bruce, M. and Bessant, J. (2002) *Design in Business. Strategic Innovation through Design.* Gosport: Ashford Colour Press.

Bruce, M. and Jevnaker, B. (1998) *Management of Design Alliances. Sustaining Competitive Advantage.* Chichester: John Wiley & Sons, Ltd.

Chesbrough, H. (2003) 'The Era of Open Innovation', *Sloan Management Review* 44(3): 35–41.

Christensen, L.-T. (1992) 'Marketing som organisering og kommunikation. En kulturteoretisk analyse af markedskommunikationens organisering og betydning i den marketing-orienterede virksomhed' ['Marketing as Organizing and Communication. A Culture-theoretical Analysis of the Organisation and Meaning of Market Communication in the Market Oriented Enterprise'], PhD thesis, University of Odense.

Clark, K. and Fujimoto, T. (1989) 'Reducing the Time to Market. The Case of the World Auto Industry', *Design Management Journal* 1(1): 49–57.

Cooper, R. and Press, M. (1995) *The Design Agenda. A Guide to Successful Design Management.* Chichester: John Wiley & Sons, Ltd.

Digerfelt-Månsson, T. (2009) 'Formernas liv i designföretaget – om design och design management som konst' ('The Life of Forms in the Design Company. On Design and Design Management as Art'), Doctoral dissertation, Stockholm University, School of Business Research Reports.

Dumas, A. and Mintzberg, H. (1989) 'Managing Design Designing Management', *Design Management Journal* 1: 37–43.

Dumas, A. and Mintzberg, H. (1991) 'Managing the Form, Function, and Fit of Design', *Design Management Journal* 2(3): 26–31.

Edvinsson, L. and Malone, M. (1997) *Intellectual Capital: Realizing Your Company's True Value by Finding its Hidden Brainpower.* New York: HarperBusiness.

Feldman, J. and Boult, J. (2005) 'Third-Generation Design Consultancies: Designing Culture for Innovation', *Design Management Review* 16(1): 40–7.

Freeze, K. (1998) 'Design Management Lessons from the Past: Henry Dreyfuss and American Business', in M. Bruce and B. Jevnaker (eds) *Management of Design Alliances. Sustaining Competitive Advantage.* Chichester: John Wiley & Sons, Ltd.

Gaver, B., Dunne, T. and Pacenti, E. (1999) 'Design: Cultural Probes', *Interactions* 6(1): 21–9.

Gemser G. and Leenders, M. (2001) 'How Integrating Industrial Design in the Product Development Process Impacts on Company Performance', *The Journal of Product Management* 18: 28–38.

Gorb, P. (1988) 'Introduction: What is Design Management?', in P. Gorb (ed.) *Design Management, Papers from the London Business School.* London: Architecture Design and Technology Press.

Gorb, P. and Dumas, A. (1987) 'Silent Design', *Design Studies* 8(3): 150–6.

Ind, N. (2001) *Living the Brand. How to Transform Every Member of your Organization into a Brand Champion.* London: Kogan Page.

Jevnaker, B. (2000) 'Championing Design: Perspectives on Design Capabilities', *Design Management Journal, Academic Revie*, pp. 25–9.

Johansson, U. and Svengren Holm, L. (2008) *Möten kring design* [*Encounters around Design*]. Lund: Studentlitteratur.

Johansson, U. and Woodilla, J. (2009) 'Towards an Epistoemological Merger of Design Thinking, Strategy and Innovation', *Design Research Journal* 1(2): 29–33.

Junginger, S. (2009) 'Design in the Organization – Parts and Wholes', *Design Research Journal* 1(2): 23–5.

Kapferer, J.-N. (1992) *Strategic Brand Management. Creating and Sustaining Brand Equity Long Term.* London: Kogan Page.

Kelley, T. (with Littman, J.) (2001). *The Art of Innovation. Lessons In Creativity from IDEO, America's Leading Design Firm.* London: Profile Books Ltd.

Kim, W. C. and Mauborgne, R. (2004) 'Blue Ocean Strategy', *Harvard Business Review* 82(10): 76–84.

Kotler, P. and Rath, G. (1983) 'Design: A Powerful but Neglected Strategic Tool', *Der Unternehmung* 37(3): 203–21.

Kristensen, T. (2004) 'The Physical Context of Creativity', *Creativity and Innovation Management* 13(2): 89–96.

Kristensson, P., Gustafsson, A. and Archer, T. (2004) 'Harnessing the Creative Potential among Users', *Journal of Product Innovation Management* 21(4): 4–14.

Leonard-Barton, D. (1992) 'Core Capabilities and Core Rigidities: A Paradox in Managing New Product Development', *Strategic Management Journal* 13: 111–125.

Lorenz, C. (1986) *The Design Dimension. Product Strategy and the Challenge of Global Marketing.* Oxford: Basil Blackwell.

Martin, R. (2007) 'Design and Business: Why Can't we be Friends?' *Journal of Business Strategy* 28: 6–12.

Marzano, S. (2005) 'People as a Source of Breakthrough Innovation', *Design Management Review* 16(2): 23–9.

Mattelmäki, T. (2006) 'Design Probes', PhD thesis, University of Arts in Helsinki.

Mintzberg, H. (1987) 'Crafting Strategy', *Harvard Business Review*, July–August: 66–75.

Mintzberg, H. (1994) 'Rethinking Strategic Planning. Part II: New Roles for Planners', *Long Range Planning* 27: 22–30.

Möller, K. and Halinen, A. (1999) 'Business Relationships and Networks: Managerial Challenges of Network Era', *Industrial Marketing Management* 28(5): 109–18.

Normann, R. (1975) *Skapande företagsledning* [*Creating Management*]. Stockholm: Aldus.

Normann, R. (2001) *Reframing Business. When the Map Changes the Landscape.* Chichester: John Wiley & London.

Olins, W. (1989) *Corporate Identity. Making Business Strategy Visible through Design.* London: Thames & Hudson.

Olins, W. (2001) 'How Brands Are Taking over the Corporation', in M. Schultz, M. Hatch and M. Larsen (eds) *The Expressive Organization. Linking Identity, Reputation and Corporate Brand.* Oxford: Oxford University Press.

Pilditch, J. (1987) *Winning Ways: How 'Winning' Companies Create the Products We All Want to Buy.* London: Harper & Row.

Prahalad, C. and Hamel, G. (1990) 'The Core Competence of the Corporation', *Harvard Business Review*, May–June: 79–91.

Prahalad, C. and Ramaswamy, V. (2000) 'Co-Opting Customer Competence', *Harvard Business Review* 78(1): 79–87.

Phillips, P. (2004) *How to Create the Perfect Design Brief.* New York: Allworth Press.

Roy, R. (1986) 'Introduction: Meanings of Design and Innovation', in R. Roy and D. Wield (eds) *Product Design and Technological Innovation.* Milton Keynes: Open University Press.

Rylander, A. (2009) 'Bortom Hajpen. Designtänkande som epistemologiskt perspective' ['Beyond the Hype. Design Thinking as Epistemological Perspective'], *Design Research Journal* 1(1): 20–7.

Simon, H. (1969/1981) *The Science of the Artificial.* Cambridge, MA: MIT Press.

Starbuck, W. (1992) 'Learning by Knowledge Intensive Firms', *Journal of Management Studies* 29(6): 713–40.

Svengren, L. (1995) 'Industriell design som strategisk resurs. En studie av designprocessens metoder och synsätt som del i företags strategiska utveckling' ['Industrial Design as a Strategic Resource. A Study of Methods and Views in the Design Process as Part of Companies' Strategic Development'], PhD thesis. Lund: Lund University Press.

Svengren, L. (1998) 'Integrating Design as a Strategic Resource: The Case of Ericsson Mobile Communication', in M. Bruce and B. Jevnaker (eds) *Management of Design Alliances. Sustaining Competitive Advantage.* Chichester: John Wiley & Sons, Ltd.

Thornquist, C. (2005) 'The Savage and the Designed. Robert Wilson and Vivienne Westwood as Artistic Managers.' Doctoral dissertation, Stockholm University, School of Business Research Reports.

Visser, E.-J. (2009) 'The Complementary Dynamic Effects of Clusters and Networks', *Industry and Innovation* 16(2): 167–95.

von Hippel, E. (2001a) 'Innovation by User Communities', *Sloan Management Review* 42(4): 82–6.

von Hippel, E. (2001b) 'User Toolkits for Innovation', *Journal of Product Innovation Management* 18(4): 247–57.

Walker, D. (1990) 'Managers and Designers: Two Tribes at War?' In Oakley, M. (ed.) *Design Management: A Handbook of Issues and Methods.* Oxford: Basil Blackwell.

Walsh, V., Roy, R. and Bruce M. (1988) 'Competitive by Design', *Journal of Marketing Management* 4(2): 201–16.

Walsh, V., Roy, R., Bruce, M. and Potter, S. (1992) *Winning by Design Technology, Product Design and International Competitiveness.* Oxford: Blackwell Publishing.

Westerlund, B. (2009) 'Design Space Exploration. Co-operative Creation of Proposals for Desired Interactions with Future Artefacts', PhD thesis, Royal Institute of Engineering, Stockholm.

# 20

# The Role of Design in Innovation: A Status Report

## BETTINA VON STAMM

A while ago, I wrote an article on 'Innovation – What's Design Got to Do With It?' for the *Design Management Journal* (von Stamm 2004). This article argued for the important contribution design could make to organizations that were trying to improve their innovation performance. The argument went something like this: companies need to grow but there are limits to achieve this through mergers and acquisitions, which leaves innovation as the main fuel for the growth engine. However, so the argument continued, many managers struggle to improve their organization's innovation performance whereby I proposed that the difficulty was a consequence of managers' prevailing mindset, namely one that is focused on cost cutting and efficiency improvements – rather than on experimentation and exploration, the fundamental cornerstones of innovation. The reason I suggested that design may have something to offer in the context of innovation was that designers, by their personal inclination, education and skill set, tend to have a set of preferences, values, tools and behaviours that closely match requirements of innovation: they like to experiment, challenge, explore and reinvent. After providing some evidence of the benefits of using design and designers I explored different attitudes towards design and closed with some comments on the location of designers (for example, whether designers are located inside or outside the organization). What has changed since–or has anything changed since? Using this article as a starting point, I will provide a report on the current situation of 'innovation – what's design got to do with it'.

Let me start with a study on the importance of and levels of satisfaction with innovation from the latest innovation survey of the Boston Consulting Group (Andrew et al. 2009). This survey found that innovation remains a strategic priority for the majority of companies and that 64 per cent of survey respondents ranked it as a top-three priority. At the same time, the survey shows that only 50 per cent of managers and 47 per cent of employees are satisfied with the results of their innovation efforts. The survey indicates that the struggle to create innovative organizations continues and that culture remains one of the major issues – it was one of the two main obstacles identified through the 2009 BCG survey. It seems that existing organizational cultures still foster a mindset that heavily emphasizes efficiency, cost-cutting, incremental changes and a focus on day-to-day business. Innovation is not likely to flourish in such a culture.

But the problem goes deeper than that. Organizations that do not value innovation tend to perpetuate such a culture by recruiting and training managers that share these traits and values. Not many organizations make a conscious effort to recruit for a willingness to experiment or take (calculated) risks. These, however, are important attributes for those who are charged with improving an organization's innovation performance. Of course, it is not all that cut and dried: people who are efficient and effective can also be innovative.

However, as a general rule, asking people who are good at cost cutting and paying attention to detail to approve projects that are characterized by high levels of uncertainty – both in the process as well as the outcome – almost feels like asking the impossible. Attitudes that are advantageous to driving out cost and being efficient include attention to detail, focus on the present, demanding certainties and predictability, being numbers driven, maintaining tight control, thriving on repetition, managing by standards and procedures, considering failure to be a disaster, evaluating and judging rationally and having a preference for preserving the status quo. On the other hand, characteristics that foster the creation of an environment that is conducive to innovation include the ability to consider the bigger picture, being future oriented, accepting (initial) ambiguity and uncertainties; being more visual and concept driven, thriving on autonomy and experimentation, being open-minded and flexible, taking failures as learning experiences, valuing emotion and challenging the status quo. People who are extremely good at one of what are considered to be polar ends of organizational skills are not likely to be good at the other. Moreover, it is often difficult for them to even see the value of the opposite approach. Nonetheless, both mindsets are essential for a long-term successful organization.

It is an important first step for companies to understand and acknowledge that certain skills are required for a successful innovator and that these skills are not the same as those required for improving efficiencies and cut costs. The next step is to bring people on board who have those skills – not to replace existing managers, but to work alongside them. This might sound quite straightforward, but is in fact not without its own challenges. Bringing in a facilitator who is equally at home in the different cultures of business and design can be necessary until individuals have come to develop an appreciation of the contributions of the 'others'. For example, Claudia Kotchka, Vice President for Design Innovation and Strategy of Procter & Gamble, very much describes her role as that of a 'translator'. Jennifer Reingold, in an article for Fast Company described Kotchka's role this way:

> Like a simultaneous translator, Kotchka must express the language of design in a way that people steeped in sales, finance or research can understand. At the same time, she needs to keep her designers motivated and clear on the fact that an idea that doesn't increase sales is meaningless in a place like P&G.

I have undertaken three rounds of interview-based research into best practices and future challenges of innovation (cf. von Stamm 2001, 2003, 2006). From these three surveys we can see a 'journey of understanding innovation' that has taken place. According to the initial survey results in 2001, most organizations then were involved with *introducing processes* to manage innovation effectively. They had introduced, for example, a stage-gate process. At the time, their understanding of innovation seemed primarily framed around technological and product innovations. Those involved in identifying what needed to be done in the context of innovation were often bright young people. Yet, due to their junior level, they generally enjoyed little authority in the organization. As a result, these young innovators felt they had to fight the perception that they were just adding to people's workload. In the second survey, conducted in 2003, it emerged that organizations had begun to review and consider the kind of *organizational structures and roles* they needed to support innovation. The follow-up survey three years later (2006) revealed that many organizations had broadened their understanding of innovation and were seeking *'new ways of doing business'*. Many participating organizations had also realized that processes and structures alone would not make a real difference. It was encouraging to see that managers became increasingly aware of the links between an innovative organization and the internal culture and mindset. Particular attention was now being paid to the role of different mindsets and a mixture of different skills have to play in the innovation process – as journalist and writer Walter Lipman once said, 'If all think alike no one thinks very much.' In the next section I am looking at innovation from a design perspective, building on the insights of these studies into innovation in organizations and add further questions. For example, what is it about the skill set of designers that contributes to innovation? And how does design thinking offer supporting tools and approaches that help an organization in achieving its innovation ambitions?

## DESIGN IN THE CONTEXT OF INNOVATION

'Firms with higher design intensity have a greater probability of carrying out product innovation', claims a government report (DTI 2005). In the context of innovation management, three interpretations of design dominate. These include (1) design as a tangible outcome – that is, the end product of the process that results in items such as cameras, cars and so on; (2) design as a creative activity; and (3) design as the process by which information is transformed into a tangible outcome. The third interpretation – design as a process – is most commonly used and this is how I, too, understand the word. For me, however, design is also an act of conscious decision making, so I would vary the definition slightly to read: 'Design is the conscious decision-making process by which information (an idea) is transformed into an outcome, be it tangible (a product) or intangible (a service).'

Design is about doing things consciously. Design questions why things are being done in a certain way and does not choose the easiest option without understanding the consequences involved. Design is about building on ideas, trying things out, comparing alternatives, exploring opportunities to finally select the best possible solution. For a long time, writings on design seemed to have been exclusively 'by designers for designers.' This began to change in the early 1980s. Today many writers and theorists tend to look at the interface of design with other business functions as well as its wider role in organizations in general and in new product development in particular (cf. Oakley 1984, Pilditch 1987, Walsh et al. 1992, Bruce and Biemans 1995). Publications such as *Business Week* and *Fast Company*, both coming out of the US, have made design one of their major topics, reporting regularly on companies such as design and innovation consultancy IDEO and Procter & Gamble (P&G), the latter a company that has made conscious efforts to bring design and designers into the heart of their organization. P&G's CEO Lafley explained in 2005, 'I want P&G to become the number-one consumer-design company in the world, so we need to be able to make it part of our strategy. We need to make it part of our innovation process.'

The aforementioned Claudia Kotchka (2006) also commented on P&G's activities: 'Designers at Proctor & Gamble historically were called at the end of the project for superficial decoration. Design thinking puts designers and several other critical personas together at the inception of the project. The power of design is leveraged at the beginning and all through the development.'

It seems, then, that design is finally receiving the necessary attention from people outside the design department – just as Philip Kotler and Alexander Rath (1984) have called for when they described design as a necessary organizational competence. In their article they declared that, 'Design is a potent strategic tool that companies can use to gain a sustainable competitive advantage, yet most companies neglect design as a strategic tool. What they don't realize is that good design can enhance products, environment, communications and corporate identity.'

While the situation remained much the same for two decades after Kotler & Rath's original article, publicized stories like the one from P&G indicate that, slowly but surely, things are changing. This is also reflected in the statement of Patrick Whitney (2006), head of the IIT Institute of Design in Chicago, that 'A titanic talent search is under way as managers scour the globe for innovators. Companies are struggling to transform themselves from cultures driven by cost and quality control to organizations that profit from creative thinking.'

Clearly, things are starting to change. The pace of change, however, remains rather slow. I believe that the role of design in innovation today is hampered by the same reasons I found in 2004:

1. A lack of shared understanding: confusion about what design and design thinking actually are remains.
2. Silent designers: non-designers continue to have a significant impact, often without being aware of it (Gorb and Dumas 1987).
3. Communication issues: despite many verbal commitments 'communicating across communities', i.e. groups of people with different mind and value sets, has not really become any easier.
4. Educational differences: associated with the above are different foci in education, for example whereas education for managers is traditionally verbal and numerical, concerned with topics such as accountancy and engineering, designers' education is visual, geometrical and concerned with craft and art. Designers are educated and trained to deal with projects that involve unfamiliar concepts, involve fuzzy problems and high levels of ambiguity and require assessments that are 'subjective, personal, emotional and outside quantification' (Walker 1990).

So far I have built the argument that design is an important strategic tool and why there might be some hurdles in taking it up. As a next step I would like to revisit why thinking about design in the context of innovation makes sense.

## A FEW DEFINITIONS

First of all, what does innovation mean? A commonly used definition is: Innovation is the commercially successful exploitation of ideas. This definition associates innovation with a tangible outcome. However, in today's fast-changing environment this is not enough. Innovation can be and has to be present in all aspects of an organization. Innovation is the art of making new connections and continuously challenging the status quo – without changing things for change's sake. Innovation is more about a certain frame of mind than a tangible product or a new technology. In an innovative organization, innovation will not be the domain of a department or a small group of people; rather, it will be everyone's responsibility. Design and designers can be a key facilitator ad accelerant in creating such a mindset.

Successful innovation is first and foremost about creating value. It does so by either improving that which exists already (generally associated with incremental innovation) or developing products, processes, services and approaches that did not exist previously (generally associated with radical innovation). Both kinds of innovation require the following:

- to challenge the status quo
- to have an understanding of and insight into consumer needs and wants
- to develop imaginative and novel solutions.

In addition, innovation is generally associated with the following:

- the willingness to take risks
- accepting high levels of ambiguity and uncertainty
- thinking outside the box
- a passion to drive the idea through to conclusion
- the ability to inspire passion in others.

Let's take a look at what David Durling (2009) has found in his research on characteristics of designers:

Designers are typically shown to be highly intuitive and analytical. This is combined with an open-ended curiosity about their world. They are similar to cognate professionals such as architects, but significantly different to business managers, engineers, computer specialists and a general population.

He refers to 'intuition' in the Jungian sense, meaning that 'intuition is a mode of perception that relies primarily on meanings, patterns or possibilities. It is beyond the reach of the senses and is sometimes referred to as insight. Intuition is future oriented and engages in explorations beyond present or past realities . . . People who have intuition as a primary preference also like solving new problems rather than routine' (Durling et al. 1996). In a further article Durling (2003) states, 'These kinds of designers also seem to like solving new problems and to seek radical solutions rather than producing more routine incremental change.' These findings indicate that designers are ideally suited for innovation. However, the fact that there is considerable overlap between designers' characteristics and what is required of innovators does not mean that innovation should be left to the designers. When everything is said and done, innovation is about teamwork. To move from idea to profits, both skill sets, adaptive and innovative, are required, all the time (Kirton 1976). I would like to point out two things. First, people with adaptive skill sets are more likely to dominate organizations that seek to improve their innovativeness; second, there is a group of people – designers – who have the skill and mind set that is required to innovate. Thus, designers have an important contribution to make to the innovation process and they can be valuable members of innovation teams.

## THE ROLE OF DESIGN IN INNOVATION

White Paper by the UK Government stated in 1995 that, 'The effective use of design is fundamental to the creation of innovative products, processes and services' (DTI 1995). This White Paper also lists these benefits of good design:

> Processes improved by gradual innovation
> Redesign of existing products in response to user needs, new markets and competitor products
> Development of new products by anticipating new market opportunities

## DESIGN AND COMPANY PERFORMANCE

Further, a study of Norwegian companies has found that companies using design consciously did not only show higher levels of innovation activity, but also (1) generate more revenue from innovation and (2) are overall more profitable than companies that do not use design (Solum et al. 1998).

A more recent study by the UK government (Department of Trade and Industry 2008) concludes that design has a crucial role in the national economy: 'Through commercializing the science base, improving business competitiveness and updating our public services so they suit the needs of users, design can help make the UK more innovative and ultimately improve our economy.'

FTSE-100 and FTSE-All Share indices over a ten-year period:
Ten-year performance 1995-2004

FIGURE 20.1    Design outperforms FTSE 100. © Design Council.

While the above statement from the 2008 White Paper by the UK Government might sound like wishful thinking, there is some hard evidence that this is indeed the case: a 'Design Index' established by the British Design Council in 1993, tracking the financial performance of the sixty-three most design-led companies in the UK, found, in 2002 that, whether in a bull or a bear market, design-led companies were outperforming the FTSE 100 and FTSE All Share index by more than 200 percent (see Figure 20.1).

There is then strong evidence of the benefits of using design in today's business environment. These benefits apply even more when we consider the changes in the economic, social, political, technical and global contexts.

- Consumers are becoming more and more sophisticated; functionality alone is not sufficient.
- Changing demographics are resulting in an older population.
- An increased need for differentiation; companies offering 'me-too' products will not survive in the medium or long term.
- Increased competition on a global basis – a balance needs to be found between maximizing economies of scale and responding to local differences in taste and needs; careful design of products and services can help to maximize standardized and shared parts, while ensuring that local or regional adaptations can be undertaken with minimum efforts.
- An increased demand for user-friendly products – even for highly innovative products. Although in the past it seems to have been acceptable for new products to require educated and technically savvy users or users who don't mind reading through reams of instructions, nowadays people place greater emphasis on ease of use and on products that are self-explanatory.
- An increased demand for products that take the impact on the environment into account.

Together, these changes open up opportunities for innovation and pose challenges that can only be addressed through the systematic application of design and design thinking, particularly if we think about Tim Brown's definition of design thinking. It is essential to understand that the key to success is to make design and innovation the core of the company's strategy. It is also worth mentioning that design and innovation can spark each other in different ways, not only directly (by involving designers in the design and development of a product) but also indirectly through the physical work environment (for example, choosing a layout of workspace that maximizes collaboration and chance encounters). It is not about bringing in a designer for a new product once every few years; it is about making design and innovation a way of life – about questioning and challenging what and how things are done and asking where there might be an opportunity to add value.

## DESIGN NEEDS TO BECOME PART OF THE FABRIC

Design applied holistically tends to be associated with product success whereas design restricted to styling is more likely to be associated with less successful products. (Roy and Riedel 1997)

Whether or not design actually contributes to the success of a product and a company's performance is critically influenced by management's attitude towards it and by how much design is embedded in an organization's thinking and culture. In understanding the degree to which design is integrated into an organization's fabric a framework developed by Angela Dumas and Henry Mintzberg (1981) is helpful; they differentiate five levels.

The first is the use of a design champion. He or she is responsible for the implementation and use of design in an organization. While this might be a necessary first step, on its own and over time it is not sufficient to realize the full potential of design in an organization – not least because everyone else in the organization is allowed to believe that thinking about design is someone else's responsibility.

A second approach is the establishment of a design policy. Dumas and Mintzberg argue that it can help to clarify beliefs that already exist in a company but that, by itself, a design policy is of little consequence. It is a declaration of intent that is not necessarily followed up with actions, thereby losing credibility and support.

The third approach is the development of a design programme. While Dumas and Mintzberg have observed that, in some cases, such a programme can trigger change, its momentum has to be maintained otherwise it will have a 'flavour of the month' feel to it.

Under the fourth approach, design as a function, performance is often associated with and measured based on the performance of marketing or production, rather than on its own independent efforts. This approach can also be problematic in that it allows design to be seen as the responsibility of only a small group within the organization – and thus irrelevant to the rest.

The fifth style that Dumas and Mintzberg observed – design as infusion – is the one most likely to lead to the successful and comprehensive employment of design. Here, the entire organization 'breathes design' and feels responsible for its consideration. Interestingly, based on my research, categories similar to the five identified by Dumas and Mintzberg can be found in the realm of innovation. Here too 'innovation as infusion', where everyone breathes and thinks innovation, is the most successful approach. Managers should be aware, though, that, be it design or innovation, the state of infusion can only be reached after a journey, not instantaneously.

To illustrate how different levels of infusion and approaches to design influence outcomes, I would like to share some case studies.

### Case Study No. 1: Celebrating the Past, Ignoring the Present

The first is the story of a former East German motorcycle producer that brought in a British design consultancy to revive its fortunes. An informal brief, a tight time frame and limited resources were the starting point for the consultancy, which developed a stunning working prototype in record time. Admired and prized by motorbike enthusiasts and journalists alike, the bike still failed to turn the fortunes of the company. Despite the fact that the designer felt he had built the bike on the foundations of the company's pre-war success – simplicity, sustainability and common sense – the company was incapable of producing the bike as designed. It lacked the skills, machinery and resources to execute the design. That meant a significant redesign, which delayed the bike's introduction and actually diluted the original design.

*Learning point.* Why did this happen? The designer did not spend enough time understanding the specifics of the company he was designing for. His starting point was the company's successful and glorious past, not its present circumstances, which involved fairly unskilled labour, out-of-date machinery and a lack of funds. It is important to understand fully existing constraints – not to work within them, but to be able to work around them.

### Case Study No. 2: Too Many Cooks

The high-speed train that connects the capitals of Belgium, France and the UK, the Eurostar, was designed and built by three consortia, one from each country. The design task, in fact, was broken down according to the level of financial commitment from each country. The British design consultancy designed the 'nose' of the train and the driver's cab; the French designers were responsible for the interior as well as the exterior design – which left the toilets, the overhead luggage racks and the seats for the Belgian design team. The result was as you might imagine; one journalist wrote, 'An excess of widely diverging advice on the design, colours and finishes in the train has left its mark. The resulting entity is a frenzy of old, new, cool, warm, kitsch and hi-tech; it's all there and the disappointment is all the greater after the excitement of the highly individual external styling' (Evamy 1994).

*Learning point.* Breaking down a design task into separate modules is likely to cause a disjointed result. If the task must be broken down, strong leadership will be required to achieve a coherent and consistent overall result. Moreover, design cannot achieve its potential impact if it is based on consensus on the lowest common denominator. In addition to strong leadership, design requires strong decision making and vision.

### Case Study No. 3: There's No Place Like Home

Ihavemoved.com was a Web-based company whose niche was providing an address-changing service. For the design of its Web site, the company brought in an external agency that was selected from a shortlist of three firms. The founders of the organization selected the finalist agency for its understanding of the opportunity and its enthusiasm for the project. However, it soon became clear that the Web design was falling short of expectations – not to mention its mandate to deliver the promised service to customers. The ihavemoved.com team commissioned a redesign but, again, the agency's work was less than satisfactory; it struggled to complete the work in time for the relaunch, despite much input and extra funds from their client. Frustrated, ihavemoved.com's team decided to look for the necessary technical and design expertise in-house. They were aware that this was quite a risky move because none of the company directors had the knowledge necessary to assess the technical capabilities needed fully. However, by using their network, they did find a recruit who turned out to be just what they needed. Not only was he able to immediately identify major flaws with the newly redesigned Web site – he was also able to fix them. Having the design resource in-house gave the company more control and also proved to be much cheaper. In addition, it turned out to be a great advantage when they started to offer customized versions of the Web site for large corporate clients who wanted to offer the address-change service as part of their own Web sites.

*Learning points.* Commissioning a design consultancy for the first time is very difficult. What is the best way to assess their capabilities and fit with the organization? How can the client ensure that they really understand the task at hand? According to research, more often than not, designers are identified by a recommendation from a trusted source (von Stamm 1998). Second and more important: If a task cuts close to your organization's core – if it represents the foundation of your offering – it is best to develop and retain the necessary expertise in-house.

### Case Study No. 4: Question Authority

The final case study is that of Black & Decker's Quattro, a multipurpose tool that, by allowing different heads to be attached, enables four different operations to be performed: sanding, sawing, drilling and screwing. After an initial design was developed externally (due to internal resource constraints), the project champion felt that the concept offered far greater degrees of innovation and differentiation than had been realized by the initial design. He handed the project to a young in-house designer for review. In-depth understanding of Black & Decker's vision and ambition, combined with a passion for improving ease of use and user comfort, resulted in a distinctive, innovative design that took the market by storm: When the product was introduced in late 1998, demand outstripped expectations. Instead of starting from the existing design, that of a battery-operated tool, the young designer had questioned what had been taken as a given by the external consultancy and by doing so was able to create a superior design in terms of ergonomics as well as aesthetics.

In this case, design and innovation were used to create an entirely new product category. It is clear that such projects are the norm rather than the exception at Black & Decker.

*Learning points.* First, having design expertise in-house is more likely to push the boundaries than using an external supplier; bear in mind that an external agency has limited selling opportunities in the commissioning organization. Second, an internal sponsor at the senior level is critical. Without being given the freedom to experiment and the opportunity to prove the value of the concept, designers and innovators have little chance of bringing something truly great and different to market. Third, it is probably more difficult to inspire enthusiasm and sense of ownership in an external design consultancy than in an internal designer. Passion and desire to succeed are essential in bringing a project to a successful completion.

### *Internal or External? Some Advantages and Disadvantages*

I firmly believe that in order to gain the full benefits of design – and innovation, for that matter – an organization needs to develop internal capability for its management and delivery (von Stamm 1998). Bringing in outsiders might help to kick-start the process, but ultimately it will remain a bolt-on, an artificial limb that is useful but not quite part of the core. In worst-case scenarios, outside help becomes a transplant that is rejected by the organism – as almost happened in the case of the motorbike project described above. However, there will always be situations where bringing in an outsider is preferable or the only option. Table 20.1 lists some of the advantages and disadvantages for each scenario.

## CONCLUSIONS

Back in 2004 I felt compelled to make the case for the use of design and designers in the context of innovation, explaining at some length what had led me to that conclusion. That conviction still holds firm. What has changed in the past six years is that more people share this belief and that more stories and even hard evidence, are available to help us believers make our case.

As my convictions remains the same, so do the reasons why more companies do not embrace design and design thinking in their endeavour to create more innovative organizations. In the end it comes down to mindset. The way of thinking, what is valued, and the language used in design and business respectively continue to differ, causing misunderstanding, misconception and confusion. To help both the business and design communities to benefit from each other more the following is needed:

- a better awareness and understanding of the differences;

- an understanding and appreciation of the differences and the value that lie therein;
- tools – and people – who can bridge the differences and bring out the best of and for both worlds.

I would like to leave you with a key message, nicely summarized by Richard Wallace (2006): 'When all that matters is the experience, design moves to the front of the bus and technology takes a back seat. Design, not technology, has captured the consumer's imagination, as anyone who has twirled a fingertip around iPod's elegantly simple dial pad can attest.'

**TABLE 20.1: Advantages and disadvantages of in-house and external designers**

|  | Advantage | Disadvantage |
|---|---|---|
| In-house | Cost-efficiency (depending on volume) Instant and continuous accessibility Easier coordination with other in-house departments Company retains control Designer develops intimate understanding of company Connected with internal networks Control and ability to prioritize | Lack of creativity/new ideas The need to keep the design team busy: e.g., the need for ongoing development work Losing touch with external developments Getting stale |
| External | New inspiration Access to specialists' expertise Relieves workload Accessibility of additional skills/staff Speed Ability to change and explore different options Can be bought if and when needed | Lack of understanding of company-specific issues Problems of ready accessibility Problems in the coordination with in-house design and/or other departments Potential lack of confidentiality Company needs skills to evaluate the design work Not-invented-here syndrome Problems with industrializing the externally developed design Loss of control Credibility gap if design is too far removed from company's own style Being made a low priority on agency's agenda |

# REFERENCES

Andrew, J., Haanæs, K., Michael, D., Sirkin, H. and Taylor, A. (2009) *Innovation 2009 – Making Hard Decisions in the Downturn*. Boston, MA: Boston Consulting Group.

Berner, R. (2006) 'Design Bridges', *Business Week*, 5 June.

Bruce, M. and Biemans, W. (1995) *Product Development: Meeting the Challenge of the Design-Marketing Interface*. Chichester: John Wiley & Sons, Ltd.

Buxton, B. (2008) 'A New Mantra for Creativity – Executives Should Apply the "Order of Magnitude" Rule to Any Problem that Demands a Creative Solution', *Business Week*, 12 May.

Department of Trade and Industry (2008) *Innovation Nation*. DTI Publications. London: HMSO.

Design Council (2002) *Facts and Figures on Design in Britain 2002–03*. London: Design Council.

DTI (2005) *Economics Paper No. 15: Creativity, Design and Business Performance*. London: DTI.

Dumas, A. and Mintzberg, H. (1981) 'Managing the Form, Function and Fit of Design', *Design Management Journal* 2: 26–31.

Durling, D. (2003) 'Horse or Cart? Designer Creativity and Personality', Proceedings, Fifth European Academy of Design Conference, 28–30 April, Barcelona, Spain.

Durling, D. (2009) 'Virtual Personalities', http://nelly.dmu.ac.uk/4dd/synd4e.html (accessed 22 September 2009).

Durling, D., Cross, N. and Johnson, J. (1996) 'Personality and Learning Preferences of Students in Design and Design-related Disciplines', in J. Smith (ed.) Proceedings of IDATER 96 (International Conference on Design and Technology Educational Research), 2–4 September, Loughborough University, pp. 88–94.

Evamy, M. (1994) 'Call Yourself a Designer?' *Design* (March): 14–16.

Gorb, P. and Dumas, A. (1987) 'Silent Design', *Design Management Journal* 8: 150–6.

Government White Paper (1995) *Competitiveness: Forging Ahead*. London: DTI Publications.

Kirton, M. (1976) 'Adaptors and Innovators: A Description and Measure', *Journal of Applied Psychology* 61(5): 622–9.

Kotchka, C. (2006) 'Design Evangelist, Procter & Gamble Company', www.innovativeye. com/front-end-of-innovation/2006/5/24/claudia-kotchka-design-evangelist-procter-gamble-company.html (accessed 16 June 2010).

Kotler, P. and Rath, A. (1984) 'Design, A Powerful but Neglected Strategic Tool', *Journal of Business Strategy* 5: 16–21.

Martin, C. (2006) 'Claudia Kotchka, Design Evangelist, Procter & Gamble Company', 24 May, www.innovativeye.com/front-end-of-innovation/2006/5/24/claudia-kotchka-design-evangelist-procter-gamble-company.html (accessed 16 June 2010).

Oakley, M. (1984) *Managing Product Design*. London: Weidenfeld & Nicolson.

Pilditch, J. (1987) 'What Makes a Winning Company?', in D. Walker and J. Henry (eds) *Managing Innovation*. Thousand Oaks, CA: Sage Publications.

Reingold, J. (2005) 'What P&G Knows about the Power of Design', *Fast Company* 95: 56.

Roy, R. and Riedel, J. (1997) 'The Role of Design and Innovation in Competitive Product Development', Proceedings of Contextual Design – Design in Contexts, 23–5 April 1997, Stockholm, Sweden.

Solum, N., Karlsen, E. and Smith, K. ([1996] 1998) 'Design and Innovation in the Norwegian Industry. Revised', STEP Group report R–12. Prepared for the Norwegian Design Council. Oslo: STEP Gruppen.

von Stamm, B. (1998) 'Whose Design Is It? The Use of External Designers', *Design Journal* 1: 41–53.

von Stamm, B. (2001) *Innovation Best Practice and Future Challenges, A Study into Innovation Best Practice in Innovation Exchange Member Companies and the Literature*. London: Innovation Exchange, London Business School.

von Stamm, B. (2003) *Second Innovation Best Practice and Future Challenges Report*. London: Innovation Exchange, London Business School.

von Stamm, B. (2004) 'Innovation – What's Design Got to Do With It?' *Design Management Journal* 15(1): 10–19.

von Stamm, B. (2006) *Third Innovation Best Practice and Future Challenges Report*. King's Lynn: Innovation Leadership Forum.

Walker, D. (1990) 'Managers and Designers: Two Tribes at War?', in M. Oakley (ed.) *Design Management, A Handbook of Issues and Methods*. Oxford: Blackwell, pp. 145–54.

Wallace, R. (2006) 'Design not Technology', *EETimes* (27 February), www.eetimes.com/op/showArticle.jhtml?articleID=180207156 (accessed 16 June 2010).

Walsh, V., Roy, R., Bruce, M. and Potter, S. (1992) *Winning by Design, Technology, Product Design and International Competitiveness*. Oxford: Blackwell Publishing.

Whitney, P. (2006) 'Design Visionary', 19 June, www.businessweek.com/magazine/content/06_25/b3989416.htm (accessed 16 June 2010).

# 21

# Connecting Marketing and Design

## MARGARET BRUCE

Marketing and design share the same ambitions – they are both customer focused and wish to improve the quality of the consumer's everyday life through the products and services they offer. They also want to attract consumers to their products and services and enhance consumer experience. Paying attention to function, style, price and quality and emotive value can differentiate products and services and draw consumers to them. Consumers may pay a premium price for products and services that are perceived to be higher value and perhaps may be considered to be a status purchase. Brand loyalty may also be engendered if the consumer is satisfied with the product or service. Consumers may act as brand ambassadors and inform others about its value. They are engaged with co-creation of meaning and value. User engagement is enhanced with social media. Marketing and design are in a symbiotic relationship, but have different insights into consumer behaviour and approaches to meeting consumer needs. Leveraging creative design principles is a challenge for marketing across manufacturing, service and retail environments. As more and more wicked problems appear – sustainability, downturn in consumer spending, youth unemployment, caring for a growing number of elderly people and so forth – that require a multidisciplinary approach, then marketing and design connections need to be considered at different levels from product to process to transformational change. This chapter explores the shifting landscape of the connections between marketing and design.

## THE VALUE OF DESIGN FROM A MARKETING PERSPECTIVE

A market orientation is defined as: 'a culture in which all employees are committed to the continuous creation of superior value for customers' (Narver et al. 1998: 242).

This view of marketing emphasizes the focus on customers and the continual drive to improve the product or service that is being offered to fulfil consumer needs. Design shares the endeavour to create superior value for consumers and can combine the requisite technical skills and creative principles with an empathy towards user needs (Molotoch 2003). The focus on 'continuous creation of superior value for customers' encompasses product and service innovation and environments. It is this connection between marketing and design, which is explored in this chapter. The salient question being: How can marketing and design create and deliver value?

Indeed, whatever the marketing goals – retaining or increasing market share or entering a new market – it is investment in design expertise that produces the new or updated/redesigned products or services, packaging and communications that consumers want to purchase. Design is a tangible asset through the products and services generated and an intangible asset through the emotive value, meaning and experience embedded or associated, with the offerings, the X factor of design. Designers can create iconic and classic items that survive their own era: we can think of Chippendale chairs or Chanel's No 5 perfume. Some of these designs may define a 'moment in time' and so capture the essence of a period, for example Biba the mini skirt of the 1960s or the minimalist furniture and interiors of the 1980s. Hence, design is entwined with visual culture. Star or celebrity designers permeate and apply their visual approach across different media and sectors, for example Starck's hotels, cars, fashion, graphics. Design signatures can be an asset and the UK retailer Debenhams has used design signatures across its home furnishings and fashion ranges to give prominence and value to these items in store. Other celebrity designer and brand owner alliances occur to provide limited editions in high street fashion, for example Jimmy Choo and H&M, Stella McCartney and Gap for children. In the luxury market, designer signatures can create a 'cult' following of consumers who seek out the brand and communicate its value to others, perhaps through social network sites. The surprising death of Alexander McQueen, a UK celebrity fashion designer, meant that the value of his products soared because of their iconic and rarity value.

There is increasing interest in the application of design thinking to services to understand the interaction between users and service providers across public and private sectors. Services are intangible and the consumption and production of the service occur at the same moment, which is why the experience at the moment of consumption matters. Services need to be planned and considered and so creative design principles can be brought to bear to this process to help to ensure that they deliver what customers really want. The NHS has used service design to improve clinical processes, such as post-diagnostic services for cancer sufferers, and elsewhere it has seen better community strategies for the police, for example. 'Service design involves a high level of user engagement ... I like the idea of bringing those design principles to public services. Good design process brings in a good understanding of the context: financial constraints, processes and how human beings behave' (Taylor 2010).

McDonald's is an example of a company that has had to improve its customer service experience. In France, McDonald's outlets are individualized and stylish to create an atmosphere and a service experience that is compatible with the local culture and allows for family eating or quiet spaces for other customers. Airlines have long paid attention to the service needs and experience consumption by different classes of users from business to economy and first class. The design of the cabin for each category varies in terms of space, expression of luxury as do the passenger waiting areas and the whole user journey from leaving home to arrival at the airport to getting on the plane and arrival at the destination. People want an all-embracing experience as they interact with a service or product. 'Owning an i-Phone is just the beginning: it's what you can do with it – the "apps" – that matter' (Aaltonen 2010). Marketing has a long established tradition with service and retail sectors and so design can learn from the tools that marketing has developed, for example service blueprint of the user journey to document points of user engagement with the service encounter (Baron and Harris 2003).

## DESIGN IN A BUSINESS CONTEXT

For business, in general, good design differentiates companies and makes products 'stand out from the crowd'. Innovative design opens up new markets. Clever design rekindles interest for products in a mature market. Well designed products communicate quality and value to the consumer, so enhancing the product's appeal and attractiveness, making selection easier (Kotler and Rath 1990). One way of thinking about design is illustrated by Crawford and Di Benedetto's (2003: 278) definition of design as 'the synthesis of technology and human needs into manufacturing products'. This means then that design connects technology, on the one hand and the consumer, on the other. In this view, design is also connecting consumers to new products, services and experiences. A design-led approach to marketing could draw upon the heritage, the place and the craftsmanship of the product, such as is the case with luxury goods (Beverland 2005). The value of Prada is immersed with its heritage, its craftsmanship and its ability to be innovative in a changing market place. Beverland (2005) refers to: 'being in, rather than of, the marketplace'. In other words, being deeply sensitive to user needs and being able to innovate to address these.

An experience-based approach to design management is that of Apple. A powerful brand based on design leadership across all the points that touch the consumer. Kester (2010) refers to Apple in the following way:

Why Apple? Because a company like that epitomizes great innovation ... they were the first company to take a mouse to market, but they did not invent those things ... Technology is just ideas. Design is about taking those ideas and making them work for people. Apple placed design at the heart of the business ... with Apple, the product becomes emblematic of a system and a service.

Apple products have a distinctive aesthetic and are ergonomically effective. Consumers have strong emotive ties to the brand – not 'I like Apple, but I love Apple', which has been reinforced through Apple's 'think different' campaign. Their behaviour is tribal and owning an Apple product can be regarded by its consumers as conveying status. Consumers are in a social community that enriches the meaning of the 'coolness' of the brand. Apple has a memorable identity through its name and logo and its personality is expressed through its youthfulness. The fun and iconic Apple products cover aspects of everyday life that are becoming increasingly personal and essential – iPhone, iPad, iMac, iPod. Apple staff act as effective brand ambassadors and are trained to give personal service and be attentive to customers. They run events, provide training and reinforce the sense of belonging to the Apple family. Some events are targeted to young consumers, for example music-making events for children to encourage them to be brand aware and create a desirability of the brand. It's a two-way street, so that by hosting such events Apple staff observes consumers and are empathetic to emerging consumer needs. The company's storefront members can provide consumer insights to internal marketing and design experts who can use these insights to develop enhancements and new products. This example shows the interconnectivity of product, innovation and service. Apple innovates and develops wholly new products and categories of products, such as the iPhone and the user interface, which makes it easy to navigate the applications offered. But, crucial to the brand is the service delivery and the experience that reinforces the brand values. Service refers not only to stand-alone service enterprises, such as restaurants and holidays, but also the service that is delivered in retail stores and to support products. For companies like Apple, the service elements cover: after sales support, repairs, customer relationship management, store environment and sales staff. These service elements have to be considered and planned. Design plays a part in this process to create and plan the elements to deliver a meaningful brand experience. Baron and Harris (2003) provide a 'service blueprint' to underpin the planning of the service and to record a user journey that reveals the key touch points of where users have service encounter. This is similar to 'mapping the customer experience' as described by D'Esopo and Diaz (2009). Pine and Gilmour (1998) consider the service environment as a stage and the actors creating the experience are all those engaged in the performance. Customers are involved in the performance too – they co-create the experience, for example getting dressed up to visit a luxury store or having family fun at a Disney theme park.

Consumers are involved with the co-creation of brand meaning through their experiences with the products, services and experiences. This deepens the relationship that consumers have with the brand and also builds brand loyalty. Design connects consumers with brands through a sensual experience – making the connection exciting and relevant to the consumers' needs. Nike ID is an example where consumers were engaged with making design decisions to personalize the product and then they took photographs via their mobiles and downloaded these on a website. They were acting as brand ambassadors for Nike (Kim et al. 2009). Other examples of this engagement, co-creation of value and creating brand meaning is Christian Dior, which had a campaign to source 'Dior Divas' amongst young adult consumers of Dior. Consumers could experiment with cosmetic products in store, then send an image from their mobiles to a web site and then votes were taken on the site to choose a Dior Diva.

The brand experience wheel shown in Figure 21.1 expresses the connections that are being made between the consumer and the brand. Design enables these deep connections to be made (Kim et al. 2009) through providing a meaningful experience to the consumer and then this can be communicated through the different touch points whereby the consumer connects with the brand, for example in-store, Internet, product. In fact, Crosby and Johnson (2007) argues that the brand experience is fundamental to developing brand loyalty. And, from a marketing perspective, brand loyalty is crucial in customer retention and leveraging revenue. Losing customers can damage reputation and recruiting new customers is expensive, so retention is a key concern for marketing.

FIGURE 21.1   Touchpoints of brand experience (Kim et al. 2009). © DMI, 2009.

## LEVELS OF INTERACTION

Design contributes positively to sustained business success. This contribution can be considered as:

- *financial* – impacting on sales, turnover, profit;
- *strategic* – to exploit new markets, to retain or recapture their market position, producers have to develop new products and services that consumers want;
- *reputational value* – provision of offerings that consumers desire and wish to purchase – design equity.

It is sometimes forgotten that investment in design is a low-risk investment. Research has shown investment in design leads to an increase of sales of over 41 per cent and that 90 per cent of new design projects are profitable (Potter et al. 1991). The researchers also discovered that projects paid back their total investment costs in less than 15 months from the time of project implementation. Joziasse and Selders (2009) show the different types of value added by design in Table 21.1. These include: profit, brand equity, innovation and faster change.

A major contributor to achieving such levels of performance is the approach taken to design management, particularly the interface between marketing and design. Indeed, effective design management entails the integration of design with marketing and production (Walsh et al. 1992; Bruce and Bessant 2002). Attention given to 'front-end' activities, for example market research to identify consumer needs and market gaps, can help gauge the likelihood of success for new products and for product improvements (Cooper and Kleinschmidt 1988). Arguably, one of the major challenges

**TABLE 21.1: Different types of value added by design**

| | | |
|---|---|---|
| 1. More profit | Prestige | 1. More sales transactions |
| | | 2. Higher premium price |
| | Costs | 3. Lower production costs |
| | | 4. Lower marketing costs |
| 2. More brand equity | Awareness | 5. Higher distinctiveness and user awareness |
| | Loyalty | 6. Better reputation and user loyalty (emotional bond) |
| 3. More innovation | Time | 7. Shorter time to market |
| | Amount | 8. More opportunities and intellectual property |
| 4. Faster change | Company | 9. Faster and smoother internal change |
| | Society | 10. Lower level of environmental degradation |
| | | 11. More solutions for social issues (aging, literacy, etc.) |

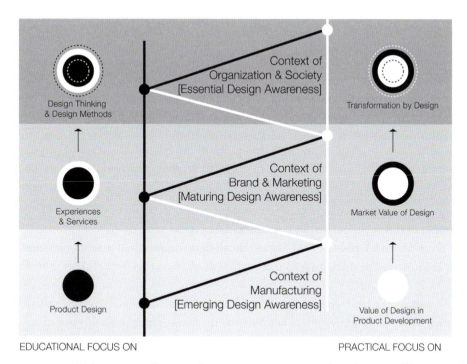

FIGURE 21.2   Making sense of current design management research and practice (Cooper et al. 2009). © DMI, 2009.

facing companies is to improve their management of design. To establish a leading position in the market, marketing executives need to understand the strategic and commercial value of design to fully leverage this resource. For design, the challenge is to connect with the marketing in its entirety and not to be treated as a functional tool or source of expertise to simply execute marketing decisions and strategy. Design and marketing need to work in partnership to influence business strategy. Indeed, design-marketing led companies are leading businesses – Ikea, Tesco, BMW, Apple and Disney are global companies in retail, manufacturing and service that come to mind. These companies can be considered as exemplars of what Cooper et al. (2009) consider as design-led businesses. They argue that design in business has shifted from a product-centric view of design – the expertise to create innovative products – to creating experiences and services to a situation where design thinking and methods enables 'transformation by design'. This shift is shown in Figure 21.2.

## SYNERGIES

Much of the research that considers the role of design from a marketing perspective focuses on design as a kind of tool within the marketer's toolkit. In other words, design has been considered typically as a resource for marketing to execute marketing's plans. From this viewpoint, studies have addressed the interface between marketing and design within the new product development process (Perks et al. 2005); considered design as part of a multidisciplinary team (Jevnaker 2000); as the creative source and energy for new products (Bruce and Bessant 2002); as providing a new and empathetic approach to consumer research (Leonard-Barton 1991); and as a process that marketing can manage through briefs and effective project management (Bruce and Daly 2007). It is clear that design plays a valuable role in business in these ways and also that marketers need to understand in detail how to leverage design expertise for specific activities, such as market research, etc. However, others suggest that a design-led perspective for strategic management (and marketing) may be a role that fully utilizes the ways of thinking and the knowledge that design embraces. And so, design becomes an enabler of business transformation (Cooper et al. 2009). Firstly an example of this is shown below. Apple could be considered another.

Trueman and Jobber (1998) argue that design is more than a functional activity and is connected to marketing in three different ways, as they express here, namely:

> At the product level, good design leads to the production of quality goods and services [that] can improve image and increase the consumer's perceived value of new products, so that they command a higher retail price in the marketplace.
>
> At the process level, designers can interpret, integrate and communicate new ideas ... [so] likely to speed up and increase efficiency ... and reduce time to market. At a strategic level in terms of brand building ... and the development of a corporate design culture.

Joziasse (2000) also identifies the need to integrate design into corporate strategy, particularly in firms where design is associated with the firm's core competitive advantage. Thus, these researchers reinforce the notion of symbiotic ties between marketing and design and suggest that these are complex and operate at different levels from product to process to strategic. We will come back to these levels elsewhere in this chapter.

## WICKED PROBLEMS: SHIFTING BUSINESS LANDSCAPE

A number of design disciplines exist with specialisms ranging from product design, interiors, furniture to textiles, fashion and corporate identity. So, there may be different technical design skills devoted to particular products at one level of the connection between marketing and design. But design-led thinking is strategic at another. This entails being able to handle complex problems, consider the longer term view and develop a creative approach to addressing such strategic problems. 'Wicked problems' are complex problems and these present challenges to design and marketing. These include: sustainability; social responsibility; recession and a downturn in consumer spending and the impact of this on particular sectors, like automotives; the lack of trust and increasing lack of loyalty to brands; and an uncertain job market for young people; obesity and healthy eating. All of these wicked or complex and challenging problems also create new opportunities and generate new markets to explore. Innocent is a UK company that specializes in healthy beverages targeted at children. The drinks are made up of fresh fruit and so can be one of the recommended 'five a day' fruit intake, the packaging is made from recycled material, and the logos and slogans are colourful, bold, distinctive and appeal to children. Innocent has exploited a niche in a highly competitive market and been successful because of its design-led approach to marketing that connects with its target market.

## TENSIONS

Another strand of research regarding the marketing and design interface argues that marketing and design are in conflict because they use different technical languages and have different goals. On the one hand, marketing is focused on commercially driven outputs, whereas, design is focused on creative and aesthetics outputs (Walker 1990). It is true that design has been rather neglected by the marketing literature. This is because design is regarded as a distinct profession. Similarly, marketing does not comment particularly on engineering or life sciences or other disciplines that are important for technological innovation. For example, in the pharmaceutical business, marketing is involved with the innovation process, as is science, technology, medicine and other disciplines. Another concern, typically expressed from a design perspective, is that marketing is essentially trying to curtail the creative process, which leads to outstanding design. In a business world, tough decisions are made about cost of materials, time spent on idea generation and execution that inevitably lead to tensions between marketing and design. One strategy to achieve closer integration between design and the firm is to give designers access to a range of business functions. Lorenz (1986) suggests that: 'for these connections to be made successfully requires a team effort in which the industrial designer's imagination, synthesizing skills and entrepreneurial drive are given as much weight as the tools of the engineer, the financial controller and the marketer'.

In contrast, however, others recommend that designers be protected from business functions (Bangle 2001; Wetlaufer 2001). BMW appointed an official mediator between design and the rest of the firm who was responsible for interfacing with business activities and giving the design the freedom to create and remain focused on their design role (Bangle 2001). In some companies, marketing may not be as powerful as the accounting or sales department. In fact, there may be a tension between marketing and sales and these two activities may not be conjoined effectively. So, design needs to have connections with a wider business community, in addition to marketing. However, Filson and Lewis (2000) identify organizational culture as the main barrier to achieving this integration. Leveraging design requires product champions, resources, constructive relationships between design and other business functions, communication and strategic input and a design-based culture (Jevnaker 2000).

Despite some indisputable tensions, marketing and design have one crucial point in common. Both disciplines are focused on the consumer and on satisfying consumer needs. This common focus fuels the relationship and also means that marketing and design are symbiotic. Marketing and design do work together and produce effective outcomes for the consumer and for business. Marketing provides an ongoing customer interface and ensures that design innovation delivers value customers find appealing (Blaszczyk 2000; Leonard and Rayport 1997).

## CONNECTING AT PRODUCT AND PROCESS LEVELS

As consumer tastes change, there is a need for new design to fill market gaps and to take advantage of new opportunities, such as long-lasting bags made from recycled material for shopping, hybrid fuel cars or easy-to-wear garments for the elderly. Technological developments in materials, information technology and so on are captured in new designs and open up new market possibilities. Design can apply existing technological knowhow to provide new products. Apple's approach to devising the iPod is one example of this.

From a strategic marketing position, when a company decides to enter a new market, it has to have the goods and services to offer. If it is seeking to increase its market share, it may restyle its existing products. If two companies have merged, then a new corporate identity may be required. Such strategic changes initiate investment in design. If these market changes are executed effectively, then there will be a commercial reward; if not, then the company could lose resources, lose prestige or miss a gap in the market. The strategic role of design is illustrated by three starkly different examples. Honda created the hybrid car to meet the growing need to cut fuel consumption and Honda gained tremendous reputational value for its innovation. Apple used existing technology to produce the iPod to create a global, new market and established leadership in this market. General Motors was complacent and believed that its large, gas-guzzling cars would continue to sell. It was wrong and the company nearly went out of business.

Design cuts across the four 'Ps' of the 'marketing mix': product design, price, promotion and placement. When managers talk about 'integrated design' they refer to the interrelatedness of the product design, with its packaging, its advertising and how all of this reinforces the overall company's image and its price and distribution. 'The driving experience' of BMW is demonstrated through the engineering and performance of BMW cars, the comfortable and stylish interiors of the cars, the promotional literature and the distinctive nature of BMW showrooms.

Companies configure products to meet criteria of price and/or differentiation. Design effort is influenced by whether the firm is focusing on a specialist, premium niche;

aiming for the mainstream; or, offering cheaper goods and competing on the basis of cost. The choice of materials, the configuration of products affects the cost of the product. Take the plastic packaging of a computer, where the least costly plastics for a given quality of appearance are selected. The toy company, LEGO, designs products for the premium end of the mainstream toy market. Its products are highly priced but offer 'play value'; keeping children busy and absorbed for long periods of time. Its emphasis on good design has contributed to its superior image and ability to sustain a long-term presence in the market.

Negative associations are given to poorly designed products. Such products may be regarded as unsafe, as damaging to the environment or unwieldy to use. They can affect the company's reputation and just not sell, so affecting business performance. It may be that certain attributes have been over-emphasized, for example, a highly stylized chair may look attractive, but may be uncomfortable to sit on.

## CORPORATE COMMUNICATIONS

Design decisions are made about corporate communications too. The company's logo, brochures, signage, packaging, Internet site and advertisements reinforce the identity of the organization. A market repositioning needs to be communicated to the target market, so that its new situation is clearly understood both by existing and potential clients. Brand image is critical in conveying meaning of the brand values to its target market and to begin to build a relationship with consumers and to drive this through brand communities. Concern about sustainability has encouraged consumers to be involved with brand lineage to be assured that the brand is ethical. Newer brands encourage cult followers, who may be averse to a corporate takeover and the core market may move on to something else. Does the corporation hide its connections with the new brand, ignore it or herald its alliance? Managing the brand image is a task for marketing and design to ensure that the meaning of the brand is conveyed consistently and clearly to its constituent markets. All aspects of corporate communications – logo, signage, brochures, promotional material, stationery and advertisements – have to work together to give a consistent message to the consumer wherever the consumer connects with the company. Consumers are channel surfers and so may connect with the company through its online and offline channels and so the company needs to present a harmonious image and identity through its various channels. Consumers may be loyal to the channel – for example, eBay or Amazon – and so the brand value and hence brand meaning may be built around the channel, as well as the products and services offered by the channel. ASOS is an incredibly successful e-tail fashion channel. Managing the brand meaning, keeping consumers interested in ASOS and cultivating brand loyalty is due to a mix of marketing and design. Marketing will identify the target market and develop a segmentation strategy with a blend of lifestyle, usage patterns and demographic profile and design will provide the creative strategy and execute this to enrich the brand value and encourage loyalty.

## STRATEGIC LEVEL: BUSINESS TRANSFORMATION

Global leading businesses and other stellar operators integrate design and marketing at all levels of the business – BMW, Disney, Apple, etc. 'Companies that are good at design are good at most other business activities' has long been recognized (Walsh et al. 1992). Fundamentally, these businesses pay attention to the consumer at all levels, it's their DNA. Increasingly, a partnership between design and marketing at a strategic level can be discerned, as shown in the company case, Firefly.

### A Real Example: Firefly

Firefly was the world leader in laminate flooring but, over time, the competition in this area has grown fiercer and the company's global position was under threat. The focus of the company has been on technical development of laminate flooring and sales to the trade, such as builders, interior designers and architects. Design was regarded as a functional activity and 'icing on the cake', rather than an integral strategic driver within the business.

As the marketplace became more challenging, with intense price competition, the company's product range began to lack lustre and was regarded as 'too European' for the global market. Senior management decided to change focus and develop a design and marketing team. The vision was: 'to become the highest brand recognized in the flooring sector' in the mid-to-top level of quality and to achieve this positioning through offering an attractive collection. Promotional activities directed at consumers would also help to 'pull through' the product to the marketplace. Consumer tastes had shifted and DIY had become a fashion industry, so to reposition itself, the company had to consider consumer behaviour, trends and lifestyles and reflect these in its product ranges. To achieve this change meant internal organizational changes to move away from technical and sales to embracing design and marketing. A design-led company means generating ideas and taking these from initial concept to the consumer and being able to predict trends and become more fashion aware, in order to move the collection away from traditional flooring products.

A vice-president of global design was appointed and set about transforming the internal processes and structure of the company, as well as commissioning market research and investing in design to develop the product ranges. To stimulate the market, the company developed four distinctive collections and each of these was underpinned by lifestyle propositions relating to each theme. These ranges were launched globally with the expectation that particular styles would be more attractive to certain regions, than others. For example, northern Europe would like pale, clean styles, whilst southern Europe would show a preference for richer and darker colours.

A storybook was created for each lifestyle theme and this contained images of each collection in different environments, such as home, office and restaurant and with complementary paints, wall-coverings furniture and accessories. This provided consumers with design ideas and gave them sources of inspiration to attract them to the collections. The storybook was a major communication platform to the trade and consumers and was the first time that a flooring company had begun to consider the consumer context for their products. A design and installation service was also provided direct to the consumer.

## DESIGN-LED STRATEGY

For design to have a strategic function within the company, significant organizational changes were required. A 'products to market group' was established, to connect design, marketing and technology. This group manages the whole development process from:

- concept development;
- presentation of the business case to the senior team;
- sample launches to country managers.

The multidisciplinary group was responsible for identifying trends and interpreting them through product development. This group hosted workshops of regional sales and marketing teams globally to assist with the activity and then met with retailers to identify their views of customer requirements. Retailers ensured that the timing of the product launch was correct to optimize product launch. Retailers have different launch dates based on their own seasonal calendars and so the flooring company had to fit in with these launch periods, on a global scale. Design consultants were also involved in this activity. The group's overall remit was to: 'deliver products that fit the market, guiding and directing the brand and reducing time to market'. Technical input reviewed the products constantly to find areas of improvement, such as new gloss finishes. Where possible, patents were obtained to protect the proprietary rights and reduce the risk of copying. This combination of technology push and market pull reduced time to market to eighteen months and reduced the risk of failure. This connection between design and marketing was important for the 'strategic direction and to guide and direct the brand'.

This reorganization to achieve a design-led approach to strategic decision-making was fundamental to the successful repositioning of the brand and for the company to strengthen its business leadership in the global market.

## CONCLUSIONS

Businesses are in a dynamic and challenging environment and have to deal with complex problems that can be described as 'wicked problems' – sustainability, social responsibility, youth unemployment, less consumer spend, new technology and so forth. New ways of thinking are required to deal with these and it has been advocated that design methods could be appropriate. This view of design in business has been described as 'transformational design' (Cooper et al. 2009) and 'knowledge design' (Konno 2009). It does seem that a shift in the connection of marketing and design has occurred, which reflects this broader shift in the business landscape. Marketing and design interface at various levels from product to process to strategic and share a common orientation by their focus on consumer needs. Whilst tensions may be observed between the marketing and design because of a different emphasis on commercial versus aesthetic endeavours, distinct solutions to this are evident. These cover: allowing greater creative freedom, a connection at a strategic level and developing a design-led business (Filson and Lewis 2000; Joziasse 2000; Crawford and Di Benedetto 2003; Cuffaro et al. 2002).

Design is not confined to product or service design, but is involved with experience. Social media means user engagement and interesting approaches to embracing this engagement in design have been outlined with the example of Nike ID. Another is that of LEGO Factory whereby users can download software, design virtual models and upload designs onto a public Web gallery and LEGO users can select the best designs. LEGO produces these.

For a sustainable competitive business, marketing and design are both necessary to create value. They understand and are empathetic to consumers and so are at the forefront of change.

## REFERENCES

Aaltonen, G. (2010) 'Service Design: Commodity In, Commodity Out', *Guardian*, 15 March.

Bangle, C. (2001) 'The Ultimate Creativity Machine: How BMW Turns Art into Profit', *Harvard Business Review* 79(1): 47–55.

Baron, S. and Harris, K. (2003) *Services Marketing: Text and Cases*. London: Palgrave.

Beverland, M. (2005) 'Managing the Design Innovation – Brand Marketing Interface: Resolving the Tension between Artistic Creation and Commercial Imperatives', *Journal of Product Innovation Management* 2: 193–207.

Blaszczyk, R. (2000) *Imagining Consumers: Design and Innovation from Wedgwood to Corning*. Baltimore, MD: John Hopkins University Press.

Bruce, M. and Bessant, J. (2002) *Design in Business: Strategic Innovation Through Design*. London: Prentice Hall/Financial Times.

Bruce, M. and Cooper, R. (1997) *Marketing and Design Management*. London: International Thomson Business Press.

Bruce, M. and Daly, L. (2007) 'Design and Marketing Connections: Creating Added Value', *Journal of Marketing Management* 23(9–10): 929–53.

Bruce, M. and Jevnaker, B. (1998) *Management of Design Alliances: Sustaining a Competitive Advantage.* Chichester: John Wiley & Sons.

Cooper, R., Junginger, S. and Lockwood, T. (2009) 'Design Thinking and Design Management: A Research and Practice Perspective', *DMI Review* 20(2): 46–54.

Cooper, R. and Kleinschmidt, E. (1988) 'Pre-Development Activities Determine New Product Success', *Industrial Marketing Management* 17: 237–47.

Crawford, M. and Di Benedetto, A. (2003) *New Products Management*, 7th edn. New York: McGraw-Hill.

Crosby, L. and Johnson, S. (2007). 'Experience REQUIRED' (cover story). *Marketing Management* 16(4): 20–8.

Cuffaro, D., Vogel, B. and Matt, B. (2002) 'Why Good Design Doesn't Always Guarantee Success', *Design Management Journal* 13(1): 49–55.

D'Esopo, M. and Dias, F. (2009) 'Mapping the Customer's Experience through Brand Design', *DMI Review*, 20(4): 38–47.

Filson, A. and Lewis, A. (2000) 'Barriers between Design and Business Strategy', *Design Management Journal* 11(4): 48–52.

Jevnaker, B. (2000) 'Championing Design: Perspectives on Design Capabilities', *Design Management Journal Academic Review* 1: 25–39.

Joziasse, F. (2000) 'Corporate Strategy: Bringing Design Management into the Fold', *Design Management Journal* 11(4): 36–41.

Joziasse, F. and Selders, T. (2009) 'The Next Phase: Laying Bare the Contributions of Design', *DMI Review* 20(2): 29–37.

Kester, D. (2010) 'Service Design: Commodity In, Commodity Out', *Guardian*, 15 March.

Kim, J., Koo, Y. and Chang, D. (2009) 'Integrated Brand Experience Through Sensory Branding and IMC', *DMI Review* 20: 73–81.

Konno, N. (2009) 'The Age of Knowledge Design: A View from Japan', *DMI Review* 20(20): 6–14.

Kotler, P. (2000) *Marketing Management Millennium Edition.* Sydney: Prentice Hall.

Kotler, P. and Armstrong, G. (2008) *Principles of Marketing.* Upper Saddle River, NJ: Pearson-Prentice Hall.

Kotler, P. and Rath, G. (1990) 'Design: A Powerful But Neglected Strategic Tool', *Journal of Business Strategy* 5(2): 16–21.

Leonard, D. and Rayport, J. (1997) 'Spark Innovation through Empathetic Design', *Harvard Business Review* 75(6): 102–13.

Leonard-Barton, D. (1991) 'Inanimate Integrators: A Block of Wood Speaks', *Design Management Journal* 2(2): 61–7.

Leonhardt, T. and Faust, B. (2001) 'Brand Power: Using Design and Strategy to Create the Future', *Design Management Journal* 12(1):10–13.

Lorenz, C. (1986) *The Design Dimension: Product Strategy and the Challenge of Global Marketing.* Oxford: Basil Blackwell.

Molotch, H. (2003) *Where Stuff Comes From.* New York: Routledge.

Narver, J., Slater, S. and Tietje, B. (1998) 'Creating a Market Orientation', *Journal of Market-Focused Management* 2(3): 241–55.

Perks, H., Copper R. and Jones C. (2005) 'Characterising the Role of Design in New Product Development: An Empirically Derived Taxonomy', *Journal of Product Innovation Management* 22: 111–27.

Pine, II. B. and Gilmore, J. (1999) *The Experience Economy*. Boston, MA: Harvard Business School Press.

Potter, S., Roy, R., Capon, C., Bruce, M., Walsh, V. and Lewis, J. (1991) *The Benefits and Costs of Investment in Design: Using Professional Design Expertise in Product, Engineering and Graphic Projects* (report). Milton Keynes: Design Innovation Group, The Open University and UMIST.

Taylor, M. (2010) 'Service Design: Commodity in, Commodity Out', *Guardian*, 15 March.

Trueman, M. and Jobber, D. (1998) 'Competing Through Design', *Long Range Planning Journal* 31(4): 595–605.

Walker, D. (1990) 'Two Tribes at War?' In Oakley, M. (ed.) *Design Management: A Handbook of Issues and Methods*. Oxford: Blackwell.

Walsh, V., Roy, R., Bruce, M. and Potter, P. (1992) *Winning by Design: Product Design, Technology and International Competitiveness*. Oxford: Basil Blackwell.

Wetlaufer, S. 2001. 'The Perfect Paradox of Star Brands: An Interview with Bernard Arnault of LVMH', *Harvard Business Review* 79(9): 117–23.

# 22

# Design Economy – Value Creation and Profitability[*]

## TORE KRISTENSEN AND GORM GABRIELSEN

It is often claimed that investing in design constitutes 'good business'. Legendary IBM CEO Tom Watson made this claim at one of the first Aspen conferences on design in the 1950s. Watson had learnt about design from IBM's Italian competitor Olivetti. Realizing the potential design could bring to IBM, he commissioned a number of design projects by famous designers. Paul Rand's IBM logo and other corporate design artefacts that resulted from these commissions count among the classics in design history. They also demonstrate that design and corporate affairs go hand in hand. Today, it is broadly accepted that companies like Apple, Alessi or Sony have built their competitive advantages around design. In fact, these companies have demonstrated superior performances for a substantial time. Yet, even if we acknowledge that design does have role in the success of some companies, we still lack conclusive data about the specific contribution of design to their overall business success. A wide range of questions remains unanswered. For example, we do not yet understand when and why design should support which aspect of a business' performance. We do not know if all kinds of design benefit the organization in equal proportions or which design efforts and activities support specific performances. To investigate these questions, we conducted a quantitative study with a random sample of twenty-five of the one-hundred largest Danish companies. Our aim was to find out whether the measures taken by these companies co-vary with the assumption that good design equals good performance.

We begin by developing three hypotheses that we then test with a survey. Hypothesis one states that the quality of product design is positively co-varied with financial

---

[*]*Source:* Adapted from Rusten, Grete and Bryson, John R. (eds) (2010) *Industrial Design, Competition and Globalization.* Basingstoke: Palgrave MacMillan.

performance. Hypothesis two claims that the logo design is associated with financial performance. Hypothesis three finally argues that a product's web design is positively co-varied with the financial performance of a product. We explain the samples and the research design before we present our findings and offer an analysis of their relevance to our three hypotheses.

## PREVIOUS STUDIES ON DESIGN AND BUSINESS PERFORMANCE

An investigation conducted by the Danish Ministry of Commerce and Construction (Erhvervs- og Boligstyrelsen 2003) claimed that the turnover for firms using designers over a period of five years exceeds the turnover of companies not using design by fifty-eight billion DKK. Moreover, companies that increasing their spending in design have a gross-income growth of forty percent more compared with those spending less on design. This Danish study is based on self-reported data gathered at one point of time. No information is given about the nature and quality of the design. Unfortunately, the report lacks information on how the data has been analyzed. Therefore it is difficult to judge the validity of the reported findings.

The British Design Council (2002, 2006) generated further insights into the way design contributes to business performance:

> More than eight out of ten design-led companies have offered a new product or service in the last three years, compared to just 40 per cent of UK companies overall. 83% of companies in which design is integral have seen their market share increase, compared to the UK average of 46%. Design is integral to 39% of rapidly growing companies but to only 7% of static ones. 80% of design-led businesses have opened up new markets in the last three years. Only 42% of UK businesses overall have done so. A business that increases its investment in design is more than twice as likely to see its turnover grow as a business that does not do so. 34% of businesses that see design as integral or significant say added value have had a great impact on their turnover. Where design plays only a limited role, the proportion is just 21%. Businesses that add a lot of value are half as likely to compete on price and more than twice as likely to compete on innovation as businesses that rely on their core product or service. (Design Council 2006)

But this study, too, does not address the character and quality of the design and the research methodology is only vaguely described making it impossible to evaluate the validity of the present findings.

The Design Innovation Group at the University of Manchester and Open University (Potter et al. 1991) investigated the relationships between payback time and the application of design and found that investments in application of design had a fast payback compared to alternative investments. This study distinguished between products, packaging and graphic design and found that graphic design achieved the fastest payback. However, generalizations of the findings of this study are difficult, as the study relates to the use of a specific governmental support system. The findings may be influenced from self-selection and support effects.

Gemser and Leenders (2001) have looked into the multiple meanings of design and what product design can actually do for a company. Their work includes a discussion of the advantages and challenges of utilizing design. In their empirical analysis, Gemser and Leenders (2001) focussed on the expenditures manufacturers of furniture and instruments spent on professional design expertise during product development. The respondents were asked to rate their own performance vis-à-vis competing firms. Performance data were measured as rank correlation. The hypotheses about a positive co-variation between design and performance were partly confirmed.

Hertenstein, Platt and Veryzer (2005) have reported a number of studies in industrial design on the relationship between 'effective design' and business performance. Their investigation makes use of an expert panel from the council of the Design Management Institute to rate designs from the best to the worst in nine industries. In their analysis they distinguish between 'effective design' and 'less effective design'. However, their study does not allow for examination about the nature of design and what distinguishes design from other factors of value creation. The authors have reported some evidence for their hypothesis that the firms most effective at demonstrating good industrial design have a higher stock market return than the firms with less effective design.

The above review of past studies leaves us with many questions. For instance, we need to know if firms are performing well because at some earlier time they made the decision to invest in design. Could it be that these companies have already performed well prior to using design and now use design to reveal their original strength? Could it be the 'peacocks tail effect' that a company that performs well demonstrates this through good design? If so, it would mean that good design is an indication that everything else works well in a firm (Dumas and Mintzberg 1989). Good design would be more like a symptom than a cause. But how are we to know whether design was the cause or the effect? In order to find answers to these questions, we need studies that track data over several years of performance.

## THREE HYPOTHESES FOR HOW DESIGN CREATES ECONOMIC VALUE FOR THE FIRM

To understand the economic value design creates for a firm, it is useful to discuss three distinct elements of design that are relevant to business. Product, logo (i.e., communication) and web design represent different design competencies a company needs to possess or employ. Each of these design competencies affects the business performance in a different way.

### Product Design as a Design Competency

Product design is an integrated issue where form and function are in harmony. For a consumer to sacrifice her money to acquire the product it is usually evaluated both functionally and in its ease of performance. The product is usually evaluated as a result of its perceived strict functionality, expressivity and the credibility. A good product design is positively correlated with a higher price and turnover, relative to the intensiveness of competition.

The product is mostly understood in a strategic context and usually the responsibility of top management. Product design tends to be horizontally integrated because it requires the concerted action of marketing, production and R&D (Griffin and Hauser 1996). Also the cash flow of the company operations must ultimately come from its market transactions, where the product is the target. This means that product design is a major issue and one that is considered vital by both the management of a firm and its stakeholders in general. Products are used for signalling the intentions of a company. The recent iPhone and iPad were the topics of commercial magazines like *Business Week* because they were seen as vital to the survival of the company and to the progress in the industry. A new product is always the object of investigation by financial analysts, because they regard the product as an asset, whose value cannot be manipulated in the same way as accounting data or new contracts. There have been attempts to use new products to signal superiority to the markets and when these signals have been found less than credible, the financial markets have punished the firms. The product is the subject of positioning of the company. The argument for the hypothesis is that a higher consumer valuation due to the higher quality of the design leads to a higher price and or increased sales. This in turn leads to better business performance.

> H1: the quality of product design is positively associated with financial performance.

### Logo Design as a Design Competency

The company's visual representation, the logo is used as shorthand for communication design. The logo is an identity asset, which is given particular attention and constantly modernized to signal other changes in a company. In a market, the logo is a major source of identification and tool for establishing and maintaining a consistent image. This concerns brand communication, where the assumption is that the marketing communication may influence the consumer, but the point of purchase is where the decision is taken offering opportunity to affective effects.

The logo also is used as a tool to express the organization's values towards internal and external stakeholders, as a visualization of projects related to corporate communication and identity. This is seen as important ways to realize effectiveness in human resources, to reduce waste and sub-optimality due to unclear values. Not the least in connection with mergers and acquisitions, communication design, may give integrative benefits and 'esprit de corps'. Also on the financial markets logo and its adjacent items are seen as important, for example in annual reports as effective organizational communication. Moreover, capital market signalling is seen as a matter of using logos. For instance, when large companies merge, the analyst look for which logo dominates or how are two logos integrated. Important inferences concerning who owns whom come out of this. Therefore, we expect a positive correlation between the evaluation of a well-designed logo and a company's financial performance.

A pertinent question concerns the valence of the co-variation for logo. In the sense that it is recognized as a valid symbol for the present organization, the correlation can be expected to be positive. On the other hand, logos are often used to set a new direction or strategic intent for the company. In that context the logo is a sign, not for what the company is, but what it ought to be (Henderson and Cote 1998). A new logo here may be used to signal a new direction and new values. If the logo is new, the valuation (Janiszewsky and Mayvis 2001) may be negative because recognition and acceptance take time, and credibility may be missing. In such instances, a negative valence should be expected, as the consumer initially does not recognize the logo to be a credible expression of what the company represents. Only a lot of additional communication may achieve that consumers understand that the company wants to change and that the new logo functions as a guide for the company' s new aspirations or that the company intends to remain what it is now. In such situations, a discrepancy between product and logo can be expected.

H2: The logo is positively associated with financial performance.

As the analysis shows, there may also be a case for negative associations. The reason is that firm may face to very different scenarios. One is the conventional, where the product design is performing well and dominates the situation. An alternative scenario is where the product does not perform as anticipated and the financial strategy (value of the shares) dominates, but where the logo does not deliver credibility to the general impression.

## Web Design as Design Competency

Although web design has been available for barely twenty years, it is now essential in the communication of products. The much acclaimed iPad success is to a large degree a question of good web design. Important parameters are simplicity, accessibility of navigation and fluency of operation. Since most users on average stay focused on a web page only for a few seconds, the visual elements of a website must be immediately accessible (Horn 1999). Increasingly, interactivity and the sense of being in contact with the firm are important. Online services and support are an important issue where users expect to get immediate responses to their questions by a call centre or automatic answers to standard questions ('FAQs') or live chats. Online communication is a two-way street. Customers' e-mail addresses can be used to send newsletters and announcements of new products, events and other forms of 'buzz marketing'. *Amazon.com* is a company that continues to advance its online interactions and services for its customers. One of their smart design features is that connect items a person bought simultaneously and to allow this information to new customers that customers who bought a particular book in many circumstances also bought particular other books. Access to networks may also have economic repercussions like network externalities (Shapiro and Varian 2000). The third hypothesis is argued by the web design may influence interactivity and services to serve the users and benefits of information efficiency and network externalities to the business.

H3. A web design is positively associated with financial performance.

## HOW WE TESTED THE THREE HYPOTHESES: SAMPLE SELECTION, SIZE AND RESEARCH DESIGN

To test our hypotheses, we had to identify a set of appropriate companies, develop a framework for evaluation of their financial data and be able to assess the quality of their product designs. We took the following steps to set up the study:

### Research Sample

The selection of firms came out of the '100 largest Danish firms' organized in a database by the Danish newspaper Børsen. These data were corroborated by public databases for the period 2001 to 2005. The companies selected for the investigation were randomly selected from the list by picking every third company from it. Unfortunately, some companies were missing essential information and could not be used in the investigation. Some of the companies were parts of larger conglomerates or local sales departments of multinational firms. Others were publicly owned and finally some did not have a legal form that requires a public account. This left us with twenty-five companies we identified as dominant local players within the industries of food, transportation, pharmaceutical, manufacturing industry, and energy.

### Financial Data

We apply different proxies for profitability, growth and size, respectively. We apply three accounting based proxies for profitability; return on investments (ROI), profit margin (PM), and return on equity (ROE). Penman (2007) regards ROI as a superior performance measure. We define ROI as operating profit before tax divided by total assets. PM is defined as operating profit before tax divided by net turnover. ROE is defined as net earnings divided by book value of equity. In addition, we apply two market based profitability measures; growth in market value of equity and market to book value. For example, Penman (2007) finds that a high market to book indicates a high future level of profitability. We measure growth as growth in turnover, net assets, ROI and PM, respectively. As proxy for size we apply net turnover, total assets and market value of equity, respectively. These proxies were for the 25 firms were recorded for the five years from 2000 to 2004.

Accounting data imposes numerous challenges and fluctuates from year to year due to a variety of reasons such as investments, mergers and acquisitions, changes in accounting principles, divestment of assets, which has little to do with the basic value creation or design. To reduce this problem, we summarize the five years observations in two ways: as a measure of the financial values, for which we calculate the mean over the five years from 2000 to 2004 and as a measure of growth, for which we calculate the mean growth rate over the same five years. The financial data are summarized in the appendix, Table 22.A1.

### Design Data

To capture the meaning of 'good design', we could have used data on the investment in design or on the number of designers, to mention a few possible indicators. We could also have selected the criteria ourselves.

Instead, we selected an expert panel with the help from the Danish Design Centre and the Association of Danish Designers. We asked both to suggest five designers they felt had a reputation for being among the best designers and the best referees in contests and legal settings. All designers had extensive industrial experience. Out of the ten designers originally selected, some were not available at the time of the experiment. Thus, seven design experts participated in our research.

### The Study

The present study includes product, logo and web designs. Samples of design for each of the twenty-five firms were downloaded from the company websites and set up as a paired comparisons experiment with three parts: one part compared products; the second part compared logos and the third part compared web designs. The lack of actual physical designs may be partly compensated by the fact that the experts are all experienced and have all been exposed to the product designs of the large companies under study. To ensure that the judgment was not influenced by conflicts of interest, we asked all experts beforehand to indicate if they were personally involved with any of the design work conducted in these firms. None was.

The method of paired comparisons is a basic tool for measuring preferences for stimuli, as well as for scaling responses to objects or perceptual attributes. The power of this method lies in its ability to make sharp discrimination of scaled objects, such as shown in Figures 22.1 and 22.2. (For a review of paired comparisons and the conjoint analysis, cf: Gabrielsen 2000, 2001.)

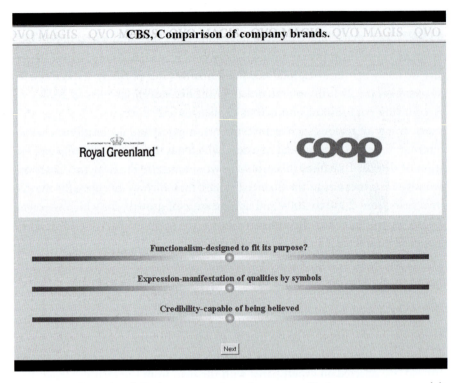

FIGURES 22.1 Example of the logo comparisons by experts: Design experts compared logos from Royal Greenland and COOP with respect to 'Functionality', 'Credibility' and 'Expressivity' and marked their preferences on the three scales below the images. © Kristensen and Gabrielsen, 2011.

For the first part of the experiment i.e. comparing products, each of the experts was sitting before a computer screen, which showed two pictures of products – one picture to the left and one picture to the right. To provide a better impression of the actual products, a projector was used to the design up in larger scale (as exemplified in Figures 22.1 and 22.2).

We instructed the experts to compare the products shown with respect to three distinct product qualities: (1) 'Functionality – designed to fit its purpose', (2) 'Expression – manifestation of qualities by symbols' and (3) 'Credibility – capable of being believed'.

## 1. Functionality

Above all, functionality, i.e., the performance of the object, seems to be a major concern for the consumer, given the meaning and the context of a product (see below). Convenience and easiness in use are important criteria. This is evident when looking at fast-moving consumer products, where the packaging is the product (Orth and Malkewitz 2008). Functionality is inseparable from the performance of a product: a coffee maker is expected to produce coffee of a certain quality when operated; a pair of scissors is supposed to cut paper and/or textiles. A coffee maker may be difficult to

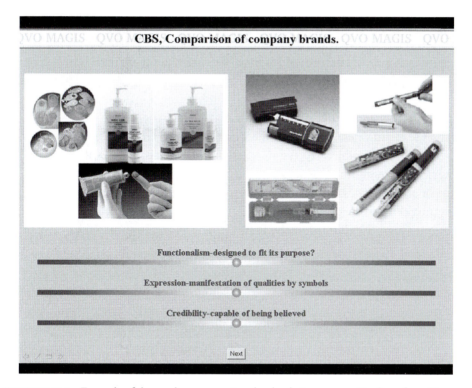

FIGURES 22.2   Example of the product comparison by the design experts: Products from Nomeco and Novo Nordisk were compared with respect to 'Functionality', 'Credibility' and 'Expressivity' and then ranked on the three scales below the images. © Kristensen and Gabrielsen, 2011.

operate or may not produce an enjoyable cup of coffee and thus fail to fulfil consumer expectations. In this case, the performance of the coffee maker is not adequate.

Sometimes, functionality is not performed in an obvious manner, but is deeply embedded. An object may just look right or feel right to a person without deliberation (Churchland 2002). In this case, the relation between the user and the object is 'seamless' and the user or consumer does not think of the thin line between the hand and the object. The actual functionality only becomes evident to the user when the object breaks down (Winograd and Flores 1985). The feeling of seamlessness and value points to cognitive processes that are tacit and take place unconsciously.

## 2. Expressivity

Good design is defined according to meaning and context. A good design must be easy to understand. This can be defined strictly or in a broad sense. Take a lemon squeezer. Most of them cost three to ten Euros, and they may not appear particular well designed. Yet, they perform the job of transforming a lemon into the juice. So they must be described as functional. Then take the Juicy Salif squeezer shaped as a long-legged spider by Philippe Starck. Many say it looks good, but attempts to squeeze lemon means splashing the kitchen and usually only a few drops reach the glass you must place under the squeezer. If the criterion is functionality, this product cannot be well designed, and yet the sales are tremendous. The question then is what is it? Surely, it is expressive. One may assume that this object is not really a lemon squeezer, but rather a piece of art intended for the kitchen. As such it is functional, but not as a squeezer. In general, we assume a well designed product to be functional, but functionality in a strict sense as indicated above – not the whole function. The squeezer is a piece of art rather than a design object. As the above example show, people are willing pay for expressivity and surely, the Starck squeezer is more expressive than most squeezers. The example also indicates a 'trade-off' between functionality and expressivity; does the expressivity enable the user to understand well the strict functionality or does it serve its own purpose, as indicated by the Rams' criteria above. In the latter case, we prefer to label such design 'art' (Spivey 2005).

### 3. Credibility

Finally, the criterion credibility is a reflection similar to the feeling of value and the term 'reliability' seems to cover what the object of design must be. Credible means that the spectator easily comes to the conclusion that this is good design (Reber et al. 2004).

A logo is intended for effective identification, typically of the company brand. For the consumer, a logo means identification of the company delivering products and services. In this sense, it is 'shorthand' for a number of design points which may include corporate architecture and even storytelling. The logo is a 'prime' for expression of promises if the consumer purchases (Orth and Malkewitz 2008). The quality of the logo is dependent on its ability to signal distortion free, enabling the awareness of the consumer and to raise the intended image of the company. There may be a source of confusion, because the logo itself is usually a symbol, which may or may not be a good design – independent of how the consumer judges the company. The confusion may come from a judgment of the logo as a judgment of the company. A company with a bad reputation may actually have a good and well-designed logo.

Web design concerns the Internet interactions between the company and its stakeholders. Its functionality concerns effective communication and interactivity. A major issue is that the stakeholder with a minimum of 'clicks' can reach the wanted information and interact with the relevant department in the firm. For that to happen it must be reliable and fast. In addition, the design must be logical and expressive of the firms' intentions as well as the stakeholder intuition of how it works. Easy access and to understand also adds to the credibility and experienced reliability of the Web design.

We then asked the experts to mark their preferences concerning each pair of products on a 30 cm linear scale running below the two pictures, each scale representing one of the three above criteria (as shown in Figures 22.1 and 22.2). The rating of each quality was performed electronically by moving a cursor from the middle to the left or to the right, indicating which design the expert favoured. A rating mark close to the middle indicates that no strong preference exists, and the two designs were considered equal. A rating mark close to the right end of the scale indicates a high degree of favour for the right picture compared to the left picture – and vice versa. After the preferences were rated, the expert pressed a button marked 'next' and a new screen with two new pictures of products appeared and the scores registered electronically.

No numeric values or labelling were added to the scales. However, to ensure the proper understanding of the comparisons, the expert were given instructions and some warm-up comparisons, were performed to train the procedure. After fulfilling the first part of the experiment – product comparisons – the experts conducted the second part of the study by comparing company logos. In the third and final part, the design experts compared company web designs.

In each part of the experiment, the two pictures they compared were presented in a random order. To reduce the demands on the designer, the number of comparisons in each part of the experiment was limited to twenty-seven comparisons following an 'experimental statistical design' allowing for preference scales to be estimated. The time to value and respond to the designs was approximately one hour per expert.

## STATISTICAL ANALYSIS OF THE DESIGN EXPERT COMPARISONS

At the end of each part of the experiment, for each quality the ratings provided by the design experts were converted into numbers. The distance from the middle of the scales to the noted mark (i.e., the entry made by the design expert) was measured electronically, positively to the right and negatively to the left. In any event it was assumed that the numerical scores increases with the strength of the preference of the one firm over the other firm, and that equal but opposite preferences corresponds to equal but opposite scores. If, e.g. the numerical score with firm $i$ to the left and firm $j$ to the right is positive, then firm $j$ is preferred over firm $i$, if the numerical score is negative, then firm $i$ is preferred over firm $j$.

The purpose of the paired comparisons is for each of the qualities to arrange the firms into a rating scale, i.e. that the assessed preferences of two firms can be attributed to the difference of the two firms on a continuous scale. The usual way of expressing a rating scale is that the signed preferences are 'additive' in the sense that the preference of firm $i$ over firm $j$ is the same as the preference of firm $j$ over firm $k$ plus the preference of firm $k$ over firm $i$ for any firm $k$.

The model is a general linear model and the estimation of parameters, therefore, is straightforward.

To obtain a comprehensive definition/measurement of design the experts were asked to compare the three distinguishing qualities functionality, credibility and expressivity. However, it was expected that the three distinguishing qualities reflected only one basic quality – 'good design'.

### Descriptive Statistics

The arrangement of the 25 firms into a rating scale for each of the three qualities is shown in Table 22.A2 and is also illustrated in Figure 22.3.

What we see is that, for the same firm, there may be considerable differences in product, logo and web design. Rockwool demonstrates the best product design, but its logo and web design perform less well. DSB scores high in its web design but low on logo.

### Similarities of the Three Scales

It was expected that the ratings of the product, logo and web design were correlated, Table 22.1, showing that the rating of product was significantly positively correlated with the ratings of both logo and web design, whereas, logo and web are not correlated, $p = 0.2724$.

This structure was moreover supported by the fact that also the rating of web and logo is conditional independent conditioning on the product, $p = 0.794$. This may indicate that the design of the product plays a more crucial role compared to web and logo design. The interpretation is that the product design seems to be the 'driving factor'. It affects both logo and web design, but it does not seem to work the other way, from logo or web design to product design. This may be interpreted that the product, which is the most difficult design to get right because of the managerial complexities related to concerted action between multiple business functions, may positively affect the rating of the other forms of design.

### The Relation between Design Quality and Financial Performance

The data generated are shown in Table 22.2. The vertical axes show the correlation coefficients for product, logo, and web design respectively.

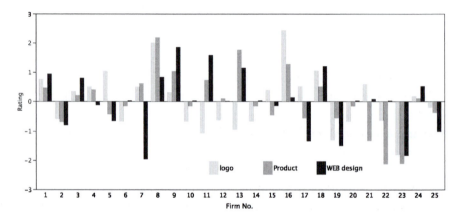

FIGURES 22.3   Ratings of the twenty-five firms with respect to product, logo and web design. ©
Kristensen and Gabrielsen, 2011.

**TABLE 22.1: Pairwise simple correlations**

| Variable | By Variable | Correlation | P |
|---|---|---|---|
| Product | Logo | 0.4849 | 0.0140 |
| Web | Logo | 0.2283 | 0.2724 |
| Web | Product | 0.5551 | 0.0040 |

**TABLE 22.2: Correlation, Spearman's rho**

| | | Product | Logo | Web |
|---|---|---|---|---|
| **Size** | | | | |
| Net turnover mean 5 years | $N = 25$ | 0.433* | 0.136 | 0.35 |
| Net assets, mean | $N = 25$ | −0.085 | 0.497* | 0.019 |
| Market value of equity, mean value | $N = 13$ | −0.456 | −0.258 | −0.06 |
| **Growth** | | | | |
| Net turnover growth rate over 5 years | $N = 25$ | 0.182 | −0.343 | 0.106 |
| Net assets growth | $N = 25$ | 0.171 | 0.161 | 0.061 |
| Return on investment[1] growth rate | $N = 25$ | 0.205 | 0.123 | 0.08 |
| Profit margin[2], growth rate | $N = 25$ | 0.442* | 0.145 | 0.188 |
| **Profitability** | | | | |
| Return on investment[1] mean 5 years | $N = 25$ | 0.07 | −0.398* | 0.063 |
| Profit margin[2] year mean | $N = 25$ | 0.104 | −0.313 | 0.209 |
| Stock market equity growth rate 5 years | $N = 13$ | −0.379 | 0.221 | 0.363 |
| Market to book, mean | $N = 13$ | −0.104 | −0.555* | 0.39 |

*Correlation is significant at the 0.05 level (2-tailed).

[1]Return on investment is defined as operating profit before tax divided by total assets.

[2]Profit margin is defined as operating profit before tax divided by net turnover.

The horizontal axes show values for various economic performances. The correlations are between the rankings of three forms of designs and a variety of performance measures.

Based on the correlations in Table 22.2 there are some indications that size of the firm is associated to the quality of the design. The correlation between net turnover and product design is positive and significant at the 5 percent level. Also net assets are negatively correlated with logo design at the 5 percent level.

Furthermore, the signs of the correlations differ; they are positive for product and negative for logo. Again, this is seen as an indication of, that the nature of product design and the other forms follow different rules. It is quite possible for product design to be positive and graphic design to be negative. This is because the product is what is offered to the user or consumer at the sacrifice of the price, while the graphic design may communicate what the firm aspires to become, but is not yet there. This finding also strengthens the claim that the relation between product and other forms of design is dominated by the product design and not vice versa.

Growth only explains the quality of the design in a modest way. The correlation coefficient between growth in profit margin and product design is positive and significant at the 5 percent level. Other growth related correlation coefficients remain insignificant.

The correlation between profitability and logo design is negative and significant at the 5 percent level. Both the correlation coefficients of the market to book and return on investment are negative and significantly different from zero (5 percent level). The explanation may be that the logos are judged significantly, but negatively due to lack of credibility. The credibility may be low due to the use of logos as expressions of the values and intentions the company is pursuing and not what it has already realized in its products. The quality of web design is neither explained by size, growth and profitability.

## ANALYSIS AND DISCUSSION

Some correlations were significant for product and logo design which means that H1 and H2 were supported with the important qualification that the correlation for the H2 was negative, but the H3 hypothesis was rejected. The reason may have been that the three scales lacked in their ability to express different qualities (functionality, expressivity and credibility of the designs). It was therefore decided to merge the three scales into one, representing the 'quality of design' including functionality, expressivity and credibility.

The design experts' valuation was treated as a single scale as the expert's opinions were very similar and they could be ranked on a single scale. The performance data were analyzed as correlations, using both Spearman's rho and Kendall's tau. They matched perfectly, and Spearman's rho was used subsequently.

For the product the correlation coefficient between its valuation and return on equity was 0.442, which was significant at the 0.05 level. Also net turnover showed a significant correlation, 0.433, which was also significant at the 0.05 level. For the logo the correlation with the net capital growth was −0.555 and significant. Correlation with net capital growth was −0.497 and return on investment −0.398, both significant.

Hypothesis H1 was supported in that there was a clear correlation between good product design (as defined by our experts) and the ranking of and total turnover performances (0.442, statistically significant at the 0.02 level and 0.433, statistically significant at the 0.01 level).

H2 was accepted although its correlation was negative. Logo is negatively correlated to stock market value/internal value (KIV) (–0.555), net capital (–0.497) and return on investment (–0.398). The explanation seems to be that the expert interpretation of what the company aspires to be was different from what the experts observe the company to be and they do not see the expression as credible.

H3: the third hypothesis concerning web design was also shown to be false. No significant co-variation was found.

The reported findings show that the main driver is product design. If this is valid, the logo and web design may follow, but not the other way around. It seems impossible to show, that logo or web design dominated product design and therefore is able to let the product piggyback on say a strong brand – just to take the full consequence. The outcomes may be a direct reflection of the fact that the three forms of design were organized in a different manner in the firms. There were different ways of managing the forms and that seems reflected in the outcomes.

It is a remarkable observation that the correlations were negative for the logo, while positive for the product. The explanation could be that they follow different rules indicating that it is accepted that a logo does not have to be a truthful impression of what the company is at the present time, but that experts accept that the logo signals what the company aspires to. The difference itself, as was explained above, follows from varying organizational rules and strategic intentions of products and other forms of design respectively.

## CONCLUSION

Unlike previous studies, we distinguished between product design, logo design and web design. By using a design expert panel that compared product pairs, we created a more 'realistic setting' and a finer grading of the quality of design.

Product and logo design may reveal very different things. A good product design is a sign that everything else works well (Dumas and Mintzberg, 1989). This is because a good product design requires all cross-functional operation to be well integrated (Griffin and Hauser, 1996). The logo, on the other hand, can be designed by design consultants with limited access to the cross functions of the company. If the company wants to cover financial problems, or just boost its figures, it may support this by a new logo, cooperative branding, etc. But in this situation, the design may not credibly show that everything else works well – in particular if the product design fails to impress.

Web design was not found to be a significant factor in this study. That does not mean that it is not significant in other firms that were not part of this study. A study including e-commerce and even iPod or Amazon.com, for example, might demonstrate a different view.

In our opinion it is safe to claim that Tom Watson's words are true, and that companies may use product design to improve their business. Yet there are more unanswered questions. Future studies of design in business must deal with all these issues and the challenges are complex. Studies until now have been limited in scope. They all added to a common body of knowledge, but more comprehensive approaches will support this. Also 'qualitative studies' like case studies can provide valuable insight since they actually go deep into the value creation of the companies and may offer inputs into the strategic considerations of design.

## REFERENCES

Barney, J. (1991) 'Firm Resources and Sustained Competitive Advantage' *Journal of Management,* 17(1): 99–120.

Blaich, R. with Blaich, J. (1993). *Product Design and Corporate Strategy: Managing the Connection for Competitive Advantage.* New York: McGraw Hill.

Bolan, Richard J. and Collopy, F. (1994). 'Design Matters for Management'. In Bolan, R.J. and Collopy, F. (eds.). *Management as Designing.* Stanford, California: Stanford Business Books.

Churchland, Patricia Smith (2002). *Brain-Wise Studies in Neurophilosophy.* Cambridge, Massachusetts: A Bradford Book MIT Press.

Design Council (2006) 'Design in Britain 2005–06' http://www.designfactfinder.co.uk/design-council/pdf/DesignInBritain2005–06.pdf downloaded 10 June 2007.

Design Council (2002) 'Competitive advantage through design' http://www.karo.com/portfolio/images/ideaspdf/Competitive%20Advantage%20Through%20Design.pdf downloaded 15 May 2007.

Dickson, T. (2006). *Dansk Design.* København: Gyldendal.

Dirickx, I. and Cool, K. 1989. 'Asset Stock Accumulation and the Sustainability of Competitive Advantage' *Management Science,* 35(12): 1504–1513.

Dumas, A. and Mintzberg, H. 1989. 'Managing Design Designing Management' *Design Management Journal,* 1(1): 37–43.

Erhvervsi- og Boligstyrelsen (Danish Ministry of Commerce and Construction). 2003. *Designs Økonomiske Effekter* [Economic Effects of Design]. Available online: http://www.ebst.dk/file/1638/designeffekter.pdf (accessed 10 June 2007).

Gabrielsen, G. (2000) 'Paired Comparisons and Designed Experiments', *Food Quality and Preferences,* 11: 55–61.

Gabrielsen, G. (2001) 'A Multi-level Model for Preferences' *Food Quality and Preferences,* 12: 337–344.

Gemser, G. and Leenders, M. (2001) 'How Integrating Industrial Design in the Product Development Process Impacts on Company Performance' *Journal of Product innovation Management,* 18: 28–38.

Griffin, A. and Hauser, J. 1996. 'Integrating R&D and Marketing: A Review and Analysis of the Literature'. *Journal of Product Innovation Management,* 13: 191–215.

Hatch, M.J. and Schultz, M. 2003. 'Bringing the Corporation into Corporate Branding' *European Journal of Marketing*, 37: 7–8.

Henderson, P.W. and Cote, J.A. 1999. 'Guidelines for Selecting and Modifying Logos' *Journal of Marketing*, 62: 14–30.

Hertenstein, J., Platt, M. and Veryzer, R. 2005. 'The Impact of Industrial Design Effectiveness on Corporate Financial Performance', *Journal of Product Innovation Management*, 22: 3–21.

Heskett, J. 2002. *Logos and Toothpicks*. Cambridge: Cambridge University Press.

Heskett, J. 2005. *Shaping the Future Design for Hong Kong: A Strategic Review of Design Education and Practice Design Task Force* [Report]. Hong Kong: The Hong Kong Polytechnic University.

Janiszewski, C. and Meyvis, T. 2001. 'Effect of Brand Logo Complexity, Repetition and Spacing on Processing Fluency and Judgement' *Journal of Consumer Research*, 28: 18–32.

Keller, K.L. 1993. 'Conceptualizing, Measuring, and Managing Customer-Based Brand Equity' *Journal of Marketing*, 57: 1–22.

Kotler, P. and Rath, G. 1988. 'Design: A Powerful but Neglected Tool' *The Journal of Business Strategy*, Autumn: 16–21.

Maltz, E. and Kohli, A.K. 2000. 'Reducing Marketing's Conflict with Other Functions: The Differential Effects of Integrating Mechanisms' *Journal of The Academy of Marketing Science*, 28(4): 479–492.

Orth, U. and Malkewitz, K. (2008) 'Holistic Package Design and Consumer Brand Impressions' *Journal of Marketing*.

Penman, S. 2007. *Financial Statement Analysis and Security Valuation*. 3rd ed. New York: McGraw-Hill.

Potter, S., Roy, R., Capon, C., Bruce, M., Walsh, V. and Lewis, J. 1991. *The Benefits and Costs of Investments in Design: Using Professional Design Expertise in Products Engineering and Graphics* [Report]. Manchester: OU/UMIST Design and Innovation Group.

Rams, D. 1995. *Weniger, aber besser* [Less but Better]. Hamburg: Jo Klatt Design+Design Verlag.

Reber, R., Schwarz, N. and Winkielman, P. 2004. 'Processing Fluency and Aesthetic Pleasure: Is Beauty in the Perceivers Processing Experience?' *Review of Personality and Social Psychology*, 8(4): 364–382.

Rumelt, R. 1984. 'Towards a Strategic Theory of the Firm.' In Lamb, R. (ed.). *Competitive Strategic Management*. Englewood Cliffs New Jersey: Prentice-Hall.

Sanchez, R. 1999. 'Modular Architectures in the Marketing Process' *Journal of Marketing*, 63: 92–111.

Simon, Herbert A. 1996. *The Sciences of the Artificial*. 3rd ed. Cambridge Mass: MIT Press.

Schwarz, N. 2004. 'Metacognitive Experiences In Judgment And Decision Making' *Journal of Consumer Psychology*, 14: 332–348.

Spivey, N. 2005. *How Art Made the World a Journey to the Origins of Human Creativity*. New York: Basic books.

Winograd, T. and Flores, F. 1985. *Understanding Computers and Cognition: A New Foundation for Design*. Norwood, NJ: Ablex.

# APPENDIX

**TABLE 22.A1: Financial data**

| Proxy | N | Mean | Std. Dev. |
|---|---|---|---|
| Net turnover, mean | 25 | 13516.9 | 12819.03 |
| Net turnover, growth rate | 25 | 0.092 | 0.247 |
| Profit margin (PM), mean | 25 | 0.063 | 0.07 |
| Profit margin (PM), growth rate | 25 | −0.004 | 0.016 |
| Return on investments (ROI), mean | 25 | 0.06 | 0.06 |
| Return on investments (ROI), growth rate | 25 | −0.005 | 0.019 |
| Return on equity (ROE), mean | 25 | 0.088 | 0.15 |
| Return on equity (ROE), growth rate | 25 | −0.022 | 0.066 |
| Growth in stock price | 12 | 4.51 | 45.92 |
| Market to book, mean | 13 | 2.715 | 2.764 |

Mean is mean of the years 2000 to 2004

Growth rate is the mean annual growth rate in the years 2000 to 2004

Profit margin (PM) is defined as operating profit before tax divided by net turnover

Return on investments (ROI) is defined as operating profit before tax divided by total assets

Return on equity (ROE) is defined as net income divided by book value of equity

Market to book is defined as market value of equity divided by book value of equity

**TABLE 22.A2: Ratings of the twenty-five firms with respect to logo, product and web design**

| Firm | No. | Logo | Product | Web Design |
|---|---|---|---|---|
| Danisco A/S | 1 | 0.76 | 0.47 | 0.96 |
| Danish Crown AmbA | 2 | −0.59 | −0.70 | −0.81 |
| Arla Foods AmbA | 3 | 0.35 | 0.21 | 0.81 |
| DLG Service A/S | 4 | 0.52 | 0.41 | −0.13 |
| Royal Greenland Seafood | 5 | 1.05 | −0.43 | −0.66 |
| Coop Denmark A/S | 6 | −0.69 | −0.16 | 0.04 |
| DFDS Seaways A/S | 7 | 0.51 | 0.63 | −1.97 |
| SAS | 8 | 2.01 | 2.19 | 0.85 |
| DSB A/S | 9 | 0.32 | 1.04 | 1.86 |
| Sophus Berendsen A/S | 10 | −0.69 | −0.16 | 0.04 |
| H. Lundbeck A/S | 11 | −1.08 | 0.74 | 1.59 |

*(Continued)*

**TABLE 22.A2: Ratings of the twenty-five firms with respect to logo, product and web design (Continued)**

| Firm | No. | Logo | Product | Web Design |
|------|-----|------|---------|------------|
| Coloplast A/S | 12 | −0.64 | 0.11 | 0.02 |
| Novo Nordisk A/S | 13 | −0.96 | 1.77 | 1.16 |
| Nomeco A/S | 14 | −0.69 | −0.16 | 0.04 |
| FLS Industries A/S | 15 | 0.40 | −0.47 | −0.15 |
| Rockwool A/S | 16 | 2.43 | 1.28 | 0.15 |
| Aktieselskabet Nordi | 17 | 0.52 | −0.57 | −1.34 |
| MT Højgaard A/S | 18 | 1.06 | 0.52 | 1.21 |
| Flextronics International | 19 | −1.32 | −0.57 | −1.51 |
| Danfoss A/S | 20 | −0.69 | −0.16 | 0.04 |
| NESA A/S | 21 | 0.59 | −1.34 | 0.09 |
| Aarhus Oliefabrik A/S | 22 | −0.64 | −2.14 | 0.04 |
| Vestas Wind Systems | 23 | −1.82 | −2.12 | −1.85 |
| Kuwait Petroleum | 24 | 0.18 | 0.12 | 0.53 |
| NEG Micon A/S | 25 | −0.20 | −0.38 | −1.03 |

# 23

# Design Transformations: Measuring the Value of Design

## DAVID HANDS

### INTRODUCTION

It is commonly agreed that design can offer real tangible benefits to the organization. Design is both an integral and intrinsic part of a variety of business cultures that provides a fertile seedbed for strategic growth and sustainable development. However, for this to be fully realized into a cohesive vision, a framework of organizational planning is vital to ensure its successful implementation and value.

Since its fledgling beginnings in the late 1970s, design has undergone considerably change both in the way it is understood at an organizational level and the way it is used at a project level. However, the business landscape has undergone radical change with the erosion of global boundaries, new emergent technologies and societal transformation all demanding reconfigured ways of working and pervasive negotiation. One thing does remain – design is still notoriously difficult to quantify in terms of its impact and many benefits. Design has many roles to play within the organization and, on reflection, it has varying levels of application; this ranges from minimal impact, mostly being used as a tool to cosmetically improve the appearance of a product, right up to being at the core 'DNA' business operation. Through its many-faceted roles, design management has the ability to shape business decision making and operation by influencing key individuals within the organization. With end consumers and customers demanding greater transparency and honesty in organizational activity, extra demands have been placed on the organization to fulfil these obligations. By having a broader understanding and appreciation of design, its value and impact could be more widespread and crucial to organizational activity.

It could be argued that design goes far beyond 'doing' and the final embodiment of that process. It is more akin to an all-encompassing philosophy that permeates every aspect of business operation. This chapter investigates how design can be measured or 'audited' to provide the organization with a unique insight into its true value to obtain new and strategic horizons for continual growth and sustainability in uncertain times.

## AUDITING DESIGN – QUANTIFYING DYNAMICS AND TENSIONS

The value of design 'enhancement' on the organization has therefore been clearly demonstrated. Wide-ranging and detailed surveys across many diverse and different industrial sectors have demonstrated the increase in turnover and profitability that can result from design investment (Roy and Potter 1997). At an individual organizational level, case studies have explained how individual products; product families and indeed services have significantly benefited from the consideration of design issues (Thackera 1997). However, very often, for an organization, this message can be frustrating, despite the evidence of research into and investment in adopting a design-centric approach; it can be difficult to know how to value design within the company; where to start to augment its application and, inevitably, how to manage it effectively. The generally accepted solution to this problem is to engage the services of an external design consultant. Although this is often successful, evidence suggests that there can be problems with this approach.

There is still continual debate as to the role and definition of design within an organizational context. Often, design is posited within the context of new product development and attendant development processes; in the company context, design is defined as a process in which products and services are developed by combining design with other core competencies. While design competence is creative in its basic nature, the role and importance of design management has a strategic character. Design management has emerged into an important stage of corporate recognition, placing greater demands on the skills and competencies of design managers, whereby they have to utilize their innate knowledge, interpersonal skills and forecasting strengths. Design-trained personnel are now in an enhanced position to lead organizations along with their management counterparts in fast changing commercial environments.

As the impact of internationalization and the emergence of overseas new markets take on a more strategic importance, design is increasingly regarded as a means to handle knowledge and complexity with the aim of attaining competitive businesses, products and services. To counter these conflicting and often contradictory demands, design managers must adjust their current thinking and understanding of design and the services they offer. A greater attention to the design process, its constituent elements and methods is required as a basis for the integration of knowledge and requirements. Design services must continually adapt and adjust according to client needs. Tacit knowledge must be described, communicated and shared with the client and constituent stakeholders to create a shared basis. Walton (2002) offers a comprehensive definition of both design and design strategy within this context, suggesting that:

> design is both process and product. It is a way of thinking synthetically, of looking at problems in unexpected and creative ways and seeking innovation. It is also about specific outcomes – products, graphics, communications and business settings. More broadly it is about how consumers and stakeholders experience the outcomes – the design interface must be all these things – a strategy for action, as well as the results of those actions and decisions.

The design process can be considered as the 'immaterial creation' of the product, that is, at the end of this process the product does not yet exist but everything is defined beforehand. As a consequence, design in its current conception incorporates a significant amount of complex information and knowledge that must be properly managed. Therefore, a mature, sophisticated and continuously improving series of processes is becoming increasingly important for organizations. This, in turn, requires a sound and robust management of the whole development process, integrating the appropriate tools and techniques and with the necessary supporting infrastructure. These problems highlight the need for a set of flexible tools, which can help in assessing the whole process and provide the opportunity to identify specific aspects in order to later focus on them during the improvement phase.

One notable series of guidelines to assist the organization in the design and development process of new innovative products is BS 7000. The standard offers a comprehensive framework on the management of design, addressing issues of responsibility and consideration throughout all levels of organizational operation. The series includes:

BS 7000-1: 1999 Guide to managing innovation
BS 7000-2: 1997 Guide to managing the design of manufactured products
BS 7000-3: 1994 Guide to managing service design
BS 7000-4: 1996 Guide to managing design in construction
BS 7000-5: 2001 Guide to managing obsolescence
BS 7000-6: 2005 Guide to managing inclusive design
BS 7000-10: 1995 Glossary of terms used in design management.

One of the key contributors to the series of guidelines, Alan Topalian (Alto Design Management) argues that the 'emphasis is placed on ensuring close coordination during the development process so all disciplines contribute effectively at all stages, before and after introduction to market'. The series of guidelines emphasizes the importance of financial planning; key decision pathways; roles and responsibilities of design and management functions from the boardroom to the designer at project level. Initially, BS 7000 was considered unwieldy and inflexible in operation; however, after a series of modifications and amendments, industry has warmly received the guidelines commonly adopting its application in practice.

In addition to the BS 7000 series of published guidelines, the Design in Business Organisation in conjunction with the Design Council have produced a three-part framework focusing on how the organization can develop innovative product and service offerings. The 'Design Atlas: A Tool for Auditing Design Capability' provides straightforward and highly applicable support information that measures design capability, processes and planning considerations in three parts. Part 1 describes the frame of reference covered by the audit tool and outlines how it can be utilized to review design capability within an organization. Part 2 provides a detailed description of the questions that constitute the entire three-part framework. Finally, Part 3 provides examples of actions that can be taken to develop and augment organizational design capability in response to the audit (Roberts 2001).

Although the Design Atlas framework provides a more accessible and easy-to-use approach to understanding design capability within the organization, its industry acceptance has yet to be fully investigated and understood. However, anecdotal evidence suggests that its popularity and application is increasing; the benefits of its implementation remain unclear.

Cooper and Press (1995: 187–222) provide a broad account and definition of what constitutes a design audit, offering methods and design auditing techniques to the reader. They offer a flexible framework whereby the organization could undertake a full strategic design audit, taking into consideration four dimensions of investigation: the wider environmental context in which the organization operates; physical manifestations of design, focusing upon visual identity and communication material; the internal corporate culture and understanding of design; and, lastly, design management procedures and processes. The authors emphasize and signpost aspects of organizational activity to investigate but do not provide an exhaustive list of considerations to adhere to. They provide a cautionary note that using audits 'will help organisations to define current practice and benchmark themselves against other organisations, but relying on audits to measure practice, stimulate improvements in practice and for overall change is inappropriate'. Predominantly, the audit framework has been and remains hugely popular with both design and management postgraduate students negotiating the complexities and nuances of auditing design for the first time.

## WHAT IS AN 'AUDIT'?

Auditing refers to a range of activities aimed at improving how an organization operates. This may be in a strategic area where current and past plans are evaluated against results. Alternatively an audit may apply to a management area where any aspects of an intangible process may be considered. Operations may also be 'performance re-searched', particularly from a compliance viewpoint where a complex range of goals both precise and imprecise may be located against what actually goes on.

Overall, an audit is concerned with structure, process and their relationship al-though the effective audit value varies considerably depending on the ability of the auditors to focus on the issues and the ability of the organizations to facilitate the audit process. A cycle classically describes most audits where a process of improvement is used that involves research. However, whereas the hypothesis-stimulated process called research may be concurrently evaluated, audits tend to have greater specificity.

The general benefits of the audit process are:

- the provision of knowledge of current practice;
- the improvements in standards of process;
- the highlighting of non-compliance;
- the identification of potentially damaging practice;
- the rewards which are intrinsically linked to accurate knowledge;
- the securing of future cohesion;
- the promotion of higher standards;
- the facilitation of further planning;
- the stimulation of learning.

An audit works best when self-improvement is identified as a goal. This may be achieved by recognizing that knowledge is lacking in the day-to-day operations of an organization. An audit should provide a structure through which change may be identified and implemented. Audits also encourage the airing and discussion of prob-lems and the shared critical evaluation of practice. The objective position of the audit should promote construction and encourage team debate and decision making rather than management-led change. In this way, change may be self-developed where man-agers may use the audit process to guide the decision-making process. Therefore, if viewed in an entirely positive light the audit process may become transferable and re-usable, so a team may be empowered by the process, effecting change internally.

Audits act to improve or to facilitate self-improvement. This is sometimes described as 'stewardship', which relates to 'capital' – financial capital, human capital, intellec-tual capital and the responsibilities and power associated with them. The development of this process of stewardship may be attributed to the complexity and density of professional life, particularly where necessary targets are present but where they may be imprecise. As a first principle, stakeholder audits may represent the development of standards and quality and, indirectly, their sharing. Certainly the impartiality and 'fact-based' nature of audits ought to elicit both comparisons and exemplars.

The audit process, within the context of design, (as with all audits) reflects the na-ture of the activity. As design is often represented by a variety of activities it is perhaps natural to focus on a qualitative audit methodology. This, as it may relate to people in design, draws upon much activity from social communication and the recognition of cooperative working. In this way a design-led organization or department may usefully be described as a collaborative enterprise but perhaps regularly celebrating its particular diversity. Collaboration should actually expect surprise, acknowledge the chaotic nature of creativity and avoid a philosophy of predictability, particularly in assessing outcomes. This, from an audit perspective, presents some difficulty; successful collaboration at work will involve the relatively unpredictable contribution of participants. This could be called 'creative collaboration', which in design creativity is encouraged but not entirely measurable through audit as it often occurs through a silent and discreet form of engagement. In fact, attempts to adopt audit compliances within creative environments reduce task variety just when its breadth may be a key business success factor. The audit therefore has to acknowledge human conditions and aspirations related to communication, trust, authority and tacit knowledge. The demands for business efficiency if narrowly defined, therefore, would actually develop predictable activity and produce, through audit, negativity. Social audit therefore proposes trust as a key aspect of control.

Auditors are required to balance their methodology. The nature of autonomous organizations (in particular certain aspects of the design office) is to serve their own local interest. This should not be audited in isolation. The power of individual design and creativity within organizations resided within a reciprocal framework of self-feeding and business delivery. This may be difficult to acknowledge from the perspective of the audit as detailed processes are often reviewed only with exclusive reference to larger organizational systems. Creative design groups 'self-reference' prior to recognizing an organizational system. In relation to this, a first-level audit defines creativity as a value within the context of the overall aims of the organization. A second-level audit is one that assesses focused operational work referred locally. An auditor is therefore required to perceive both levels at the onset but acknowledge the recursive nature of most organizations. Reinterpretation of larger scale compliance through audits is often central to the audit task. This is ultimately a qualitative process if one acknowledges the responsibility of creative units in the development of goods and the associated qualities of experience. An audit that centres on control should therefore measure the effect of control at both levels. A bureaucratic organization will centre on verifying adherence to negotiations; creativity naturally questions such control and so audits may represent reconciliation between each level.

Western management processes often appear to increase bureaucratic control; 'managing' creative design may subscribe to this perception. However, there is a growing acknowledgement of the need for internal self-control and responsibility suggests that our competitive future will be increasingly reliant on fostering creativity in organizations. The traditional audit view of capacity should increasingly be related to self-control. Intrinsic control (Beer 1966) therefore should be applied – that is control more focused on tacit knowledge and less on normative control. Alternatively the auditor as an extrinsic controller, taking a 'Fordist' view, will adjust systems to exterior performance criteria. This, in turn, may actually reduce effective control of creative processes, as trust between levels is lost and individual contribution denied. Business pressures naturally increase the control dilemmas focused in hierarchical relations. Furthermore, the increased complexity of creative process linked to new product development actually calls for more focus on compliance and detail (for example, product safety standards). This compliance naturally attacks the required flexibility to operate creatively in a design development. The decision to audit, particularly in recursive environments, should be taken with the expectation of change for all. Paradoxically audits may arrive for new areas like design from a development of 'bureaucracy', which naturally assumes predictable behaviour.

Audits should aim to develop understanding between autonomous units and to reduce the expectation that individuals may have of unchallenged implementation of their personal policies. In this way, the first step that an organization may take, particularly involving creative elements like design, is to demystify the complexity surrounding tacit knowledge and its counterparts. Change from a mutual understanding between autonomous groups could therefore be seen as a rather more natural process than one that is brought about by a compliance audit, potentially based on organizational norms. Mutual understanding is a product of agreement through communication; therefore audits may be used to:

- form strategic alliances between units;
- develop deeper level communication;
- understanding complementary competences.

This dispels the traditional image of auditing as solely concerned with non-compliance. Audits therefore develop knowledge; it is important to work with what you have (designers, managers, salespeople, account handlers and so forth) but a deeper level audit ought to inform the company of the precise nature of such people and resources.

## INTRODUCING THE DESIGN AUDIT

The individual designer is at the heart of the assessment of design competence within the firm via an audit. Designers will not see themselves as 'measurable' even though their efforts plainly relate significantly to the operation of the company. Design groups tend to reference value within themselves; this is typically found in all professional groups. Indeed design solutions which themselves vary from company expectations may be viewed, by designers, as independent. Traditionally the designer is the free 'irreverent' employee who alone appears to have specific personal engagement with his or her work. In directing design one is aware of the necessity to encourage design-ers to question and think (unlike many other employees) beyond the current scope to the company. The environment fostering the design and development of new prod-ucts and services is complex and involves and variety of risks at a number of levels in a wide range of situations. The literature on design management is equally varied from that which is calculable to that which surrounds the perceptions of adventure in de-sign (Jerrard 2000).

How then can design be audited? Do we take a social audit model or economic one? Or is design more closely related to the development of alternative solutions as in ecological/environmental auditing? Are the metrics just too complex? Plainly aes-thetic judgements cannot be 'measured' and the benefits of good design might relate to spiritual benefit (both the designer and customer) as well as an economic benefit. Designer's motivations, therefore, are always likely to be partly covert. Consequently a range of assessment criteria for a design audit may include the following broad cat-egorization both strategic and operational levels:

- aesthetic policy and achievement;
- the working environment;
- professional (peer) evaluation;
- financial management;
- training;
- education;
- marketing;
- design research;
- client orientation;
- company alliance strength.

These categories may lead to quantitative, 'reductionist' conclusions and as such would move the audit away from the issue of design as being intrinsically valuable.

However, assuming that a design department is not assessable would isolate it. One approach has been to socially audit popular product processes as in the Body Shop and Ben & Jerry's Ice Cream (Johnson 1996). Also a 'stake holder' view (Jackson and Carter 2000) may objectify the view of design departments. This is echoed by the Audit Commission in its expectations in arts auditing (Matarasso 1997). Design in it-self may be seen as profit making and a number of accounts suggest that it is pivotal to the fortunes of many companies (Design Council 1998) without specific rules to use successful processes in evaluating design are likely to be socially based, for example, focus groups of users, consumers and manufacturers would contribute significantly to design audits. Involving such stakeholders is well known in social auditing, figure re-quirements a potential audit cycle.

## METRICS OF MEASUREMENT

Within any organization design will involve complex activities and a variety of links to other activities. Furthermore the associated activities within most definitions of design (in the humanities) may even defy description. However there are a variety of starting points for describing design activities within companies so they may be assed or even measured. Initially the design department may be viewed as a series of informal and formal networks. Common characteristics of networks include (a) social measures and (b) performance measures. Social measures will include: participation, communication, trust, professional norms and common purpose. Performance measures will include: business contribution, product or service market innovation. Clearly an audit of design (and the design function) is likely to have particularly broad but interrelated measurement criteria. Some common themes from a wide approach include:

- a mixture of qualitative and quantitative data;
- an ability to equate formal measurement with assessment and derive accurate 'hybrid' conclusions;
- measurements of aspects of value systems (for example, social capital) with elements of structure (networks inside and outside the company);
- comparison, involving contextual references within the company, between design and other developments and functions.

There are, however, specific problems in attempting to measure design performance. These include:

- an understanding of social and economic measures;
- a conceptualization of generic questions to specific situations;
- the process of obtaining and categorizing diverse information;
- difficulties in equating innovation and change with compliance and benchmarking;
- the absence of precise performance indicators;
- the interpretation of audit results.

The tension involved in promoting the design audit as a specific process is ever present. Design as with other functions may be viewed as essential to the company but valueless if viewed through traditional measurement/assessment techniques. What makes a strong or weak design department? Is the design department able to be commonly ranked against departments? Rigorous measurement is one challenge, another related aspect is demystification of design in order to allow for audit results to be understood and potentially acted on. Evaluation of the design function appears to predicate measurement of it even though inputs and outputs may be particularly imprecise in design. What then makes a good design audit, one that provides valuable and accurate 'opportunity-cost' results, which may be used as a measure of effectiveness? The required indicators would include:

- *specificity* – target areas to be measured;
- *measurability* – access to information and people;
- *reliability* – features which are both representative but also are consistent within the criteria;
- *rigour* – real life information rather than wholly anecdotal interpretation;
- *comprehensiveness* – the features which represent day to day activities in design;
- *continually* – common, transferable audit techniques.

Design, as well as involving complexity, is internationally diverse throughout a variety of sectors. The ability to audit design universally (Roberts 2001) is hampered by sector specific practices and geographical locations. Plainly design practice involves common values and common business ideals but how design effectiveness is measured in one industry or country may be vastly different to another.

## AUDITS FROM MODEL DEVELOPMENT

Management models have always provided representation of process success and failure. Although many such models exist, their effectiveness in design environments should perhaps be questioned. An examination of a simile seems often typical – companies as machines, human bodies, journeys, etc. Such glib similes add little to the analysis of a design management problem. The purpose of all audits is related to the development of 'accurate' picture as opposed to generic problem representation. However the day-to-day terminology of management (for example, performance targets, quality assurance and so forth) appears too general or even inappropriate for design. Modelling design is often viewed as inaccurate and, like all such systems, usually represents scientific convenience and reductionism. How, then, can we model design processes 'accurately' as part of a process of problem (or success) diagnosis within an audit? A rather more isomorphic approach is called for – if the process resembles spaghetti it may well be to describe (or model) it so. Therefore isomorphic models are perhaps best for applications in design departments where it is essential to preserve the operational characteristics in the design department within which a novel (specific) process terminology exists. How, then, can design audits gain currency against audits in other areas of the company?

It is evident that there needs to be a degree of common process and terminology in all management models within the same organization. Individual departments may be best modelled at an operational level when conceptual models would apply generically. However this suggests denial of the strategic role design often has within an organization. The best way to model design is therefore to integrate its isomorphic model with a homomorphic company model. The model of the company will acknowledge design and the design department will acknowledge the company. This, at the level of operations, recognizes synergy as a component of business success. This account illustrates that design audits allow for the application of management and process models to areas of the company that are often overlooked.

## CAPTURING DESIGN 'KNOWLEDGE'

Knowledge of all kinds exists within organizations at a number of management and operational levels; it is often considered intangible and 'unmeasurable'. However, the process by which knowledge is acquired and subsequently used is via individuals in often discrete and tacit ways. In recognition of the interdisciplinary nature of design in area like NPD (Jerrard et al. 1999), the study of the associated knowledge is important.

Networks formed by individuals are unlikely to provide all of the day-to-day specific knowledge that the organization requires; informal networks, however, continue to be the lifeblood of small competitive companies. These networks contain social capital and as such often add tremendous value to the life of companies. The successful integration of information from different sources therefore may be said to provide 'informational advantage'. Knowledge has long been thought of as a key commercial resource but one that is so often elusive to many and frequently misunderstood.

For such knowledge to be advantageously retained and embedded by the company, then, a process to evaluate it needs to be developed; refined and embedded within day-to-day practice. However, for a small or medium-sized enterprise it is often extremely difficult for a number of reasons. These include the fact that small companies necessarily have to deal with simultaneous information for complex concurrent working whilst operating on tight deadlines and limited financial and human resources. The personal nature of networking often also reduces the perceived value of knowledge as it may appear subjective and may not be shared in full. There often appear to be significant differences between organizations that are easily able to generate and use knowledge and those that find it difficult. This is partly geographical; it is not surprising that national or even global competition is derived from regionally clustered groups of companies. Such networks may be considered in relation to one firm, where employees may see themselves as occupying a more central aspect of knowledge dissemination. However, such networks are often viewed to be more dynamic, with third parties relating to fourth parties and so on. In this way, knowledge can produce specific and valuable capacity, which in turn may be generally viewed as 'social capital'. Plainly, assessing social capital poses a problem of scale; networks are often defined through an individual and that individual's efforts. Some observers (see Burt 1992) have attempted to measure the strength and diversity of an individual's network. However, this may only be possible in generic terms; an lone individuals may not be able to contextualize what they know through others.

In assessing the knowledge potential of a small organization, in design, it appears that networks may be particularly important. Membership of professional bodies for designers (for example the Chartered Society of Designers or the Design Business Association) has always aided business although many would admit to learning little through membership; this is unlikely to be true.

In a knowledge-creating company there appears to be a strong emphasis on internal entrepreneurism where all may be considered knowledge workers (see Nonaka 1991). This may be perceived as a loan activity but providing velocity for the knowledge appears linked to a process of sharing. Designers seem to do this all the time, a community of designers may be observed regularly making tacit knowledge explicit and indeed this may serve as a metaphor for the design process itself. The promotion of design knowledge may be an evangelistic pursuit where such tacit 'behaviour' is regularly reviewed and integrated within day-to-day work. Functional knowledge is often viewed as complementary to design knowledge and may be evidenced by the provision of appropriate business outlets for the complex skills of a designer. Within many design-centred organizations such divisions may not yet be detectable, which may explain why the auditing design appears complex and so often contradictory.

## REFERENCES

Beer, S. (1966) *Decision and Control*. Chichester: John Wiley & Sons.

Burt, R. (1992) *Structural Holes: The Social Structure of Competition*. Cambridge, MA: Harvard University Press.

Cooper, R. and Press, M. (1995) *The Design Agenda: A Guide to Successful Design Management*. Chichester: John Wiley & Sons.

Corfield, K. (1979) *Product Design*. London: National Economic Development Office.

Design Council (1998) *Design Council Red Paper 1*, April. London: The Design Council.

Jackson, N. and Carter P. (2000) *Rethinking Organisational Behaviour*. Harlow: Pearson Education.

Jerrard, R. (2000) 'Researching Designing: Cycles of Design Research', *Doctoral Education in Design: Foundations for the Future, La Clusaz, France*. Stoke-on-Trent: Staffordshire University Press.

Jerrard. R., Trueman, M. and Newport, R. (eds) (1999) *Managing New Product Innovation*. London: Taylor & Francis.

Johnson, J. (1996) 'Social Auditors: The New Breed of Expert', *Business Ethics*, March/April.

Kotler, P. and Rath, G. (1990) 'Design: A Powerful but Neglected Strategic Tool', *Journal of Business Strategy* 5: 16.

Matarasso, F. (1997) *Use or Ornament, The Social Impact of Participation in the Arts*. Stroud: Comedia.

McGoldrick, P. (1990) *Retail Marketing*. Maidenhead: McGraw-Hill.

Nonaka, I. (1991) 'The Knowledge-Creating Company', *Harvard Business Review*, November–December.

Roberts, P. (2001) 'Corporate Competence in FM: Current Problems and Issues', *Facilities* 19(7/8): 269–75.

Roy, R. and Potter, S. (1997) 'The Commercial Impacts of Investment in Design,' in M. Bruce and R. Cooper (eds) *Marketing and Design Management*. London: International Thompson Business Press, pp. 175–203.

Thackera, J. (1997) *Winners! How Today's Successful Companies Innovate by Design*. Aldershot: Gower.

Walton, T. (2002) 'Design and Knowledge Management', *Design Management Journal* 2(1): 5.

# 24

# Major Challenges for Design Leaders over the Next Decade

## ALAN TOPALIAN

Thinking forward ten-plus years may be a long time, however 1995 now seems just a few years ago.

*Jouko Peussa*

We are bad at looking decades ahead ... We're clever, but seldom wise.

*Ronald Wright*

Chance favours the prepared mind.

*Louis Pasteur*

## INTRODUCTION

The contributions of design and design professionals within business and society evolve continuously. Design leaders need to be at the forefront of such changes and, ideally, should drive them. This chapter encapsulates and builds on twenty-two respondents' views of major challenges anticipated for design leaders over the next ten or so years. Challenges were interpreted as serious demands, concerns, hurdles, limiting factors, influences and opportunities that mould crucial decisions.

Building on best practices within design, respondents were encouraged to distil from perceived challenges distinctive roles design professionals might fulfil to expand perceptions of design contributions (perhaps in unexpected areas), demonstrate the benefits of involving design professionals and raise aspirations to lead through design.

Data from sixteen respondents were collected in early 2005 followed by data from a further six respondents in 2009. Two respondents gave their views in writing; the others through interviews, most lasting around 45 minutes.

The first part of this chapter lists significant aspects of design leadership. The second sets out critical factors that provide some background from which identified challenges arise. The third part outlines the challenges that emerged from respondents' views, presented under five broad headings.

## ASPECTS OF DESIGN LEADERSHIP

Where should we look for leadership in design? The following are significant aspects:

- *Distinctive and innovative design approaches and solutions* that change our perceptions, set new benchmarks and trends, enhance the quality of life and create new opportunities.
- *Leading through design*, using the 'designerly approach to problem-solving' to upgrade corporate performance substantially, particularly where design professionals take the lead to drive change and deliver against the odds (Topalian 1990).
- *Leaders in design*: individuals at different levels in organizations who are outstanding 'creatives', nurture creative environments and/or inspire exceptional creative performance. Not all will be design trained, nor will their titles and job descriptions feature the term 'design' (British Standards Institution 2008b).
- *Design leadership over time*. First, where successful solutions are generated quickly and immediately 'feel right' to clients and users. Second, where the effectiveness of solutions and overall contribution of design are sustained over the longer term, even when attention is distracted or priorities change. And third, by maintaining momentum in design investments and building on experience.
- Finally, *gaining acknowledgement for achievements through design* (or reclaiming them for design when credit is assigned elsewhere). This encompasses conveying the 'design message' effectively to prime audiences and providing ammunition for other stakeholders to promote design on to colleagues. A key characteristic of such leadership is ensuring that the term 'design' is never buried, sidelined or otherwise ignored.

## CRITICAL FACTORS UNDERLYING CHALLENGES

A number of factors relating to professional design practice emerged from discussions with respondents that provide added context and background to the challenges in the following part. These are outlined below.

**TABLE 24.1: Critical factors underlying challenges**

1 Total experiences encompassing all senses.

2 Expansion of boundaries to encompass integrated systems.

3 Resonant platforms.

4 Personalization and sharper targeting.

5 Maintaining integrity of design solutions despite dispersal of work.

6 Multi-sourcing and co-creation through alliances.

7 Waste of unfulfilled potential.

8 Legislative demands and uneven playing fields.

9 Multiplier effect of intellectual property.

10 Full discharge of design responsibilities.

## 1. Total Experiences Encompassing All Senses

Attention now focuses on creating and managing experiences for customers and users of what organizations stand for, offer as products and services and how they operate (Topalian and Stoddard 1997). Designing total experiences could harness all senses: taste, smell and sound in addition to sight and touch.

Additional benefits would accrue if attention were paid to designers' and their clients' experiences during the complete span of design projects.

## 2. Expansion of Boundaries to Encompass Integrated Systems

Boundaries demarcating 'tangible' products are eroding as the nature of artefacts undergoes radical transformation. Increasingly, products are complemented by services and services incorporate tangible components that can be patented and branded. These offerings will not deliver satisfactory experiences unless they are conceived and work as integrated systems (Elliot and Deasley 2007). Where boundaries are set is critical when deciding the components to bring together into seamless, functionally and aesthetically attractive solutions.

Interfaces and operating systems constitute the familiar 'faces' and 'languages' that present distinctive characteristics, hide technical complexities and enable interoperability between different products. These provide the glue to securing the ultimate prize: repeated use and loyal ownership.

## 3. Resonant Platforms

Products and services tend to become more complex over generations (Arthur 2009). However, enhancements are not valued when they cannot be comprehended quickly, used intuitively or tailored to the way users operate.

'Un-bundling' and stripping back to fundamentals are sound strategies to distil the essence of offerings (Stern et al. 2006). Such 'core platforms' are more likely to generate emotional resonances and loyalty among users through easier, effective use. These could be reinforced through complementary enhancements targeted at the specific needs of different market segments.

## 4. Personalization and Sharper Targeting

Organizations rarely have the last word on the products and services they offer because customers effectively 'complete' them with experience through use, amendments, impressions and emotions. In the process, they 'personalize' those offerings and the brands they are marketed under (Topalian 2003).

How people use products and services is seldom predictable: much can be learned from observation and feedback, especially when disappointment or satisfaction is beyond expectation. Communications tend to be more resonant when organizations are clearly knowledgeable about how customers use their products and provide options that accommodate the idiosyncrasies of users and help boost performance.

Identifying target audiences and potential customers accurately, then reaching them effectively when necessary, to inform and deliver product experiences, requires harnessing an expanding range of channels to their full potential.

## 5. Maintaining Integrity of Design Solutions despite Dispersal of Work

Design work tends to be pigeonholed into predetermined silos, then chopped into bits that are often parcelled out for execution around the world – offshore or outsourced – before delivery to target markets. In many instances, no-one has a meaningful overview or formal responsibility for all this work. Compartmentalization and fragmentation of design work can lead to a loss of control and debilitating compromises, just when it is crucial to ensure that the different strands and outcomes work seamlessly together. Solutions reaching markets may be stripped of the distinctive characteristics and coherence that commended their initial concepts for development.

Pressures to commoditize design suck executives into a piecemeal, short-term perspective with a blinkered overemphasis on cutting costs. Investment is curtailed rather than using design intelligently to build the expertise and design management infrastructure to enhance performance and ensure the quality of products and integrity of brands are not compromised. In such circumstances, it is difficult to envision a strategic role for design. For example, few grasp the fact that all innovations, bar none, are designed and, often, it is design professionals who drive innovation.

All this distances design professionals from key decisions, severely limiting their influence. Design leaders need to ensure that proper attention is paid to design professionals' contributions and recommendations. They should also insist that design professionals always have direct contact with and feedback from, all parties, especially clients and end users (Topalian 1989).

Design leaders and their clients need to understand clearly the business problems underlying projects and how design solutions should perform to deliver planned experiences to target audiences. These help to avoid ill-informed shortcuts and compromises, so defend the integrity of solutions through development to launch, ensuring that all involved are familiar with the approaches taken and why solutions are formulated as they are.

It is unlikely that wisdom accumulates or that resources are used effectively when work is chopped up inappropriately. The resulting loss in corporate memory limits an organization's capacity to generate effective solutions. Care should be taken to ensure that expertise does not degenerate when, for example, sales and marketing rarely visit design studios and service workshops, R&D is relocated offshore and manufacturing is out-sourced.

## 6. Multi-sourcing and Co-creation through Alliances

As the complexity and costs of projects increase, a wider range of specialists and suppliers contribute to the design process and execution of solutions. They bring with them different skills, values and networks that need to be mastered.

The principles of inclusive design and open sourcing provide strong catalysts to 'co-create': harnessing the expertise and energies of users, staff and supply chain partners to formulate and validate design solutions. Strategic alliances help spread workloads, expenditure and risk; they can also increase agility to respond to opportunities and compatibility through common standards.

Consequently many projects have several 'clients' and teams, some dispersed around the world. 'Noise' and heat generated can claim significant proportions of ever-shortening work programs and budgets as wants, approaches and cultural factors vie for attention.

## 7. Waste of Unfulfilled Potential

Organizations whose products are used to their full potential have a competitive advantage: customers perceive higher value and feel greater satisfaction, hence goodwill. These factors provide a foundation for higher margins, growth in revenues and market share.

By contrast, ineffective use of products represents a waste, perhaps resulting from a poor understanding of product capabilities or a lack of confidence to demand more of them. Overspecified offerings with features and benefits that customers do not see or take advantage of are unlikely to be valued. Better guidance and support planned as an integral component of products would help, perhaps with simulations of product operation provided online before purchase and demonstrations of effective use within after-sales support (Topalian 2003).

## 8. Legislative Demands and Uneven Playing Fields

Organizations in advanced nations can be handicapped by higher standards imposed in home markets. Competition is rarely on a level playing field, especially when established players have difficulty protecting against intellectual property infringements by new entrants from less regulated emerging economies.

## 9. Multiplier Effect of Intellectual Property

As creators of intellectual property (IP) with a special grasp of design and certain technologies, design leaders are in an advantageous position to apply an organization's IP to best effect, internally and through licensing, in familiar and unfamiliar fields. Considerable value could be added for all parties if licenses – say, of a new technology – incorporated design management support to help partners achieve desired outcomes through enlightened applications.

Moreover, design leaders could advise on IP that should not be shared (say, in strategic alliances), then ensure these are appropriately protected to avoid compromising an organization's competitive position.

## 10. Full Discharge of Design Responsibilities

No business will attain its long-term potential unless design is taken seriously. Without enlightened and stretching design leadership, expenditure that precedes and follows design cannot be optimal – a distinct disadvantage in highly competitive global markets.

Design may be too important to leave to designers yet, all too often, senior executives fail to discharge their design responsibilities appropriately – through a lack of awareness, misjudging the difficulty in leading design or disinterest. Consequently subordinates, who rarely have the skills or authority, flounder without direction and support (Topalian 1990).

Selecting ideas to sponsor is tough; these choices subsequently affect levels of change and investment. They can be infectious across organizations, especially when a clear vision influences choice. Sometimes, asking for what seems impossible opens up new possibilities to pursue.

## ANTICIPATED CHALLENGES

The remainder of this chapter is devoted to the challenges anticipated for design leaders over the next decade or so. These are set out under five headings – see Figure 24.1.

| Prospect beyond the horizon | Upgrade to business design | Manifest corporate conscience through design | Nurture future generations of leaders | Join top table |
|---|---|---|---|---|
| Escape the present to envision valued futures | Establish design as the key unifying discipline at the heart of business | Do the right things right | Sustain centres of excellence in design | Step up to lead in business |
| Map strategic terrain for exploration | | Seduction carries responsibilities | Revamp training to lead through design | 'Borrow with pride' best practices from other fields |
| Wise management of transitions | Cultivate a distinctive voice and expertise | Embrace diversity, promote inclusion and sustainability | Value staff by championing their creativity | Acknowledge diversity in design leadership |
| | Develop deeper practical knowledge of users | Achieve a better work / leisure balance | Target new skills for development | |
| | Insights into competitor outputs through creative analyses | Build on achievement through virtuous cycles of development | Infiltrate 'non-design' jobs to widen perceptions of design | |
| | | | Augment shared experience of professional practice | |
| | | | Ensure continuity by grooming successors | |

FIGURE 24.1   Overview of key challenges. © Alan Topalian, 2005–10.

## PROSPECT BEYOND THE HORIZON

### 11. Escape the Present to Envision Valued Futures

Design leaders are responsible for developing their organizations' ability to explore territory 'beyond the horizon' and envision needs, customers and markets that do not exist yet (Topalian 2000; British Standards Institution 2008a). Organizations that are able to see further into the future have more time to prepare for whatever the future presents: they can create opportunities to renew themselves, become more agile and change direction as necessary – all through design. Creative friction, especially between new groupings of different individuals, is vital to bring about new thinking and solutions. However, staff need to break free from the present and conventional thinking.

> One doesn't discover new lands without consenting to lose sight of the shore for a long time.
>
> *André Gide*

Design leaders should encourage colleagues to be different, change mindsets, expand networks, update their ways of working and challenge top executives, even in adverse circumstances. In sum, to operate as opportunistic innovators to turn perceptive ideas into new realities.

> Marketing colleagues tend to provide us with extrapolations from where we are now. By contrast, designers offer us leaps forward in our thinking ... One of our designers' special contributions has been the uncanny accuracy of their forecasting. Many design proposals that seemed far-fetched proved to be just right when the time came ... The leaps forward enabled through design are the catalysts that make all those other advances possible. They build the confidence to envision different ways forward and by-pass 'me too' strategies and solutions.
>
> *Harry Rawlinson*

## *12. Map Strategic Terrain for Exploration*

It is important that design leaders generate a continuous stream of proposals for developing their organizations, transforming ideas into attractive outputs and revenues. Introducing design into planning cycles and corporate plans will go a fair way to reinforce design's strategic role in business. This will highlight design leaders' contributions in fields far wider than design – particularly in setting the agenda, influencing priorities and the allocation of resources in business and society (Topalian 2000; see also, Topalian 1995; Cooper and Press 1995).

Design leaders can also provide 'roadmaps' to track trends in behaviour and identify emerging needs (British Standards Institution 2008a). These will help to make sense of the 'fuzzy front-end' when exploring new opportunities and solving problems generally. The confidence that comes with thorough preparation and sharing information is a significant factor in opening minds to different, more adventurous approaches and solutions.

## *13. Wise Management of Transitions*

Virtually all design leads to change, and managing change is a design exercise in its own right. Project outcomes, however attractive, do not always commend themselves and the value attached to final outcomes is influenced by the journey traversed (Topalian and Stoddard 1997). However, just as anticipated journeys are rarely mapped out, neither are design professionals involved to help make those experiences more tangible at the outset or prepare for smoother transitions. Quality improvement programs decades ago illustrated these shortcomings, as do the majority of design projects whose scopes rarely encompass the effective management of transitions (Topalian 1994).

Design leaders need to ensure responsibility for transitions is assigned formally and prepare their organizations to accept and exploit fully the solutions generated. Typically this would involve explaining project configurations, justifying strategies pursued, protecting the integrities of journeys, then championing the solutions generated.

## UPGRADE TO BUSINESS DESIGN

### *14. Establish Design as the Key Unifying Discipline at the Heart of Business*

Design involves devising effective means to achieve desired ends. All organizations use design: making decisions when creating, running and expanding businesses are all fundamentally design exercises. Frequent use of the term 'design' in business media attests to this fact. Consequently, failure to take design seriously and use creative design professionals extensively to stretch aspirations and enhance performance, is tantamount to corporate sabotage.

Despite superficial acceptance of its importance, business executives find design difficult to exploit. Their poor grasp of design management leads to difficulties in evaluating design's contribution. A common view is that design requires long-term expenditure but yields uncertain outcomes.

Design leaders need to steer non-design and design colleagues through the 'design management universe' and communicate core messages about design – within and outside the profession, particularly government – so its potential is exploited fully in business and society. This should include raising awareness of the design process and contributions across business, dismantling organizational silos that constrain and undermine design work, and integrating the administration of design so that it is handled as a powerful strategic resource that measurably enhances performance and warrants long-term investment. The principles that guide design analyses and recommendations need to be explained, too. Exploring 'soft' issues around problems can help to understand better the 'hard' issues at their core. The inspiring vision formulated for Terminal 5 at London's Heathrow Airport illustrates this point (Turner and Topalian 2002).

Another challenge is to persuade venture capitalists to look for a 'design strand' in submissions received to judge how effectively applicants will turn their propositions into real businesses, through design.

### 15. Cultivate a Distinctive Voice and Expertise

The 'designerly approach to solving problems' provides the holistic perspective and motivation to tackle complete problems and create total solutions, co-ordinating the contributions of different disciplines to serve shared visions of desired outcomes.

Acute observation, creative analyses and interpretation enable design professionals to see opportunities that escape others. Empathy with users – promoting 'the voice of customers' – yields fresh insights and possibilities, and helps to manifest user experiences more tangibly. Solutions tend to be more rounded with components that are integrated seamlessly to work in user-friendly ways, often addressing a wider range of needs and delivering beyond expectation (Topalian and Stoddard 1997).

The use of visual imagery and data visualization helps to surmount language and cultural hurdles, especially when working in cross-disciplinary teams drawn from several nations. Rapid testing of propositions through a succession of models and simulations provides further vivid bases for members to spark creatively off each other, gauge reactions and validate options for development (Topalian and Stoddard 1997; Brown 2008; Hambrose 2009).

Aping the client's language may facilitate routine communication. However this often reinforces an impression of subservience unless design leaders find a distinctive voice to highlight design's distinctive contributions. Using business language in fresh ways can surprise and inspire, particularly in terrain not normally associated with design. Potent ways to demonstrate a distinctive approach is to ask unexpected questions that help to restructure and express problems in fresh ways: these bring different terrains into play, sometimes working with what appear to be mutually-exclusive factors to generate novel outcomes that competitors find difficult to match.

### 16. Develop Deeper Practical Knowledge of Users

Designers gain the trust of users and a deeper understanding of their needs when adopting the 'voice of customers' role, often proposing the means of enhancing user performance before the demand emerges. Such insights help create better-focused products and services.

> MK Electric had great difficulty in getting to the 'voice of the customer'. To bridge the gap and introduce greater customer orientation, we started using an industrial design consultancy ... One of their key roles is to help us communicate with customers in a 'futuristic' way.
>
> *Steve Amphlett*

Little is taken for granted. For example, research during a washing machine design project sought to determine what customers mean by 'clean clothes' and how they strive to achieve such cleanliness. Findings revealed customers' values and priorities in different laundering tasks – vital inputs into the design process. In another organization, when designers took over the groundwork from engineers relating to kitchen furniture design, the research undertaken changed dramatically not least as designers observed how people used their kitchens and asked questions to reveal the reasoning behind their habits. The findings gave rise to a novel range.

### 17. Insights into Competitor Outputs through Creative Analyses

As prime creators of products and services, design professionals are in a special position to ascertain how competitors' offerings are put together. These provide insights into product configuration, fabrication, how technologies are harnessed, distinctive features, strengths and weaknesses, as well as cost structures and marketing strategies. All of these could have a significant influence on competitive position.

## MANIFEST CORPORATE CONSCIENCE THROUGH DESIGN

### 18. Do the Right Things Right

Design professionals can help organizations to fulfil their corporate social responsibilities through efficient and effective use of resources, not least the enlightened management of creativity. After all, quality, continuous improvement and innovation are all designed into business activities.

Design leaders often act as the conscience of business enterprises when championing customers' interests so as to create better integrated and targeted offerings. They also ensure that their organizations 'do the right things right' and are accountable for their actions, pushing for the impact of an organization's operations and outputs on the environment and society to be transparent, measurable and appropriately documented.

## 19. Seduction Carries Responsibilities

A primary role of design professionals is to make organizations and their offerings attractive by helping to 'seduce' target audiences: customers to purchase, financiers to invest, suppliers to collaborate, media to praise and so on. Some suggest this seduction is 'purer through designers' as they have a cleaner, more trustworthy image compared with, say, advertising agencies.

As competition mounts and the powers to seduce develop, design professionals will be under considerable pressure to manipulate and push the boundaries of ethical conduct. Design leaders need to deter those exerting pressure and shield colleagues from them, ensuring that creative expertise is exercised sensitively while highlighting the professional and ethical values underlying design work with an openness that does not mask the downsides.

## 20. Embrace Diversity, Promote Inclusion and Sustainability

Inclusion and sustainability are key factors in the discharge of corporate social responsibility. Design strategies and solutions should not disadvantage potential users or other stakeholders.

Sustainability encompasses achieving more with less in the design, fabrication, delivery and use of offerings, not least by cutting out what is unnecessary, hence wasteful. Moreover, products and environments should be easy to access, process, assemble, service, recycle and so on.

Other contributions include planning products with sensible lifespans. Backward compatibility and the potential to upgrade will help in this respect.

Adopting the principles of inclusive design may encourage senior executives to evolve an inclusive approach to *management* with work integrated across an organization using diverse teams that operate in a truly *inter*-disciplinary manner with due sensitivity to the range of communities served, internally and externally (British Standards Institution 2005).

## 21. Achieve a Better Work/Leisure Balance

Design leaders could have a significant role in re-designing the future of work by seeking to achieve a better balance between private and work lives – not least in devising modes of work that draw out creativity and enhance people's sense of wellbeing. Reconfiguring work environments and processes could also have significant impacts.

### 22. Build On Achievement through Virtuous Cycles of Development

Introducing a successful innovation is no easy matter; evolving the culture and infrastructure that nurture serial innovation constitutes a quantum increase in difficulty (Topalian 2000; British Standards Institution 2008a).

Enterprises are poor at extracting the most from past experience, building on achievements from successive cycles of development. Yet, many ideas for improvement and development are revealed by carefully observing how products and services fit into and enhance customers' operations and users' lives (Topalian and Stoddard 1997).

'Stop/start' strategies – often resulting from a lack of confidence and short-termism in corporate and product development – are prime inhibitors of innovation. They sap enthusiasm, waste resources and throttle the vitality of organizations. Continuity of investment is necessary as major creative breakthroughs do not occur at the flick of a switch but rather as a series of projects unfold (Robertson and Schybergson 1995).

Design leaders need to promote longer term perspectives and continuity of investment to develop the infrastructure and skills necessary to maintain momentum in their development programmes. This will be considerably easier with greater transparency and understanding of how design contributes to customer satisfaction and facilitates the achievement of organizational objectives without imposing unreasonable burdens on any party. Moreover, investment decisions would be easier if interim benefits were evident during long-term projects.

## NURTURE FUTURE GENERATIONS OF LEADERS

### 23. Sustain Centres of Excellence in Design

The UK's reputation as a centre of excellence in design – in both education and consultancy sectors – is constantly under threat, not least as developing nations increasingly contribute to higher value work. Nevertheless there is a fair gap for others to bridge given the ability of British design professionals to think independently in unfettered ways within absolute constraints. Much the same applies to other nations with reputations for design excellence.

As work is redistributed around the world, how will creative enterprises grasp business opportunities to enhance their nations' reputations in new markets? Perhaps this may be through fresh services, sensible activities off-shore, staff and student exchanges, building new long-term relationships and developing new clusters of skills.

## 24. Revamp Training to Lead through Design

Serious concerns are expressed about the output of design schools. The change from vocational courses to a quasi-academic approach has resulted in graduates who are not adequately prepared for the rigours of professional practice. Graduates lack basic skills – observation, drawing, analytical thinking, etc. – and direct experience of all stages of design. Concept generation is of limited value without knowledge of the real world (characteristics of materials, standards relating to the built environment and so forth), limitations of processes and experience of implementation.

Design leaders have a central role in bringing academe and practitioners together. They could act as mentors to faculty members and help them expand their networks within and outside design. They could also provide clear briefs, plus guidance, for the refreshment of courses offered by design, engineering and business schools so they remain relevant to the future needs of client organizations and design practices – especially relating to continuing professional development. These would be based extensively on documented professional design practice that informs, inspires and challenges students as well as key decision-makers (Topalian 2002).

## 25. Value Staff by Championing their Creativity

> Only by encouraging people to be creative can they become truly human.
>
> *Rowan Williams*

One of the best ways organizations can demonstrate that employees are their greatest asset is to harness their creativity and place them on the front line to contribute fully to corporate performance. As creativity is at the core of leading through design, design leaders could drive initiatives that involve actively seeking and respecting staff views on organizational values and operations, then building effectively on their expertise to derive maximum benefit (Reid and Oliver 2009).

Design leaders can help introduce creative team practices across organizations. Often, their relative ease with young individuals who work in 'creative' ways in and out of teams is an advantage although, superficially, these may not look like working at all. They entrust colleagues with tough challenges (without concession), demand novel outcomes, then allow appropriate freedom, applying a light touch within clear operating guidelines.

## 26. Target New Skills for Development

Design professionals need to embrace and co-ordinate a wider range of specialisms to undertake their work competently. Relevant skills are in constant flux as new technologies, materials and procedures feature in increasingly complex projects.

Design leaders need to target and nurture appropriate skills within their organizations. Being innovative about the expertise brought to bear on problems can help generate novel solutions. New skills clusters may emerge that enable them to offer distinctive services, extend the range of problems they can tackle, enhance their efficiency, so raising their competitive advantage. Knowledge management systems can help by incorporating details of employees' professional and personal expertise and interests. Other skills can be located externally through a range of networks to respond to emerging requirements with an agile, opportunistic approach.

### 27. Infiltrate 'Non-design' Jobs to Widen Perceptions of Design

Considerable benefits should derive from trained designers gaining experience in 'non-design' jobs during their careers. In such positions, they can demonstrate the value of their transferable skills and may have greater influence on design outcomes than practicing designers (Topalian 2002). Design experience on both client and consultancy sides also helps broaden perspectives and raises confidence.

Desirable outcomes are a heightened understanding of design among non-design colleagues and enhanced expertise among design professionals to address higher-level, wider-ranging issues in business – not least through a greater understanding of client motivations and decision-making. Design leaders should promote this model of career progression and make special efforts to facilitate 'non design' appointments.

### 28. Augment Shared Experience of Professional Practice

Little of the diversity of design management experience has been documented or shared. Without such references, design is unlikely to be taken seriously by top executives when addressing problems.

Academic institutions have failed as guardians, builders and disseminators of the body of knowledge in design management. The reality of professional practice is reflected poorly in courses. Published references do not cover all key issues – a critical shortcoming when corporate memory is constrained by the fragmentation and dispersal of design work. Practitioners rarely turn to academics for new thinking and independent guidance on leading design practice. The paucity of relevant research and publications leads practitioners to search outside design, without checking (and thus undermining) references generated from within.

Design leaders need to adopt the discipline of evaluating progress and outcomes of all their projects then documenting experiences in ways that can be shared within and outside their organizations. This is still a rarity, in part because clients realize that, more often, such evaluations reflect on their own performances, rather than that of commissioned specialists. Projects could be offered as bases for academic research with guidelines for preparing rigorous case studies – complete with teaching and learning objectives – to build a 'bigger picture' of professional design management practice. This will ensure that documented cases address a common set of issues so approaches and outcomes can be compared, leading to better informed decisions and the means to evaluate actions taken.

### 29. Ensure Continuity by Grooming Successors

Grooming future generations of design leaders is vital to capitalize on design achievements and raise the discipline's profile. Yet succession planning is alien territory for the vast majority of organizations, as is creating opportunities for design staff to lead through design then publicizing their achievements. Design leaders should detail this requirement in executives' job descriptions to remedy this shortcoming (Topalian 1989).

## JOIN TOP TABLE

### 30. Step Up to Lead in Business

Design leaders need to raise their own and clients' aspirations so they are seen as *leaders in designing businesses* rather than just leaders of design in business. This huge shift requires design – whose essence is to devise effective means to achieve desired ends – to be acknowledged as *the* core discipline in business.

Design leaders need to establish credibility among top executives as worthy partners when addressing higher level, systemic problems. Ideally this would be achieved by initiating discussion, setting challenging agendas and demonstrating that including design professionals in strategic teams yields benefits beyond distinctive solutions. For example, helping to refresh the way organizations formulate and communicate business strategies internally and externally (Topalian 1995), planning how a nation will continue to operate when the 'swine flu' pandemic reaches its most virulent phase, creating resource-efficient buildings and designing corporate identities (in the true sense) for multinational organizations are current realities in professional practice.

Design leaders could act as creativity mentors to senior executives, building trust to share their deepest concerns within satisfying, productive relationships in which equally demanding parties spark off each other to mutual benefit.

### 31. 'Borrow with Pride' Best Practices from other Fields

Design leaders tend to rely on experience accumulated within their organizations; few are aware of best practices in other organizations and disciplines. So widening perspectives and networks, monitoring experiences elsewhere and 'borrowing with pride'[1] practices that have proved effective for others should be adopted as normal practice.

Design leaders should devote equal energy to publicizing appropriate practices and achievements within design, then facilitating their adoption in other disciplines, ideally extending the use made of design professionals for wider problem solving in business and society.

### 32. Acknowledge Diversity in Design Leadership

Design leaders are needed at several levels of organizations; not all will be design trained. 'Friends of design' and potential leaders need to be identified across organizations (within and outside the professions), entrusted with design responsibilities then supported to discharge those responsibilities effectively, with confidence. Such responsibilities need to be clarified and guidance provided on aspects that might be delegated to enhance overall performance (Topalian 1990).

### CONCLUSION

The challenges described above remain relevant today, five years after most respondents were interviewed. Relatively few will surprise leading design practitioners and many may appear to be of a lower order than challenges put forward in the engineering and business fields (US National Academy of Engineering 2008; Hamel 2009). However, those challenges could not be addressed without enlightened design, in both the traditional and wider senses. All feature a myriad of design processes that need to be integrated and aggregated over time to create the breakthroughs envisioned: sparks that occur during extended periods of hard graft. Design leadership is vital to get details right within the essential building blocks that reveal transformative 'bigger pictures'.

By contrast, the potential for creating distinctive roles for design professionals may be more discomforting – perhaps even tinged with fantasy. However, design leaders are already fulfilling several of these roles. As such, they demonstrate the obverse of 'silent design' (Gorb and Dumas 1987): the non-design contributions of design professionals to business success. They become involved in far wider fields than design, are often more rigorous than their clients when tackling the most intractable problems in business (for example, starting with unblinkered definitions of the business problems that underlie design projects) and, as integrators, are frequently the key drivers to enhance performance in associated areas where they might not be considered to have any role (Topalian, cited in Stevens et al. 2008).

Where do we go from here? One strand of research relates to extraordinary achievements through design: investigating how design professionals contribute in unusual ways to deliver outstanding outcomes. Surprisingly, design professionals find it difficult to interpret 'extraordinary achievements'; top performers can be modest and what is relatively common in design practice may appear exceptional to those in other disciplines.

So here are a few indicators of what 'extraordinary achievements' might encompass:

- more problems are solved (and solutions remain effective longer) than planned at the start of projects;
- design professionals provide a lead in areas where they are not considered to have any interest, let alone competencies;
- outcomes are well beyond expectation and could not have been anticipated beforehand;
- considerable time and resources are saved by redefining problems through which new perspectives emerge;
- guidance provided ensures a total solution is implemented, with greater impact;
- further opportunities are created for innovation to generate competitive advantage;
- outcomes set new standards which competitors find very difficult to match;
- clients' thinking, practices and organizations are transformed;
- the status quo of markets is disrupted with, say, novel interpretations of needs, new price points and trendsetting solutions.

The best way to predict the future is to create it.

*Peter Drucker*

## ACKNOWLEDGEMENTS

I am grateful to the following whose views contributed to this chapter: Brian Ashwood, Quality Manager, Cambridge Consultants; David Bernstein, Principal, Kelland Communication Management; Richard Eisermann, Director of Design and Innovation, Design Council (now Partner, Prospect); Rob Fielding, Divisional Director – Technical, Domnick Hunter; David Griffiths; Patrick Goff, Founder, hoteldesigns.net +; Richard Hill, Brand and Design Manager, Eurostar UK; Yo Kaminagai, Responsable de l'Unité Management du Design, RATP; Jeremy Lindley, Global Category Design Director, Diageo +; Matt Lyons, Head of Construction Design Services, Boots; David Mercer, Head of Design, BT; Peter Peirse-Duncombe, Managing Director, Bisque Radiators; Jouko Peussa, Vice-President Technology, Compair Group; Harry Rawlinson (ex Managing Director, Aqualisa Products) +; Patrick Reid, Regional Engagement Director, IS Innovation, AstraZeneca +; Charlie Rohan, Director of Consumer Experience, NCR; Richard Seymour, Creative Director, Seymour Powell; Brian Smith, Managing Director, FeONIC +; Graham Telling, Industrialist, Department of Trade and Industry; Ivor Tiefenbrun, Managing Director, Linn Products; Paul Warren, Group Design Director, Glen Dimplex; Les Wynn, Manager of Industrial Design, Human Factors & Covers Engineering, Xerox Europe +. '+', 2009

Any shortcomings in the integration of the wide range of views within a framework are entirely my own.

## NOTE

1. This phrase is believed to have gained currency in General Electric.

## REFERENCES

Arthur, W. (2009) *The Nature of Technology: What it Is and How it Evolves*. New York: Penguin/ The Free Press.

British Standards Institution (2005) *BS 7000-6 Guide to Managing Inclusive Design*. London: British Standards Institution.

British Standards Institution (2008a) *BS 7000-1 Guide to Managing Innovation*. London: British Standards Institution.

British Standards Institution (2008b) *BS 7000-10 Vocabulary of Terms used in Design Management*. London: British Standards Institution.

Brown, T. (2008) 'Design Thinking', *Harvard Business Review*, June: 84–92.

Cooper, R. and Press, M. (1995) *The Design Agenda*. Chichester: John Wiley& Sons.

Elliot, C. and Deasley, P. (eds) (2007) *Creating Systems that Work: Principles of Engineering Systems for the Twenty-First Century*. London: The Royal Academy of Engineering.

Gorb, P. and Dumas, A. (1987) 'Silent Design', *Design Studies* 8(3): 150–6.

Hambrose, H. (2009) *Wrench in the System: What's Sabotaging Your Business Software and How You Can Release the Power to Innovate*. Chichester: John Wiley & Sons, Ltd.

Hamel, G. (2009) 'Moon Shots for Management', *Harvard Business Review*, February: 91–8.

Rawlinson, H. (2003) *Impact of Design Leadership on the Valuation of Innovative Companies*. London: Design Leadership Forum.

Reid, P. and Oliver, D. (2009) 'Creativity and the Practical Innovation Process, A Driver of Competitive Advantage', *Management Online Review* (May), online publication accessed through www.morexpertise.com.

Robertson, A. and Schybergson, O. (1995) 'Multimedia Complexity: A Pathfinder Product Development Strategy', in P. McGrory (ed.) *Proceedings of 'The Challenge of Complexity' Third International Conference on Design Management*, University of Art and Design, Helsinki, pp. 14–19.

Stern, Y., Biton, I. and Ma'or, Z. (2006) 'Systematically Creating Coincidental Product Evolution: Case Studies of the Application of the Systematic Inventive Thinking® (SIT) Method in the Chemical Industry', *Journal of Business Chemistry* 3(1): 13–21.

Stevens, J., Moultrie, J. and Crilly, N. (2008) 'How Is Design Strategic? Clarifying the Concept of Strategic Design', *Design Principles and Practices: An International Journal* 2(3): 51–9.

Topalian, A. (1989) 'Organizational Features that Nurture Design Success in Business Enterprises', Proceedings of the Second International Conference on Engineering Management, Toronto, pp. 50–7.

Topalian, A. (1990) 'Design Leadership in Business: The Role of Non-executive Directors and Corporate Design Consultants', *Journal of General Management* 16(2): 39–62.

Topalian, A. (1994) 'The "Design Dimension" of Quality Improvement Programmes', *Creativity and Innovation Management* 3(2): 115–23.

Topalian, A. (1995) 'Design in Strategic Planning', in McGrory, P. (ed.) *Proceedings of 'The Challenge of Complexity' Third International Conference on Design Management.* Helsinki: University of Art and Design, Helsinki, pp. 5–13.

Topalian, A. (2000) 'The Role of Innovation Leaders in Developing Long-term Products', *International Journal of Innovation Management* 4(2): 149–71.

Topalian, A. (2002) 'Promoting Design Leadership through Design Management Skills Development Programs', *Design Management Journal* 13(3): 10–18.

Topalian, A. (2003) 'Experienced Reality: The Development of Corporate Identity in the Digital Era', *European Journal of Marketing,* MCB UP, 37(7/8): 1119–32.

Topalian, A. and Stoddard, J. (1997) '"New" R&D Management: How Clusternets, Experience Cycles and Visualisation make more Desirable Futures Come to Life', Proceedings of the 'Managing R&D in the Twenty-First Century: Theory and Practice, The Tools of the Trade' Conference, Manchester Business School.

Turner, R. and Topalian, A. (2002) *Core Responsibilities of Design Leaders in Commercially Demanding Environments.* London: Design Leadership Forum.

US National Academy of Engineering (2008) *Introduction to the Grand Challenges for Engineering,* www.engineeringchallenges.com.

## FURTHER READING

Better Place (2010) *Better Place*, www.betterplace.com.

# 25

# Design Leadership: Current Limits and Future Opportunities

## FRANS JOZIASSE

After a century of industrial development, management now recognizes design as a tool to create competitive advantage. The management of design is under control; it is time for design leadership!

Chiva and Alegre (2009) argue that the next level of design will see the need for design leadership being congruent with the corporate objectives. Likewise, design management will focus on the bottom line and on getting more bang for each buck in design. According to de Wit and Meyer (1999), combining exploration with exploitation of the design function within organizations is a natural issue for leaders to tackle. The managers and leaders of future design functions will have to find the right balance between the exploration (leadership) of new possibilities, future directions and the exploitation (management) of old corporate certainties (March 1991). Before analysing the current state of the art of design leadership, let's look at the origin of design leadership and design management and explore their similarities and differences.

Design leadership does not have a long history. In the early 1920s, design leadership was distinguished from design management and it included the following key responsibilities (Turner and Topalian 2002):

- envisioning the future;
- manifesting strategic intent;
- directing design investment;
- managing corporate reputation;
- creating and nurturing an environment of innovation;
- training for design leadership.

One could argue that this is similar to the definition of strategic design management (Joziasse 2000), which differs from design leadership in that its main occupation is the translation of vision into strategy. But design management and design leadership depend on each other heavily. Design management needs design leadership in order to know *where to go* and design leadership needs design management to know *how to get there*. This also means that (design) leaders and (design) managers have to cooperate and continuously assess the level of agreement about what to do and how to do it (Christensen et al. 2006) to be effective and efficient. Although Turner and Topalian (2002) clearly separate the two and even argue that one can't fuse a design manager and design leader in one person, they do not specifically mention the conditions to create this good relationship between design management and design leadership.

## CONTEMPORARY DESIGN LEADERSHIP

How does contemporary design leadership take place? An interesting view on design leadership comes from the Next Design Leadership Institute in New York. The Institute was founded in 2002 by G. K. VanPatter and Elizabeth Pastor as an experiment in innovation acceleration. Its self-pronounced mission is to raise our awareness of the radical changes in cross disciplinary innovation leadership at the leading edge of the marketplace and to explore how those changes are affecting designers. Next Design works at the level of foundational pattern language. This means that it distinguishes between pattern optimization (perfecting the known), which is connected to design management, and pattern creation (seizing the unknown), which is connected to design leadership According to Next Design, these two languages and their interconnectivity can be adapted to any design challenge in different circumstances, including architecture, industrial design, communication design, environmental design and experience design. The circumstances under which design leadership currently takes place are described on the Next Design web site and include the following:

- work is done in cross disciplinary teams;
- tasks include the development of interconnected inbound (towards the internal stakeholders) and outbound (towards the customer and other external stakeholders) co-operation and collaboration skills;
- projects demand user participation;
- observation of user behaviours is vital;
- it includes development of visual processes for navigation;
- it includes development of adaptable design management tools;
- the design frames not only problems but also opportunities;
- work has to adapt to scale (small and large) and complexity (low and high).

## DESIGN LEADERS

Now we have identified the circumstances in which design leadership flourishes, it is important to note that we also need the right people – those who can bring design leadership into organizations. Kevin McCullagh (2008) argues that design leaders tend to share three general qualities:

- they are good at envisioning the future; they constantly ask what is changing and what opportunities flow from this change;
- they think strategically; they question existing core competencies and ask which we need to build in the future and how we are going to set ourselves apart from the competition;
- they lead others; they know how to develop, inspire and maintain teams.

This seems to enforce Turner's argument that you cannot find a design manager and a design leader in one person because managers are more focused on controlling an existing situation. But how can one become a design leader? As yet there is no education that focuses on producing design leaders. Looking at the traditional business environment, great leaders simply work very hard and are extremely persistent in reaching their goals. This often comes in combination with a high level of self-awareness and self-management as well as being able to manage conflicts and being empathic. These skills are referred to by Daniel Goleman (1998) as the emotional intelligence of all leaders. Goleman is convinced that these skills can be part of one's talents but he insists that they can also be learned and monitored along the way, for example by using Trait Emotional Intelligence training (Petrides and Furnham 2001). Despite the common characteristics of leadership personalities, it is generally accepted that design leadership can come in different styles and with different faces. McCullagh (2008) identifies ten types of design leaders who have pushed the boundaries of the profession in recent decades. He distinguishes between:

- 'the Maestro' (Jonathan Ive, famous for his Apple designs);
- 'the Entrepreneur' (personified by UK designer Terence Conran);
- 'the Entertainer' (exemplified by Philippe Starck, often used as an example of a 'Star designer');
- 'the Visionaries' (Bill Moggridge of IDEO);
- 'the Manager' (Chris Bangle former BMW design director);
- 'the Ambassador' (Tim Brown most notably at the World Economic Forum in Davos);
- 'the Scholar' (Patrick Whitney, director of the Institute of Design at IIT, Chicago);
- 'the Provocateur' (James Woodhuysen with his exposes of design's muddled thinking and low expectations);

- 'the Scribes' (who have pioneered new forms of design journalism like Bruce Nussbaum);
- 'the Curator' (museum or gallery directors who have presented design in new ways to new audiences like Stephen Bayley at the V&A Boilerhouse and the London design museum.

## CAN DESIGN LEADERSHIP FACE THE FUTURE CHALLENGES?

If we take Next Design as an indicator, we might say that design is moving away from design management, which focused on the integrative nature of design, towards design leadership, which sees design as a transformative power. Simultaneously we see a shift from solving discipline-specific and simple problems towards addressing complex and non-discipline-specific problems. Others argue that design's future challenges are in organizational, social and business design, referred to as 'design thinking' (Brown 2008, Martin 2009). Is design ready for to take these challenges? Of course it is not enough to have a group of design leaders to lift a whole profession to the next level. Design leaders as Stefano Marzano of Philips Design have strongly contributed to envisioning Philips future, to building brand reputation and maintaining a culture for innovation. This kind of design leader is a great inspiration for many designers and design managers. But this best practice is rare and is not yet anchored in design education or in the daily practice of organizations.

Of course it is not enough to know the responsibilities – what counts is the act of defining the responsibilities on the one hand and being assigned to them on the other. The latter is linked to the ability to create mutual respect for design leadership among peers, which in today's global context of organizations and markets is very challenging.

I see two major issues here. First, many 'design leaders' still complain about their limited direct ability to generate substantial and continuous funding. It is still difficult for many design leaders to prove the value they generate for the business. Hence the design budget is often among the first to be cut. This often means that they need to address low-complexity challenges rather than high-complexity ones and work on more defined challenges instead of undefined ones. Even Philips Design, with a great design leader, did not pass internal reviews after the credit crunch and had to lay off 25 per cent of its workforce. Second, great leaders have the ability to hover above the detail, see the big picture and think abstractly to imagine a different direction (Roger 2007). And while integrative thinking appears to be one of the core competencies of designers (Pink 2005), possessing it alone will not put it into action. The development of interconnected cooperation tools and the work in cross-disciplinary teams, in particular, seems to still be a problem for many design leaders. Joziasse et al. (2005) found that design leaders seem to have reasonably professional links with other departments but that the frequency and the moment of contact (earlier in the process) is far below what they desire.

It could be argued that design leadership faces significant issues that might limit its ability to develop to the next stages of leading the development from product and service design into organizational transformation design, social transformational design and business design.

## THE SOURCES OF THE LIMITS OF DESIGN LEADERSHIP

If there are limits to design leadership, we need to understand where they come from. In an attempt to contribute to this discussion, I will look at the causes that I consider to have the potential to limit design leadership. They can be grouped in four different areas. The first is the psychology of a designer, the second is the economic context, the third is culture and the final area is linked to sustainability.

### Psychology

Psychology involves the designer's perspective, the designer's self-awareness, the designer's self-management and the designer's empathy and ability to manage conflicts. The importance of these psychological aspects is explained below.

*Designer's Perspective.*   Too many designers have created a reality that doesn't fit into the leadership model. From their perspective they have created a strong set of myths (which are often reinforced in traditional design education):

- designers are creative; others are not;
- designers can solve any problem in the world in a different, better way;
- process, structure and management of design reduce creativity;
- designers know the consumers and know what is best for them.

This may seem simplistic but is very much the reality still within the mindset of many designers. This in some way arrogant stance can become stronger when more designers sit within design departments, especially when – as is often the case – the design department is not really (read formally and physically) connected to other departments in the organization. It makes it very difficult to work in cross disciplinary teams and to develop interconnected inbound and outbound cooperation skills.

*Designer's Self-awareness*   Lack of self-criticism and unshakable self-confidence can also hinder a designer from becoming a leader. It can result in being perceived as arrogant and difficult to approach. Many designers still live in closed organizational communities; sometimes they are even kept apart from others in the organization so as not to disturb the creative process. BMW's former design director, Chris Bangle, believed that this was the best way to ensure a design team's creativity (see Bangle 2001).

*Designer's Self-Management.*   In the traditional management area – project management – designers tend not to do so well. They often lack the training and skills to manage complex projects. They say design is difficult to manage, it is chaotic, iterative. This is partially true and necessary but to make yourself credible and to lead design you need to control the process as well, especially in a commercial environment.

*Designer's Empathy.* Empathy as a competency or behaviour is poorly understood by those who need it most and it is difficult to train for it and to acquire it. Understanding the emotions of others is core to connecting with them and to building interpersonal relationships (Goleman 1998). The ability to empathize and communicate marks good leaders and helps them to motivate individuals and teams. It also supports them in managing conflicts. Although designers normally tend to be more 'touchy-feely' types (according to the Myers–Briggs personality test) they also have a strong tendency towards being introverts. It means that they probably have a talent for understanding other people but they lack the communication part.

### Economic

Limits for design leadership that derive from the economic side include the creative industry and its position in the business environment; the effectiveness of design, as well as metrics to measure design's impact financially and, last but not least, its ability to scale economic approaches to small and medium-sized enterprises and businesses.

*Creative Industry and its Position in the Business Environment.* The good news is that the European Union has calculated that the creative industry in Europe has grown strongly in the ten years since 2001 to approximately 850 billion euros turnover and now, according to the European web site dedicated to European creative industries (www.european-creative-industries.eu) accounts for as many workers as do the chemical or the food industries. It must be questioned, though, whether this increase in numbers is an improvement. Does 'more' mean 'better'?

Research indicates that true design leadership is not only about leading design into effective business results but also about leading the process of business innovation through design (Hargadon 2005; Verganti 2009). Thus defined, design leadership is hardly ever recognized in the business community.

*Effectiveness of Design and its Financial Support.* Over the last twenty years, advertising agencies like WPP and EURO have been growing into global companies with a strong focus on their financial management. Although one can disagree with the quality of the output of the ad creation process, one must acknowledge the way the ad world has created a platform for funding their activities. By measuring and 'proving' the effectiveness of their campaigns with widely accepted methodologies like monitoring sales, new customers, requests for information, phone inquiries, retail store traffic, web site traffic they create a gateway to the significant marketing budgets.

There are, of course, some design leaders like Claudia Kotchka, the former Head of Design of Procter & Gamble, who have been able to implement new metrics to measure the successful application of design in innovation (Lafley and Charan 2008).

But this has not yet led to an industry-wide methodology that makes it more easy and credible to direct design investments.

*Scaling from Large to Small Businesses.*    Many case studies and specifications about best practice in design leadership talk about large companies. But how far did the design community succeed in transferring this learning into the smaller scale business environment? And how are smaller budgets coped with?

Of the 19.3 million enterprises in the European Union (EU) today, the World Bank classifies 99.8 per cent as SMEs. These SMEs employ some 75 million people (source: http://siteresources.worldbank.org/CGCSRLP/Resources/SME_statistics.pdf (accessed 30 March 2011). In the last decade, SMEs were the primary generators of new jobs, whereas big companies downsized and reduced employment. Yet SMEs provide products and services that the big competitors do not concern themselves with. The SMEs often deliver what no one else seems to want to deliver and, in many cases, they do it very well.

So what is the role of design leadership in all this? Basically, SMEs seem to be attractive for design leadership as, on average, these companies are younger and more driven by innovation, hence design often frames opportunities instead of problems. They are closer to their markets and customers and they work more often in multidisciplinary teams. A recent study of the European Union shows that design leadership has rarely entered into these companies (Kootstra 2009).

## Cultural, Learning, Education

Some limits for design leadership can be linked to culture, learning and education. This concerns specifically the education of designers and the funding of design leadership promotion. The more general field of politics is full of obstacles for design leadership.

*Education of Designers.*    Most design education has its historical roots in art schools, polytechnics and universities. A few exceptions emerged in the early 1980s. These include the Delft University in the Netherlands, which introduced an educational model built on four pillars, geared towards setting and maintaining high standards in their research programmes. The Illinois Institute of Technology in Chicago (IIT) is another example. The IIT focused on a human-centred innovation model and strong links with the industry. Unfortunately many design schools still do not believe that designers should be educated in a context that combines aesthetics and human factors with the commercial imperative.

Larry Keeley, is the managing director of the Chicago-based innovation management consultancy, the Doblin Group and also teaches at the IIT. He observed that, although designers have taken notice of design leadership matters, designers still do not understand the sheer scale of modern business transformation. Design and designers, according to Keeley, did not transform at the same pace as the design specialization and 'focused factories' in countries like India and China have changed (Keeley 2005). Shaping and leading design specializations through training is a big opportunity for the educational world, he concludes.

*Funding of Design Leadership and Politics.*    From the moment when it was noticed that countries can create competitive advantage through leadership in education, R&D and politics (Porter 1990) researchers have observed an important role for creativity in the context of economic developments as well (Florida 2002). We have already witnessed a transformation from production towards service and creativity in the Western world in recent decades. We are now watching rapid advances in terms of learning and transformations in Asian countries like China, India and Korea. Funding of creativity, innovation and R&D-related research is now a priority on the agenda of their policy makers and promoted at a much higher pace than we have ever seen in Europe.

### Sustainability

The final source that limits design leadership concerns matters of sustainability. Keywords in this context include the triangulation of economy, society and the environment; ecological design and the number problems.

*Triple Bottom Line (Economy, Society, Ecology) Dilemma?*    Design as contribution to more products in shorter lead times, for global markets and lower costs? Or design with strong ethics, a responsible citizen and responsibility. One can argue that not much has changed since Papanek (1971) published his book *Design for a Real World.*

Still many products and services around us are unsafe (for example toys), showy (for instance SUVs), maladapted (most mobile phones), or essentially useless (designer stuff). And designers are, of course, human beings as well, who once started their studies to improve the world and its artefacts. It seems that the majority of designers are still not taking the lead into balancing the triple bottom line. Some countries, like Sweden, have a strong tradition in taking their design education responsibility seriously. The US has only very recently taken cautious steps towards balancing economic growth with a care for the environment.

Design has slowly developed some methodologies and tools for designing products and services with less impact on the environment (for example, design for disassembly) with more relevance for the consumer (for example, inclusive design). But the influence of design is still limited here as long as many companies mainly do 'green washing' instead of developing clear visions and design strategies to have less impact on the environment.

The world is rapidly transforming and design should take its position in this transformation. We know the world's challenges and we know they are more complex, ambiguous and volatile. The challenge is that this will almost certainly cut across dozens of specialties about which designers may know very little.

Designers need to work in teams. Designers need to combine their talents with unfamiliar expertise from other fields. Design teams need not only to accept but actually embrace the notion of accountability. Designers need to invent and use a series of methods that bring rigour and robustness to the field. Designers must come to understand that making something hip and culturally connected is only one dimension of being strategic. (Keeley 2006)

The good news is that the world is moving towards the designer's way. So many parts of daily life now need to be humanized, reinvented and made more gracious, involving, and understandable. That is why, despite the limiting circumstances we have discussed up to this point, there is reason to remain optimistic about the future of design leadership. The question is then, what steps will we need to capitalize on these opportunities.

## NEW OPPORTUNITIES FOR DESIGN LEADERSHIP

There are several encouraging developments that serve as supporting evidence for our assessment that design leadership is making inroads in real organizations. These developments are taking place within private companies but are now supported by government policies, even at the European level. The contributions that design management is making in these contexts range from changing perspectives to setting up benchmarks that aid the evaluation of design management.

### Changing Organizational Perspectives

Although it is still rare to find designers in the company board room or those that report directly to the board, one might say that their presence is growing. Mauro Porcini at 3M, Claudia Kotchka at P&G, Han Hendriks at Johnson Controls and Adriaan van Hooijdonk at BMW demonstrate that designers can have a role in changing organizational perspectives and in influencing strategic decisions.

These particular designers have worked on changing the persistent attitude that designers alone are creative in an organization and others are not. In their respective settings, they have multiplied the creative potential of their organization by discovering and strengthening the links between various creative areas, for example by bringing together people in engineering, marketing, manufacturing and logistics. In doing so, they have raised their own credibility and clarified that design itself is not in a position to solve all the problems in the world but that design can lead and facilitate the search for new approaches and new solutions. Each of these design leaders has invested time and effort to develop and integrate rigorous design processes, structures and management tools to improve the productivity of design. The integration of their processes into the other business processes can be seen as key to their success.

## Setting Up Benchmarks to Evaluate Design

In recent years, we have witnessed new developments that aid the evaluation of design. There are now more government programmes and policies in place to promote the use of design than ever before. In addition, more money is now dedicated to funding design-related projects and research.

The EU now views design as an essential tool in creating competitive advantages for the EU countries. The EU now promotes and funds projects that foster awareness and capabilities in design management and in design leadership. Following the Lisbon Strategy of the European Council in 2006, the EU set out to become the most competitive and dynamic knowledge-based economy in the world by the year 2010. This desire was coupled with the aim to promote sustainable economic growth and the creation of better jobs, thereby contributing to greater social cohesion across the EU.

As a sign of commitment to the Lisbon Strategy, the EU established two design initiatives in 2006 that remain relevant in the context of design leadership.

The first initiative saw the creation of the Design Management Europe Award, which was presented for the first time in 2007 and is now an annual event. This recognition is intended to demonstrate the value of good design management practices and the commercial benefits to businesses. The Design Management Award is part of the ADMIRE project, funded by the EU Commission. ADMIRE targets specifically SMEs. Its aim is to inform SMEs about the possibilities and the roles of design in fostering innovation and competitiveness and to demonstrate the impact of good design management developing successful products and services. Although it can be argued that this award is still one for, by and with designers and design managers, the award demonstrates that design leadership is, in essence, about new business creation. This notion reaches far beyond the role many businesses currently assign to design management.

A second and more recent initiative to highlight the importance of design in the EU came in the form of the European Year of Creativity and Innovation in 2009. Dedicating a whole year to the role of creativity and innovation for personal, social and economic development made it clear that the EU is committed to building on the initiative and to broadening its scope. The aim of the European Year of Creativity and Innovation was to disseminate good practice, foster research, share knowledge and exchange educational approaches to inform ongoing policy debates on related issues. One example was the Creativity World Forum, a two-day international congress (1 and 2 December 2009) with renowned international keynote speakers, expert panels, a cooperation platform and an exhibition, as well as a local visitor programme on 'creative industries' in Baden-Württemberg (3 December 2009), linked with a political meeting of international delegations from the Districts of Creativity regions.

Although it is disappointing that the ambassadors for the creativity and innovation programme did not include design leaders from industry, it is still encouraging to see that the activities of the European Year of Creativity and Innovation 2009 enabled a wide range of people – educators, managers, policy makers and the general public – to become involved at the European, national and local levels.

### Demonstrating the Value of Design within Organizations

While the EU initiatives served to build greater awareness of the roles of design in innovation in general, some organizations have worked on developing means to demonstrate the value of design for their specific businesses. Companies such as P&G (see Lafley and Charan 2008) and Apple (see www.businessweek.com/technology/ByteOfTheApple/blog/archives/2009/04/almost_a_billio.html, accessed 16 March 2011) have lent credibility to the fact that good design pays off. The question of interest to other businesses is 'how do design leaders plan their success?'

This question has triggered several organizations, among them Johnson Controls and the Danish Railways, to pursue metrics that allow them to measure the success of design once it is being implemented. New methodologies are being developed and tested in the UK by the Design Business Association and in the Netherlands (Kootstra 2009). These methodologies are intended to determine the effectiveness and measurement of design efforts. It is too early to make statements about the practicality or the rigour of these methodologies. Nonetheless, they show that design is taking leadership and no longer waits for others in the organization to assign it its position.

### Merge Education!

The last decade has seen an explosion of new programmes in design management both at the undergraduate level and the graduate level. Unfortunately most of these programmes are situated within traditional design schools that understand the task of design management as being to tweak design practices to the needs of professional business processes and aims. Seldom do we find educational programmes that offer leadership training and that understand the activities of designing as the key to creating future business success.

A few new programmes are concentrating on merging design with business and science. Most of these new programmes do not arise in design schools but in business schools. Leading business schools like the Rotman School of Management, Lancaster University, Wharton University of Pennsylvania and MIT Sloan Executive Education have discovered the innovation and design arena and are busy integrating design into their business programmes. Time will tell whether this is a trend or a hype that will pass. Regardless of this, educational institutes in design must be conscious about this development and advance their own courses before losing ground. Some have already started to do so. The Aalto University in Helsinki, Finland, is among the institutions that have taken this challenge head on. It was created by merging three established universities: Helsinki's School of Economics, the University of Art and Design, and the University of Technology into one institute. Named after the legendary Finnish architect and designer, Alvaro Aalto, the new institution pulls together various fields of study to offer a holistic approach to design studies. Already, the university is referred to locally as the Innovation University. 'In bringing together experts from disparate arenas including design, media and technology to develop new approaches to common problems, the university manifests the Finnish government's conviction that innovation will come from cross-disciplinary efforts' (Kao 2009).

### Develop Design Leadership into Value Networks for SMEs!

So far, the lessons of international experience show that very few government and donor initiatives have succeeded in implementing sustainable strategies for SME development. To succeed, sustainable SME development will require concerted efforts among the various parties concerned including commercial and rural banks, leasing companies and equity providers, consulting and training firms, internet providers, as well as local business associations. (Lukács 2005)

Many members of the design community are part of the SME environment themselves. They are often fragmented and badly organized. It is difficult for a SME to identify and to approach a proper designer. It follows that designers themselves need to become more active in SME networks and use these channels to educate SMEs about their contributions. Only a few of the larger design consultancies – for example, IDEO and Frog design, both based in the US – enjoy steady financial support from their mother companies (Steelcase for IDEO, Aricent for Frog). This support allows them to finance projects for a large range of SMEs. Another model is the venture capital model promoted, for example, by NPK products (NL) and Q Innovation Capital Group (NL), which create and manufacture innovative solutions for SMEs.

### Become Aware of Your Role in a Sustainable World!

Finally, designers must stand up and become accountable to some moral and commercial standard. A designer must be professionally, culturally and socially responsible for the impact his or her design has on the citizenry. A designer cannot afford to hire investigators to compile dossiers about whether a business is ethical or not (see Heller's interview with Allan Chochinov at www.aiga.org/content.cfm/to-design-or-not-to-design-a-conversation-with-allan-chochinov). Yet certain benchmarks must apply, such as knowing what, in fact, a company does and how it does it. And if a designer has any doubts, plenty of public records exist to provide for informed decisions. However, each designer must address this aspect of good citizenship as he or she sees fit. And this must definitely go beyond the creation of visions for a better world!

The best company that has used design to enable good citizenship is Philips in the Netherlands. With a long heritage, Philips was one of the first global organizations to publish an environmental annual report. Top management has integrated design in its attempt to create long-term sustainable solutions.

In its publication 'Seeds for Growth' Philips concentrates on solutions that focus on health and wellbeing and energy-efficient lighting (source: www.Philips.com). This is the result of a long-term shift in Philips' corporate goals as it changes a technology-driven company into a human-centred company (something that was confirmed with their new corporate message 'sense and simplicity', which started with an initiative from the design department that encouraged the board to use Philips products for a weekend long experience).

## DISCUSSION

In concluding this chapter there are two issues that must be discussed relating to the further development of real design leadership:

- Do we need designers as design leaders?
- Should design leaders move into other areas like the design of management?

### *Does one need to be a designer to become a design leader?*

The answer to the question is simple: no! The following examples will illustrate this point.

Steve Jobs, the famous CEO of Apple, Inc, is a great leader who is not a classically trained designer. RitaSue Siegel, the founder of the design-focused executive head hunter agency RitaSue Siegel Resources, says it takes a leader who recognizes that design is critical to reaching a company's strategic goals. A. Lafley, P&G's CEO is another good example of a design leader. After his intermediate marketing positions in Japan, he became convinced that design could become one of the game changers for P&G. And ever since he has acted to support design within his organization:

> The roots of me being a 'design believer' actually lie when I lived in Japan from 1972 through 1975. I ran retail and service operations. Years later, I moved back to Japan in 1994 to lead P&G's Asia business. Japan was an incredible experience. Design is important in Japan – not only outstanding product design, in everything from automobiles to electronics and exquisite packaging, but also in the design of everyday experiences. (Lafley and Charan 2008)

Lafley used design as a catalyst for moving a business from being technology centred to one that is more consumer-experience centred. Moreover, he installed a direct reporting line for his design director – the company's former packaging design director. This means the design director is reporting directly to Lafley himself. Finally, Lafely insisted on integrating design into the innovation process right from the beginning.

There's probably no one way towards design leadership and it does need more than one. The best leaders are those who stay close to their business, are passionate about it and have a great wisdom, which they truly believe in and communicate. To create design leadership, organizations need both design champions (design believers – people with a strong passion for design) in the top management and great and strategic design managers who have a direct line to them.

## DESIGN THINKING

Design thinking has become the latest buzz word. It is often used to describe how design competencies, methodologies and skills can be used in other arenas like new business models, services and politics. A number of new educational institutions and programmes – like the Danish 180° Academy, which educates 'concept makers' – have tried to leverage this buzz into new educational offerings that combine methods from design, the human sciences, marketing and business strategy.

> Creativity, therefore, isn't a thing that magically appears, but a process you work through.
>
> *Hammersley 2009*

So design leadership promises to play a role in design thinking where it can apply its best practices to broader and more complex tests. The challenge will be not to fall into the same pitfalls and limitations as the ones they are trying to leave behind!

## REFERENCES

Bangle, C. (2001) 'The Ultimate Creativity Machine: How BMW Turns Art into Profit', *Harvard Business Review* 79: 47–57.

Brown, T. (2008) 'Design Thinking', *HBR Review*, June: 1–9.

Chiva, R. and Alegre, J. (2009) 'Investment in Design and Firm Performance: The Mediating Role of Design Management', *Journal of Product Innovation Management* 26: 424–40.

Christensen, C., Marx, M. and Stevenson, H. (2006) 'The Tools of Cooperation and Change', *Harvard Business Review* 84(10): 73–80.

Clark, Matthew, A. (2006) *Mastering the Innovation Challenge.* New York: Strategy + Business Books.

De Wit, B. and Meyer, R. (1999) *Strategy Synthesis; Blending Conflicting Perspectives to Create Competitive Advantage*. London: Thomson Learning.

Florida, R. (2002) *The Rise of the Creative Class. And How It's Transforming Work, Leisure and Everyday Life*. New York: Basic Books.

Goleman, D. (1998) *Working with Emotional Intelligence*. New York: Bantam Books.

Hammersley, B. (2009) 'Reinventing British Manners the Post-it Way', *Wired UK*, December: 130–5.

Hargadon, A. (2005) *How Breakthroughs Happen, The Surprising Truth about how Companies Innovate*. Boston, MA: Harvard Business School Press.

Joziasse, F. (2000) 'Corporate Strategy, Keeping Design Management into the Fold', *Design Management Journal* 11(4): 36–41.

Joziasse, F., Selders, T., Voskuijl, W. and Woudhuysen, J. (2005) 'Innovation, branding and organisation: what international design managers think about their performance'. DMI eBulletin, June 2005, www.dmi.org/dmi/html/publications/news/ebulletin/ebvjunepk. htm (accessed 10 April 2009).

Kao, J. (2009) 'Tapping the World's Innovation Hot Spots', *Harvard Business Review*, March: 112.

Keeley, L. (2005) 'Interview by G. K. Van Patter', *The Business of NEW*. New York: NextD Journal Rethinking Design, Conversation 4.

Keeley, L. (2006) Interview, www.nextd.org (accessed 10 April 2009).

Kootstra, G. (2009) 'The Incorporation of Design Management in Today's Business Practices. An Analysis of Design Management Practices in Europe'. Rotterdam: Centre for Brand, Reputation and Design Management, Rotterdam: INHOLLAND, University of Applied Sciences.

Lafley, A. and Charan, R. (2008) *The Game Changer*. New York: Crown Business.

Lukács, E. (2005) 'The Economic Role of SMEs in World Economy, Especially in Europe', *European Integration Studies* 4(1): 3–12.

March, J. (1991) 'Exploration and Exploitation in Organizational Learning', *Organization Science* 2(1): 71–87.

Martin, R. (2009) *The Design of Business: Why Design Thinking is the Next Competitive Advantage*. Boston, MA: Harvard Business Press.

McCullagh, K. (2008) The Many Faces of Design Leadership, www.core77.com/blog/ featured_items/the_many_faces_of_design_leadership_by_kevin_mccullagh_ 9962.asp (accessed 10 April 2009).

Papanek, V. (1971) *Design for a Real World*. Chicago: Academy Chicago Publishers.

Petrides, K. and Furnham, A. (2001) 'Trait Emotional Intelligence: Psychometric Investigation with Reference to Established Trait Taxonomies', *European Journal of Personality* 15: 425–48.

Philips.com. 2008. 'Seeds for Growth', www.philips.com.

Pink, D. (2005) *A Whole New Mind: Why Right-Brainers Will Rule the Future*. New York: Riverhead.

Porter, M. E. (1990) *The Competitive Advantage of Nations*. New York: Free Press.

Roger, M. (2007) *The Opposable Mind: How Successful Leaders Win Through Integrative Thinking*. Boston, MA: Harvard Business School Press.

Turner, R. and Topalian, A. (2002) 'Core Responsibilities of Design Leaders in Commercially Demanding Environments', Inaugural presentation at the Design Leadership Forum, 2002.

Verganti, R. (2009) *Design Driven Innovation – Changing the Rules of Competition by Radically Innovating What Things Mean.* Boston, MA: Harvard Business School Press.

## ONLINE SOURCES

www.businessweek.com/technology/ByteOfTheApple/blog/archives/2009/04/almost_a_billio.html (accessed 16 March 2011).

www.european-creative-industries.eu/ (accessed 16 March 2011).

www.nextd.org (accessed 16 March 2011 ).

http://siteresources.worldbank.org/CGCSRLP/Resources/SME_statistics.pdf (accessed 4 April 2011).

www.core77.com/blog/featured_items/the_many_faces_of_design_leadership_by_kevin_mccullagh_9962.asp (accessed 16 March 2011).

http://grips.proinno-europe.eu/ministudies (accessed 16 March 2011).

# PART IV

# Into a Changing World

# Editorial Introduction

## SABINE JUNGINGER AND RACHEL COOPER

James Pilditch opened his book *Talk about Design* with a chapter titled 'Into a Changing World'. We think this is a fitting title for the seven contributions of this last section. As design management is moving forward, the world continues to change. Broadened concepts of what constitute a product or design practice; changing management approaches and a rise in cultural values continuously challenge our understanding of design management and its boundaries. All the while there is a growing understanding of what constitutes a design problem. The first intent of this last section is to reflect on some of the lingering issues that have the potential to shape the course, the role, the practices and the methods of design management in the near future. The second intent is to articulate some of the directions and turns design management may take in the future.

One of these turns is sure to come in form of a geographical and cultural shift. Until now, much of the literature, most of the research and certainly most of the practices of design management have been shaped by Western and capitalist models of management. The case studies we have available today tend to involve design consultancies in developed industries. Only very little is known about the role or the forms of design management in many parts of the world, including Asia and Africa. Among the most watched developments in Asia are the efforts by the Chinese government to establish design itself as a profession and practice and to educate Chinese businesses about the role design can have in their organizations within their cultural context. **Cai Jun** provides us with key insights into the rising demand for design management in mainland China. His chapter 'The Evolution of Design and Design Management in China' traces the chronology of the development of design management and lays out some of the future directions for design management in China.

Little has been said about the role of design management in organization design. Although the relevance of this area of research has been pointed out before (see for example Oakley 1984), design management has yet to expand into this territory. This matter seems to gain urgency, as other fields are rapidly filling this gap. The field of organizational studies has emerged as an active area of research and practice for issues of design, many of which imply aspects of design management. In an effort to point out the directions research is taking here, we have taken the unusual step of reprinting a paper by **Georges Romme** and **Gerard Endenburg,** renowned scholars in the field of organizational studies: 'Construction Principles and Design Rules in the Case of Circular Design'.

This turn towards organizational studies is accompanied by a more nuanced approach towards management. Researchers in the field are now openly questioning and examining which aspects of management have relevance to design. **Ulla Johansson** and **Jill Woodilla** make the case that much of design management research and practice operates under a functionalist management paradigm. Bringing to bear a Scandinavian management perspective, they offer a critical assessment of the way knowledge from design merges with knowledge from management. In 'A Critical Scandinavian Perspective on the Paradigms Dominating Design Management', they call for design management research to expand into what they describe as the 'radical humanist paradigm' complemented by interpretative and reflexive methodologies and a critical management approach.

While management paradigms have influenced the field of design management from its inception, the growing role of design in improving and changing public services call for a reflection on their relation to policy making and public management, argue **Sabine Junginger** and **Daniela Sangiorgi**. In their chapter, 'Public Policy and Public Management: Contextualising Service Design in the Public Sector', they lean on Buchanan's Four Orders of Design to demonstrate how many current service design projects in the public realm remain focused on Third Order design activities. This means that service designers have yet to embrace their role in addressing Fourth Order design problems. Using examples from three case studies, they illustrate the consequences for service design's role in public policy, public management and organizational change.

The changes in the cultural and global focus, the fact that more and more design 'business' has actually been derived from the public sector also has consequences for the ways in which design consultancies operate. In 'Future Business Models of Design Management' **Rachel Cooper**, **Martyn Evans** and **Alexander Williams** point to the need for designers to develop new approaches to conducting design business. Their chapter is based on findings from the Design 2020 Project, undertaken from 2006–8. For this study, the research team consulted with a wide range of stakeholders that together constitute 'the design business'. The study documents trends in the national design industry and show future prospects that have implications for design business and design businesses around the globe. One of the key findings that emerges from the dialogue with design practitioners, design consultants, design buyers, clients, design policymakers and design educators is the need for new design business models. Contemporary business models fail to embrace the rapidly changing design practices and roles that are currently reshaping the field of design. These new business models, in turn, call on design managers to develop their skills and broaden their understanding of management.

**Mike Press** turns our attention to the methodological shifts that are beginning to affect the practices of design management and that are driven by fundamental shifts in the relationships between users and producers. In 'Working the Crowd: Crowd sourcing as a Strategy for Co-design' he uses a set of examples to demonstrate diverse practices that 'place the crowd in the driving seat of innovation'. He goes on to map the historical precedents of participative and user-focused design strategies before he summarizes some of the key theoretical perspectives that are useful in understanding the changes taking place. Press builds on work by Charles Leadbeater who sees the orchestrating of creative conversations and the enabling of crowds to come together and to share their ideas productively as one of the central tasks in the future. If this is so, the focus of design management may change profoundly. In the least, we can anticipate new methods and new areas of practice and research.

For the closing chapter, we have asked **Richard Boland**, who together with **Fred Collopy** coined the term 'Managing as Designing' in 2004, to reflect on the implications of their interpretation. We learn that both philosophical and practical concerns have led them to object to the role of the manager as decision maker. The 'mindset of managing as decision-making' has been problematic in many areas but its principal problem, Boland states, is that it does not reflect the reality of managing. Instead, he explains that 'managing as designing is a call to see that what most managers do, most of the time, is design.' We think these insights into management practice have the potential to create ripples through, if not shake, the foundations of design management. It is as much a change in perspective of who and what a designer is as it questions the characteristics and the roles assigned to the position and practice of a manager. If the manager is already a designer, where does this leave the designer in the organization? And what about the skills and characteristics of the 'design manager'? If managing is a design activity, where and how does the design manager, who has studiously learned about decision making over the past decades, fit in?

We are off into a changing world. Exciting it is. And uncertain. It will be interesting and fascinating to observe how the field of design management will cope with these new turns and shifts.

# 26

# The Evolution of Design and Design Management in China

CAI JUN

The progress of China since the 1980s has marked a crucial chapter in the modern history of global economic and social development. China's economy is influential in the world today, yet it was not worth mentioning in the 1980s. China has experienced rapid social evolution and economic transformation, from which modern Chinese design has gradually emerged. The role of design within China is closely related to the nation's development into a major global manufacturing base. By joining the World Trade Organization (WTO), China bolstered its position as a low-cost supplier to the world market. The ubiquitousness of products 'made in China' and Chinese production is a direct result of the global transfer and integration of the manufacturing supply chain since the 1980s. When the Chinese economy became global, design turned into a recognized and respected business tool that both the state and privately owned companies in China strive to harness for their respective purposes.

The Original Equipment Manufacturer[1] Model (OEM) is no longer sustainable as it does not allow for continued business development. There is now a call and a need for more ODM, OBM and OSM models within Chinese business. What we see is a shift from 'made in China' to 'created in China'. This shift marks a transformation in the way Chinese companies perceive design. In fact, it represents a milestone in the development of design management in China. It is predicted that design management as a practice will facilitate the gradual movement of Chinese companies to the centre stage of the world market.

## DESIGN IN THE CONTEXT OF THE CHANGING CHINESE ECONOMY

Tremendous changes occurred in China between 1978 and 2008. It developed from a poor country into the third largest economy in the world. China has been transformed from an enclosed, backwards and autocratic society into a more open and increasingly global community. From a traditional, agricultural country in the preliminary stages of industrialization, it has progressively become an industrial society in which modern information technology is widely used. In thirty years, China has completed a process of industrialization that took Western countries more than a hundred years to accomplish. The 'China model' has turned out to be the most important event in the history of international economic development since the late twentieth century. It was this economic phenomenon that gave rise to contemporary design in China.

In view of the global changes in the twentieth century, the government of China was deeply aware of the importance of economic reform and social revival to uplift the country from years of stagnation.

On the one hand, the purpose and function of industrial design became noticeable to society when the country started to reform and change from a government-planned economy to a market economy. Industrial design was gradually introduced and became influential. On the other hand, economic development relied heavily on export manufacture and markets. Hence multinational companies held the advantage in technology and management; under these circumstances it was extremely difficult for local design to survive.

The progress of the Chinese economy reflected the growth of local companies. In the initial stage of economic development, local companies had to struggle for capital accumulation through very low cost mass production. Cost, market share and technology were therefore the major concerns of those companies. Design was thought to be an additional production expenditure and increased the risk of sales of a new product. To 'design' a new product was, in practice, the reproduction of an overseas buyer's prototype or the mass production of a buyer's sample, hence the end result was the imitation of an existing design.

## ATTITUDES TO DESIGN IN THE ERA
## OF THE PLANNED ECONOMY

If we want to understand current developments and the progress of design in China and the Chinese economy fully, we need to reflect briefly on the condition of design in China prior to 1978. Before 1978, Chinese society functioned strictly according to the model of planned economy. Private businesses were scarce as most companies were state owned. Hence a free market and competition hardly existed. The country was ruled according to the principle of equality. Under this code, consumers had no privileges. Heavy industry was the main basis of the economy. In such situations, the government viewed arts and crafts as a means to sustain the export of traditional handicraft products in exchange for foreign currencies. Design was more for government architecture, political propaganda and showing off the country's achievements. As a tool for ornamenting products, design seems to be part of traditional arts and crafts. There were no freelance designers, no design companies and definitely no design businesses. The design disciplines in colleges were limited to ceramic arts, textile and drying, architectural decorations, book binding and the like. Many design activities were performed around the topics of government buildings, huge political gatherings and propagandist exhibitions. The government slogan addressing design restricted its role and practice to the decoration of 'light' products (Deng Ying 2007): 'Turning arts and crafts into daily use items and daily utility items into arts and crafts' – as the government slogan connoted, design was merely for the decoration of light products.

## 1978–85: INDUSTRIAL DESIGN ENLIGHTENMENT IN CHINA

Since the late 1970s, manufacturing industry in China began to import foreign advanced technology. This technology introduced concepts and thinking related to modern industrial design. In the period between the late 1970s and early 1980s, China changed from government planned economic monopoly to an economy characterized by the coexistence of and competition between, private businesses and state-owned companies.

The implementation of the economic reform policy triggered the large-scale importation of foreign capital and technology. In 1979, China started to build Export Processing Zones and Bonded Areas along its coasts to attract foreign companies, which would bring materials, parts and designs and assemble their products in China before re-exporting them to other countries. At this point, China attracted foreign investment in the construction of Chinese factories. This opened the era in which China's business model rested on its ability to offer a vast and cheap supply of labour that could handle mass manufacturing of products for foreign companies.

Simultaneously, China began to accept foreign capital by avoiding import taxes for providing local manufacturing through Semi Knock Down (SKD) and Complete Knock Down (CKD) for assembly parts of products designed and sold overseas. By providing a cheap labour force for manufacture, this model provides private enterprises with greater development opportunities.

At the same time, China imported finished assembly lines for the production of household electrical appliances from foreign countries. The aim here was to manufacture in order to supply the Chinese domestic market. The local Chinese market, at that time, had never been exploited and the supply of household appliances was inadequate. Thus it was a seller's market and sales were determined by productivity. The products coming off the state owned production lines compared poorly in their quality, design, function and make with products from foreign imported production lines. Many Chinese companies needed manufacture according to the foreign production lines' constraints and the finished products were readily acceptable to the domestic consumer at that time.

Through the 1970s to the early 1980s, manufacturing was mostly supported by production lines and technologies that originated from Japan, Germany or Italy. In 1979, the State Council approved the foundation of the Preparatory Committee for China Industrial Design Association. This was the most important event in the field of design in China and remains significant to this day. The government became aware of the function of industrial design in business innovation and development. However, there were more important issues from the economic reform process to resolve first. The government's purpose in promoting industrial design was more to facilitate the development of the nation's light industries and to increase the range of products, styles and variations available in the market. Design was not elevated or promoted as a tool for nationwide industrial innovation. The government's policy towards the development of design had not been as advanced as that of Japan and Korea until recent years. This may explain Chinese society's lack of the concept of and respect for, intellectual property, which in turn became a hurdle to the promotion of industrial innovation. It was not until 2000 that the situation for design in China improved significantly.

## DESIGN EDUCATION SHIFTS FROM
## ARTS TO INDUSTRIAL DESIGN

Systematic education in industrial design began to surface between 1977 and 1985 in China. The design academies in China first initiated contacts with academies already established in the western hemisphere and in Japan. For example, a number of Chinese scholars went to study in German or Japanese institutions such as Staatliche Akademie der Bildenden Künste, Stuttgart and Tsukuba University. Modern industrial design, system thinking and concepts were introduced after they returned. Moreover, they wanted to contribute to their country through a vitalization of industrial design. Meanwhile, educators from Japanese and German academies also visited China, often conducting workshops that promoted industrial design to local college professors. Together, these events and activities served to kick start a development process that would eventually result in the establishment of an industrial design influence in China. In 1984, the Central Academy of Arts and Design located in the capital city of Beijing and the Hunan University and Guangzhou Academy of Fine Arts located in the southern part of China, pioneered the first schools of design in the country. Bauhaus and other Western industrial design traditions had been introduced in their teaching programmes since the 1940s. Numerous young people with aspirations about industrial design flocked to these new programmes.

A handful of influential design scholars, foremost among them Professor Liu Guan Zhong, led the design education efforts in the 1980s. Liu was the founder of the Department of Industrial Design at the Central Academy of Arts and Design, which is now the Academy of Art and Design at Tsinghua University in northern China. Liu studied at Staatliche Akademie der Bildenden Künste in Stuttgart of Germany, where he was exposed to the concepts of German industrial design and the fundamental design education system of the HfG Ulm. His work created a big impact in the field of design education in China when he returned. Liu indicated that design is not only a technology but also a culture. Beyond that, design is the 'way to create a more reasonable existence (use)' (Liu Guang Zhong 1995). In the south of China, the Guangzhou Fine Art Academy under the leadership of Professor Yin Ding-Bang quickly emerged as a design education force. Yin linked design with market revival and commercialization. He promoted a range of design professions (interior design, advertising, product design and others) and emphasized the essential link between design and business that stimulated the development of the design profession.

From the beginning of the growth of industrial design in China, there existed a conflict between the utopia design ideals and the practical commercial values in the design thinking. From 1978–85, for example, design was linked to mass production but its role was to apply styling to products. Designers were not able to innovate and to stimulate culture. During this time period, the growth of the Chinese economy was highly unpredictable. Conflicts between planned economy and market economy still existed. Social resources were still controlled by the government. A business had to devote all its efforts to acquiring initial capital. Design was of lesser concern because domestic market consumption had yet to be exploited and the average consumer had little information or opportunity to compare products. There was primarily no long-term business development strategy.

## 1986–95: INDUSTRIAL DESIGN ACCEPTANCE IN CHINA

There were complications and changes in government policies toward the economic reformation throughout the 1980s and in their implementation. The progress of society and industrial development was affected by all of this. It was a particularly difficult time for business. However, those companies that emphasized marketing and management continued. By the end of the 1980s, household appliances had become one of the major industries in China and competition was strong. Individual companies began to realize the importance of brand building and the role of design in this context. Corporate identity quickly rose to strategic significance. Graphic design firms began to provide branding services to corporations but many companies were under the illusion that visual identity alone could achieve market success for them.

The first freelance design office in China was set up in Beijing in 1986. It did not survive for more than six months. Then in 1988, Southland Industrial Design was founded in Guangzhou by three staff from the Guangzhou Fine Arts Academy. Southland Industrial Design is today considered to be the first successful and officially registered industrial design firm in China. It focused on providing design services to the emerging private household appliance manufacturing companies in the Pearl River Delta region of the Guangdong province.

Six years later, the Haier Group of China and the GK Company of Japan co-founded QHG. This consultancy focused on the development of Haier products. QHG was the first Chinese–overseas joint venture design company. GK brought the processes and methods of modern industrial design and management to support Haier in advancing new product development and innovation. Haier, in turn, was aggressive in strategic marketing and in the promotion of its product and services. The joint venture had a role in quickly establishing Haier as one of the most famous household appliance brands in China. Other companies, for example, Konka, Changhong, Hisense, TCL and Skyworth, also developed important brands and had grown into the largest manufacturers of household appliances in China by the end of the 1990s.

Beginning in the 1990s, Chinese companies started to establish their own design departments. Initially, industrial design was positioned and managed within engineering and production departments. 'Design' was mainly the replication of product samples and styles from overseas and 'product development' was merely reproduction or partial alteration of an existing design. Most companies were close followers of overseas designs. There were few concepts or strategies for original design development.

In 1992, *Design,* published by the China Industrial Design Association, was the first magazine in China focusing on industrial design. The magazine was very important for the dissemination of industrial design knowledge to the society of the time. Now it is one of the most active magazines for design in China.

Industrial design education started to bloom in the 1990s. Design programmes began to develop both in arts and in engineering schools. We can count the Central Academy of Arts and Design, the Guangzhou Academy of Fine Arts and the Zhejiang China Academy of Fine Arts among those offering design education in the arts and crafts tradition. Tongji University, Hunan University, Zhejiang University and the Beijing Institute of Technology offer design education in the contexts of engineering and architecture. The developments in the arts and in engineering often took place independently from each other and there were few collaborations between these schools until recently. Today, design education in China has solid roots both in the arts and in the engineering disciplines.

Also in the 1990s, a couple of designers set up industrial design firms in cities, such as Dragonfly Design in Shenzhen and Zhu Jian Hua Industrial Design in Shanghai. These individuals were eager to exploit the field of transportation design, particularly motorcycle and car design. In 1993, Dragonfly Design launched a first family car 'xiao fuxing' (little lucky star) in China. Unfortunately, as the transportation industry itself was still underdeveloped and government policies and license were not open to private industries, the effort was failure. However, this event had an important impact on Chinese design in that period.

FIGURE 26.1 'Xiao Fuxing' mini family car design by Dragonfly Design, 1994. *Source:* selected works of China Art Grand Exhibition 1997 (volume of Arts & Design) by Shanghai Book & Painting Press (Duoyunxuan) 1997.

Although they may have been premature in their timing, design overall has grown in influence in China during these years. The success of Sundiro's motorcycle which sold over 1.5 million is a good story from Zhu Jian Hua Industrial Design. Its design integrated engineering and production in the local market. During this period, consumers became aware of product design elements and began to recognize its value. Design of the international name brands, imported by joint venture companies such as Panasonic and Philips, is also becoming influential in China.

## 1996–2002: THE CHINESE DESIGN EXPANSION

For a long time, Chinese industrial design has advanced only slowly and in small steps. As in other countries, the initial role of design was limited to providing the aesthetics for a product – that is, to beautify and lend style. During the phase of early industrialization in China, companies did not see a need to invest in design. Instead, design was a function of either engineering, advertising or packaging. The Chinese term used for 'designers' at that time literally translates into English as the 'form maker'. The debate about the meaning of design and modern art and crafts in academia reflected this. Some report shows that before the end of the 1990s, there were less than fifty industrial design firms in the whole of China (Ye Zhen Hua 2008).

Despite the seeming design ignorance in businesses, design educators in Chinese design academies were at this very time, developing design capabilities and training future designers. The professors and teachers also held freelance positions and worked as design professionals with businesses and industry. This may have contributed to the increasing awareness of industrial design in society. Eventually, individual local governments implemented policies to stimulate and raise companies' industrial design competencies. The Bejing municipal government, for example, founded the Beijing Industrial Design Promotion Organization (BIDPO) and the Beijing Industrial

Design Centre (BIDC) in 1995. The BIDC aims to promote collaboration in design policy making between the government, academies and industries. The aims of BIDC are to promote the acceptance of industrial design in the business sector through actual participation in collaborative government projects. The BIDC has successfully brought together government support and the development of design in business, which set an inspiring model for the other local governments in China to promote industrial design. In the few years, between 1996 and 1999, through demonstration projects and other activities, BIDC has penetrated the medical and computer industries to push forward understanding and acceptance of industrial design in these companies. Notably, it has stimulated the Lenovo Group to set up its own industrial design team and operating system for innovation design.

The market success of some new products in 1997 led Lenovo to fully appreciate the value of industrial design in business innovation. The company founded the Lenovo Industrial Design Centre in 1998 and it has become an important organization for business innovation. In just a few years the Lenovo Industrial Design Centre built its innovation design system and attracted design talents to join. It has established

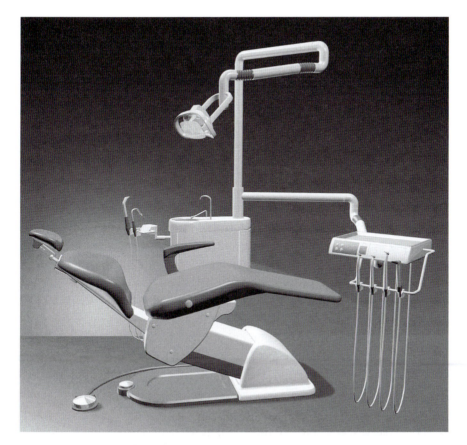

FIGURE 26.2   Dente unit design by BIDC, 1997.

1997-1999

2000-2003

2004-2005

FIGURE 26.3    Lenovo computer design development process from 1997 to 2005.

collaborative relationships between international design consultancies and rapidly adapted international design and research methodologies. On the other hand, it also creatively managed design and engineering collaborated interactively at each stage. Led by the customer experience strategy, the design management process covered from user research, trend analysis until engineering prototype. Through the success of a series of computer products especially developed for the consumers in China, Lenovo became a predominant brand in China and also gained a global reputation after it merged with IBM computer in 2004. Other companies in the household appliances and computer fields also established their own design team in the same short period of time. Haier, Konka, Hisense and Changhong illustrate this point, each having built their own design centres.

In 2002, the Academy of Arts and Design at Tsinghua University organized and hosted the Tsinghua International Design Management Symposium. This was the first international symposium on design management held in China. The symposium gathered together leaders in the design field, including Earl Powell, Dieter Rams, Patrick Whitney, Harald Leschke and Tetsuyuki Hirano. Over 200 participants across China and overseas joined it. The symposium accentuated international collaboration between colleges. In 2002, together with eight schools, Tsinghua University and University of Salford, UK, initiated the Asia Link-Design Management Network program, which was funded by the European Union.

The design market in China grew rapidly after 1999. One of the reasons was keen market competition, which increased the demand for newly designed products. The other reason was that overseas design companies began to enter the Chinese markets to develop joint businesses with the big companies, such as the start up of collaboration in design between Lenovo and the American design company Ziba. In this period of time, the rapid emergence of the mobile phone market has also nurtured many design houses, which specialize in providing full services from styling design to mockups for the IT industry.

At the same time, the capacities in all aspects of manufacturing further advanced in China. More multinational companies began to set up production for the more sophisticated and precise products.

In 1993, Motorola was the first multinational company to set up a R&D centre in Beijing. Numerous foreign companies followed. According to research by the United Nations Conference on Trade and Development in 2004, among the world's 700 largest multinational companies, thirty-five percent had R&D setups in China. This percentage falls just below that of the US and UK (Shi Xing Hui 2006).

However, in the joint-venture companies in China, design is often caught up in internal conflicts. From China's perspective, instead of systematically importing foreign designs from joint partners, they wish to develop more suitable products for the domestic market and strengthen their own innovation ability. But foreign partnering companies are not willing to develop products in China. They stipulated that all design and development has to be done abroad. Some do not even allow the Chinese team to be involved in their design and development processes. Design became a source of discontent between the joint venture partners because of values, intellectual property rights and different interests of the parties. For these reasons, the progress of design was stifled. As an example, the Volkswagen Santana model produced in China has taken over fifteen years to upgrade.

On 11 December 2001, China eventually became a member of the WTO, which signified not only its success in the transformation from a planned economy to a market economy but also that it had become a major world economy. Since then, the economic development of China has been closely connected with the global economy. China's manufacturing industries have become a part of the global economy and the consumption power of the Chinese market is emerging in the eyes of the world. From this point on, the whole community has had a fresh new look on design.

## 2002–8: DESIGN ENRICHMENT IN CHINA

The evolution of design in China has accelerated since 2002. There are more medium and small design enterprises appearing in the Yangtze River Delta and Pearl River Delta regions. In 2003, private enterprise accounted for 59.2 per cent of China's GDP (OECD 2005: 125). A highly competitive design market and an expanding design business landscape emerged. International design consultants have also started to enter the Chinese market. This wave was set off by Korean design firms that specialized in mobile phone design. American, European and Japanese design firms such as IDEO, Ziba, GK, quickly followed. Their main clients are large manufacturers. In this competitive market, design businesses gradually diversified; both foreign and local design companies focused on their own market segments leading to a differentiation of service models. Further evidence of the brisk advance of design in China is provided by the following:

- In 2003, US *Business Week* reported for the first time the progress of Chinese design and its influence and future prospects. The publication drew the world's attention to the growth of the Chinese market.
- In 2004, Lenovo announced the acquisition of IBM's worldwide business for PC – a milestone for the internationalization of Chinese enterprise. The incident indicated the merging of Chinese design with the international market as part of its economic globalization. Also in 2004, Wuxi Municipal People's Government conducted the China International Design Summit, which set an example of a high-level design conference organized by a local government in China. The programme attracted hundreds of renowned participants from the fields of design, education, business and government bodies from China and abroad. It was also the first time such program was broadcasted by CCTV (China Central Television). The summit was a large 'variety show' of design. Nonetheless, the summit was a good promotion to attract business investment. Since then, many cities in China followed, promoting design together with their city's image. Through this design promotion, the Chinese design activities became more widely known abroad.
- In 2006, Lenovo won the IDEA grand award with its Opti multimedia desktop computer. Since then, Chinese companies have continued to receive other international honours such as the G-mark, IF and Red Dot Design Awards. These are indicators that Chinese design is becoming increasingly recognized in the international design community.
- The *China Industrial Design Annual,* published by the China Industrial Design Association in 2006, revealed that there were 328 industrial design firms in China in 2005. The number increased after 2007. Most design firms were located in cities like Shanghai, Beijing and Shenzhen of Guangdong.
- In 2006, the China Red Star Design Award was launched by the China Industrial Design Association (CIDA), BIDC and other institutions (China Industrial Design Association 2006). It is the most important award for industrial design and innovation in China. The Red Star Award was endorsed by the International Council of Societies of Industrial Design (ICSID) in January 2009. This award is offered annually and it attracted 400 design submissions from

200 participating companies in the beginning. The numbers increased to 758 companies and 3,821 design entries by 2009. It is proof of the rapid changes in design in China.

- Since 2004, the Chinese automobile industry entered into booming market, increasing by 20 per cent increase every year. The domestic automobile enterprises market share in the Chinese market has been over 28.7 per cent since 2006. This is the first time that it has been ahead of international companies (cf: Brandt and Rawski 2008). Proprietary brands such as Geely, Chery, Great Wall and Brilliance Auto quickly recognized the value of design in market competition. The huge consumer demand has made the policy of importing technology and developing research and design a success.

The year 2008 represents a major milestone in the history of contemporary Chinese design progress. China successfully hosted the twenty-ninth Olympics in Beijing. For the first time, design was received and recognized at the higher governmental levels. Design was not only an essential communication tool but also a strategic instrument for the promotion and enhancement of the China's global image. Through the process of Olympics, the government was able to understand design in more depth, which in turn helps in future policy making for the further development of design and

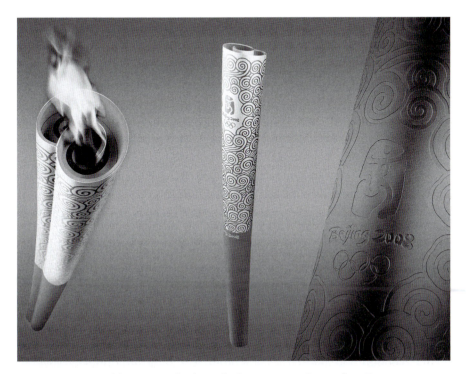

FIGURE 26.4 Beijing Olympics torch, design by Lenovo team. *Source:* from Lenovo.

innovation. The Beijing Olympics also brought an awareness of design to the general public. It helped them to understand and appreciate the significance of design in business and brand landscape. The whole design system of the Beijing Olympics with its logo, mascots, pictograms, opening and closing ceremonies had special meaning to the world. It signified the nation's determination to bring about a cultural revival and it has left an indelible impression on the hearts of the Chinese people.

The 2008 Chinese cultural revival represented the return of Asian culture after China's long-term pursuit of Western modernization. It was also a reflection of Chinese economic strength after China's rapid growth into an influential market in the global economy and the return of the people's cultural confidence along with this growth. Multinational companies – whether Nokia or Samsung, Ford or Volkswagen – have started to look towards Chinese culture and its uniqueness for design ideas that could generate greater product value and greater advantage in penetrating the Chinese consumer market and so have local companies. The characteristics and lifestyles of Chinese consumers are being explored and influenced globally.

## THE EVOLUTION OF DESIGN MANAGEMENT IN CHINA

The growing recognition that design can add value to businesses and products raises interest in design management. The evolution of design management in China, like the evolution of design itself, was in every stage influenced by the political organizational, technological and cultural aspects of the country's reformation.

Industrial and economic developments have been the prime objectives for policy making by the China government since 1978. Emphasis has been placed on technological development, which was considered the core advantage of the country, but design and intellectual innovation were not part of this. However, the importance of design to market economy and industrial competence is gradually being understood from the bottom up. Due to pressure from academia and industries alike, the government's National Development and Reform Commission eventually started to evaluate and debate design topics for policy making in 2003. To the government, industrial design is part of the creative industry and many local authorities have already recognized the importance of creative industry. Industrial design was emphasized again by Wen Jia-bao, the Premier of the State Council, in 2007. Soon after, local authorities began building creative industrial parks. This wave of building is now rolling rapidly throughout the country. The conventional industrial parks and technology parks are being quickly replaced by these creative parks. Various design firms often group together in these creative industrial parks to form a cluster of design services that can provide local companies with total design services from conceptual design to product engineering. In cities like Shenzhen, Ningbo and Shunde in Guangdong, where the economy and industry is more prominent, local authorities have set objectives and policies for advancing the design industry. Industrial design companies are supported in many ways, including through tax breaks, bank guarantees and governmental funding sources.

In Hangzhou, the Creative Hangzhou Industrial Design Competition – organized annually by the government since 2008 – attracts several hundred enterprises, many colleges and numerous design companies every year. It is effective in enhancing the level of local design quality. In 2008, more than fifty companies came together to establish the Shenzhen Industrial Design Profession Association. This is an example of a non-profit-making and non-governmental organization that is actively promoting the development of industrial design and design management in the region.

From the industries to the government, the important shift from service manufacturing to self-initiated innovation and brand building is gradually being recognized. This change requires not only originality in aesthetic design but also intellectual property (IP), which includes the advancement of new concepts, technologies, methodologies and management. The challenge is to differentiate Chinese offerings from the existing designs of the West.

While currently only a few large companies like Lenovo and Haier have committed to industrial design, we see an increasing investment by small and medium-sized businesses. From technological development to product differentiation, from production to brand management, from fashion to utility, from electronics to heavy industries, the change is happening everywhere. It is evident that the shift from 'made in China' to 'created in China' is finally underway.

Based on intensive production and low labour costs, imported technology and highly competitive markets, Chinese enterprises gradually learned how to succeed using market strategies, brand building and strategic planning. The market offered

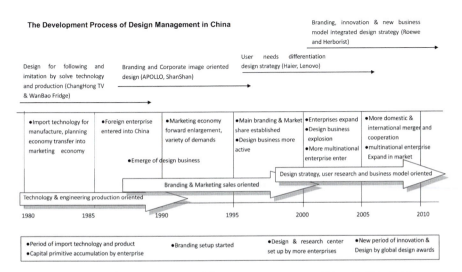

FIGURE 26.5  The development process of design management in China. © Cai, 2011.

a different model for enterprises with a good opportunity for success. Following this transformation, design management has played a more important role in the competitiveness of enterprises. For those enterprises, promoting the value of design and strengthening competitiveness through good products, services and brands have become common objectives. The strategic management of competition in design can be traced from a policy of following the market and imitating technology in the early 1980s (for example, both the TV manufacturer ChangHong and the fridge producer, WanBao, were following and imitating Japanese products), to branding and corporate image-oriented design in the 1990s (for example the company Apollo succeeded by promoting its brand image in the market and the fashion company ShanShan came to be well known because of its corporate identity). We can see, then, a shift towards user needs and the differentiation design strategy (Haier, Lenovo) in the early 2000s and currently an effort to link branding, innovation and new business models through integrated design strategy.

## THE POST-GLOBAL FINANCIAL CRISIS DEVELOPMENT OF DESIGN MANAGEMENT

The global financial crisis, which began in September 2008, is another important driving force in the rise of design management. Businesses in China are highly affected by financial world markets. Along the coastal areas, thousands of companies that provide manufacturing services to the overseas markets went bankrupt. The business model of export-oriented mass production, which has been the main driver of economic growth since the 1980s, is now in question. Noticeably, companies that developed original brands and designs for the domestic market have survived. Some of these companies seized opportunities arising from reduced competition and even found new areas for expansion stemming from the crisis. Many design companies that provide services to these companies have had the chance to grow, too.

This showed when Tsinghua University and The University of Salford, UK, co-organized the D2B 2009 Tsinghua International Design Management Conference in April, 2009. Despite the financial crisis, there were very positive responses from academia and industry all over the world. The conference signified that design management is being studied in China on a consistent and long-term basis and that the development of the Chinese market remained of interest to the world, even during its recession period. A few months later, in November, the city government of Beijing together with the Central Academy of Fine Arts, hosted the Icograda World Design Congress Beijing 2009, which was one of the more important international design symposia in China these years. At the same time, seven Mainland China design colleges joined the Cumulus International Association of Universities and

Colleges of Art, Design and Media, ensuring future collaborations between design education in China and the world. We can therefore say that an international basis for the academic development of design and design management in China has been formed.

## CHINESE DESIGN MANAGEMENT OF THE FUTURE

Significant changes are occurring in the design businesses and design management in manufacturing companies. For the OEM service manufacturers and export contract manufacturers, product designs were often simply the reproduction of an existing prototype and the design concerns were more about product form, colour, material, ergonomics and production process. As the market grows, design differentiation in styling alone is no longer sufficient. Today, manufacturers need to face not only the challenges presented by downstream development of new technology but also upstream supply chain concerns such as customer service, market analysis, sales forecasts and business strategy, in order to exploit any blue ocean market potential arising from design innovation (Kim and Mauborgne 2005). The design industry is also extending its capacity into original concept creation including technological and market innovations as well as conventional aesthetic and engineering designs, model making and tooling services. The Shenzhen Newplan Industrial Design Co. Ltd is a good example of a design consultancy fully utilizing the well-established supply chain of the Pearl River Delta region as a platform to provide the market with a one-stop service from product development to manufacturing.

At the upstream, user studies are increasingly popular. Design and management consultants often team up to address market integration and to provide a company with strategic product design and brand building proposals. Design diversification and innovative integration are happening everyday and this is just the beginning. Design management plays an important role in the diverse supply chain and this trend is sure to continue.

## CONCLUSION

The growth of the Chinese economy since the 1980s and during the financial crises has been surprisingly stable. China has been transformed from a planned economy to a free-market economy. Its industries have been transformed from low value-added mass production to integrated brand marketing and manufacturing. The Chinese market will merge further with the global supply chain. More Chinese enterprises are emphasizing international brand building for long-term development.

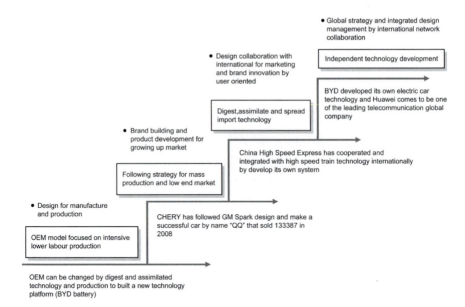

FIGURE 26.6   The strategy of design management in China. © Cai, 2011.

Design in China has started to spread its wings. Design management has been influenced by technology, marketing and culture. It has also been transformed from traditional manufacture and reproduction into marketing original innovation and integrated brand-building strategy. Enterprises have started to engage with and in design. They may begin with brand identity and style or focus on quality and business management for production. In this process, traditional business models of the planned economy have been discarded. In their place we find modern management and manufacturing ideas. Clearly, the time for design management has come.

Design management will be integrated with technology and business as it becomes essential to integrate product and service innovations to penetrate of the local markets. To do so, both local and foreign companies have to gain a better understanding of Chinese consumer culture. User studies and case studies have become important topics in corporate design management. New government policies support this by favouring intellectual property rights and original innovation.

Along with economic growth we see an emerging cultural confidence in the people. Cultural inspiration becomes the crucial approach to design innovation for stimulation of sales in the China market. The Chinese market is becoming more influential in the world economy and Chinese cultural elements and styles are becoming more influential in international design concepts. Designers from China, who are already prominent in both local and international companies, will soon achieve world recognition. The recent Red Dot design awards have showed evidence of the change and progress.

FIGURE 26.7 Red dot design award 2010: (top) Huawei air modem, (bottom left) Skyworth TV, (bottom right) Newplan elderly phone. *Source:* http://www.red-dot.org.

## NOTE

1. OEM (original equipment manufacture), ODM (original design manufacture), OBM (original brand manufacture), OSM (original strategic manufacture).

## REFERENCES

Brandt, L. and Rawski, T. (eds) (2008) *China's Great Economic Transformation*. Cambridge: Cambridge University Press.

China Industrial Design Association (CIDA) (2006) *China Industrial Design Annual*. Beijing: Intellectual Property Press, pp. 350–4.

Deng Ying (2007) 'Make a Wing of Art for Consumer Goods by Modern Arts and Crafts', www.cnci.gov.cn.

Kim, W. and Mauborgne, R. (2005) *Blue Ocean Strategy*. Boston, MA: Harvard Business School Press.

Liu GuanZhong (1995) *Apple Collection – Discourse on Design Culture*. Haerbin: Heilongjiang Science & Technology Press, pp. 21–2.

OECD (2005) Economic Survey of China. Paris: OECD, www.oecd.org/dataoecd/10/25/35294862.pdf (accessed 5 April 2011).

Shi XingHui (2006) 'Creative Engine-Evaluation Report on-25 China Best Research Institute of Enterprises', *Global Entrepreneurs* 125.

Ye ZhenHua (2008) *Report of Survey for Intellectual Property in Creative Industrial Design Firms*. Beijing: National Intellectual Property Bureau, pp. 62–72.

# 27

# Construction Principles and Design Rules in the Case of Circular Design*

A. GEORGES L. ROMME AND GERARD ENDENBURG

*This paper proposes science-based organization design that uses construction principles and design rules to guide practitioner–academic projects. Organization science implies construction principles for creating and implementing designs. These principles serve to construct design rules that are instrumental in developing organization designs. Testing and implementing designs require pragmatic experimentation in complex, dynamic settings. The authors explore a circular design process as an example of science-based organization design. Tests of this circular process in over 30 organizations suggest that construction principles are important for creating new design rules as well as for a deeper understanding of the systems and practices created from these rules. In addition, explicit principles and rules for organization design appear to facilitate the transfer of learning between different projects. As such, they can help reconnect organization research to ongoing design work.*

Several authors argue that organization studies should venture out from its adolescence – characterized by observing and explaining natural events in a rather disengaged manner – toward a more mature science that also engages in engineering organizational structures and processes (e.g., Boland and Collopy 2004a, Schein 1987, Starbuck 2003). Indeed, organizations need all the help they can get from organization researchers to create humanly satisfying and sustainable organizations. For example, many corporations and other organizations have great difficulties in creating

*Source: Reprinted from Romme, A.G.L. and Endenburg, G. 2006. 'Construction Principles and Design Rules in the Case of Circular Design.' *Organization Science*, 17(2): 287–297. Courtesy of the Institute for Operations Research and the Management Sciences (INFORMS).

commitment to strategies and policies, particularly when crafted at the top level. Large organizations also have difficulties in ensuring that vital information reaches senior management promptly and accurately (e.g., Kim and Mauborgne 1993, Milliken et al. 2003, Moss and Sanchez 2004). Moreover, attempts to fundamentally redesign organizational processes frequently fail to produce the intended outcomes (e.g., Foss 2003, Hardy and Leiba-O'Sullivan 1998, Moss and Sanchez 2004).

These observations suggest the need to rethink the nature of organization design as a scholarly field. In this paper, we advocate a *science-based design* approach to organization design. This approach draws on a research cycle involving organization science, construction principles, design rules, organization design, and implementation and experimentation. It focuses on developing construction principles and design rules grounded in organization science as well as on organizational solutions implemented and tried out in real-life settings.

To illustrate this approach, we explore the development of the circular design method. This particular method involves several construction principles inferred from the notion of circular process and a set of design rules with regard to designing, producing, and implementing a circular organization. Our work in this area involves collaboration between an engineer trained in design methodology and an organization theorist trained in the social sciences. As such, we believe organization studies can overcome the divide between the academic and practitioner worlds by adopting an integrated approach drawing on the social sciences as well as design methodology.

In this respect, the social rather than the natural sciences may actually be the 'hard' sciences, because social processes are not neatly decomposable into separate subprocesses, but are closely interrelated and therefore inherently complex (Simon 1996, Warfield 1994). Controlled experiments are therefore hard to conduct in the social sciences (Starbuck 2004). Moreover, even if controlled experiments could be conducted, the creation of a firm or hospital does not permit testing for years before finding its way into our lives: 'The test arena is our life situation, not a laboratory' (Warfield 1994: 8). Thus, organizational design and development involves pragmatic implementation and experimentation (Wicks and Freeman 1998).

The paper proceeds as follows. The first part of the paper describes the role of construction principles and design rules in a science-based approach to organization design. In the second part, we turn to the example of circular design. Subsequently, we will discuss the implications for researching and practicing organization design.

## CONSTRUCTION PRINCIPLES AND DESIGN RULES AS BOUNDARY OBJECTS

Simon (1996) defines *science-based design*, or the *science of design*, as the entire body of intellectually tough, analytic, partly formal, and partly empirical knowledge for the design process. Over time, the status of design and intervention research in organization studies has changed from a core activity to a relatively minor project outside the mainstream (e.g., Baligh et al. 1996). Moreover, this research tradition in our field has been largely disconnected from the development toward a generic design science (cf. Warfield 1994).

McMaster (1996) suggests the first step in designing is to create the principles from which the rest will develop. This implies the need for a coherent set of construction principles for creating new organizational designs or redeveloping existing ones. We explore the critical role of principles and rules serving as *boundary objects* between organization science in the academic community and organization design work by practitioners. Boland and Collopy (2004b) define a boundary object as

> An artifact (for example, prototype, report, pictorial representation, model) that serves as an intermediary in communication between two or more persons or groups who are collaborating in work. The phrase connotes that the object is a symbolic carrier of multiple meanings, which the parties to the interaction can use as a basis for productive interaction. A boundary object allows parties with diverse interests to raise the possible meanings it has for them in a conversation of discovery. (268)

We suggest that a science-based approach to organization design includes the following five components (see Figure 27.1).

*Organization Science.* This is the cumulative body of key concepts, theories, and experientially verified relationships useful for explaining organizational processes and outcomes. As such, organization science provides, or should provide, the theoretical foundation for construction principles (defined in the next paragraph). We deliberately refer here to 'experientially' verified relationships, to include both empirical evidence (e.g., survey data) and pragmatic evidence obtained by engaging directly in real-life settings. Whatever the nature of the data, the prevailing mode of thinking in organization science currently is, and should be, propositional in nature (e.g., 'given conditions C, if A occurs, then B is likely to follow').

*Construction Principles.* Any coherent set of imperative propositions, grounded in the state-of-the-art of organization science, for producing new organizational designs and forms and redeveloping existing ones. Given the plurality of organization science as well as the diversity of organizational practices, different sets of construction principles can be developed. Construction principles also serve to bridge the descriptive nature of scientific propositions and the prescriptive nature of design rules. In this respect, a construction principle emphasizes the importance of a certain type of solution in view of certain values or goals (e.g., 'to achieve A, do B'). If those engaging in a design project develop some awareness of construction principles used, their learning capability as well as the effectiveness of their actions in the project tends to increase. In this respect, construction principles outline the deeper meanings and intentions behind design rules; as such, they help to reconstruct design rules, for example, when try-outs and implementation outcomes suggest the need to change these rules.

*Design Propositions or Rules.* Any coherent set of detailed guidelines for designing and realizing organizations, grounded in a related set of construction principles. Design propositions are preliminary design rules that are not yet (sufficiently) tested in practice and grounded in science (Romme 2003). We will only use the term 'design rule' in the remainder of this paper. Design rules are elaborate solution-oriented guidelines for the design process (e.g., 'if condition C is present, to achieve A, do B'). These rules serve as the instrumental basis for design work in any organizational setting. It serves as a heuristic device that describes ideas and intentions underlying a particular organization design, and as such helps to make sense of the processes produced by this design and any changes the design needs to undergo. An individual design rule can typically not be applied independently from other rules. Given the integrated nature of organizations and their designs, design rules are therefore developed and presented as part of a coherent set of related rules.

*Organization Design.* Developing representations of the intended organizational system being (re)designed with help of the design rules. Simon (1996) suggests design involves actions in a virtual world using representation (e.g., drawings, models, narratives). Any particular organization design arises from the interaction between design rules, the contingencies of the design situation (e.g., history, size, technology), and the preferences of the people engaging in the organization design effort. Design rules help to develop a tailor-made design; they cannot be directly tested in practice, only organization-specific solutions can. Representations of the intended design initially tend to be highly visual in nature (e.g., a diagram depicting communication flows). Over time, when the design is implemented and practiced, people increasingly use narrative representations (e.g., anecdotes illustrating who is in charge of a particular domain).

*Implementation and Experimentation.* Implementing the organization design and trying out the processes caused by that design. Implementing and testing a preliminary design in the dynamic complexity of a particular organization is a highly pragmatic process. This type of experimentation acknowledges the importance of experimenting with new ways of organizing and alternative forms of discourse, while being tempered by the pragmatic commitment to develop practices that are useful (Wicks and Freeman 1998). In this respect, pragmatic experimentation is about finding out whether the design 'works.'[1]

The science-based design cycle is completed by observing, analyzing, and interpreting the processes and outcomes generated by the design, and where necessary, adapting

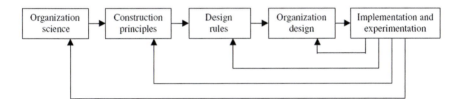

FIGURE 27.1   The research and development cycle in organization design.

existing organization theories or building new theory. In addition, experiences and observations regarding implementation and experimentation may lead participants to rethink the design as well as the rules and principles used (see Figure 27.1).

As boundary objects, construction principles and design rules provide themes 'on which each organization must write its own variation. It is not cast in concrete' (Ackoff 1999: 181). In other words, many different organization designs can be created and implemented, even if we would assume and start from one coherent body of scientific knowledge supporting a single set of construction principles (which is, of course, not a valid assumption).

The process flow in Figure 27.1 produces design solutions implemented and tested in organizational settings as well as construction principles and design rules grounded in organization science (cf. Van Aken 2004). Grounding in science serves to develop a robust understanding of how and why the design (rules) operates. The primary interest of the practitioner (e.g., executive) is in whether the design works, that is, whether it produces the intended organizational processes.

Many organizational processes may complicate, undermine, or rather reinforce the organization design process. For example, any organization design that is actually practiced and evolving is socially constructed, with cognitive frameworks, power distribution, and other aspects of the broader system either undermining or reinforcing the design project (e.g., Bate et al. 2000). In these dynamic and complex settings, the process flow in Figure 27.1 provides a framework that executives, consultants, and researchers engaging in a design project can adopt to create a shared language for discussing processes and outcomes. Figure 27.1 may create the false impression that organization research drives design. However, complex organizational problems, related to the current organizational design and processes, trigger most (re)design work (cf. the arrows in Figure 27.1 from implementation to organization design, design rules, and so forth).

We focus here on organization design projects in which practitioners team up with researchers to create, try out, and implement designs that are grounded in research. Evidently, many organization design projects are not grounded in organization science and, implicitly or explicitly, draw on the conventional wisdom about organization design shared by the practitioners involved. These projects, however, are outside the scope of the argument in this paper.

If organization design develops into a mature science based discipline, practitioners can choose from and draw on different sets of construction principles and design rules firmly grounded in organization science, whereas exceptional projects may focus on developing new principles and rules. In the adolescent state of organization studies (Starbuck 2003), we have to engage in developing and codifying principles and rules or any other boundary objects that help to align organization science and actual design work.

## CIRCULAR ORGANIZATION DESIGN[2]

The case study in this section involves a sustained effort to develop construction principles and organization design rules around the notion of circularity. As such, it illustrates how construction principles and design rules can serve as boundary objects between organization design and science.[3]

This section first explores the thinking that produced the initial construction principles and design rules; subsequently, we explore critical experiences and applications that served to redefine and extend these principles and rules. For a preview of the results, the appendix contains an overview of the most recent version of the circular design.

One author of this paper pioneered the circular design, in both executive positions and consulting roles, for more than 35 years. The other author has been engaged in researching this design practice as an academic outsider for more than nine years. Bartunek and Louis (1996) suggest that joint insider-outsider research produces the quality of *marginality*, that is, being neither altogether inside or altogether outside the system. In insider-outsider partnerships 'the outsider's assumptions, language, and cognitive frames are made explicit in the insider's questions and vice-versa. The parties, in a colloquial sense, keep each other honest – or at least more conscious than a single party working alone may easily achieve' (Bartunek and Louis 1996: 62). The remainder of this section draws on joint insider-outsider research, in the form of case studies, reports of experiments, (internal) evaluation studies, and participant observation; some of these studies have been reported elsewhere (e.g., Van Vlissingen 1991, Endenburg 1998, Romme 1999).

### Developing Construction Principles
### Based on Cybernetics

The creation of the circular design started with the development of several construction principles from cybernetics in the late 1960s. Cybernetics, as the science of steering and control, served as an important framework for organization design at the time (e.g., Beer 1959). In this period, Gerard Endenburg, who previously studied cybernetics as an engineer, took over his parents' firm in the Dutch electrotechnical industry. This firm had been struggling for some time with the implementation of a works council, a consultative body that was required by a new Dutch law. In the first few years of operating this council in combination with a conventional administrative hierarchy, participants grew increasingly dissatisfied with this consultative body. Instead of providing genuine consultation between management and workers' representatives, it frequently produced conflict. Endenburg therefore decided to completely redesign this consultative system.

Drawing on the notion of circularity from cybernetics, he first developed a number of *construction principles* that would have to apply to any kind of system 'capable of maintaining a state of dynamic equilibrium' (Endenburg 1998: 65). In this respect, cybernetics suggested the purpose of any circular process 'is to detect the disturbance of a dynamic equilibrium and to take steps to restore it. It is a process which is unnecessary in a static equilibrium, because the factors influencing a static equilibrium are not variable' (Endenburg 1998: 65). As such, Endenburg (1974) identified the following construction principles for building a self-regulating system:

- *'Weaving' must be possible.* In driving a car from A to B, the driver follows a certain road. While driving, he must continually correct his course to keep the car on the road. Figure 27.2 illustrates the process of continually correcting course to get from A to B, in the simple case of a completely linear route (the dotted line).
- *The circular process makes it possible to search.* The car driver has to engage in several different steering and feedback processes simultaneously. We define this process as a circular process that consists of three basic components: an operational, measuring, and directing (i.e., comparing and instructing) component. The driver's feet and hands do the operational work; measuring is done by the driver's sensory organs (mainly the eyes) taking in signals that the brain converts into mental images; comparing and instructing activities

FIGURE 27.2   Weaving must be possible to get from A to B.

involve another part of the brain that compares images and may issue 'instructions' to correct the speed level or steering. If the driver shuts his eyes for a longer period, the circular process is broken and it will be impossible to reach the destination or avoid an accident. Thus, a system is only able to search and find its way toward a target under the following conditions:

- The steering and feedback (circular) process is closed.
- Performance is being measured.
- The system has sufficient scope to weave from side to side.
- The circular process enables the system to search (e.g., to locate the part of the road that has the fewest potholes).

- *'Mistakes' must be made.* Assume the steering wheel has been mechanically fixed. The driver will then not be able to continually correct his course. In other words, to realize an objective one has to deviate from the optimal route. It follows that, if we wish to reach a destination, we can only do so if the direction in which we travel is at variance to the desired direction.
- *Exploring and setting limits: Do more with 'more or less.'* There is no such thing as the 'correct' route. However, one can define the route 'more or less,' that is, explore and set acceptable limits within which to move toward the destination.
- *In the case of collaboration, agree on acceptable limits.* If two or more people (e.g., cyclists) agree to pull a trailer between them toward B, the situation shown in Figure 27.3 results. To keep balance, each person will have to weave from side to side. But this 'oscillation' must not be so great that one is forced to let go of the trailer. The fact that both individuals are continually weaving implies they will experience substantial levels of tension (or 'conflict') between them as well as with the trailer. To reach the destination, they will have to keep these tensions within acceptable limits (see Figure 27.3). Moreover, the collaboration can only function properly if it is managed in such a way that each participant is able to make informed objections to any proposed (changes in these) limits.

### Developing and Testing Preliminary Design Rules

The construction principles described in the previous section provided the framework from which design rules were constructed. First, these principles serve to construct the rule of *informed consent* as follows:

For example, a heating or cooling system has a thermostat that regulates temperature, an operational device for heating or cooling the air, and a sensor device for

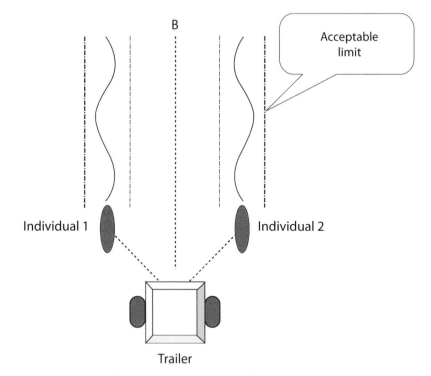

FIGURE 27.3   In case of collaboration, agree on acceptable limits.

measuring the actual temperature. These components must be designed and linked to one another such that the system as a whole functions properly. The three components must 'decide' together whether they are able to maintain the set temperature, given their acceptable limits. A single component or a majority of components cannot 'take' this decision. If, for example, the temperature is set to 100°Celsius (or 210°Fahrenheit) and the thermostat instructs the boiler to raise its water temperature to that level, the boiler will be destroyed. Thus, even a simple technical system can only function properly if it is directed in such a way that *each* component can raise informed objections to proposed changes. By consequence, the rule of informed consent must also apply to organizations in which the circular process of operating-measuring-directing is essential to performance. (Endenburg 1974)

Another design rule, *double linking* between circles, was constructed as follows:

For practical reasons, a circle of people generally consists of only a limited number of people. The design of a larger organization will therefore have to incorporate multiple circles. Circles then have to be linked in a way that maintains the principles of circular process. The main challenge here is how to link circles that have a hierarchical relationship (e.g., the business unit circle and an operational

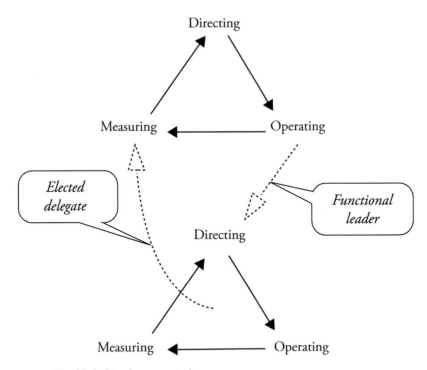

FIGURE 27.4   Double linking between circles.

circle). [Figure 27.4] shows two circles – each responsible for directing, operating, and measuring its activities – that are hierarchically linked but maintain a circular flow of information and power. There are two connecting links. The first one, the functional leader, connects operating activities in the upper circle with the directing activities in the lower circle; the functional leader is appointed (by consent) in the higher circle. This link is similar to Likert's (1961) linking pin. The second link, the delegate, serves as the feedback connection between the lower circle and the measuring task in the higher circle; the lower circle appoints the delegate(s) by consent. In practice, the lower circle selects a delegate to represent its directing, operating, and measuring tasks in the measuring process in the upper circle. This serves to extend Likert's single connection to two connections. Moreover, by linking both pins to the same (upper or lower) circle, the problem of creating two separate communication channels is avoided. (Endenburg 1974)

The first draft of the circular design, involving decision making by informed consent and double linking between circles, was discussed broadly in Endenburg's firm. One question raised was how people would be appointed to positions. The informed consent rule implicitly assumed that appointing people to positions is a key policy decision. To make this assumption explicit, a third rule was defined to emphasize that 'people' decisions are policy decisions that should be taken openly. In sum, the pre-liminary design involved the following three rules:

1. Informed consent, defined as 'no reasoned and paramount objection,' gov-erns all decision making on policy issues in circles. This means a policy deci-sion can only be made if nobody raises a reasoned and paramount objection against it. (Policy is defined as the limits set to a certain operational or other process.)
2. Vertical connections between two circles are formed by way of double link-ing, that is, the participation of at least two persons in both circles – including the functional leader and at least one elected delegate from the lower circle.
3. Persons are elected on the basis of consent, after open discussion.

These design principles implied that a circular organization is added to the exist-ing administrative structure, and in addition, each member of the organization par-ticipates in at least one circle – a policy-making unit composed of people sharing an objective. The coexistence of the circular and administrative structure implied that circles were authorized to decide on policy (in terms of 'acceptable limits'), whereas the administrative hierarchy remained responsible for getting things done within the policies decided upon in circles.

While the initial versions of the circular design were being drafted and dis-cussed, Endenburg set up the first experiments in his firm. That is, several circles were formed and trained. These experiments and their outcomes have been de-scribed in detail elsewhere (Endenburg 1998, Romme 1999). A key finding was that the three design rules defined earlier failed to define what a circle exactly was. The first experiments suggested several key elements of a fourth design rule: *Every member of the organization belongs to at least one circle, a unit of people with a com-mon work objective. A circle serves as a policy-making unit. That is, it formulates and updates its objective(s); performs the three functions of operating, measuring, and di-recting (i.e., comparing and instructing); and maintains the quality of its resources by means of integral education.*

### Learning from Implementation and Experimentation in Other Organizations

Starting in the mid-1980s, an increasing number of other organizations started de-signing and implementing new organizational structures with help of the circular design rules. These opportunities for implementation and experimentation in other organizations accelerated the process of redefining and extending the construction principles and design rules.

Since the mid-1980s, 32 organizations implemented the design. These organi-zations are as diverse as agribusiness and industrial firms, IT firms, consulting and accounting firms, and educational and health care organizations. The size of these organizations varies between small and medium sized; the largest firm in this population currently employs about 1,500 people. Most organizations that have implemented the circular design are Dutch; in addition, firms in Canada, the United States, and Brazil have implemented the design. In addition to these successful implementation cases, many other organizations have also explored and, in some cases, tried out the design. We discuss both successful and less successful cases here, to explore critical experiences and learning with respect to construction principles, design rules, and conditions for applying these rules.

*Case 1: Implementation Driven by the CEO.* In the late 1980s, the CEO of a medium-sized American firm – a manufacturing company in the high-tech sector – decided to introduce the circular design to get his employees more involved in the decision-making process. The CEO preferred to implement the design himself; this approach was to some extent inevitable because consultants with relevant expertise were not yet available on the American continent. The CEO introduced the method in different circles and tried to show his employees how (he believed) the method should work. During the start-up period of each new circle, he acted as the chair and subsequently asked the functional leaders of each circle to take charge of this task. After one year, circles were given the opportunity to elect their own chairs. The CEO, however, continued to be present during circle meetings at lower levels in the organization.

After two years, the circular design did not appear to work. People felt they were 'forced' to have circle meetings, many therefore did not show up for meetings, and those who did tended to be silent. In 1993, the American economy moved into a crisis, and the firm started facing some severe performance problems. The CEO redirected his attention to these problems. Without the CEO's direct involvement, circle meetings were gradually dissolved. By 1997, the circular design was no longer in operation in this firm.

A detailed follow-up study of this case showed that the attempt to introduce the circular design mainly failed as a result of the 'hit and run' strategy adopted by the CEO as well as his strong need to be in control: the CEO personally facilitated and guided the implementation process, which he expected to produce direct results. This implied that people in the organization did not get a real opportunity to explore and acquire some of the skills and routines required. On a more fundamental level, the CEO exclusively 'owned' the implementation strategy, although the initial purpose was to involve people in directing and controlling processes. In early 2004, the CEO therefore decided to reintroduce the circular design, but now with substantial support and involvement of external consultants.

The learning outcome from this case was that implementation processes are not obvious and self-evident. At the time, we concluded that an explicit design of the implementation approach was needed. As a key component of the circular design approach, this implementation approach would have to be communicated up front to any organization interested in the circular design. Later in this section, we will return to the implementation approach.

*Case 2: Introducing Circular Design in a Crisis Situation.* In the early 1990s, Matrex, a medium-sized industrial firm in the Netherlands, was facing a severe problem as a result of new competitors from Eastern Europe entering their market. The situation became extremely critical when the firm's bank announced its intention to file a bankruptcy petition against Matrex. In this situation, the CEO of Matrex, who once attended a workshop on the circular design, contacted a consultant with expertise in circular design. The consultant adopted a short-cut version of the circular approach. In a few hours time, a general management circle and several unit circles were created. The consultant chaired the first meetings of the general circle, in which she explained the basics of working in a circle meeting and then immediately turned to a discussion of the performance problems. An important step involved assessing the workflow from order acquisition to delivery and customer service. As a chairperson, the consultant also facilitated the other circles in a series of six meetings each, to supervise and train people in solving problems in their own units and in the interaction between units. Within several weeks, this intervention resulted in substantial improvements in Matrex' workflow.

The introduction of the circular design also affected the leadership style of the CEO. The CEO, a hands-on manager highly involved in managing operations, started acting very differently as a manager. He refocused his attention and leadership style toward the policy-making process in the circle responsible for business policy across all units, while leaving unit-level policies to the unit circles. The result was a dramatic increase in productivity as well as the restoration of profitability within one year after the financial crisis became apparent. In the second year after the start of the circular redesign project, Matrex also decided to install a top circle (i.e., board of directors), which it had not created previously because priority was given to solving the most urgent business problems in the first year.

The learning that emerged from this case is that participative decision-making processes generated and bounded by the circular structure can be effective in a crisis situation. On a more fundamental level, the introduction of the circular design apparently enhances connectivity by bringing people, information, and knowledge together to solve problems and make policy decisions (Romme 1999).

*Toward Design Rules for Implementation.* The latter case involves a crisis situation. For organizations in non crisis situations, a more elaborate implementation approach was developed. This approach includes an experimental period involving at least one pilot (i.e., a unit that experiments with informed consent, the operating-measuring-directing process, etc.). At the end of this period, the project team that coordinates the pilot takes a 'go-no go' decision regarding organization-wide implementation of the design. This project team is, for example, composed of the CEO, all business unit managers (including the manager of the unit serving as pilot), a delegate appointed by the pilot circle, and an external consultant as the team's advisor.

A key element of the implementation approach developed in the last 15 years involves the statutory structure. The prevailing statutory structure in the private sector is the investor-owned corporation, characterized by a formal structure in which the investors have the ultimate authority in all matters – through the shareholders meeting and their representatives in the board of directors (Hansmann 1996). This implies that, in the case of implementing the circular design, the formal owners could abolish the circular organization design at any time. In several implementation projects this actually occurred.

We thus observed the need to create an *inclusive* statutory entity, in which the corporation is controlled and owned by financial stakeholders as well as by other key stakeholders such as executives and employees. A group of lawyers and consultants, including Gerard Endenburg, created a solution involving a statutory structure that safeguards the circular design, once implemented. This statutory structure draws on a synthesis of the corporation and foundation (Endenburg 1998). Eleven organizations have implemented this statutory structure, which helped them to safeguard their circular structure and processes – in some cases over several decades. In not-for-profit organizations that implemented the circular design (e.g., in education, health care) this statutory change is not feasible because governmental bodies largely determine their formal structure.

We observed earlier that all organizations that applied circular design rules are small or medium sized. Many larger organizations did explore and try out the circular design rules. A recent example is the implementation of the circular design in a unit of a large industrial plant operated by one of Europe's largest multinational corporations. Unit management decided to introduce the circular structure to provide the opportunity to voice opinions and share experiences as well as decide on policy issues by informed consent. As a result, the unit created a culture of involvement, openness, and transparency. The unit had, and currently still has, the autonomy to sustain the circular design at the unit level. However, unit management has been unable to get senior management on board, and as such, intraorganizational diffusion of the circular design has not occurred.[4] These experiences, in addition to those discussed earlier (cf. Case 1), led to an explicit set of design rules for the implementation process (see appendix).

*Hierarchy as Construction Principle.* Several attempts to introduce the circular design in organizations structured as an association uncovered the – until then implicit – role of hierarchy. For example, one of these projects was done in a cooperative association of a large number of agribusiness firms. The delegates of these firms participated in a general membership meeting constituting the ultimate (formal) authority in the organization. This general membership meeting appointed a board charged with running the activities of the association. The ambiguous nature of the relationship between members and the board created a major governance problem: members were formally in charge but in practice unable to act collectively, whereas the board was formally subordinate to the members but actually governed the organization. Thus, interpersonal relationships between members were generally rather poor and highly dependent on the board as intermediate agent. This problem caused several associations to explore the circular design.

However, all attempts to add a circular organization to these organizations failed as a result of the absence of a clear hierarchy (as a sequence of accountability levels). Moreover, in all these cases, there was in fact no one authorized to repair this fundamental problem to create the conditions for developing a circular organization for policy making. These experiences suggested an additional construction principle that defines the importance of hierarchy. In this respect, the existing organizational structure in previous cases did involve a clear hierarchy. We therefore decided to formulate an additional construction principle, drawing on the literature that defines the role of hierarchy. For example, Simon (1973) describes the role of hierarchy in large complex (self-regulating) systems as follows:

> By virtue of hierarchic structure, the functional efficacy of the higher-level structures, their stability, can be made relatively independent of the detail of their microscopic components. By virtue of hierarchy, the several components on any given level can preserve a measure of independence to adapt to their special aspects of the environment without destroying their usefulness to the system. (23–24)

In terms of the construction principles previously defined, hierarchy therefore is *a fundamental property* of all viable systems. Using this additional construction principle, we subsequently redefined and extended the design rules and conditions for applying them (see appendix).

## Overview

We have explored the development of construction principles and design rules grounded in cybernetics as well as practical experiences in more than 30 organizations. The appendix describes the current version of the construction principles, rules for designing a system of circles, key conditions that have to be created, and design rules for the implementation process.

Circular construction principles were largely constructed from cybernetics. More recently, the circular design method was also grounded in the literature on political science (Endenburg 1998), organizational learning and control (Romme 1999), unanimity rule (Romme 2004), employee commitment (Buck 2002), and property rights (Romme and Endenburg 2001).

## DISCUSSION AND CONCLUSIONS

Apparently, early choices regarding the notion of circular process drove, but also created boundaries around, subsequent stages in the project. Over time, the experimental process of developing, testing, and adapting principles and rules thus tends to become locked into the boundaries set early in the process. In the case of circular design, one new principle was defined and several design rules and conditions for applying these rules were uncovered. These changes involved corrections and extensions rather than major transformations of the initial set of principles and rules.

The design literature refers to this lock-in effect in design projects in terms of liquid and crystallized states: during the liquid state, a design problem is still open to many possible directions; once a project crystallizes, the ability to revise key elements of the design without incurring substantial extra costs is greatly reduced (Boland and Collopy 2004b). A major crystallization moment in the evolution of the circular design project apparently involved the organization-wide implementation of the circular design in the firm Endenburg managed, after a more 'liquid' period in which he developed several preliminary design rules and pilot-tested different solutions produced with these rules.

A limitation of the circular design case arises from the nature of the data obtained via interviews, document study, and participant observation. These data are largely experiential and narrative and therefore not rigorous from the perspective of mainstream organization science. This is partly due to the key role of pragmatic experimentation that, as we have argued earlier, focuses on developing practices that actually work (see also Endnote 1). Thus, we have not established the internal and external validity of our findings and observations in a more robust manner.

Moreover, the full complexity of the context in which implementation of a particular organization design occurs will always be somewhat loosely coupled to the general concepts and theories developed in organization research. In this respect, Worren et al. (2002) introduce the notion of *pragmatic validity* and argue that scholars and practitioners will interpret this notion differently. In the propositional mode of thinking that prevails in organization science, the emphasis on pragmatic validity leads to the development of more prescriptive theory – drawing on explicit causal propositions, operational definitions of constructs, and descriptions of how implementation is to proceed to achieve desired outcomes (Worren et al. 2002). By contrast, more ambiguous knowledge serves important functions in organizational practice. Thus, practitioners are more likely to rely on narrative and visual data to achieve pragmatic validity – for example, stories and anecdotes illustrating a particular concept or diagrams visualizing a conceptual model (Worren et al. 2002).

Construction principles and design rules, as instrumental boundary objects, may help to bridge these different meanings of pragmatic validity. They can help realign the perceptions and understandings by the people involved in implementation and experimentation settings with the basic assumptions and propositions researchers develop about these settings. Rather than trying to align propositional thinking directly with the ambiguity and complexity of organizational practice, the design approach advocated in this paper involves construction principles and design rules as boundary objects. These boundary objects can serve as a conceptual framework for productive interaction and collaboration between practitioners, consultants, and academics. That is, the border territories between the highly different epistemic cultures of academia and practice need to be cultivated by means of instrumental language and knowledge that helps scholars and practitioners work together productively. To advance organization design research in this area, further work on extended notions of pragmatic validity as well as boundary objects shared by scholars and practitioners is essential.

Overall, we have proposed an approach toward organization design very similar to the research and development cycle connecting the natural sciences, engineering, and technology. A key idea is that organization designs, produced from construction principles and design rules, should be extensively tested in practice. The circular design case illustrates that construction principles are important for creating new design rules as well as for a deeper understanding of the organizational designs and practices created from these rules. In addition, explicit construction principles and design rules facilitate the transfer of learning outcomes between different projects that draw on similar principles and rules.

This approach implies that principles and rules, ideally, come before design. As Figure 27.1 suggests, the actual flow of activities in a design project may also start with developing and enacting a particular design without a science-based understanding already in place. The organization then first explores whether the design works, that is, produces the expected organizational processes, and subsequently tries to develop a deeper understanding of how and why it works. This approach is more easily reconciled with the intervention culture currently prevailing among executives and their consultants. In any case, it is not realistic to expect that executives, consultants, and other practitioners will engage in deeper reflections on their designs and interventions *if* they do not team up with scholars trained in organization science.

In sum, we have advocated a science-based design approach that grounds construction principles in organization science and tests design solutions in practice. We explored the development of principles and rules for circular design that were tested in more than 30 organizations. An important finding is that construction principles and design rules can serve as boundary objects between organization design and research. This suggests any organizational practice can be better understood as well as more easily advanced by uncovering the underlying principles and rules. The pivotal role of construction principles and design rules opens up new opportunities to reconnect organization science to organizational practice.

### Acknowledgments

Earlier versions of this paper were presented at the NSF-sponsored conference on Organization Design, New York University (2004), and the European Academy of Management conference, Munich (2005). The authors express their appreciation to three anonymous *Organization Science* reviewers, an anonymous reviewer for the EURAM conference, Bill Starbuck, and Joan van Aken for constructive feedback and comments.

## APPENDIX. OVERVIEW OF THE CIRCULAR DESIGN METHOD (IN 2005)

### Construction Principles

In a system that is self-regulating or wishes to build self-regulating capacity:

1. 'Weaving' must be possible.
2. The circular process makes it possible to search. That is, a system is only able to maintain a state of dynamic equilibrium under the following conditions:

   - the steering and feedback circle is closed and performance is being measured;
   - it has sufficient scope to weave from side to side;
   - the circular process enables the system to search.

3. 'Mistakes' must be made.
4. Boundaries are continually explored and set (i.e., do more with 'more or less').
5. Acceptable limits are set and agreed upon (in case of collaboration).
6. Hierarchy is a fundamental property.

### Circular Design Rules

To build organizational capacity for self-regulation and learning, apply the following rules to create a circular design tailored to your organization:

1. Decisions on policy issues are taken by informed consent (defined as 'no reasoned and paramount objection').
2. Every member of the organization belongs to at least one circle, a unit of people with a common work objective. Each circle formulates and updates its objective(s); performs the directing, operating, and measuring/feedback

functions; and maintains its skills/knowledge base by means of integral education.

3. The double link, i.e., the vertical connection between two circles, is constituted by the participation of at least two persons in both circles – including the functional leader and at least one elected delegate from the lower circle.

4. The circular structure, defined in the previous rules, is added to the administrative hierarchy. This administrative hierarchy, as a sequence of accountability levels, contains all functional leaders that are responsible and accountable for implementation of policies made in circles.

5. Circles elect persons only on the basis of informed consent, after an open discussion.

### Conditions

The following conditions must be created (if not already present):

1. The intention to create an organization that is economically as well as socially viable (i.e., profitability *and* safety). The minimum requirement is that top management (including the board) shares this intention when implementing the circular approach; preferably, other people in the organization share this intention.

2. All members of the organization have access to information systems and flows. Exceptions to this rule can be made, but only if the 'why' and 'how' of these exceptions are transparent.

3. There is a sequence of unambiguous levels of accountability in the organization that differentiates the performance of the entire system into higher and lower level issues. Without this type of hierarchy, a circular structure cannot be designed.

### Implementation Design Rules

Design the implementation process as follows:

1. Obtain top management's commitment early in the process, by raising the 'how' as well as 'why' question regarding the choice for a circular design.

2. Set up a project team that coordinates and monitors the implementation and experimentation process. Connect this project team directly with the top team: that is, the project team should include at least one top manager (preferably the CEO or managing director of the organization).

3. Invite external experts to help the project team, if this expertise is not available in the organization.

4. Organize the implementation process as an experimental process, involving at least one pilot:

    a. Each pilot involves a unit of people that is trained on the rights and skills linked to the circular process, decision making by informed consent and integral education.

    b. These pilots are embedded in ongoing operational and management processes.

    c. After a predefined number of pilots, top management (including the board of directors, if any) takes a decision regarding organization wide implementation, on the basis of a proposal drafted by the project team.

5. Create statutory safeguards for the circular organization design to make it sustainable over time.

## ENDNOTES

1. Pragmatic experimentation is similar to the clinical approach developed by Schein (1987). Pragmatic experimentation differs from controlled experimentation in laboratory conditions, because the former takes place in social reality where conditions are given and randomized selection is not possible. Pragmatic experimentation is also different from quasi-experimentation, in which authentic (non randomized) groups are given different experimental treatments (Riecken and Boruch 1974). Applying different experimental treatments to different systems (e.g., organizations or units within an organization) is rarely feasible and also often unethical in organizational design and development. Moreover, pragmatic experiments can be conducted without a fully formed theoretical understanding already being in place. In the latter instance, one first explores whether the intervention 'works' and subsequently develops a deeper understanding of how and why it works – this is reflected in Figure 27.1 in the arrows from implementation and experimentation to organization science, construction principles, and design rules.
2. To increase readability, we will refer to one of the authors of this manuscript (Endenburg) in the third rather than first person in the remainder of this paper.
3. Please note that circular construction principles and design rules initially – in the late 1960s and early 1970s – developed from the science of steering and control (cybernetics) rather than organization science as it currently is. This raises questions about the relevance of organization theory (at the time) for organization design problems in practice. These questions are outside the scope of this paper.
4. Apparently, design becomes much more difficult if the organization has an identity that key agents in the organization try to protect (Fiol and O'Connor 2002). That is, over time an organization tends to become locked in to a certain identity and power structure. If the existing identity and structure are too different from those underlying the new design, the selfpreserving forces may be too powerful (cf. McMaster 1996).

# REFERENCES

Ackoff, R.L. 1999. *Re-creating the Corporation: A Design of Organizations for the 21st Century.* New York: Oxford University Press.

Baligh, H.H., Burton, R.M. and Obel, B. 1996. 'Organizational consultant: Creating a Useable Theory for Organizational Design.' *Management Science,* 42: 1648–1662.

Bartunek, J.M., Louis, M.R. 1996. *Insider/Outsider Team Research.* Thousand Oaks, CA.: Sage.

Bate, P., Khan, R., Pye, A. 2000. 'Towards a Culturally Sensitive Approach to Organization Structuring: Where Organization Design meets Organization Development.' *Organization Science,* 11: 197–211.

Beer, S. 1959. *Cybernetics and Management.* London, UK: English Universities Press.

Boland, R.J., Collopy, F. 2004a. 'Design Matters for Management.' In R.J. Boland and F. Collopy (eds.), *Managing as Designing.* Stanford, CA.: Stanford University Press, 3–18.

Boland, R.J., Collopy, F. 2004b. 'Toward a Design Vocabulary for Management.' In R.J. Boland, F. Collopy (eds.), *Managing as Designing.* Stanford, CA.: Stanford University Press, 265–276.

Buck, J. 2002. *Employee Commitment in Sociocratic Organizations.* Unpublished Master's thesis, George Washington University, Washington, D.C.

Endenburg, G. 1974. *Sociocratie: Een Redelijk Ideaal.* Zaandijk, The Netherlands: Klaas Woudt.

Endenburg, G. 1998. *Sociocracy as Social Design.* Delft, The Netherlands: Eburon.

Fiol, C.M., O'Connor, E.J. 2002. 'When Hot and Cold Collide in Radical Change Processes: Lessons from Community Development.' *Organization Science,* 13: 532–546.

Foss, N.J. 2003. 'Selective Intervention and Internal Hybrids: Interpreting and Learning from the Rise and Decline of the Oticon Spaghetti Organization.' *Organization Science,* 14: 331–349.

Hansmann, H. 1996. *The Ownership of Enterprise.* Cambridge, MA.: Harvard University Press.

Hardy, C., Leiba-O'Sullivan, S. 1998. 'The Power behind Empowerment: Implications for Research and Practice.' *Human Relations,* 51: 451–483.

Kim, W.C., Mauborgne, R.A. 1993. 'Procedural Justice, Attitudes, and Subsidiary Top Management Compliance with Multinationals' Corporate Strategic Decisions.' *Academy Management Journal,* 36: 502–526.

Likert, R. 1961. *New Patterns of Management.* New York: McGraw-Hill.

McMaster, M.D. 1996. *The Intelligence Advantage: Organizing for Complexity.* Boston, MA.: Butterworth-Heinemann.

Milliken, F.J., Morrison, E.W., Hewlin, P.F. 2003. 'An Exploratory Study of Employee Silence: Issues that Employees don't Communicate Upward and Why.' *Journal of Management Studies,* 40: 1453–1476.

Moss, S.E., Sanchez, J.I. 2004. 'Are Your Employees Avoiding You? Managerial Strategies for Closing the Feedback Gap.' *Academy Management Executive,* 18(1): 32–44.

Riecken, H.W., Boruch, R.F. (eds.). 1974. *Social Experimentation: A Method for Planning and Evaluating Social Intervention.* New York: Academic Press.

Romme, A.G.L. 1999. 'Domination, Self-determination and Circular Organizing.' *Organization Studies,* 20: 801–832.

Romme, A.G.L. 2003. 'Making a Difference: Organization as Design.' *Organization Science,* 14: 558-573.

Romme, A.G.L. 2004. 'Unanimity Rule and Organizational Decision Making: A Simulation Model.' *Organization Science,* 15: 704–718.

Romme, A.G.L., Endenburg, G. 2001. 'Naar een nieuwe vormgeving van het aandeel.' *Tijdschrift voor Management Organisatie,* 55(3): 40–53.

Schein, E.H. 1987. *The Clinical Perspective in Fieldwork.* Newbury Park, CA.: Sage.

Simon, H.A. 1973. 'The Organization of Complex Systems.' In H.H. Pattee (ed.), *Hierarchy Theory: The Challenge of Complex Systems.* New York: George Braziller, 1–27.

Simon, H.A. 1996. *The Sciences of the Artificial.* 3rd ed. Cambridge, MA.: MIT Press.

Starbuck, W.H. 2003. 'Shouldn't Organization Theory Emerge From Adolescence?' *Organization,* 10: 439–452.

Starbuck, W.H. 2004. 'Why I Stopped Trying to Understand the *Real* World.' *Organization Studies,* 25: 1233–1254.

Van Aken, J.E. 2004. 'Management Research based on the Paradigm of the Design Sciences: The Quest for Field-tested and Grounded Technological Rules.' *Journal of Management Studies,* 41: 219–246.

Van Vlissingen, R.F. 1991. 'A Management System Based on Consent.' *Human Systems Management,* 10: 149–154.

Warfield, J.N. 1994. *A Science of Generic Design.* 2nd ed. Ames, IA.: Iowa State University Press.

Wicks, A.C., Freeman, R.E. 1998. 'Organization Studies and the New Pragmatism: Positivism, Anti-positivism, and the Search for Ethics.' *Organization Science,* 9: 123–140.

Worren, N., Moore, K., Elliott, R. 2002. "When Theories Become Tools: Toward a Framework for Pragmatic Validity. *Human Relations,* 55: 1227–1250.

# 28

# A Critical Scandinavian Perspective on the Paradigms Dominating Design Management

ULLA JOHANSSON AND JILL WOODILLA

Whether a distinctive Scandinavian management style exists in practice has been debated within management research (see Jönsson 1996; Czarniawska and Sevon 2003). Regardless of whether the managerial praxis is different or not, typical management research in Scandinavia differs from both the US and the continental European context. Critical management (Alvesson and Wilmott 1992), institutional perspectives on management (Czarniawska-Joerges 1993; Brunsson 2000; Rovik 2000; Olsen 2007), cultural management and narratives (Corvellec 1997; Czarniawska 1999) are not exceptions, as in the US and continental tradition, but rather the norm among the most published and respected professors in Scandinavia, especially in Sweden. Critical management, in the broad sense that is exemplified in the biannual Critical Management Conference in the UK and the network of critical management scholars (www.criticalmanagement.org) has always had a strong anchoring in the Swedish community of management researchers. This anchor extends to the design community and the Scandinavian perspective on design management.

From a Scandinavian and a critical management perspective, the paradigmatic ground of much design management research is problematic due to widespread acceptance of functionalist platforms within design management. From a Swedish perspective, where management researchers tend to critique prominent scholars such as Michael Porter or John Kotler for their neo-positivist and functionalist platforms, it is disquieting to realize that these scholarly texts are accepted elsewhere without question. For the authors of this chapter, the unequivocal acceptance of the merger between the design field and the often positivist or neo-positivist management research field is especially awkward because both authors embrace perspectives that hold paradoxes and ambiguities rather than blindly accepting the orthodox empirical method focusing on causality (see Johansson and Woodilla 2005). They realize that other paradigms exist along with the functionalist one. These alternative management discourses enjoy respect within the Scandinavian management research community and might be better suited to the design discourse.

In this chapter we first provide a short overview of Scandinavian design management research and then use the critical perspective introduced above to problematize the way knowledge from design merges with knowledge from management.

## DESIGN MANAGEMENT IN SCANDINAVIA[1]

Design management is part of a long tradition in Scandinavia. Lisbeth Svengren, the first design management researcher there, joined the field in the 1980s as a doctoral student within the Triad project, which was initiated by Harvard Business School and organized by the Design Management Institute, Boston. She became the second scholar worldwide to defend a dissertation in design management successfully (Svengren 1995).

The Scandinavian situation differed strongly from that in England. In England, design management researchers had their base sometimes in management departments, sometimes in design schools, and at other times in business schools. In Scandinavia, until the early 2000s, there was basically only one person dedicated to design management in each country. Lisbeth Svengren in Sweden and her colleagues in Denmark (Tore Kristensen) and Norway (Birgit Jevnaker) all belonged to business schools. Jevnaker joined the design management area in the early 1990s and since then has published extensively in strategy and design (cf. Jevnaker 1993, 2005), whereas Kristensen has worked within the area of marketing and design in Denmark (cf. Kristensen and Lojacono 2002). Only in Finland did design management research take place at a School of Art and Design (at UIAH, now Aalto University). Art historian Pekka Korvenpää and design practitioner Peter McGrory led these efforts.

During the 1990s up to the early 2000s, close connections existed among these more-or-less independent, individual scholars, but design management did not grow into larger research groups. Practitioners generally eyed design management with great scepticism during this period, and among management scholars, design management was not considered 'cutting edge' research. There was not a single group of qualified design management researchers based at any of the many design schools in Scandinavia. This situation continues at the time of writing. However, after 2000, a shift happened in the Swedish design management group, including some important events.

At Stockholm University, where Svengren moved after her dissertation, a number of doctoral students not only joined the international design management discourse but also embraced the Scandinavian critical management tradition. The discourse of art and management also had strong support at Stockholm University at that time. A handful of doctoral students created a foundational group at the intersection between art and management, design management and critical management, with Clemens Thornquist defending his dissertation in 2005 and Theresa Degerfeldt-Månsson hers in 2009. Several other doctoral students continue within in this group.

At Växjö University and Kalmar College University (renamed Linneaus University in 2010), courses in design management were held at the master level even before they started in Stockholm. Personal ties played a role here, too. In 2001, Ulla Johansson joined Växjö University and the design management discourse in a major research project in collaboration with Svengren (Johansson and Svengren 2008), followed by her own research evaluation of the Swedish government's design programme (Johansson 2006a, 2006b).

In 2006, a strategic effort at University of Gothenburg led to the founding of a new Business and Design Lab, a cooperative effort between the Faculty of Art and the Faculty of Management at the university. In autumn 2009, Johansson was installed as the first professor in design management in Scandinavia, the Torsten and Wanja Söderberg Professor of Design Management. The group she leads consists of ten people – doctoral students, postdoctoral researchers and senior researchers – who pursue one of two research directions: action-oriented research projects (doctoral students in collaboration with the local design community) and theoretical and critical reflections (postdoctoral and senior researchers).

In Finland, where design management is integrated into design schools, design management has taken another turn. Current research focuses mostly on visual identity and branding (cf. Koskinen et al. 2003; Karjalainen 2004). This area represents the intersection between design and management but with a strong design foundation. In Denmark, where design management is closely linked to the Copenhagen Business School and Tore Kristensen, design management is an area within marketing. Research here focuses on how design and designed artefacts are valued by customers (cf. Kristensen and Lojacono 2002). In Norway, on the other hand, design management is tied to the Norwegian School of Management and led by Birgit Jevnaker who has written about design strategies for more than fifteen years. A recent cooperation between the Norwegian School of Management and the School of Design and Architecture may encourage more Norwegian researchers to focus on design management.

In summary, design management in Scandinavia has expanded from a closed network of four scholars during the 1990s to a larger and more diversified group, and continues to be an emerging and growing area of scholarly interest. This is not a homogeneous group pursuing a single discourse. However, a number of trends are emerging. We can see a shift from normative, or 'mainstream' research methods and projects towards methods based on critical reflexive methodology and innovative research designs. Within Sweden, there has been a distinct turn that embraces diverse management theories as a foundation. New doctoral students in Stockholm, Gothenburg,

Lund, and Växjö have found it impossible to use conventional management theories in relation to design – they say they are not a good match and prefer to depart from non-mainstream theoretical frameworks so as to build a new, and more relevant, conceptual platform for their empirical studies. We take this to indicate a need for a different paradigmatic framework as a useful basis for the discipline of design management, and suggest that the approach introduced by Burrell and Morgan in the late 1970s may help develop these new directions (Burrell and Morgan 1979).

## ANALYTIC REVIEW OF DISCOURSES CONTRIBUTING TO DESIGN MANAGEMENT

### Presentation of Analytic Tool

To demonstrate our thesis – that mainstream management is not a good match with design – we use the paradigmatic framework that Gibson Burrell and Gareth Morgan presented in their influential sociological paradigms and organizational analysis (Burrell and Morgan 1979). They analysed organizational/managerial research to reveal the paradigmatic grounds or taken-for-granted assumptions that guide all approaches to research and to articulate distinctions between different schools of thought. Their aim was 'to show what each of the paradigms has to offer, given the opportunity to speak for themselves' (Burrell and Morgan 1979: 395).

Burrell and Morgan's analytical framework creates four paradigms bounded by two continua, one representing the ways in which scholars view the social world and the nature of society, grounded in objectivism versus subjectivism, and the other representing whether the research presupposes to contribute to a society in social order or to social change. The objective–subjective distinction (Burrell and Morgan 1979: 1-9) relates to the debate on the nature of organizational phenomena (ontology), the nature of knowledge (epistemology), and 'the model of man' and methods for investigating the social world (methodology). The objective–subjective dimension is complex and multifaceted. The second distinction relates to the purpose of research, whether conscious strategies or taken-for-granted assumptions, as a concern for regulation in human affairs (p. 17), or research for radical change at the other end of the continuum. Together the two dimensions form a 2 × 2 matrix with four distinct paradigms that encompass a wide range of social theories. The four paradigms are labelled: (1) radical humanist paradigm (with endpoints of subjective and radical change); (2) Radical structuralist paradigm (with endpoints of objective and radical change); (3) interpretative paradigm (with endpoints of subjective and regulation); (4) functionalist paradigm (with endpoints of objective and regulation) (see the matrix structure in Figure 28.1).

Burrell and Morgan's framework provides an analytic tool for analysis of the various research streams within the domains of design, current management, and design management, resulting in 'paradigmatic patterns' of these fields. Our analysis begins

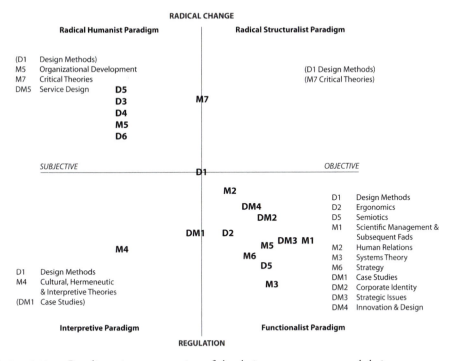

FIGURE 28.1 Paradigmatic representation of the design, management, and design management discourses. © Johansson and Woodilla, 2011.

with the design discourse, highlighting contributions from Scandinavian researchers, and discussing underlying assumptions using the Burrell and Morgan matrix. We continue with an overview of the management discourse, augmenting the classic theories discussed in detail by Burrell and Morgan with contemporary research, followed by a presentation of research in the emerging field of design management.

## The Paradigmatic Pattern of Design Discourses

The design discourse is not one but many. Design embraces various subfields each with its own research area, such as Design for All, Democratic Design, Gender and Design, or Sustainable Design. Each area reflects and focuses on different concerns and assumptions. Compared with management, the field of design is smaller in size with fewer researchers, and might be due to the field being relatively new, the epistemological discussion has not reached the same intensity. This is not to say epistemological concerns are non-existent, and indeed, similarities can be found between the two disciplines. Here we are primarily concerned with industrial design and the epistemology of design as a method and way of thinking – that is, with the design process rather than the result of this process.

The special character of design research has been discussed ever since Simon (1969) claimed that design research has an epistemology of its own, distinguishing it from the humanities, social sciences and natural sciences. Because design is creating what has never existed before, research must take this into consideration and can be neither explanatory nor descriptive, but rather experimental. Other scholars have also discussed the attributes of design research. Lawson (2005) theorizes design competence to be a specific way of making sense of problem solving in a comprehensive and synthetic way, whereas Buchanan (1992) describes the core of the design competence as handling and solving 'wicked problems'.

Edeholt (2004) identifies two design characteristics that distinguish it from both social science and technical areas: first, design is about how things 'ought to be' or about creating an alternative world and, second, design starts with problem solution rather than problem analysis. Such concerns suggest that design, in general, falls within the radical humanist quadrant of Burrell and Morgan's framework, defined as subjective (rather than objective) and interested in radical change (rather than regulation). However, our discussion below of different fields and groups within design research suggests that design research, although predominantly located in this quadrant, also has areas located in other paradigms.

The discourse on design methods and how to describe them has changed over the years in ways that can be linked to our epistemological and methodological discussion of design paradigms. The various design schools, each with different underlying assumptions, have come into prominence and for the most part have replaced one another. In the 1960s and 1970s structuralist approaches were prominent, such as Alexander's (1964) methods for complex problem solving without preset conceptions. Later, influenced by the general interest in system theory in academia, scientific and systems-theory oriented descriptions prevailed (Cross 1999, 2001). Such views were challenged by Schön's (1983) literal and reflexive approach. Over the last few years, design methods research has partially merged into the epistemological debate and has partially been influenced by broader social science methods. The design methods discourse has become a multidisciplinary forum for scholars, which we acknowledge by positioning it at the intersection of the axes of the Burrell and Morgan framework.

Within the field of industrial design, Bauhaus in Europe and Dreyfuss in the US are considered the founding fathers of research. The prominent Bauhaus group in Germany had a clear ideological base, from which the members wrote and practised as designers and architects (Droste 1990). The Bauhaus maxim was 'form follows function' and simplicity was not only beautiful but also a means to the end – to enable everyone to benefit from modernity. Dreyfuss (1967) was the Bauhaus's American counterpart with a passion for industrial design and ergonomics. He focused (not without similarities to early management theorists Taylor and the Gilbreths) on the human body. He measured and recorded the types and ranges of motion, establishing thereby research into ergonomics as a scientific field (Clark and Corlett 1984; www.hfes.org). Research into functions and ergonomics has been a substantial area ever since, with a peak in the 1960s and 1970s.

Many researchers from the ergonomics school had a specific interest in how things and the environment could be changed to be inclusive for individuals with disabilities.

The areas of Inclusive Design (Clarkson 2003) and Design for All (www.designforall. org) emerged in the 1960s and still represent active fields of research. Applying Burrell and Morgan's framework here shows how these fields have moved away from a functionalist perspective to a more radical humanistic view.

The design community has always included an ethical dimension. Eco-Design and Design for Sustainability have their roots in this important area of design research and practice. Eco-Design is a current business 'hot topic' with its emphasis on using recyclable materials (Brezet and Van Hemel 1997), while Design for Sustainability advocates radical changes in all systems (cf., Margolin 2002; Manzini and Jegou 2003; Thackara 2005; Thorpe 2007). This research falls within the Burrell and Morgan's radical humanist paradigm.

The semiotic design discourse lies at the border between design and design management and has grown since the early 1990s in research and practice (Mono 1997). Semiotic research has a different character from the schools above due to its deep philosophical roots and connections to many interdisciplinary areas. Semiotic design research (Krippendorff 1989; Vihma 1995) is concerned with how images are perceived and 'read' by different audiences and stakeholders and most research is related to regulation rather than radical change. When used in practice, it has an objective perspective – the objective is to determine that business goals are met by the way products, advertisements and other touch-points are expressed in and through images.

Interactive design (Edeholt and Lowgren 2003) represents an area that began at the man-machine interface both in mechanical and in computerized applications. Given the prevalence of information technology and human-computer interaction in all aspects of life, research in this area has grown rapidly. More an area in itself rather than a subfield of industrial design, methods of user-centred design merge with other design methods. Because many studies in interactive design consider the computer from a more holistic view, regarding it as a tool for work and pleasure, this area has come to intersect with organization studies, specifically those in the radical humanist paradigm (Ehn 2006; Björgvinsson 2007).

The various areas of design research are designated by the letter D in Figure 28.1. Our analysis indicates that while design research falls predominantly in the radical humanist paradigm, it also has areas located in the other paradigms.

## THE PARADIGMATIC PATTERN OF THE MANAGEMENT DISCOURSES

When Burrell and Morgan analysed theories of management and organization within each paradigm, they found that dominant theories belonged in the functionalist quadrant (having assumptions of objectivism and regulation). They identified some, although very little, research to fit within the quadrants that relied on assumptions of subjectivity or radical change. Today there are multiple research activities in at least three of the four quadrants of Burrell and Morgan's analytical framework, as indicated by the following overview of the current areas of management research. The multiplicity of fields and discourses existing within the management field highlight the necessity to choose carefully with whom to partner in matters of design management.

There are many ways to divide the research area of management. Most divisions fall into a historical sequence, which does not mean that one theory displaced another in chronological order. Rather, each topic area was born in the academic conversation of a particular context and emerged to take care of deficiencies and criticisms of the then-current theoretical perspective. Key components of the previous theory remained and continued to influence the practice of management. It is therefore more helpful to think of the different areas as family members born in different times but most still alive and having quite different and complicated relationships with each other. Such discussions are commonly found in management textbooks, so here we provide only an overview. Figure 28.1 includes these significant streams with the 'management discourse' with the key M.

The starting point of management research is often traced to the beginning of the last century when Frederick Taylor (1911) published his book on scientific management, thereby indicating that his treatise was not about rules of thumb but science. Taylor's theories about how to manage workers were widely accepted in businesses of that time because they created efficiency and promised a system that would standardize the work of artisans and labourers. The ways of making sense of management that characterize Taylorism – that is the belief in objectivism and the possibility of finding one optimal and universal way of managing work – have been both heavily criticized and defended. Either way, it remains one of the core platforms of management theories. There are still many active scholars who follow this tradition. Examples include business process re-engineering, lean production, and many of the fads of the 1980s and 1990s. Morgan (1986) uses the metaphor of machine to describe the way these theories picture the organization and its character – clearly very different from the design discourse and its creative foundation.

During the 1930s and 1940s Taylorism became heavily criticized, particularly after the Hawthorne experiments that could not be interpreted within the framework of scientific management (Roethlisberger and Dixon, 1961). As a consequence, relations between human beings rather than their physical actions formed the grounding for the human relations movement and a different view of an organization, characterized by Morgan (1986), using the metaphor of an organism. This perspective continues as human resource management, with new studies and perspectives, including the prevalent discourse on diversity. For the most part research is concerned with investigating functional cause-and-effect relationships but some fall into the interpretive paradigm with a concern for subjective impressions of organizational life.

During the 1950s and 1960s, management theories were influenced by a number of new fields, in particular computer science, systems theory and psychoanalysis. Emerging discourses looked at the organization as a system, and many models of an organization resembled those drawn by computer scientists (Katz and Kahn 1978).

Open systems theory led to the consideration of organizational environments and the influence of political and societal changes on the internal and external relationships of organizations (Emery and Trist 1965). Theories of power merged into management theories, leadership developed as a separate theoretical area, and organizational development – representing a merger between organization theory and pedagogical perspectives – emerged (Levitt and March 1988). Morgan's metaphors for the main theoretical developments during this period were the organization as a system and as a brain.

During the 1980s and 1990s, cultural perspectives of organizations recognized individual members within a specific context and taken-for-granted and institutionalized assumptions (for example, Hofstede 1980; Schein 1985; Powell and Dimaggio 1992). What was then labelled as 'cultural theory', however, consisted of research with different paradigmatic underpinnings, ranging from anthropological case studies (such as Van Maanen 1988; Kunda 1992) to more positivistic survey studies (for example, Hofstede 1980). Theories of organizational change emerged, ranging from structural perspectives (Hannan and Freeman 1984) to values-driven organizational development (Beckhard 1969), to ethnographic studies (Czarniawska-Joerges and Sevon 1996). Morgan's metaphors of organizations as psychic prisons or as flux and transformation capture the variety of perspectives active in organization theory during this period.

Another stream from this period, and an important one for design management, is the strategy stream that continues to be popular among researchers, educators, and practitioners. The roots of strategy can be traced back over 2,500 years to military endeavours in Greece and China but the subject of strategic management as applied to organizations and businesses draws heavily on the work of Michael Porter (1980, 1985) and his concepts of competitive strategy and competitive advantage. Strategy is clearly functionalist in the nomenclature of Burrell and Morgan, and is so dominant within the management discourse that it lays claim to being a 'paradigm' of its own (Prahalad and Hamel 1994).

During the last decade both traditional 'main stream' and more 'critical management' (for an overview see Alvesson and Wilmott 1992; Hatch 1997) perspectives have become equally accepted within the Academy of Management. Even so, the number of scholars working with a critical radical humanist perspective are far fewer than those with mainstream functionalist viewpoints, and critical scholars who desire radical structural change are even fewer in number.

## THE PARADIGMATIC PATTERN OF THE DESIGN MANAGEMENT DISCOURSES

We consider the management discourse to be about a hundred years old, the design discourse about half of that, and the design management discourse only a few decades old. The first academic masters degree course in design management was introduced in the 1970s at the London Business School, and it was not until the 1980s that the Royal College of Art in London was the first school worldwide to teach design management to designers (Johansson and Svengren 2005).

Design management is concerned with the integration of design into management and vice versa. The design management area is more coherent than either design or management; this may be due to the relative smallness of the area but also due to underlying assumptions. A basic assumption, rooted partly in research and partly in experience, is that management most often lacks sufficient knowledge about design to take advantage of the strategic potential that design promises. As a consequence, research is most often of a normative and prescriptive character.

The Triad project is viewed as the beginning of design management research. Triad, a collaborative research project over three continents, was initiated in 1988 by the Design Management Institute (DMI – formed in 1975 to 'heighten awareness of design as an essential part of business strategy') and involved the Harvard Business School. The project's purpose was to develop case studies in design management, and the thirteen completed case studies from US, European and Japanese contexts formed a platform for design management research (cf. Svengren 1995) The Triad project focused on descriptive cases from companies where design was an integrated part of the product development and innovation process and where design had played a part of the success of the company, so called 'excellence cases'. This type of research became a role model for much of the subsequent research in design management.

Research on corporate identity forms a separate, but intertwined area of research and development within design management. When at its peak in the late 1980s, 'corporate identity' was a way to integrate the different visual messages coming from the company's logo, written material (graphic design), products (industrial design), and the environment in which production and selling took place (interior design and visual merchandising) to provide the customer with a stronger and more coherent visual message. Researchers claimed that this stronger message should be in coordination with business strategy and goals, not merely as unrelated pieces of art (Olins 1989). Corporate identity can be seen as a forerunner to the brand discourse; the difference, according to Johansson and Svengren (2005), is that brand is related to marketers and corporate identity to designers.

In the 1990s the brand discourse flourished in marketing and management and the design management discourse focused on design as a strategic resource. Research questions centred on ways in which design is a strategic resource and how it should relate to and be integrated into top management's general strategies (Cooper and Press 1995; Svengren 1995; Olson et al 1998) in addition to marketing strategies (Bruce and Cooper 1997; de Mozota 2003). During the 1990s and early 2000s, design as a strategic/marketing resource was the subject of several articles and books and became a platform for education in many countries.

A related, but different, subfield of design management considers design and innovation (Von Hippel 1978; Stamm 2003; Verganti 2006), and the partnership of designers, engineers and managers. In the first part of the twenty-first century 'design and innovation' has been in vogue, replacing the area of 'design and strategy' and the discourse is now common in the general business media. In the US, innovation and creativity have become 'driving forces for the new economy', according to Tom Kelley in his bestseller about practices at IDEO (2001). One argument in the US is that with the manufacturing base lost to overseas producers, innovation is all that is left as the driver of the economy. Design here is less about aesthetics and giving form but more about creative thinking and more about novelty than improvement. Thus design thinking, a process Buchannan (1992) previously connected to taming wicked problems, becomes the servant of innovation. Design and innovation can be seen as creative alternatives or supplements to the (functionalist) operations management discourse.

Service design is another emerging area within design, both in practice and in design management research. Morelli (2002), for example, states that designers should not be focused on products, but rather on the product-service-system and its design from a user value perspective. The service design discourse thereby is expanded and transformed from the more traditional 'design of products' to include management-related research for improved cooperation with the customer in the 'fuzzy front end' of product development. This is currently an active area of research that is still expanding (Heskett 1986). This area unites the functionalist and radical humanist quadrants of the paradigms.

## HOW DOES DESIGN MANAGEMENT
## FIT INTO THE OTHER DISCOURSES?

The paradigmatic base of the design management discourse is represented in Figure 28.1 by the letters DM. Most design management research exists within the functionalist paradigm, relying on objectivity and regulation as foundational assumptions. It is scarcely represented in the radical humanist paradigm, where design has its base. This is surprising, because if design management is an intersection between design and management research, one could expect the paradigmatic platform to have its anchor in the paradigm where the two of them overlap and thereby find some paradigmatic coherence. From a design perspective this is somewhat problematic, because design thinking as such is differentiated from rational, analytical processes, rather being defined as a holistic way of creating something new and unanticipated. Instead, design management has a comfortable place within management research, relying on mainstream, functionalist management literature and gurus such as Porter (1980, 1985) and Kotler (Kotler and Rath 1984).

# CRITICAL REFLECTIONS ON THE PARADIGMATIC PARTNERSHIP BETWEEN DESIGN AND MANAGEMENT

A good partnership within an interdisciplinary area like design management requires a common paradigmatic ground that embraces the character of each discipline. Such a simple statement is obvious from a Scandinavian perspective, where paradigmatic consistency is both valued and required. Yet, as we have shown in our analysis above, this statement is problematic because of the very nature of the paradigmatic inconsistencies within the contributing disciplines.

Management research started in the functionalist quadrant, with assumptions of objectivism and regulation of the existing society, then spread out to include work in all four quadrants, but mainstream management research still has its centre of gravity in the functionalist paradigm. Leading Scandinavian management researchers do not adhere to this neo-positivist paradigm, but instead criticize the status quo and point to the many dimensions that are omitted.

Design research, although spread all over the four paradigms, has its centre of gravity in the radical humanist paradigm, where the assumptions are a combination of subjectivity and radical change. This situation reflects a change from early design research, when objectivity was taken for granted as the norm for 'real' research, and there was functionalist research. Scandinavian design research is clearly centred within the radical humanist paradigm.

Design management research does not spread over the different paradigms (with some few exceptions). Most design management research relies on the functionalist paradigm. This paradigmatic inconsistency is problematic from many points of view. However, in Scandinavia, both the interpretative paradigm, represented by the work of Svengren (1995), Thornquist (2005), Edeholt (2004) and Johansson (2006a) in Sweden, and Korvenpää and Karjalainen in Finland, as well as the critical paradigm, represented in the works of Johansson and Woodilla (2008), have relatively strong voices.

Figure 28.2 illustrates the way in which the research areas of management, design, and design management overlap. The boundaries are drawn schematically to demonstrate the relative extent of each area rather than to place coordinates exactly on the axes.

The location of the area of overlap among the disciplines is probably not a surprise to those involved in design management. However, it is worth noting because we might have hoped for a better paradigmatic alignment. Instead, given the Scandinavian attention to epistemology, this is a good place from which to interpret the discrepancy between the centre of gravity for design research and design management research.

One reason for design management research adhering to mainstream management could be that the former is a new research area that seeks legitimacy among both academics and practitioners, and therefore originates in the dominant mainstream area.

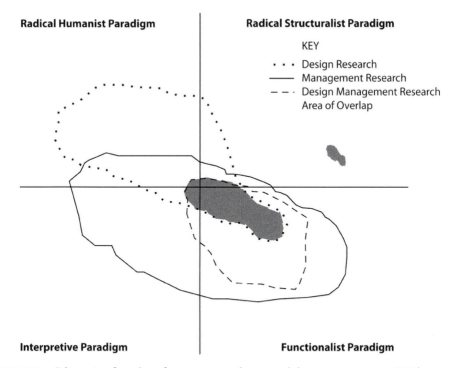

**Radical Humanist Paradigm**

**Radical Structuralist Paradigm**

KEY

· · · Design Research
—— Management Research
– – · Design Management Research
Area of Overlap

**Interpretive Paradigm**

**Functionalist Paradigm**

FIGURE 28.2 Schematic of overlap of management, design, and design management. © Johansson and Woodilla, 2011.

Or, as some have said, it is an advantage to 'go with the theories that people from top management already know' and introduce design from that perspective. Despite this, we believe it is just as important – and even required from a Scandinavian perspective – for a research area to encompass different paradigms so that the range of underlying assumptions becomes broader and deeper. In particular, a research area that does not include critical and reflexive research is in danger of being too shallow. Research in the design management area has developed to a point where it is time to look for a paradigmatic broadening. Richer methodological and epistemological debates within design management research would contribute to its intellectual development.

Our second reflection is that because the research areas of 'design' and 'management' have different paradigmatic centres, it is important for 'design management' to have a paradigmatic awareness in addition to a broad paradigmatic base for the research. A functionalist centre of gravity makes it difficult or even impossible to embrace the paradoxical and ambiguous aspects of praxis-based design knowledge. The irony might be that design management proclaims design, but in such a way that the design characteristics cannot be seen! Design management researchers need to be aware of the way in which they embrace approaches based on different underlying assumptions. There has been a lively debate among management scholars as to whether the various paradigms are incommensurable – that is, whether it is even possible for researchers based in different frameworks to work together (cf. Gioia and Pitre 1990). Rather than engage in this debate, we suggest that design management researchers consider deliberately espousing an alternative paradigm by working with research methodologies from frameworks other than functionalism.

As an initial approach to broadening the research base, and drawing from the Scandinavian experience, we suggest the following possibilities for the three underrepresented paradigms.

### Radical Humanist Paradigm

To date, most case studies in design management have had the nature of 'analysing success stories' (for example, Svengren, 1995; Johansson and Svengren 2008). Alternative methodologies could draw from action research (Baburoglu and Ravn 1992; Whyte 1991), collaborative research (Adler et al. 2004) and other approaches where the researcher takes a more active experimental role, uncovering the tensions and synergies in the relation between design and management in a particular situation. These alternative approaches mean relating more to the 'organizational development' discourse than to the 'organizational strategy' discourse.

In Scandinavia, design research is anchored in the radical humanist paradigm. Established professors in interaction design (for example, Ehn 2006) and in industrial architecture and engineering design (for example, Törnqvist and Ullmark 1989) conduct management and innovation research such as the Swedish Fenix project (a doctoral programme for designers working in industry sponsored jointly by Chalmers University of Technology in Gothenburg and The Royal Institute of Technology, Stockholm). Also, current work in the radical humanist paradigm embraces values like sustainability, gender, and inclusiveness that provide a sense of self-efficacy for participants. For example, 'Design against crime' (Davey et al. 2005) and 'Doors of Perception' (Thackara 2005) are design research and development projects with relevance to design management. They could therefore be used to inspire future design management research. Examples of Scandinavian initiatives along this vein include research for over twenty years within interaction design (Bødker et al. 2000; Koskinen 2003) as well as recently initiated, open-ended design management projects in Gothenburg (Jahnke 2009; Wetter Edman 2009) that try to find paths within the radical humanist rather than the functionalist paradigm.

### Critical Management

Critical management studies already have an established base within management research with a division within the US Academy of Management and a biannual European Critical Management Conference (www.criticalmanagement.org/). Alvesson, Scandinavia's most published management professor claims a critical management

perspective (Alvesson and Willmott 1992) and a number of the recent design management doctoral dissertations have taken a critical stance (for example, Thornquist
2005). Opportunities therefore exist for design researchers to engage with critical
management researchers to develop dynamic and fresh perspectives.

### Interpretative and Reflexive Methodology

There is also room for more reflexive research: reflexive around the practical encounter
of design and management, and reflexive around the paradigmatic differences and the
communication problems that arise. What really happens when design thinking meets
management thinking? In order to answer this question – which can be answered in
several ways – both empirical material and associated reflections are needed (Alvesson
and Sköldberg 2000), and the nature of the reflections would require careful attention.
Examples include ethnography that includes researcher reflections similar to Kunda's
(1992) study of the culture of a multinational company, but instead focusing on designers
and what is happening when they enter the world of business, and the work of Edeholt
(2006) and Buchanan (2001) where they reflect about the profession of design, but such
reflections would be about the design management discourse.

## SUMMARY

In this chapter we have discussed design management in Scandinavia and, taking a
critical Scandinavian management point of view, we have scrutinized the design management platform with the help of Burrell and Morgan's paradigms as an analytical
tool. As a result, we have two main observations.

First, there is a paradigmatic inconsistency in how design and management have
merged within the field of design management. More specifically, it is the dominant
management research from the functionalist paradigm that has merged with the
dominant design research from a radical humanist paradigm. Even though we understand how this match came about, we find it problematic for at least two reasons.
First, unintentional paradigmatic inconsistency is always problematic for academic
quality and, second, this specific partnership of design and management tends to strip
the design characteristics and values from the partnership in favour of the goal-driven
management agenda.

Second, we can see some hope despite this problematic situation. Similar to the
management area that, after Burrell and Morgan made the notion and outcomes of different paradigms explicit, has become much more diversified and richer paradigmatic-
wise, we anticipate a similar future for design management. We look forward to seeing
more interpretative studies, as well as action research and critical design management
studies. From our Scandinavian review we see signs of this happening over the last few

years. We therefore hope that Scandinavian design management will be at the forefront of continuous expansion and reflection on the paradigmatic partnerships of interdisciplinary research.

## NOTE

1. The authors gratefully acknowledge Dr Lisbeth Svengren Holms's assistance with this section.

## REFERENCES

Adler, N., Shani, A and Styhre, A. (2004) *Collaborative Research in Organizations*. Thousand Oaks, CA: Sage Publications.

Alexander, C. (1964) *Notes on the Synthesis of Form*. Cambridge, MA: Harvard University Press.

Alvesson, M. and Sköldberg, K. (2000) *Reflexive Methodology*. London: Sage Publications.

Alvesson, M. and Wilmott, H. (1992) *Critical Management Studies*. London: Sage Publications.

Baburoglu, O. and Ravn, I. (1992) 'Normative Action Research', *Organization Studies* 13(1): 19–34.

Beckhard, R. (1969) *Organizational Development Strategies and Models*. Reading, MA: Addison-Wesley.

Björgvinsson, B. (2007) *Socio-Material Mediations. Learning, Knowing and Self-produced Media within Healthcare*. Karlskrona: Blekinge Institute of Technology.

Bødker, S., Ehn, P., Sjögren, D. and Sundblad, Y. (2000) *Co-operative Design – Perspectives on Twenty Years with 'The Scandinavian IT Design Model,* http://cid.nada.kth.se/pdf/cid_104. pdf (accessed June 2010).

Borja de Mozota, B. (2003) *Design Management. Using Design to Build Brand Value and Corporate Innovation*. New York and Boston: Allworth Press and Design Management Institute.

Brezet, H. and van Hemel, C. (eds) (1997) *Ecodesign: a Promising Approach to Sustainable Production and Consumption*. The Hague: United Nations Publications.

Bruce, M. and Cooper, R. (1997) *Marketing and Design Management*. New York: Thompson Business Press.

Brunsson, N. (2000) *The Irrational Organization*. Bergen: Fagbokforlag.

Buchanan, R. (1992) 'Wicked Problems in Design Thinking', *Design Issues* 8: 5–21.

Buchanan, R. (2001) 'Design Research and the New Learning', *Design Issues* 17: 3–23.

Burrell, G. and Morgan, G. (1979) *Sociological Paradigms and Organisational Analysis*. London: Heinemann Educational Press.

Clark, T and Corlett, E. (1984) *The Ergonomics of Workspaces and Machines*. London: Taylor & Francis.

Clarkson, J. (2003) *Inclusive Design*. London: Springer.

Cooper, R. and Press, M. (1995) *The Design Agenda*. Chichester: John Wiley & Sons, Ltd.

Corvellec, H. (1997) *Stories of Achievements*. New Brunswick, NJ: Transaction Publishers.

Cross, N. (1999) 'Design Research: A Disciplined Conversation', *Design Issues* 15: 5–10.

Cross, N. (2001) 'Designerly Ways of Knowing', *Design Issues* 17: 51–5.

Czarniawska, B. (1999) *Writing Management.* Oxford: Oxford University Press.

Czarniawska, B. and Sevon, G. (2003) *The Northern Lights.* Copenhagen: Liber-Abstrakt-Copenghagen Business School Press.

Czarniawska-Joerges, B. (1993) *The Three-dimensional Organization.* Lund: Studentlitteratur.

Czarniawska-Joerges, B. and Sevon, G. (1996) *Translating Organizational Change.* Berlin: De Gruyter.

Davey, C., Wooton, A., Cooper, R. and Press, M. (2005) 'Design Against Crime: Extending the Reach of Crime Prevention through Environmental Design', *Security Journal* 18: 39–51.

De Mazota, B. (2003) *Design Management. Using Design to Build Brand Value and Corporate Innovation.* New York: Allworth Press.

Digerfeldt-Månsson, T. (2009) *Formernas liv i Designföretaget: om Design och Design Management som Konst.* Stockholm: Stockholm University Press.

Dreyfuss, H. (1967) *The Measure of Man.* New York: Whitney Library of Design.

Droste, M. (1990) *Bauhaus, 1919–1933.* Koln: Taschen.

Edeholt, H. (2004) *Design Innovation och andra Paradoxer – om Förändring Satt i System.* Göteborg: Chalmers Tekniska Högskola, Innovativ design – Arkitektur.

Edeholt, H. (2006) 'Sustainable Innovation from an Art-and-Design Perspective.' The Center for Sustainable Design, Sustainable Innovation 06, Chicago, USA, 23–24 October 2006.

Edeholt, H. and Lowgren, J. (2003) 'Industrial Design in a Post-industrial Society: A Framework for Understanding the Relationship between Industrial Design and Interaction Design', The European Academy of Design, Fifth European Academy of Design Conference, University of Barcelona, Spain, 28–30 April 2003.

Ehn, P. (2006) 'Participation in Interaction Design: Actors and Artifacts in Interaction', in S. Bagnara and G. Crampton Smith (eds) *Interaction Design.* Mahwah, NJ: Lawrence Erlbaum, pp. 137–55.

Emery, F. and Trist, E. (1965) 'The Casual Texture of Organizational Environments', *Human Relations* 18: 21–32.

Gioia, D. and Pitrie, E. (1980) Multiparadigm perspectives on theory building. *Academy of Management Review* 15(4): 584–602.

Hannan, M. and Freeman, J. (1984) 'Structural Inertia and Organizational Change', *American Sociological Review* 49: 149–64.

Hatch, M. (1997) *Organization Theory.* New York: The Free Press.

Heskett, J. (1986) *Managing in the Service Economy.* Boston, MA: Harvard Business School Press.

Hofstede, G. (1980) *Culture's Consequences.* Beverly Hills, CA: Sage Publications.

Jahnke, M. (2009) 'Design Thinking as Enabler of Innovation in Engineering Organizations', The European Academy of Design, Design Connexity: The Eighth European Academy of Design Conference, The Robert Gordon University, Aberdeen, Scotland, 3 March–3 April 2009.

Jevnaker, B. (1993) 'Inaugurative Learning: Adapting a New Design Approach', *Design Studies* 14(4): 379–401.

Jevnaker, B. (2005) 'Vita Activa: On Relationships between Design(ers) and Business', *Design Issues* 21(3): 25–48.

Johansson, U. (2006a) *Design som Utvecklingskraft. En Utvärdering av Regeringens Designsatsning 2003–2005*. Växjö: Växjö University Press.

Johansson, U. (Ed.). (2006b) *Design som Utvecklingskraft II. Fem Uppsatser om Utvalda Projekt från Regeringens Designsatsning 2003–2005*. Växjö: Växjö University Press.

Johansson, U. and Svengren, L. (2005) 'Design and Branding – a Nice Couple or False Friends?', in J. E. Schroeder and M. Salzer-Mörling (eds) *Brand Culture*. London: Routledge, pp. 122–36.

Johansson, U. and Svengren, L. (2008) *Möten Kring Design. Om Relationerna Mellan Designer, Tekniker och Ekonomer i fem Svenska Företag*. Lund: Studentlitteratur.

Johansson, U. and Woodilla, J. (eds) (2005) *Irony and Organizations: Episemological Claims and Supporting Field Stories*. Copenhagen: Copenhagen Business School Press.

Johansson, U. and Woodilla, J. (2008) 'Towards a Better Paradigmatic Partnership between Design and Management', Design Management Institute, DMI Education Conference, Cery-Pointoise, France, 14–15 April 2008.

Jönsson, S. (ed.) (1996) *Perspectives of Scandinavian Management*. Gothenburg: Bokförlaget BAS.

Karjalainen, T-M. (2004) *Semantic Transformation in Design*. Helsinki: University of Art and Design.

Katz, D. and Kahn, R. (1978) *The Social Psychology of Organizations*. 2nd edn. New York: John Wiley & Sons, Ltd.

Kelley, T. (2001) *The Art of Innovation*. New York: Doubleday.

Koskinen, I. (2003) 'User-Generated Content in Mobile Multimedia: Empirical Evidence from User Studies', *International Conference on Multimedia and Expo* 2: 645–6.

Koskinen, I., Battarbee, K. and Mattelmäki, T. (2003) *Empathic Design*. Cambridge, MA: MIT Press.

Kotler, P. and Rath, G. (1984) 'Design – A Powerful but Neglected Strategic Tool', *Journal of Business Strategy* 5(2): 16–21.

Krippendorff, K. (1989) 'On the Essential Contexts of Artifacts or on the Proposition that Design is Making Sense (of Things)', *Design Issues* 5: 9–36.

Kristensen T. and Lojacono G. (2002) 'Commissioning Design: Evidence from the Furniture Industry', *Technology Analysis and Strategic Management* 14(1): 107–21.

Kunda, G. (1992) *Engineering Culture*. Philadelphia, PA: Temple University Press.

Lawson, B. (2005) *How Designers Think*. 3rd edn. Oxford: Architectural Press.

Levitt, B. and March, J. (1988) 'Organizational Learning', *Annual Review of Sociology* 14: 319–38.

Manzini, E. and Jégou, F. (2003) *Sustainable Everyday*. Milan: Ambiente

Margolin, V. (2002) *The Politics of the Artificial*. Chicago: University of Chicago Press.

Monö, R. (1997) *Design for Product Understanding*. Stockholm: Liber AB.

Morelli, N. (2002*)* 'Designing Product/Service Systems: A Methodological Exploration', *Design Issues* 18: 3–17.

Morgan, G. (1986) *Images of Organization*. Beverly Hills, CA: Sage Publications.

Olins, W. (1989) *Corporate Identity*. London: Thames & Hudson.

Olsen, J. (2007) 'Understanding Institutions and Logics of Appropriateness: Introductory Essay 1', Working Paper No. 13, Arena Centre for European Studies, University of Olso. Available online: www.arena.uio.no/publications/working-papers2007 (accessed 9 June 2010).

Olson, E., M., Cooper, R. and Slater, S. (1998) 'Design Strategy and Competitive Advantage', *Business Horizons* 41: 55–61.

Porter, M. (1980) *Competitive Strategy*. New York and London: Free Press.

Porter, M. (1985) *Competitive Advantages*. New York and London: Free Press.

Powell, W. and DiMaggio, P. (eds) (1992) *The New Institutionalism in Organizational Analysis*. Chicago: University of Chicago Press.

Prahalad, C. and Hamel, G. (1994) 'Strategy as a Field of Study: Why Search for a New Paradigm?', *Strategic Management Journal* 15: 5–16.

Roethlisberger, F. and Dixon, W. (1961) *Management and the Worker*. Cambridge, MA: Harvard University Press.

Røvik, K-A. (2000) *Moderna Organisationer. Trender inom Organisationstänkandet vid Millennieskiftet*. Malmo: Liber.

Schein, E. (1985) *Organizational Culture and Leadership*. San Francisco, CA: Jossey-Bass.

Schon, D. (1983) *The Reflective Practitioner*. New York: Basic Books.

Simon, H. ([1969] 1981) *The Sciences of the Artificial*. Cambridge, MA: MIT Press.

Stamm, B. (2003) *Managing Innovation, Design and Creativity*. Chichester: John Wiley & Sons.

Svengren, L. (1995) *Industriell Design som Strategisk Resurs: en Studie av Designprocessens Metoder och Synsätt del i Företags Strategiska Utveckling*. Lund: Lund Institute of Technology.

Taylor, F. (1911) *Principles of Scientific Management*. New York: Norton.

Thackara, J. (2005) *In the Bubble*. Cambridge, MA: MIT Press.

Thornquist, C. (2005) *The Savage and the Designer*. Stockholm: Stockholm University Press.

Thorpe, A. (2007) *The Designer's Atlas of Sustainability*, Washington, DC: Island Press.

Törnqvist, A. and Ullmark, P. (1989) *When People Matter*. Stockholm: Swedish Council for Building Research.

Van Maanen, J. (1988) *Tales of the Field*. Chicago, IL: University of Chicago Press.

Verganti, R. (2006) 'Innovating Through Design', *Harvard Business Review* 84: 114–22.

Vihma, S. (1995) *Products as Representations*. Helsinki: University of Art and Design.

von Hippel, E. (1978) 'Successful Industrial Products from Customer Ideas', *Journal of Marketing* 42: 39–49.

Wetter Edman, K. (2009) 'Design Methods for Improved Service Innovation', Nordic Design Research, NORDES 2009 Doctoral Consortium, Oslo School of Architecture and Design, Oslo, 30 August 2009.

Whyte, W. F. (1991) *Participatory Action Research*. Newbury Park, CA: Sage Publications.

# 29

# Public Policy and Public Management: Contextualizing Service Design in the Public Sector

SABINE JUNGINGER AND DANIELA SANGIORGI

The only sense in which social is an honorific term is that in which the medium in which human living goes on is one in which human living is enhanced.

*Dewey, 'Common Sense and Science'*

The future of the public sector is imbued with tremendous challenges for governments across the globe. Even without the additional burden of a financial crisis and its costly bailout to taxpayers, which, as a consequence, eroded much cash available for public services, the efficient, effective and just management of common goods remains one of the core problems of human societies. It has been shown by now that there is a place for design practices, design thinking and design methods in the public sector. In this context, service design has attracted much attention and current explorations by service designers into public services are encouraging. Our chapter calls for a reflection on the design work that has been undertaken so far by service designers in the public realm. We wonder, if service designers are equipped to deal with the larger organizational and policy contexts they are entering now. Looking at the service design projects so far, we argue that many public managers continue to look to design as a tool for implementing existing policies, even if these policies pose problems for the kinds of services that can be designed. This, we suggest, poses a dilemma for service design: while service designers may improve the experiences of individual interactions, they may unwittingly develop and deliver services that either cannot deliver desired social outcomes or, worse, facilitate the implementation of policies that in the larger scheme of things, for example in regards to social wellbeing and social cohesion, may actually do a disservice to particular groups of people. This also might limit the abilities of service design to transform public services, one of the very goals it is currently seeking to achieve.

## BACKGROUND

Of all organizations, government agencies are often perceived to be the least likely places for design activities. Most of these agencies represent enormous bureaucratic structures that have grown over decades, if not centuries. Government agencies tend to be environments that show less flexibility and more ingrained thinking and behaviour than might be the case in a private business. We envision grey and sterile corridors, old and nondescript chairs and outdated computer software. These organizations deal with highly complex subject matters – federal laws and policies that have grown over time into enormous webs of rules and regulations. Power and hierarchy matter as much as political philosophies and economic circumstances. More often than not, government organizations have structured their operations around procedural and legal demands rather than on the needs of the people they serve. This situation can pose a problem in the administration and delivery of services. In the larger scheme of things, government agencies, although not counted as part of the 'service industry', develop and deliver a vast number of products and services to residents and citizens.

The way that most everyday people come in contact with government organizations is through public services like healthcare (hospitals and doctors) welfare (offices and case workers) or taxation (tax forms and letters). These services and the tangible products that facilitate these services give form and expression to the policies in place. Moreover, they turn the 'thoughts' that go into the making of a policy into 'action'. As a former Australian tax official observed not long ago:

> They [the users] do not readily distinguish between the policy settings of the system (the determination of what they must do and what they are entitled to expect), the legislation that gives legal support and expression to those policy settings, the administrative arrangements set up for the system, the brochures and guides that explain the system, the forms that need to be filled in to access the system or the notices they receive and to which they may need to respond. (Robinson 2001)

A recent report for the UK Cabinet Office by the Sunningdale Institute (Adebowale et al. 2009) highlights the need for a design approach that reconnects the ways in which policies are being planned and developed with the situation and experiences of the people who execute and administer these policies. In recognition of this need, the Australian Tax Office has invested in developing an integrated tax design approach, which aligns the design process from policy making to implementation. In this process, the tax office discovered a need for internal design capabilities. Others, like the United States Postal Service or the National Health System in the United Kingdom have or are employing design thinking and design methods to improve the 'client' experience.[1] These projects' aims extend from the development of services and products that are accessible, usable and useful to those people they are intended for, to the shifting organizational values and practices through scenario developments, prototyping and user research. Of particular interest in light of recent global problems is the role of design in creating sustainable organizations, in ensuring human dignity and in advancing human rights (managerial and organizational ethics). Academic workshops on these topics are being held at major management and design conferences – for example by the American Academy of Management and the European Group of Organization Studies. In the private sector, too, there is a growing movement around the ideas and the practices of Design for Social Business (cf. D4SB, Milan 2010). Design for Social Business started with the idea that businesses can be re-oriented around social purpose, in particular to reduce poverty and to enable and empower people. A social business model enables the people it employs to earn well and it reinvests any profit it makes into another social business.

These parallel developments in the private and public sectors fit with the increasing awareness of human-centred design in business and in management. Human-centred design as it takes hold in management now is influenced by the philosophies of American Pragmatist John Dewey, particularly his two works *Art as Experience* (1934) and *Logic – The Theory of Inquiry* (1938) and by the work of Richard McKeon, who studied under Dewey and had a role in the founding of UNESCO (cf. McKeon and Rokkan 1969). McKeon, who saw the role of the UNESCO as being to ensure human rights and human dignity, reflects in 'Philosophy and History in the Development of Human Rights' (McKeon 1970) on the challenges involved in bridging the gap from 'thought to action'. The gap from thought to action remains one of the key problems for public institutions and it aptly describes a core design problem for government agencies. With the rise of service design in the public sector, this design problem is now a concern for service designers. The problem, however, presents itself in reverse: how do we get from action (i.e., the implementation of services) to thought (i.e., to ask why and what particular services are meant to achieve). We argue that it is crucial for service designers to understand the intent behind the policies they are implementing through newly tailored services. How are the services they develop linked with these policies? What impact do service designers have in shaping these policies? What roles are service designers assuming in these organizational and political contexts? How transposable are the concepts, practices and methods of service design to the public sector? To understand where we see service design at the moment, we begin with a reflection of how service design emerged as a practice and field.

## INFLUENCES THAT LED TO THE
## EMERGENCE OF SERVICE DESIGN

Service design, as a distinct discipline originated in the 1990s when a group of in-formed thinkers (notably Hollins and Hollins 1991; Morello 1991; Manzini 1993; Erlhoff et al. 1997; Pacenti 1998) started to perceive and describe it as a new de-sign agenda. The rapid growth of a tertiary economy sector in many Western coun-tries posed a clear departure from traditional design practices and the design cultures, which had developed around physical and tangible outputs of the secondary sector, tied to manufacturing.

Eyed with suspicion by traditional designers and organizations alike, service de-sign pioneers sought to demonstrate their relevance through specific contributions in the area of intervention. The theoretical argument was advanced when services were understood to be 'products' in their own rights (Hollins and Hollins 1991; Erl-hoff et al. 1997) and thereby qualified as 'objects' of design to which design methods could be applied. Hollins and Hollins (1991) and Mager (2004) built the bridge to design management when they suggested that services should be designed with the same attention to processes as was common in 'product development'. Subsequently, service design embraced theories and practices of interaction. Elena Pacenti's doc-toral dissertation (1998) offers a formal explanation of the shift from the interpre-tation of services as complex organizations to one of services as complex interfaces to the user. She defined service design as the design of the 'area, ambit and scene where the interactions between the service and the user take place', opening up a liaison with the schools, research and methodology of interaction design. In the US, human-centred interaction design early on pushed for a broadening of the limiting notion of interaction design that marked early digital human–computer interaction research.

The analogy between designing 'interactions' (user–device interface) and design-ing 'service interactions' or 'service encounters' (user–service interface) is at the core of service design's identity and practice. What has been gradually changing in the last decade is the context and nature of the interactions that service design has been dealing with (Sangiorgi 2009): from one-to-one to many-to-many interactions; from sequential to open-ended interactions (Winhall 2004); from within to amongst or-ganizations. While 'scaling up', service design is also 'reaching out' and 'deepening in'; this means that when both the complexity of challenges and the objects of design become larger, design needs to collaborate with a wider number of stakeholders and professions, but also to work 'within' service organizations and user communities to provide tools and methods to deal with change and complexity on a daily basis. Buchanan (1998) talks about 'third-order' and 'fourth-order' design as a way to rep-resent this recent need to move where strategic decisions are made in order to be in a position to influence future directions. Buchanan has put forth the idea that there exist four different orders of design: the first order encompasses the design of symbols and graphic communication; the second order refers to the design and construction of things and objects; the third order concerns itself with designing interactions and strategic planning whereas the fourth order focuses on the design of environments and systems. The fourth order of design implies the understanding, analysis and design of complex systems that require both an analytical capacity to evaluate their parts and reciprocal interactions and relationships but also an integrative understanding of the ideas, thoughts and values that drive their unity and functioning as a whole (cf. Buchanan 1992; Golsby-Smith 1996). In an organization, fourth-order design problems concern visions, values and strategies. In a complex social or public system, for example, a public health care system, policies determine and influence how the different parts and organizational elements can relate to each other and interact together. We find the four orders useful to examine the changing role of service design in the public sector.

## SERVICE DESIGN AS A SECOND-, THIRD- AND FOURTH-ORDER DESIGN ACTIVITY

Organizations are places with a wide range of design problems. They cannot ignore aspects of communication and therewith have to address first-order design problems. They cannot ignore tangible products that are indispensible for internal and external purposes. This means they have to concern themselves with second-order design problems. But foremost organizations have to understand the kinds of interactions that they want to have with their customers and design appropriate services.

If we were to take inventory of the service design projects undertaken in the public sector so far, we would find a wide range of second- and third- order design problems: the main focus of these projects has rested on extensions of the product (by creating 'touch points') and on service interfaces that looked into how people interact with and experience these service products (exemplified by 'moments of truth' and 'service encounters'). There exists, now, an array of models, toolkits and frameworks, all of which situate design as a resource within existing business strategies and visions. At the same time, service design practitioners increasingly face fourth-order design issues as they enter organizational systems and find themselves confronted with fundamental assumptions, values and beliefs held by people in these complex environments.[2] For example, service designers have begun to discuss their role as drivers for territorial transformations.[3] Via scenario-building activities and service design processes, designers are increasingly working to redefine the ideas, thoughts and values that shape territories identity and development. This all suggest a need to shift from third-order issues to fourth-order design as an area of practice.

As we have argued elsewhere, services are always deeply embedded in organizational systems and inherently possess great potential to become vehicles for organizational change (Junginger and Sangiorgi 2009). The observation by Yoo et al. (2010) that 'a shift in the architecture of a product causes shifts in the organizing logic of a firm' also holds for the redesign and the development of services in public organizations. The problems in public management and public policy that manifest themselves in public services and products present an opportunity and a necessity for service design to embrace questions of organizational change – that is, to embrace fourth-order design problems. The shift is not radical but pointed. It means that we have to look beyond designing services, to look beyond instances of interfaces or touch points or the moment of truth as someone using a service is experiencing it. Instead, we have to begin to look at the overall design or framework of the policies that enable or deter the kinds of services that are being offered. Looking beyond service design means inquiring if the policies that are in place are useful and desirable in the sense that they are suitable for developing desirable services that improve the life of people and achieve the policy's intent. Does it make sense to design specific services, however usable and desirable or useful they may be in themselves, when the actual policy turns out to be flawed? Can service design go beyond 'informing' policies?[4] Or do we need to focus on bringing in the general principles and methods of design to policy-making in order to achieve transformative changes?

It is not surprising that service design has been invited into the public sector at a time when new public management (NPM) was calling for private business practices and methods to be applied to the management of public resources and affairs. The theories and methods of NPM aim to improve the ways in which public organizations administer, develop and deliver public services. Yet, this can be done with two very different objectives in mind. Pollitt and Bouckaert (2004) distinguish between 'NPM Marketizers' and 'NPM Modernizers'. Marketizers 'see a large role for private sector forms and techniques in the process of restructuring the public sector' whereas Modernizers 'place greater emphasis on the state as *the* irreplaceable integrative force in society, with a legal personality and operative value system that cannot be reduced to the private sector discourse of efficiency, competitiveness and consumer satisfaction' (Pollitt and Bouckaert 2004: 98). The Marketizer and the Modernizer represent two strikingly different philosophical positions. Service design can have a role in both. In fact, a fourth-order approach to service design satisfies the demands of Marketizers and Modernizers without losing focus of human rights or human dignity and has an eye for social wellbeing.

We argue that for design to have a role in public institutions and in public life, we need to have a better understanding of the connections between service design (which focuses on interacting and experiencing elements of a system), public policy (which concerns the thoughts and values that shape a system) and public management (which involves changing and organizing by implementing policies through products, processes and procedures). To reach this understanding, Service Design needs to develop a reflexive capability.

## THREE EXAMPLES

To illustrate the links between service design, public policy and public management, we present three service design situations from three real public organizations. We discuss each of these 'service design' situations to show the links between people's interpretation of this service, the design methods and approaches involved and the policies that triggered the need for the specific service. We then show how each of these situations fit within third- and fourth-order design.

### Integrating Services for People in a Human Welfare Department

Different human welfare offices and case workers often provide different services to the same person. Because these services are provided by different offices and administered by different case workers, people receiving these services have to spend a significant amount of time and cost to get to their various appointments. We can think of a mother who is out of work and receives unemployment benefits. If she is raising her children on her own – for example, because their father is absent – her children might get some support from another welfare programme. Dealing with these different programmes and offices, she may find herself on the bus for three different days to visit these different human welfare offices and to receive the support she needs and is entitled to. Does this make sense? And does this help get this woman back into a job? An American human welfare department thought not. Rather than having this woman and other people they serve spend their days and their money on different costly bus trips to various agencies, they thought to improve on their service delivery by integrating them. Someone receiving various services would then be able to get everything done 'in one stop.' The head of the agency brought in a well-known management consulting firm, which persuaded him that the problem of integration was one of centralizing data and of making all data about one client available to all case workers. The consulting firm set up an intranet web portal and a year later, it still did not work. Instead, one of the large rooms in the human welfare was reserved for ten MBA graduates who continued to struggle with implementing the software and with getting the interface right. There were many complaints about the web site and rather than helping to streamline services for 'clients,' the web site development had turned into a beast that devoured financial and human resources. What had gone wrong? Master students from Carnegie Mellon University who looked into this problem found that caseworkers' needs and views were ignored during the design process. In fact, unlike the agency head, the case workers felt it ethically improper or questionable that they should know everything about the person they assist, without that person having control over who gets information about her. Caseworkers thought that this approach violated both the human rights and the human dignity of the person they were assisting. In addition to these fundamental problems, which were anchored in core beliefs and values people (here the caseworkers) held, the web site also showed a lack of understanding of a range of interface and interaction design principles and methods. Of course, it was these secondary problems – the issues of visual presentation, verbal choices and interaction tools (pull-down menu or not?) that caught the attention of the designers who worked for the consulting firm. These things they could test and generate data on. Their honest aim was to deliver a tool that would be useful, usable and desirable to achieve the goals the department head had in mind. And yet, their efforts, even if they were to achieve these initial aims, would not have produced the outcome the agency was looking for. They had focused on second- and third-order design questions, ignoring the fourth order issues that involved fundamental assumptions, values and beliefs held by people.

### Redesigning Interactions and Experiences of Jobseekers

Employment is one of the key issues in national governments. The more people are employed, the lower are the demands on social services. People in employment tend to be more satisfied with their lives and also have a greater sense of social wellbeing as they feel more integrated into society. The role of national employment agencies is to keep people employed, help those who face redundancy and unemployment to shift gears and find new jobs and to offer training and support to people who are already out of a job and seek to enter the job market. A design research project in a German national employment agency revealed the role of service design in this context. Observing a set of consultation talks between job seekers and job consultants, the following picture emerged: Many job seekers have a rich and varied background. They have, as job consultants say, a 'diverse and discontinuous job biography'. What this means is that a person might have spent three months sailing on a boat working as a waiter, then entering into an educational programme in the US, then moving back to Europe but without finding employment in his field. Instead, this person might have entered a third and maybe a fourth kind of industry. The job consultant in the agency has few means to harvest and work with this rich range of experience and skills if they lie outside of traditionally approved job paths. Another common problem is for the job agent to seek placement for someone who is qualified and trained in a traditional job but whose field is changing so radically and quickly that these traditional jobs are disappearing. We can think of the photographer who has been made redundant in her past four jobs she received through the employment agency. Now she is back again, wanting to work and, at the same time, hoping to leave her field to find a career opportunity that fits her aspirations and motivations. The information and tools at the hand of the employment agent as well as the performance targets he is supposed to meet work against her goal. In the end, the employment agent signs her up for yet another training session to learn about the latest digital photo software. She has little choice but sit through the classes, however futile she thinks this might be in addressing her employment issues and her future career. A second-order design approach does little to help this person or to fulfil the agencies goals of placing people in jobs and of keeping people employed. We can think of a range of possible interface and touch-point issues that can make the interaction between the two – the job consultant and the job seeker – more pleasant. But unless we get to the underlying intents and policies and contextualize them with public management issues, which deal with the implementation of these policies, we will not achieve innovative solutions that lead to the desired outcome for both, the job seeker and the job consultant.

### Understanding Design Practices in Primary Care in the UK

Practice-based commissioning (PBC) is a strategic framework established in England in 2003 to shift responsibility for the commissioning of healthcare services from more centralized managerial organizations, such as primary care trusts, to front-line general doctors. The argument goes that doctors in general practice have the closest contact with the public and will therefore be better able to commission appropriate, tailored and locally based services. In doing so, they can improve the effectiveness, efficiency and equity of the health services provided. Clinicians are also thought to be in a better position to deal with the complex health situation of an individual than consultants in hospitals. They can imagine services that cross current organizational boundaries and professional domains. ImaginationLancaster carried out research into PBC groups' activities and governance models as part of a wider innovation centre called HACIRIC (Healthcare and Care Infrastructure Research and Innovation Centre). This study has revealed the wide gap existing between the 'thought' driving the PBC framework and the 'action' of PBC groups trying to implement it. Most of the barriers for an effective PBC were related to structures, mechanisms and professional practices that resist and conflict with collaborative and integrated modes of commissioning and delivering healthcare services (Carr et al. 2009, 2010). These conflicts were partly inherent in professional cultures and commissioning processes and partly embedded in existing infrastructures and policies (such as payment and reward systems, professional contracts or ICT platforms) that tacitly protected the status quo. The research team worked with doctors to explore how design methods such as persona, design games and scenario-building exercises could be integrated in their commissioning processes to enhance human-centred design approaches as well as collaborative and creative design practices. At the same time this work has been necessarily integrated with general considerations and recommendations on the overall PBC framework to try to fill the gap between the policy intention and the contradictions manifested in its implementation. As designers are increasingly engaged in redesign projects in healthcare, they need to develop this higher level understanding of commissioning policies and infrastructures to better apply their change strategies, linking third-order with fourth-order design issues. This would mean a shift from informing policies to shaping policies.

## THIRD- AND FOURTH-ORDER ISSUES INVOLVE
## PARTICIPATION, FACILITATION, VISUALIZATION

There is growing recognition that services, the design of services, and the service delivery have a crucial role in public life. There is growing criticism of the ways in which new public management has been applied so far. The reasoning is that both the 'server and the served' model that reduces the interactions between people to the exchange of symbols (money) and things (goods) and the new public management model have failed to provide us with useful concepts of what a service is, could be or ought to be. Neither offers a base for sustainable and desirable public services, as neither clarifies the value of a specific service. This is one of the findings of the final report of the *Public Services 2020 Project* conducted in the UK. Referring to research by Vargo and Lusch (2004, 2008), the report points to the need to understand how value is being created and highlights the idea of co-creating relationships:

> Rather than viewing public services as though they were goods – complete 'things' that are presented to service users – services might better be seen as value propositions', where actual value is co-created in the relationship between provider and user.

The co-creation of value in the relationship between provider and user cannot take place unless both the provider and the user are ready to engage, ready to abandon existing beliefs, challenge assumptions and realize emerging ideas by developing and implementing new service models. The question, though, is: which value are we talking about and for whom?

We have seen from the three examples above that service designers, in this process, run into issues of policy making. The problem of public organizations does not begin with the administering, developing and delivering of public services. They begin with the policies that provide guidance and direction through the formulation of intent and purpose. Ian Ayres and John Braithwaite (1992: 20) argue 'sound public policy must and can speak to the diverse motivations of the regulated public'. If we return to the three examples above, what would be the intent and purpose of these three agencies? The human welfare office would seek to provide dignified services in an efficient manner to the people it serves. It would rethink itself not merely as a service dispenser but also as one of enabler. The employment agency, similarly, would not focus on dispensing and assigning but on developing new career pathways with individual job seekers. The healthcare agencies, like the welfare office, would seek to work more closely with the different people involved in the use and delivery of healthcare services.

In short, service designers would not only implement services and would not merely inform policies in retrospect – design would actually have a role in policy making.

For this to happen, service designers have to engage more deeply with organizational issues. They have to develop means to reveal and address existing institutional

values and beliefs. In short, they have to engage in matters of organizational and social change. Service designers who are not aware and not equipped to deal with more
substantial issues of public life will merely succeed in changing the expectations of
everyday people about the role of government in their lives. The bigger success and the
more promising application of service design, we find, is to point to the problems that
current policies impose on current services, to identify opportunities for policymakers
to shift their focus from one of allocating resources to one of addressing real issues of
real people. If, like in the UK, we are faced with a new proposition for a new kind of
relationship between the citizen, society and the state, what should this new relationship look like? How can service designers contribute to this? This brings us to questions of the role of public services in how people experience their societies. Here, the
issues of social coherence, social cohesion and social wellbeing enter into our minds.

Social coherence is described as 'the perception of the quality organization and
operation of the social world and it includes a concern for knowing about the world'
(Keyes 1998). Social coherence depends on people's sense of being treated fairly and
appropriately and therefore has a significant impact on how individuals perceive themselves: 'Social coherence is analogous to meaninglessness in life and involves appraisals
that society is discernible, sensible and predictable' (Seeman 1959, 1991; Mirowsky
and Ross 1989). Research further shows that 'individuals who find the social world
more unpredictable, more complex and more incoherent also tend to view their own
lives as complex affairs, full of insurmountable obstacles and unpredictable contingencies' (Keyes 1998: 130).

Is there a broader role for service design in public policy and in public management, based on how people experience services, based on how people understand their
role within their societies? In looking to the works of Erving Goffman (1959, 1961),
we can say absolutely. If, as Dewey (1948: 198) stated, 'the only sense in which social
is an honorific term is that in which the medium in which human living goes on is one
in which human living is enhanced', service designers have their work cut out.

## (PUBLIC) DESIGN MANAGEMENT, SERVICE
## DESIGN AND ORGANIZATIONAL CHANGE

So far, we have focused on the implications for service design, as more and more opportunities for design arise in the public sector. We have tried to point to the need of
service design to familiarize itself with concepts of public policy making and public
management. We have not discussed design management much, yet it is obvious that
the activities and efforts of design in the public realm require new ways of design management. Rather than remaining the silent designers (Gorb and Dumas 1987) who
produce invisible design, designers who engage in projects that include or concern issues of public policy making and public management, design management needs to
be articulate and create the spaces and environments in which such explorations can

take place. Design management has mainly focused on managing design in the context of corporate services and products.

The transformative nature of design in regard to organizational logic is an aspect that was highlighted early on by scholars like James Pilditch (1976). Yet, subsequent developments in design management have focused more on aspects of managing products and projects. With that, the management paradigm of control, contain, sustain (profit/status quo) dominated over a wide range of design activities. As a consequence, design assumed a specific role and place within the existing organizational management logic. Design turned into a resource that could be used or not, employed in instances and cut like any other resource, too. In this capacity, design was not able to bring to the fore its ability to change.

Only now do we see efforts to make the inherent change capacity of design the unifying theme of design theory and design practice. Experience design, interaction design and human-centred design have all focused on different aspects of transformation: experience design concerned the transformation of the experience a person can have with a product, interaction design pointed out how experiences influence relationships people can have, whereas human-centred design emphasized that the purpose of any meaningful change or transformation links the individual experience with collective human aims and concerns: sustainability, human dignity and human rights.

The design of everyday life has been mostly explored in the context of material culture and through the examination of physical artefacts (cf. Norman 1990; Shove et al. 2007). The invisible aspects of these designs refer to the messages they convey, the actions they encourage or deter and their implications for social status. Design research is just beginning to look into how organizations and the demands they impose shape the life of everyday people. It is here that design research can make significant contributions and apply its capacity for meaningful change. It is time to broaden our understanding of service design as part of the activities of designing, changing, organizing and managing.

## CONCLUSION

However young the discipline of service design might be, its growing involvement in the public sector situates it at an important crossroad. Rather than arguing about the kind of activity that is needed to improve public services – i.e., pushing for marketizing or pushing for modernizing – we could begin to reframe these points of contention and focus on bringing the best of each of these approaches to the fore. Rather than looking at the impossibilities and mutual exclusivities, we could employ design concepts, methods, principles and practices to give expression and form to functions, manner and materials of services and products. Service design can contribute to social coherence, social cohesion and social wellbeing but only if it embraces and understands fourth-order design issues. It is important for service design to inform policies (as illustrated by DOTT Cornwall 2010). It is even more important to find a way to bring design as a general capability into the public realm. One way to achieve this is to educate people in public organizations, bureaucrats, administrators and public managers about design. It is interesting in this context, perhaps, to note that the research into the employment agency mentioned above was carried out by public policy Master students, not by service designers.[5]

## NOTES

1. The National Health Service Institute for Innovation and Improvement (NHSi) in 2005, with the support of IDEO design consultancy, developed a model of work based on the design innovation process. To test that process, NHSi developed a pilot project in collaboration with the social design agency Think public, applying design methods focusing on patients' experiences – Experience-Based Design (Bate and Robert 2007); since then they have organized a series of training workshops and pilot projects to support the adoption of this approach on a wider scale across the NHS.
2. One example of this is the work by engine service design agency that has been gradually moving from the design of experiences, to the design of service architectures to the shape of service propositions and organizational capabilities (see: www.engine group.co.uk).
3. Examples are the ongoing project 'Feeding Milano' developed by DIS at Politecnico di Milano, aiming at reconfiguring the food system in the Agricultural Park area south of Milan or the experiences by La 27e Région in France (see www.la27ere gion.fr). European research into these new kinds of design interventions is ongoing and available at www. sustainableeverydayexplorations.net.
4. As an example, Design of the Time, a ten-year programme coordinated by the Design Council, driving design-led solutions to economic and social challenges throughout the UK, seeks to work at three main levels: Projects level, People level and Policies level (Dott Cornwall, 2010). At the policy level, their aim is to develop new insights to inform national policies.
5. Hertie School of Governance, Berlin, Master in Public Policy Programme.

## REFERENCES

Adebowale, V., Omand, D. and Starkey, K. (2009) 'Engagement and Aspiration: Reconnecting Policy Making with Front Line Professionals', Cabinet Office, 31 March, www. nationalschool.gov.uk/downloads/EngagementandAspirationReport.pdf (accessed 5 April 2011).

Ayres, I. and Braithwaite, J. (1992) *Responsive Regulation*. Oxford: Oxford University Press.

Buchanan, R. (1992) 'Wicked Problems in Design Thinking', in V. Margolin and R. Buchanan, R. (eds) *The Idea of Design: A Design Issues Reader*. Cambridge, MA: MIT Press, pp. 3–20.

Buchanan, R. (1998) 'Branzi's Dilemma – Design in Contemporary Culture', *Design Issues* 14(1): 3–20.

Carr, V., Sangiorgi, D., Büscher, M., Cooper, R. and Junginger, S. (2009) 'Clinicians as Service Designers? Reflections on Current Transformation in the UK Health Services', First Nordic Conference on Service Design and Service Innovation, Oslo, 24 November 2009.

Carr, V., Sangiorgi, D., Büscher, M., Cooper, R. and Junginger, S. (2010) 'Creating Sustainable Frameworks for Service Redesign at Practice Level in the NHS', Proceedings Haciric International Conference, Edinburgh, Scotland, 22–24 September.

Dewey, J. (1934) *Art as Experience*. London: Allen & Unwin.

Dewey, J. (1938) *Logic – The Theory of Inquiry*. New York: Henry & Holt.

Dewey, J. (1948) 'Common Sense and Science: Their Respective Frames of Reference', *The Journal of Philosophy* 45(8): 197–208.

DOTT Cornwall (2010). 'Big Society by Big Design. Working with Citizens and Communities in a Collaborative Process of Innovation and Enterprise', Dott Cornwall, Redruth, UK, http://dottcornwall.s3.amazonaws.com/big_society_by_design_b7029e33901507c0.pdf (accessed 1 November 2010).

Erlhoff, M., Mager, B. and Manzini, E. (1997) *Dienstleistung braucht Design, Professioneller Produktund Markenauftritt für Serviceanbieter*. Herausgeber: Hermann Luchterhand Verlag.

Goffman, E. (1959) *The Presentation of Self in Everyday Life*. New York: Doubleday.

Goffman, E. (1961) *Asylums: Essays on the Social Situation of Mental Patients and Other Inmates*. New York, Doubleday.

Gorb, P. and Dumas, A. (1987) 'Silent Design', *Design Studies* 8(3): 150–6.

Golsby-Smith, T. (1996) 'Fourth Order Design: A Practical Perspective', *Design Issues* 12(1): 5–25.

Hollins, G. and Hollins, B. (1991) *Total Design: Managing the Design Process in the Service Sector*. London: Pitman.

Junginger, S. and Sangiorgi, D. (2009) 'Service Design and Organisational Change. Bridging the Gap between Rigour and Relevance', IASDR09 Conference, Seoul, Korea, 19–22 October 2009.

Keyes, C. (1998) 'Social Wellbeing', *Social Psychology Quarterly* 61(2): 121–40.

Mager, B. (2004) *Service Design. A review*. Köln: Köln International School of Design.

Manzini, E. (1993) 'Il Design dei Servizi. La progettazione del prodotto-servizio', *Design Management* 4: 7–12.

McKeon, R. (1970) 'Philosophy and History in the Development of Human Rights', reprinted in McKeon, Z. (ed.) (1990) *Freedom and History and other Essays*. Chicago: University of Chicago Press, pp. 37–61.

McKeon, R. with Rokkan, S. (eds) (1969) UNESCO. *Democracy in a World of Tensions*. New York: Greenwood Press.

Mirowsky, J. and Ross, C. (1989) *Social Causes of Psychological Distress*. New York: Aldine.

Morello, A. (1991) 'Design e mercato dei prodotti e dei servizi', PhD thesis in industrial design, Politecnico di Milano, Milan, 11 December.

Norman, D. (1990) *The Design of Everyday Things*. New York: Doubleday.

Pacenti, E. (1998) 'Il progetto dell'interazione nei servizi. Un contributo al tema della progettazione dei servizi', PhD thesis in Industrial Design, Politecnico di Milano, Milan.

Pilditch, J. (1976) *Talk about Design*. London: Barrie & Jenkins.

Pollitt, C. and Bouckaert, G. (2004) *New Public Management: A Comparative Analysis*. Oxford: Oxford University Press.

Robinson, V. (2001) 'Rewriting Legislation, Australian Federal Experience', www.opc.gov.au/plain/docs.htm (accessed 2 April 2011).

Sangiorgi, D. (2009) 'Building Up a Framework for Service Design Research', Connexity, Eighth European Academy of Design Conference, Aberdeen, Scotland, 1–3 April.

Seeman, M. (1959) 'On the Meaning of Alienation', *American Sociological Review* 24: 783–91.

Seeman, M. (1991) 'Alienation and Anomie', in J. Robinson, P. Shaver and L. Wrightsman (eds) *Measures of Personality and Social Psychological Attitudes*. San Diego: Academic Press, pp. 291–74.

Shove, E., Watson, M. and Ingram, J. (2007) *The Design of Everyday Life*. Oxford: Berg.

Simon, H. (1968*) The Sciences of the Artificial*. Cambridge, MA: MIT Press.

Van de Ven, A. and Poole M. (1995) 'Explaining Development and Change in Organizations', *Academy of Management Review* 20(3): 510–40.

Vargo, S. and Lusch, R. (2004) 'Evolving to a New Dominant Logic for Marketing', *Journal of Marketing* 68(January): 1–17.

Vargo, S. and Lusch, R. (2008) 'Service-dominant logic: Continuing the Evolution', *Journal of the Academy of Marketing Science* 36(1): 1–10.

Weick, K. (1984) 'Small Wins: Redefining the Scale of Social Problems', *American Psychologist* 39(1): 40–9.

Weick, K. and Quinn, R. (1999) 'Organizational Change and Development', *Annual Review of Psychology* 50: 361–86.

Winhall, J. (2004) *Design Notes on Open Health*. London: Design Council.

Yoo, Y., Henfridsson, O. and Lyytinen, K. (2010) 'The New Organizing Logic of Digital Innovation: An Agenda for Information Systems Research', *Information Systems Research,* Twentieth Anniversary Special.

# 30

# New Design Business Models: Implications for the Future of Design Management

RACHEL COOPER, MARTYN EVANS
AND ALEX WILLIAMS

Design management in the UK has developed significantly since the early 1990s, with design managers now operating across both design consultancies and within businesses. These organizations operate in a dynamic global environment; as such it is necessary to respond to changes, especially in the business environment in which the design industry itself operates. This chapter draws on the results of a study conducted by the authors and colleagues into the future of the UK design industry, which considers the implications for design management in the design consultancy sector as well as design management in its client base.

The study described in this chapter, Design 2020, identifies challenges and opportunities that the UK design industry will face in the second decade of the twenty-first century and presents a framework to signpost and support change.[1] The remainder of this chapter explains the research context, the findings of each stage of the study, followed by discussion of the results with implications and recommendations for those involved in design management.

## WHY IS IT IMPORTANT TO STUDY THE FUTURE OF THE DESIGN CONSULTANCY INDUSTRY?

Since the early 1990s the UK's move from an industrial-based to a knowledge-based economy has been accompanied by changes in the design industry, especially the design consultancy sector. There have also been concerns within the UK design industry regarding issues such as a blurred identity for the industry, the commoditization of design, the loss of specialisms and shifting patterns of client demand. The sector has been subject to significant change, with the emergence of alternative providers of design and innovation services, diversification in the range of design services offered, greater integration of design and responsibility for implementation, the emergence of the service sector and growing consumer awareness of corporate social responsibility issues such as eco-sustainability.

This has been recognized in a number of key reports – e.g. *The Cox Review of Creativity in Business* (Cox 2005); the DTI's *Creativity, Design and Business Performance* (DTI 2005); and the Department of Culture, Media and Sport's *Staying Ahead: The Economic Performance of the UK's Creative Industries* (The Work Foundation 2007). It has also accompanied by much rhetoric yet few evidence-based propositions for the future have been put forward. Indeed research in the field is largely fragmented.

Many designers recognize that the industry is in a state of transition and that it needs to become highly adaptive to allow it to manage rapid, continuous and disruptive change. Internally, the design industry would appear to be polarizing between commoditized design and high-value strategic design. Design management as a discipline and profession needs to understand the context of this change and position itself in a responsive manner.

Most research has focused heavily on the business of design. The UK Design Council's research (Design Council 2010) represents perhaps the most comprehensive picture of the UK design industry. Whilst the Design Council has other programmes that have shed light on commissioners or buyers of design and developments in design education, there is still little objective work on design industry structure and operation. The design industry cannot be viewed in isolation; indeed industry structure and trends are influenced by a wider range of stakeholders, encompassing clients, higher education institutions (HEIs), supply industries, consumers and trade associations.

Much of the difficulty and challenge facing the sector appears to lie in the business models adopted by design consultancies and the extent to which these are recognized or valued by current and potential clients. Little knowledge exists about the business model possibilities for design consultancies compared with other professional service firms (PSFs), such as management consultancies.

## THE DYNAMICS OF THE DESIGN INDUSTRY

Current research shows a clear myopia where the design industry is viewed and investigated from the inside-out and that surveying the business of design from the standpoint of its practitioners alone provides a scenario that only perpetuates the industry's own myths and aspirations. There is a need to outline the dynamics amongst all of the various stakeholders, specifically those relating to the design industry and identify appropriate future(s) for the sector. Literature from the PSF field, in which the significant power of clients and other stakeholders over the PSFs is widely recognized, can

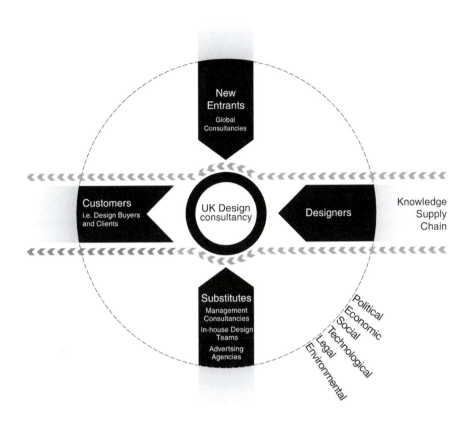

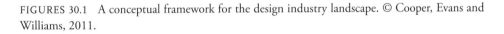

FIGURES 30.1   A conceptual framework for the design industry landscape. © Cooper, Evans and Williams, 2011.

be applied to the design industry. The use of Porter's five forces theory (Porter 1979) and a PSF conceptual framework developed by Scott (1998) provides a starting point upon which to develop a conceptual framework to identify key stakeholders mapped against their interactions with the design industry.

Applying Porter's framework to the design industry (see Figure 30.1) helps to identify key relationships within the industry, envisaging this as a knowledge economy and exploring the impact of various stakeholders on the business of design including knowledge buyers (design clients), knowledge suppliers (technology developers and consumers), alternative design providers and design associations (gatekeepers).

### Applying the Framework to the UK Design Industry

This framework was used as a basis to investigate the system in which UK design consultancies operate and draw out from the stakeholders in the system, current and future opportunities and challenges.[2]

Research was conducted in two stages. The first involved a review of literature and focus group research, which identified key issues and concerns within the sector both now and likely to emerge in the second decade of the twenty-first century and informed the development of a conceptual framework and scenario tools. A second stage involved interviews and focus groups with three sets of stakeholders: (1) design practitioners and design consultancies; (2) design buyers/clients (including both private and public sectors); and (3) design policy makers and design educators. These stakeholders were consulted in order to establish the nature of the transactions between all parties in the knowledge supply chain and sought to identify new, more appropriate models of practice relevant to each of the scenarios developed.

Focus groups were used to establish, through discussion, whether the framework and scenarios had validity and to elicit information that would enrich the model. The focus group used the framework's five dimensions to identify the following current concerns and issues:

*Design Consultancies: Competition Level.*   Many design consultancies appear not to be proficient at or interested in, business development, starting off as loose groups of colleagues and ending in the establishment of break-away businesses with the appropriation of clients. Despite their creative ability, many such businesses are not good at managing change, lacking the time or resources to devote to business development, risk management and sustainability. More often, consultancies do not possess well developed design management mechanisms.

Differentiation is dependent on the quality of the intellectual capital of the firm, embodied in its reputation and the collective ability of its senior people, a factor that is not scale specific. Small-to-medium-sized design groups appear to be more susceptible to closure, being neither big enough to attract a portfolio of new clients nor small and flexible enough to weather economic downturn. Pressure on smaller consultants is compounded by free pitching and commercial pressure to lower fees. A number of consultancies are considering whether to take ownership/equity stakes in projects, making the transition from fee to royalty-based services. However, only the larger agencies appear to have the cash flow to support the investment required.

None of these issues are new; this has been a concern of the industry for decades. However placing this within the Porter framework and comparing the profession of design consultant to that of other professional services clearly indicates weaknesses and where the industry can learn from other high fee earning industries such as management consultancy.

*Suppliers of Knowledge.*   The supply of design knowledge is an issue challenging the industry. For the suppliers of design knowledge – for example: design educators, design graduates or the lone designer – the impact of open innovation (von Hippel 2005) is yet to be seen. However, clients are increasingly aware of the power of social networks in forming and gathering opinion. In addition, there is a growing emphasis on experience and service design. Therefore, there is a recognized need to embrace more innovative means of engaging self-selecting social groups both globally and regionally – the latter is evident particularly in relation to healthcare, an ageing population and long-tail niches – and in facilitating their participation in design.

At the same time, an escalation in the rate of such technological developments reinforces the need for specialization, particularly in terms of systems, materials and applications. This results in increasing levels of contracting out to alternative providers. In addition, the respondents felt that there is an oversupply of graduates and that skill gaps between education and design practice had increased. Rewarding and retaining talented designers is increasingly difficult as the urge to work for themselves results in a high staff turnover as designers themselves gain experience and move on. Design management has an important role with regard to the development of design resource strategies and the recruitment, retention and sustainability of design and design related knowledge within organizations

*Customers: Design Buyers and Clients.*   The focus group respondents believed that it is common for design services to be seen not as knowledge providers but as other commodity suppliers. Clients tend to be unwilling to pay a premium. Even where clients understand that design adds value to a business, creativity and innovation are perceived to increase financial risk, especially in smaller businesses. At the same time, there is a reluctance for middle management (where responsibilities often lie) to make decisions on design whereas PSFs operate predominantly at board level.

Clients have the power to choose among various competing design resources, either in the UK or abroad. However, where they see design as critical to their business, there is a tendency to invest in in-house design capability instead of outsourcing, as a means of maximizing value. Moreover, clients are able to play design agencies off against each other, as in reality over capacity has lead to the development of a buyer's market. Whilst many designers believe design services can move to a more strategic level, clients tend not to share this perception, failing to view design as a business strategy and tending to bring it in to solve problems at the end of the process. Also, to date, designers have had a narrow view of what constitutes a client and tend to overlook the public sector, whereas much of the growth dynamic regionally is in the blueprinting and outsourcing of public sector services (despite cuts in public spending, this trend is likely to increase through the transfer of such operations to the third and private sectors). Design management has had a significant role in facilitating and developing relationships between organizations and their design needs; however, this is still a marginal activity in the industry. Not unless or until it is a mainstream role will it influence the customer client relationship.

*New Entrants/Alternative Service Providers.*   Much Asian design appears to be concentrated on product development; its low cost base has proved particularly attractive, as has its co-location with production. A number of Western design groups have been able to compete successfully in Asia because of their value-driven insights, simultaneously addressing Asia's growing resource issues and increasing the cultural relevance of products imported back to the West, particularly with regard to eco-sustainability issues. However, the focus groups in our study report a notable shift from 'A-Z' to 'A-G' product development or the front-end concept development, particularly within commodity product sectors, as clients have sought increasingly to commission detailing and prototyping with designers co-located at the point of production. However, diversification in design has also created niche opportunities, particularly in areas related to strategic design and design thinking, where many smaller organizations lack individual authority in this field and are subject to extensive competition from other PSFs.

*Substitutes.* The consensus is that design is essentially a cottage industry with the majority of design companies having less than five employees. Whilst the barriers for the PSF are well defined in terms of client relationship, credibility and the ability to hire talent and keep it – these are absent in the design industry and there are few perceived or indeed real barriers to entry. However, perceived value appears to be lower in design than other PSFs, resulting in few sustainable client relationships. The competition between designers is likely to be price based rather than value based. As such, many design associations are now championing the introduction of accreditation and regulation as a means of raising quality standards. These associations are, however, factional and design representation is poor with few designers on policy bodies.

In terms of the design sector, at one extreme end of the spectrum, highly consolidated segments are dominated by larger consultancies that closely track the globalization of client firms and the shift of economic activity away from mature economies. At the other, consultancies are fundamentally regional, competing to service the needs of local clients. As a professional discipline, this sector essentially provides two types of service, direct design and design management consultancy.

Looking at this sector it is clear that whilst changing business issues and opportunities facing clients have led to new management techniques and concepts, the adoption of alternative design business or client relationship models is yet to be seen.

## DESIGN 2020: THE FUTURE OF THE UK DESIGN INDUSTRY

In projecting some ten years ahead, the authors recognize that it is not possible to predict a single future and that analyses can only indicate a series of possibilities. Scenarios do, however, represent a vehicle for testing the relevance and responsiveness of current and emergent practice within a possibility space and are key to this study. The four alternative political, economic, social and technical (PESTLE) scenarios described here were developed in collaboration with cutting-edge thinkers – comprising futurologists, sustainability experts, businesses futurists, consumer experts, sociologists, technologists and designers – and build on an extensive review of national and global futures. The scenarios themselves consist of:

- *BRIC dominance* – in which Brazil, Russia, India and China maintain their trajectory as powerful economic forces within the global economy, resulting in a shift from globalization to 'glocalization'.
- *Global flow* – in which wider global agreements and connectivity emerge as a response to common crises, resulting in enhanced trading, wider cultural acceptance and open innovation.
- *Eco-imperialism* – in which the supply of energy and other essential resources assumes new significance, carbon trading is common practice and corporations seek to control the supply of resources and by extension, access to R&D.
- *Special Interest Groups* – in which the emergence of long tail economics and logistical solutions allow niche groups of consumers to become more instrumental in determining, resolving and satisfying their own requirements.

Subsequent, focus-group consultation with design practitioners and consultancies, design buyers/clients, design policy makers and design educators, using the conceptual framework as a basis for considering the four scenarios, identified the following insights into the future of the UK design industry.

### Design Practitioners and Design Consultancies

Designers believed that there would be:

- More of a need to specialize, to become fully conversant with one or two areas of importance; for example, there would be a need to be able to specify and have knowledge of new materials, particularly those that would make products more recyclable or renewable.
- Simplification of products and services to make them usable for all – all ages, all cultures – as these became accessible in the global marketplace.
- Development of products and services to meet 'global' niche markets, with more personalization to meet the cultural needs of different markets and the needs of the individual consumer.
- Longer lasting/more sustainable products via 'cradle to cradle' design, which produce less waste, reuse materials and use energy efficiently.
- Global design proliferation, which would create a highly competitive design marketplace, with more global products/services developed with a local perspective.
- Pick 'n' mix design, in which clients will be spoilt for choice and be able to choose amongst a variety of design services.

- A prevalence of localized design and production.
- A growth of in-house product and service design.
- An increase in design through consultation and collaboration.
- A greater need to take shared financial responsibility – taking a risk on new ideas, alongside clients or other collaborators.
- A movement to develop more meaningful brands/products/services would prevail.
- A prevalence of design with responsibility (and away from profligacy).

In response to these challenges the participants considered the competencies designers will need in the future. These are:

- To have a high level of knowledge and experiences to influence design decisions.
- To be effective at the management of design hubs/collaborative processes.
- To have an ability to filter large amounts of information.
- To be able to develop a two tier design structure, encompassing (i) leadership/strategic thinking and (ii) specialization and detail thinking.
- To develop a depth and quality of knowledge that has value for and is recognized by clients.
- To be thinking as well as, or instead of, jobbing designers.
- To have the ability to consult at all levels and create collaborative relationships.
- To design by looking out to the world rather than looking introspectively to the design world.
- To provide clear leadership and thinking, moving away from application and activity.
- To act as 'spongy connectors' – absorbing information and connecting clients to the right knowledge, design response and people

Policy initiatives necessary to support the development of the industry in the future include:

- The introduction of design company accreditation – providing clear standards to ensure and enhance design value.
- The introduction of national policy of grants and investments. This might include government investing positively in design, for example:
  - Finance support for design consultancy business development, particularly small consultancies.
  - Subsidies for design consultancies with international work experience.
  - Public sector commissioning and promotion of high profile design projects, to give design thinking higher profile.

- Public funding for design that has higher social value.
- Government providing designers with briefs to tackle serious design issues rather than setting up flippant design challenges with small budgets.
- Tax breaks for design R&D, not manufacturing in Asia, positive business carbon footprint and ethical production.

Clearly, there is a desire amongst designers to look at models of business that bring value and enhance their professional credibility. While representatives of design support agencies and industry associations might argue that many of the actions that are being put in place now are enabling designers to prepare for a future design industry that can participate in a rapidly changing global marketplace, it is still unclear how designers as a collective group are supported by the various industry bodies. However, it is evident that designers want one body headed by an elected design advocate who promotes the value and awareness of design. Such a body would have a lobbying role to reinforce the value of design and a responsibility to raise public awareness of design value and design training.

### Design Buyers/Clients

Using the same future business scenarios, clients considered their future needs for design, potential working patterns and what competencies they might require in designers to support them in their business within such contexts.

For clients there were some overriding themes emerging with respect to what competencies they required from designers to help them operate effectively in the future. These were:

- cultural understanding;
- willingness to collaborate;
- strong leadership;
- expertise in, for example technology, remote working, project management, designing data management systems, global warming, sustainability;
- agility;
- ability to create services, not just products;
- broad based innovation capabilities;
- biochemistry, engineering and science-knowledge;
- an ability to challenge clients and their thinking.

Expectations of the design industry in general were that:

- consultancies should remain ahead of the curve, be proactive, more visionary and lose their 'small-minded perspective'; they should open global offices or work with design satellites to bring the cultural perspective;

- designers should be less artefact focused and more able to design systems, services or experiences;
- consultancies should have the courage to confront issues of sustainability (environmental factors, etc.) and to take responsibility for the effect that design can have on the environment and society;
- consultancies should be bolder in developing uniqueness and should not try to become a homogenized oneness.

The client view of the future is a challenging one. To achieve the support that they envisage, it seems that we need a different type of education to supply the designers of the future – that designers themselves should be highly educated and more ambitious in their aspirations; that designers should be able to understand and perhaps work with clients to develop new business models and should go beyond designing artefacts. The design consultancy sector faces challenges from new entrants, such as in-house design teams and global competition as well as the emergence of the citizen designer. Whilst there are still relatively few barriers to substitutes, fragmented accreditation and professional body representation and low brand presence are set to continue.

### Design Policy Makers and Design Educators

A number of issues relevant to the design industry and practising designers were identified within the study. Firstly, 60 per cent of UK design companies comprise less than five employees; this raised questions as to whether the small design agency model is obsolete. Such micro-businesses cannot afford the time to develop their own capabilities. On the other hand the design industry could be considered analogous to the music industry, albeit needing to develop its own positive approach to high-risk, high-reward environments. It was perceived that the design industry is unable to understand its opportunities, pays lip service to collaboration and that the industry lacks designers with requisite competencies; indeed less than 15 per cent have the skills to undertake strategic thinking. There is a need for designers to move from having drafting skills to having business acumen. On the other hand this may not be possible and perhaps many designers might remain as technicians. Design managers have a role here, if only it was recognized by both the clients and their designers.

In the light of all the scenarios it was felt that new models of design business are necessary, that there should be greater focus on co-specialization but that design is poorly geared to exploit these. Indeed other businesses are more likely to seize such opportunities. Similarly the respondents in the study believed that other opportunities such as corporate social responsibility often fail to be addressed by designers.

Finally the respondents' view was that government still needs to recognize the value of design and support its use in business.

The roles of design associations and policy bodies (government/regional business support agencies) in formulating and implementing support strategies for the sector were explored through consultation with design policy makers and design educationalists. Rather than consider support for the design sector as a homogeneous whole, this study predicted continued fragmentation, the extent and shape of which is dependent on which future scenarios emerge and that policy development should be focused on supporting appropriate business models to ensure fitness for purpose.

Strategy in this sense is more about the ability for design businesses to react to rapid change and policy might best be focused on equipping them to achieve this. For nations seeking to compete, the national design policy allows a means of global positioning and branding for the local design and manufacturing industry, supporting brand growth by adding marks of good design and demonstrating national support. However, it could be argued that not every nation can join the race to position itself as the design resource for the global economy. There is a need to look inwards to see how design can empower and enrich its own economy and culture and address its internal imbalances.

Consideration needs to be given to distinguishing between what should be undertaken by a government-appointed independent body – assuming the role of government is to draw up policy and commission it – and trade associations representing the interests of the industry.

## NEW MODELS FOR THE DESIGN BUSINESS

Consultation with stakeholders across the design industry through focus groups resulted in the generation of ten new models for the design sector over the coming decade. These are described in Table 30.1.

It is recognized that the emergence of particular models is largely dependent on which of the scenarios or parts thereof, will predominate in the second decade of the twenty-first century. However, there are clear patterns likely to materialize, as illustrated in Table 30.2, which the authors will continue to monitor.

## DISCUSSION

Rather than consider support for the design sector as a homogeneous whole, this study has predicted continued fragmentation, the extent and shape of which is dependent on the future business scenarios that emerge. Research indicated that the UK design industry is likely to partition into discrete sub-sectors representative of the specific needs of the various engagement models outlined above. Indeed, since completion of the study, many of the patterns alluded to here have begun to emerge.

**TABLE 30.1: New models for the design business**

| Model | Description |
| --- | --- |
| Design strategists | Designers providing services in strategic innovation and change management. This extension into non-design and service sectors that means greater engagement with other business disciplines, probably operated by small independent, loose affiliation business models |
| Global design NGO | This is an international design NGO with a board of top design directors. The focus is a socially responsible model and supports Corporate Social Responsibility (CSR). Fees are earned from UN, DFID, MNCs with a remit to address significant problem hotspots around environmental, social, conflict-related contexts |
| IP investors/speculators | Designers invest in their own and/or others innovations using equity share models. This model exploits open innovation, works with blue chip brands, and seeks to secure venture capital |
| Mega design corps | This is a large multinational design organization serving large organizations, globally based, operating in all sectors. Offers creative and organizational development and is driven by growth and acquisition |
| Own-brand entrepreneurs | Design-led entrepreneurs who operate design-manufacturing collaborations around luxury/craft/homeware/apparel sectors. Collaboration with manufactures to develop own brands often in niche areas |
| SIG (special interest group) niche network | Exploits social media as essentially a C2B2C model. The structure involves co-design and participation between design communities and special interest groups. Designers' role is as facilitator and mediator. Fees would be based on scale of contribution and would be reliant on long tail economics, outsourcing production and distribution. High public sector engagement such as the redesign of services, empowered communities, and local authorities |
| Small independents | Micro enterprises and freelancers (1–10 employees), regionally focused, personality led, generalists with a cottage industry mentality. However, there may be high clustering and high levels of competition with consultancies continuously searching for new clients |

*(Continued)*

**TABLE 30.1: New models for the design business (Continued)**

| Model | Description |
|---|---|
| UK design centres in BRIC economies | UK design centres in each of the BRIC (Brazil, Russia, India, and China) economies, with a UK Trade and Investment (UKTI) referral mechanism for accredited UK design services. Would be able to provide specialists in cultural insight and design adaptation for EU markets |
| UK export engine | Design consultancies focus on supporting UK clients exporting overseas. Networked local design affiliates in key export markets who provide regional cultural insight. Provide and support 'glocal' design services |

**TABLE 30.2: The likelihood of business models by scenario**

| New Models for the Design Business | Scenarios | | | |
|---|---|---|---|---|
| | BRIC Dominance | Global Flow | Eco-imperialism | Special Interest Groups |
| Design strategists | | * | * | |
| Global design NGO | | | * | |
| IP investors/speculators | | * | * | |
| Mega design corps | | * | * | |
| Own-brand entrepreneurs | | * | | * |
| SIG (special interest group) niche network | | * | | * |
| Small independents | * | ** | * | ** |
| Specialized innovation services | ** | ** | ** | * |
| UK design centres in BRIC Economies | ** | * | * | |
| UK export engine | | * | ** | |

*Key*: * likely; ** highly likely

Despite the sector's aspirations, it is dependent on client demand and the need to refine and redefine its offerings in light of new client business models, configurations and locations. A key area of interest is therefore the emergence of new client sectors, such as public sector procurement, communities and special interest groups. The greatest growth is anticipated in the service and strategic design fields, although these will remain comparatively small.

Design specialization will increase, particularly within the health, corporate social responsibility and eco-sustainability fields but these are likely to become the domains of larger non-design-specific consultancies, themselves specializing in other forms of innovation. Whilst there is a role for design here – associated with design thinking – cultural dissimilarities are unlikely to result in long-term relationships and many small businesses may find themselves consumed.

Competition remains a key driver and, unlike other industry sectors, collaboration is unlikely. Although the business models proposed provide some differentiation, the different formats do impact on one another, many of those interviewed believing that freelancers and small independents, including early-stage start-ups, represent the greatest threat to the credibility of the sector. There are many calls for more professionalism, but awareness of the problems associated with accreditation, arbitration, training and the identification of professional pathways is limited and, as such, presents challenges to any initiatives to professionalize the design sector. There is a possibility that these challenges are likely to result in a two-tier design economy in which the large grow larger and the small become smaller. As these are unlikely to be serviced by one-size-fits-all policies, the various sector-facing trade associations may well become champions for discrete subsectors.

There is still debate as to whether to subsidize the design sector directly or encourage indirect uptake via funded innovation. The latter approach has not favoured design consultancies and agencies, the vast majority (more than 80 per cent) being too small to qualify for and engage in, tenders under regionally administered procurement processes. Whilst national design and innovation agendas have focused on economic development, governments themselves have often failed to engage with design as demonstrated in the lack of a public sector design procurement policy. Very often these agendas also confuse the indirect funding of demand for design through industry grants, with advocacy for and funding of the design sector.

This has led some designers, albeit those of a critical size with access to capital, to pioneer the emergence of a new category of design and intellectual property entrepreneur, based on models of shared risk and royalties. This value-added perspective is not, however, new and is evident in many developing economies where no dedicated support agencies or government-subsidized initiatives exist.

Thus our study identified that there will not be one single model for design businesses in the future and, very much like today, there will be a multilayered design industry in which a range of business models will exist. What was evident is that as client business models evolve, design will need to be both able and willing to respond to this changing environment. Moreover, whilst the design business models we identified in this study were based on the UK, they were embedded in a global context and at least half were conceived as operating in countries other than the UK.

## IMPLICATIONS FOR THE FUTURE OF
## DESIGN MANAGEMENT

This study set out to place the UK design sector as a professional service within a theoretical model based on Porter's five forces to assess its readiness for the future. In doing so we characterized the buyers and suppliers of design knowledge, skills and the competition from alternative service providers and substitutes within the model. We identified that the design sector does not exhibit some of the defence and leadership mechanisms that other PSF sectors have built up, such as the growth of the sector versus the supply of graduates or seasoned professionals, the ability to use value billing or the prevention of substitutes through the possession of propriety materials such as brand.

The research identified quality of life, sustainability and localization as dominant forces now and for the future, they saw the need for two types of designers, product designers and design facilitators, whilst design clients saw a further demand for a breadth of skills and competencies, including cultural understanding, leadership, the ability to create a service as well as products and a wider world view with greater interdisciplinarity. They believed they would require consultancies that were global in vision, less artefact focused, able to confront issues of sustainability and who were bolder in developing uniqueness.

Clients provided extensive insight into what responses each theme would mean for the industry. These included radical change to the education of designers and the challenge of new entrants and global competition as well as the emergence of the citizen designer. They predicted new forms of collaboration, global agreements, IP arrangements and broader based consultancies with multiple discipline specialisms.

In their response to the future of the design industry, designers saw more specialization, the need for experts to understand technology and materials, both globalization and personalization of product development, more emphasis on environmental issues in a highly competitive marketplace and increasing substitution from in-house design and service design. They also saw alternative business models such as shared investment shared risk in design. The designers on identifying these issues also identified a need for both design leaders and design specialists and a design industry that had a broader vision of the world.

It is clear therefore that design management as a discipline and as a profession has some challenges and opportunities:

*New Business Models; New Design Management Roles.*  The majority of our business models demand a level of competence in strategy, finance, international relations, general management and specific cross-cultural knowledge, far outweighing that which exists in most nationally based design consultancies. Obviously it is easy for design businesses to bring in a general manager, an accountant and a local specialist, but as has too often been recognized, such individuals/groups have little sympathetic understanding of the role and value of design. In addition many of the business models suggest that designers need to become 'strategic designers' in order to avoid becoming undervalued as many design activities become 'commoditized'.

Are we therefore looking for designers to turn into design managers or for a new generation of skilled design managers conversant with the global opportunities of design in business and the skills to facilitate these new business models?

*The Rise in 'Design Thinking'.*   There has been some debate around the idea of designers becoming involved very early on in the formation of an organization's strategy and that their ability to problem solve is a much needed requirement to provide organizations, NGOs and governments, with solutions to problems as far reaching as the global warming crisis, how to get the NHS working better or how to develop the next new product or service offer for commercial business. The experience of companies such as IDEO, who have been leading in this way, would indicate that only 15 per cent of their time is spent on doing this form of work (Brown 2008), although this is on the increase. Others have commented on how it was still an uphill struggle to get clients to consider design in this way, yet our overall findings indicate that design thinking and designing beyond the artefact will be a future alternative for the design sector.

*Beyond Design Management.*   If we are not careful, design is in danger of becoming ghettoized within the educational system, with graphic or product design categorized as craft-based disciplines removed from a wider intellectual sphere. Design management might also remain on the boundaries of design and management, yet there is evidence to show that design as both a process and instrument plays a significant role in addressing complex problems and as such, could also become a more integral part of other subjects, such as science or social science. This may result in the emergence of new disciplines or professions such as design psychology, design ethnography, humanitarian design or ecological design.

So perhaps we should return to examine design management within Porter's model. Design thinking may be a potential substitute; design ethnographers may constitute new entrants and knowledge suppliers are still trying to understand their role in a challenging global landscape. Whereas the new business models proposed offer opportunities for the development of design management, they also suggest a possible diversification away from it. Which future predominates is largely dependent on how the discipline chooses to define itself and how responsive it is to external market forces.

## ACKNOWLEDGEMENTS

This research was funded by the Arts and Humanities Research Council and Engineering and Physical Research Council as part of the 'Designing for the 21st Century' initiative. The project team was led by Professor Alex Williams, Principal Investigator,

University of Salford and comprised Professor Rachel Cooper and Dr Martyn Evans, Co-Investigators, Lancaster University; Nick Hall, Research Assistant, Lancaster University; Linda Hodgson (seconded to the BDI) and Dr Qian Sun, Research Assistants, University of Salford; and Maxine Horn, Collaborating Partner, British Design Innovation (BDI). The authors would like to thank all the participants and participating organizations. Special thanks to PDD and the Design Council, which hosted our workshops, future panel and focus groups.

## NOTES

1. Design 2020 was a research project undertaken by the University of Lancaster and the University of Salford, in partnership with the UK design industry body, British Design Innovation (BDI). The research was funded by the Arts and Humanities Research Council under their Design in the Twenty-First Century Initiative. The research focuses on the UK design consultancy sector, with specific reference to brand and corporate identity, multimedia, new product development, packaging and service design. It does not consider designer-makers or craft-based designers.
2. The study was undertaken in 2007–8.

## REFERENCES

Brown, T. (2008) 'Design Thinking', *Harvard Business Review*, June: 84–95.

Cox, G. (2005) *Cox Review of Creativity in Business: Building on the UK's Strengths*. London: HM Treasury.

Design Council. (2010) *Design Industry Insights 2010*. London: Design Council.

DTI (2005) *Economics Paper No. 15: Creativity, Design and Business Performance*. London: DTI.

Porter, M. (1979) 'How Competitive Forces Shape Strategy', *Harvard Business Review*, March–April: 93–101.

Scott, M. (1998) *The Intellect Industry: Profiting and Learning from Professional Service Firms*. Chichester: John Wiley & Sons.

von Hippel, E. (2005) *Democratizing Innovation*. Cambridge, MA: MIT Press.

Work Foundation (2007) *Staying Ahead: The Economic Performance of the UK's Creative Industries*. London: Department for Culture, Media and Sport.

# 31

# Working the Crowd: Crowdsourcing as a Strategy for Co-design

## MIKE PRESS

Barack Obama forged a movement from it. The Arctic Monkeys harnessed it to launch the UK's fastest selling album since The Beatles. Apple has used its inventiveness to provide the killer apps for its iPhone. Muji uses it to design its products. Many of us rely on it to work our computers. Proctor and Gamble uses it to solve its complex R&D problems. The music industry is taking it to court.

The crowd. Until very recently it was seen as a source of irrational behaviour, of the 'lynch mob' mentality, to be harnessed and controlled by propaganda and the force of law. In the age of the Internet it is the driver of new political movements, the collective brains behind open-source software, the source of new innovations and a network of communities who share products, culture and ideas.

Every year since 1927 *Time* magazine has nominated a 'Person of the Year'. Chosen largely from the world of politics, representatives from innovation and industry are rare. An early exception was 1928's choice: Walter Chrysler, founder of the Chrysler motor corporation. He had pulled the corporation together from smaller firms and made it one of the world's top three motor giants alongside Ford and General Motors, expressing this achievement through his daring contribution to New York's skyline. Great corporations, new innovations and the driving force of change were the product of visionary corporate bosses who harnessed innovation and capital to make markets. In this view, consumers were – well, just people who consumed the products of this corporate power and vision. Innovation was clearly a top down thing. *Time*'s Person of the Year in 2006 was a challenge to this view. They nominated … 'You'. According to Time, Web 2.0 has placed ordinary people at the centre of change, innovation and culture. Today we are all bloggers, creators of our own culture on YouTube, open software participants and contributors to product innovation. When it comes to innovation, users are now doing it for themselves.

Recent research reveals how product ideas and innovative applications of technology are arising directly from users and this appears to be particularly pronounced in the digital sphere such as gaming and the development of new Web technologies. This would seem to challenge not only accepted models of innovation theory and knowledge transfer but also the very methods by which business is conducted, the relationships that it develops with consumers, how it carries out design and how it protects intellectual property. In short, we are witnessing a fundamental shift in the relationships between users and producers; indeed there is an accelerating blurring of these very definitions, the implications of which are profound.

This chapter explores some of the implications of these shifts for design management by drawing on relevant research in the field. The first part provides a number of selected examples that demonstrate the diverse practices that place the crowd in the driving seat of innovation. The second part maps out the historical precedents of design strategies that are participative and user focused, and summarizes some of the key theoretical perspectives that are useful in understanding the changes taking place. The third part identifies the issues and challenges that arise from these perspectives. The most notable challenge is the subject of the fourth and concluding part – the future role of design. If the design of the industrial age was led and managed by testosterone-fired visionaries such as Walter Chrysler, the design of this new age requires a very different set of skills and attitudes. To borrow an idea from Charles Leadbeater, the central task now is to 'orchestrate creative conversations' that enable crowds to come together to share their ideas productively (Leadbeater 2008: 118).

## THE IN CROWD

Crowdsourcing appears to cover all the bases in terms of research, innovation and design from the classification of galaxies to the design of t-shirts, from financing football teams to the design of consumer electronics, from solving complex R&D problems to enhancing the value of existing products. One wiki that was set up to document as many examples as possible cited over 230 crowdsourcing projects by 2009. A few cases will illustrate the varied dimensions and nature of crowdsourcing initiatives.

Galaxy Zoo is a voluntary project established in 2007 that invites amateur astronomers to classify the galaxies shown in over one million images available from the website. In the first year of the project over fifty million classifications were received and at least one significant astronomical discovery made – Hanny's Voorwerp. Named after Hanny van Arkel, the Dutch primary school teacher who first observed the strange blue blob on a distant galaxy, the new phenomenon is probably the first stellar object to be discovered by a person who does not own or operate a telescope (Cho and Clery 2009). Crowdsourcing shifts astronomy out of the observatory and into the living room.

Threadless is a Chicago-based company employing twenty people, selling 50,000 t-shirts and earning over $100,000 in profit per month (Piller and Ihl 2009).

Customers, who submit designs online that are subsequently voted on by the 'Thread-less community', do all the design work. The most popular designs go into production and their designers win cash prizes of $2,000. The enthusiastic and committed community commits itself to purchase its chosen designs, thereby reducing the market risk. Crowdsourcing takes unpredictable fickleness out of the fashion market.

In 2008, the English football club Ebbsfleet United became one of the world's first businesses to be taken over by an online community as 28,000 club fans were mobilized by myfootballclub.co.uk to contribute £35 each in order to secure a 75 per cent controlling interest in the club. Three months later, Ebbsfleet United won the FA Trophy at Wembley, its first trophy in 118 years. Crowdsourcing can provide financial alternatives to the rich billionaires who increasingly dominate the game – and can win the cup.

Crowd spring describes itself as 'a marketplace for creative services'. Companies post creative challenges on the website and offer cash prizes for the winners, inviting creative solutions for everything from collateral design to consumer electronics. In 2009, the Korean LG company offered $20,000 for the design of a mobile phone, additionally providing entrants with access to Autodesk SketchBook Pro for use in producing the designs. Crowdsourcing provides a low cost alternative to source design work.

InnoCentive is the eBay of scientific R&D. Companies such as Boeing, DuPont and Proctor & Gamble post on the website scientific or technological problems that they need solutions for. People submit their ideas in a bid to get a cash prize of usually between $5,000 and $100,000 given to the most effective solution. In November 2006 the first $1 million prize was announced. For the forty or so Fortune 500 companies that use InnoCentive, it provides a far more flexible way to doing R&D than relying wholly on in-house teams. For scientists and technologists – and they could be freelance, retired and students – there is the promise of financial reward and employment flexibility. As Brabham observes: 'InnoCentive solvers win very large awards, but the bounties pale in comparison to what the equivalent of that intellectual labour would cost seeker companies in in-house R&D' (Brabham 2008: 83).

Apple has demonstrated how the crowd can add value to an existing product. The iPhone's success is in part due to the massive library of applications that can be downloaded for it. Many of these are authored by individuals who, after validation, can sell them through Apple's iTunes webstore and retain most of the selling price. 'The iPhone is not just a device but a platform on which thousands of third parties have developed phone applications. This has multiplied the ways an iPhone can be used making it more valuable' (Leadbeater 2009: 12). Apple freely provides the programming tools to develop applications. To date, the youngest 'app' author is nine-year-old Lim Ding Wen from Singapore whose Doodle Kids was released just weeks before the billionth app download was recorded at the iTunes site.

As the examples above illustrate, crowdsourcing does not represent a single strategy, rather 'it's an umbrella term for a highly varied group of approaches that share one

obvious attribute in common: they all depend on some contribution from the crowd'
(Howe 2008: 280). Journalist Jeff Howe was the first to coin the term 'crowdsourcing'
in an article for Wired:

> Technological advances in everything from product design software to digital video
> cameras are breaking down the cost barriers that once separated amateurs from pro-
> fessionals. Hobbyists, part-timers and dabblers suddenly have a market for their
> efforts, as smart companies in industries as disparate as pharmaceuticals and televi-
> sion discover ways to tap the latent talent of the crowd. The labor isn't always free,
> but it costs a lot less than paying traditional employees. It's not outsourcing; it's
> crowdsourcing. (Howe 2006)

In Howe's later, more detailed analysis of the phenomenon he provides a fourfold
categorization:

- *Collective intelligence:* based on the principle that groups of people have expert
  knowledge, the challenge is to provide mechanisms to capture this expertise,
  as exemplified by InnoCentive and Crowd spring.
- *Crowd creation*: Wikipedia is one example of how the crowd can collaborate
  in creating an entire repository of knowledge. Apple has also harnessed crowd
  creation to add value to its products and brand.
- *Crowd voting*: Threadless uses voting as a means of identifying consumer preferences and thus reducing the risks of production.
- *Crowdfunding*: the crowd can be mobilized financially to raise funds, as in
  the example of Ebbsfleet United, providing an alternative to banks and other
  financial institutions.

As Howe explains in his wide-ranging survey of crowdsourcing, various combinations of these approaches are evident in most initiatives, with Threadless using
them all. An alternative typology can be based on the business model that they are
pursuing:

- *Corporate*: Apple and Dell are among many companies that have crowdsourcing strategies. Dell's IdeaStorm invites the crowd to post suggestions for product improvements and to vote on them. In the small print of the website it
  explains how 'You grant to Dell and its designees a perpetual, irrevocable,
  non-exclusive fully-paid up and royalty free license to use any ideas, expression of ideas or other materials you submit.' In common with many corporate
  initiatives, the crowd is being sourced but not rewarded. Crowdsourcing here
  is being used to provide customers with active engagement with the brand
  and to enhance the company's market intelligence.

- *Intermediary*: InnoCentive and Crowd spring are intermediaries that invite design and research challenges from business clients, who in turn provide prize money to attract participation from the crowd. The motivation on the part of clients is to source design more cost effectively and flexibly, through exploiting expertise of fast growing 'pro-am' communities (see below). The challenge for the intermediary is to build a committed and expert community that can rise professionally to the challenges set them.
- *Public*: ibridgenetwork.org is an online network of university research departments that aims to develop a global audience for their work and facilitate collaboration with the commercial sector. Galaxy Zoo is also rooted in publically funded university research and seeks to outsource, on a voluntary basis, relatively low level but time-consuming research tasks. Through the change.gov website, President Obama effectively used crowdsourcing methods to connect citizens with his emerging policy agenda.
- *Peer*: Linux software and Wikipedia are both produced through networks of co-operators on a voluntary basis. While the open-source movement and peer-to-peer production are both distinctive phenomena, they can share some characteristics with crowdsourcing and, indeed, the latter is evolving to enable more collaborative and equitable forms of production. French startup Crowd spirit initially appears similar to the Dell model – people submit product ideas and designs covering the whole gamut of consumer electronics and the user community votes on them. But the key difference is that partner manufacturers are sought and community investors secured. Royalties are then carefully apportioned throughout the product development chain. In so doing, this model takes some aspects of the nineteenth century co-operative movement and places them within high-tech manufacture in a fully networked economy.

Kleeman et al. (2008) differentiate between the following types of crowdsourcing: participation of consumers in product development and configuration; product design; competitive bids on specifically defined tasks or problems; permanent open calls; community reporting; product rating by consumers and consumer profiling; customer-to-customer support.

In terms of definitions, crowdsourcing is an ever-evolving and at times disputed territory, with subterritories being defined within it. For example, 'citizen journalism' is a specific field within the crowd-creation category, while 'citizen science' is a term often preferred for initiatives such as Galaxy Zoo. Crowdsourcing methods have also opened up what Kittur, Chi and Suh (2008) describe as 'micro-task markets'. For example, Amazon's mechanical turk web site, offers people small fees (usually less than one dollar) to undertake simple tasks that can be completed on screen. In East Africa, txteagle has developed a similar service for simple voice-based tasks using mobile phones. As Whitla argues

Despite the abundance of these tasks on sites such as mechanical turk, there is some question whether this is truly a form of 'crowdsourcing'. Certainly the tasks are being opened to a crowd and users are paid for completing them, however in a strict sense the tasks being undertaken are not replacing something that was previously done in-house by the firms' employees. (Whitla 2009: 23)

Crowdsourcing is a web-enabled process that accesses the collective intelligence, creativity and resources of the public and applies it to specified needs of enterprises or communities. This chapter has a central interest in three of its forms:

- *Design outsourcing*: the outsourcing of design work using open competitions and other mechanisms (for example, Crowd spring).
- *Co-creation*: consumers are involved in the design of the product or service (for example, Threadless and Apple app store).
- *Peer production*: the development of 'end-maker' communities that create value through open design systems (for example, Crowd spirit).

It is argued here that these three processes will fundamentally transform the nature of design and innovation practices in the twenty first century and the business environments that support them. This becomes clearer when we consider some of the theoretical frameworks that account for and explain the rising power of the crowd in design.

## SEEING THE CROWD

We all consume, many of us work and a few of us are paid to create. These neat distinctions are being challenged by two fundamental ideas that lie at the heart of most discussions on the new forms of forms open design that are emerging today. The first is that many consumers can be empowered to create, thereby contributing to design and innovation processes and the second is that there will be a realignment of the work/leisure, produce/consume and amateur/professional distinctions, creating new forms of self-activity to sustain and fulfil us. There is, however, nothing new about these ideas. Writing in 1749, Adam Smith observes the following in his book *The Wealth of Nations*:

A great part of the machines made use of in those manufactures in which labour is most subdivided, were originally the invention of common workmen, who, being each of them employed in some very simple operation, naturally turned their thoughts towards finding out easier and readier methods of performing it ... In the first fire engines, a boy was constantly employed to open and shut alternately the communication between the boiler and the cylinder, according as the piston either ascended or descended. One of those boys, who loved to play with his companions, observed that, by tying a string from the handle of the valve which opened this communication to another part of the machine, the valve would open and shut without his assistance and leave him at liberty to divert himself with his playfellows. One of the greatest improvements that has been made upon this machine, since it was first invented, was in this manner the discovery of a boy who wanted to save his own labour. (Smith 1993: 17)

Labour historians such as Thompson and Hobsbawn have further documented the inventiveness and design skills of 'common' working people. But it was Karl Marx who evoked a self-determined future that provided us each with our own 'portfolio' of self-activities, unconstrained by professional distinctions. In *The German Ideology*, written between 1845 and 1847, Marx and Engels maintained that labour – forced, unspontaneous and waged work – would be superseded by self-activity in which: '… nobody has one exclusive sphere of activity but each can become accomplished in any branch he wishes … to hunt in the morning, fish in the afternoon, rear cattle in the evening, criticize after dinner, just as I have a mind without ever becoming hunter, fisherman, shepherd or critic' (Marx and Engels 1998: 231). The principles of today's open design communities therefore find their roots within the writings of Adam Smith and Karl Marx.

According to Charles Leadbeater, 'Most of the internet depends on software developed by volunteer user-programmers. Computer games now outsell Hollywood films, in part because players can add to the games, en masse. The initial release of a computer game is just the starting point for an extended process of user adaptation, which makes the game richer and extends its life' (Leadbeater 2006: 8). Leadbeater uses the term 'Pro-Am' to describe 'innovative, committed and networked amateurs working to professional standards', clearly regarding them as an essential driving force of digital culture:

In mobile phones, media, computer games and software, ideas are flowing back up the pipeline from avid users to the technology producers. Pro-Ams should play a much larger role in innovation policy. Lead users should play a larger role in foresight exercises to chart the future course of innovation and policies to deregulate markets should also open up spaces for Pro-Am innovations. Pro-Am communities are the new R & D labs of the digital economy. (Leadbeater and Miller 2004: 64)

It is not just an issue of some 'consumers' having creative talents equal to that of professional 'producers', but that collectively their talents could be greater than the sum of their parts and thus exceed that of the professionals. French cyberspace theorist Pierre Lévy shares Leadbeater's optimistic view on the capacity of the crowd to be empowered creatively through new web technologies, describing it as 'collective intelligence'. Such ideas gained considerable currency in the first decade of the twenty first century, popularized especially by New Yorker journalist James Surowiecki. The French intellectualism of collective intelligence was translated into the populist title of his book: The Wisdom of Crowds. Surowiecki contends, 'Under the right circumstances, groups are remarkably intelligent and are often smarter than the smartest people in them ... When our imperfect judgments are aggregated in the right way, our collective intelligence is often excellent' (Surowiecki 2004: xiii).

When we turn to academic research within design and innovation we also discover perspectives and methods that focus on the involvement of consumers or 'users' in the design process going back over a number of decades. According to Sanders and Stappers, since the 1970s there has been increasing attention focused on how to involve users in the design process. They differentiate between two broad approaches: the user-centred design approach (or 'user as subject') that has emerged particularly in the US and the participatory approach (or 'user as partner') that finds its roots in Northern Europe.

The former approach has been particularly notable in high technology sectors and has been associated with the integration of ethnographic methods into design. A UK report that examined user-centred design in the US argued that the term 'people-centred design' is more appropriate to the approach increasingly adopted in high technology sectors: 'people in their social context rather than task-centric users should be considered a fundamental source of innovation' (Wakeford 2004: 7). However, the key emphasis of user-centred or people-centred design remains that of better understanding the needs of users as a means of informing the professional design process, rather than involving users in that process as active creative participants.

Sanders and Stappers have sought to define co-design and co-creation as collective acts of creativity that transcend the limits of the 'user-as-subject' approach. In their discussion they define co-creation as any act of collective creativity (i.e. shared by two or more people), while co-design is collective creativity as applied to the span of a design process:

> Thus, co-design is a specific instance of co-creation. Co-design refers, for some people, to the collective creativity of collaborating designers. We use co-design in a broader sense to refer to the creativity of designers and people not trained in design working together in the design development process. (Sanders and Stappers 2008: 6)

The contemporary roots of co-design can be found in the community design movement from the early 1970s that aimed to politicize the design process. According to Toker, 'community design stood for an alternative style of practice based on the idea that professional technical knowledge without moral and political content is often inadequate' (Toker 2007: 309). Involving people in the shaping of their environment was the critical objective in a movement that gave form to community architecture.

Commenting on this movement, Cross argues: 'Although the experiments have not always been successful – in either process or product – there is at least a recognition that the professionals could and should, collaborate with the non-professionals. Knowledge about design is certainly not exclusive to the professionals' (Cross 2006: 20). From the mid 1970s this movement extended its influence from the community to the workplace with the emergence of workers' planning and socially useful production. As part of a strategy to oppose factory closures and to reduce reliance on military production, workers at Lucas Aerospace in the UK developed their innovative plans for alternative production (Wainwright and Elliot 1982). However, the greatest progress in workplace-based participatory design emerged in Scandinavia where there was a more willing political context for such initiatives. Focused on the issue of the introduction of new technologies into the workplace 'several projects in Scandinavia set out to find the most effective ways for computer-system designers to collaborate with worker organizations to develop systems that most effectively promoted the quality of work life' (Sanoff 2006: 133).

Perhaps most influential in these initiatives was the UTOPIA project, which engaged Nordic trade unions in the printing industry in a unique participatory process to redefine work design in the context of the early digitalization of the industry. This proved highly influential in the field of HCI in terms of developing methods for participatory design. Nearly twenty years after UTOPIA, four of the project's original researchers reflected on its legacy:

> Our 'political' focus on worker participation and the development of new co-operative design methods have in the 1990s become a 'success' in the USA as 'Scandinavian participatory design'. The reason is simple: participation is not only a political and emancipatory category, it is also a basic epistemological (knowledge theoretical) principle. Participation is a fundamental process, not only for democracy, but also for learning. It would certainly be to overestimate our political impact to confuse the two. Today we are more at home in the academic world, than on the political arena. The researchers are no longer dissidents, but for good and bad pretty main stream socio-technical researchers and designers. (Bødker et al. 2000: 7)

Similarly, participatory design has long since made its transition from a 'dissident' politicized design philosophy, to a mainstream method that enables companies to get 'closer to the user'. In so doing it is developed a range of well-established techniques, methods and practices (see Muller 2002).

Eric von Hippel, of MIT's Sloan School of Management, has for over three decades researched the ways in which consumers influence and shape new product development, most notably through his lead user theory that has proven particularly influential. Lead users are those consumers at the leading edge of market trends, having a propensity to innovate themselves. According to lead user theory, needs experienced by lead users today will be experienced by the rest of us tomorrow, thus their inno-vations will have a wider latent appeal. Most recently in Democratizing Innovation, von Hippel marshals further evidence of how and why lead users innovate, identifying the factors that lead to variance between different product markets. His quest for lead users has taken him into hospital operating theatres, libraries, corporate IT depart-ments as well as the domain of extreme sports. Whether a lead user is abseiling down a waterfall or developing a search facility for a public library, what they have in com-mon are heterogeneous needs and thus dissatisfaction with commercial products on offer that are developed from a 'one-size-fits-all' ethos. Users and producers develop different types of innovations, which can be accounted for through information asym-metries: users are rich in information about needs and context-of-use, whereas pro-ducers are well resourced in generic solution information. Both types of information are 'sticky' making them costly to move from the site where the information was gen-erated to other sites. As a result, users tend to develop innovations that are functionally novel, whereas manufacturers tend to 'develop improvements on well-known needs that require a rich understanding of solution information' (von Hippel 2005: 8).

The vision presented by von Hippel is one in which professional design activity be-comes increasingly marginalized: 'In physical product fields, product development by users can evolve to the point of largely or totally supplanting product development – but not product manufacturing – by manufacturers' (von Hippel 2005: 14). In fore-seeing manufacturers providing toolkits for user innovation and custom design, von Hippel argues how this will 'democratize the opportunity to create'.

While lead-user theory focuses on the potential of a relatively elite, forward-focused group of users to be incorporated within the new product development process, the idea of mass customization suggests that a wider community of consumers can be en-gaged in a process of co-design. Frank Piller has long championed and researched mass customization, describing it thus:

> Customer co-design or mass customization … is a particularly promising way of serving individual customers both individually and efficiently. The objective of mass customization is to deliver goods and services that meet individual customers' needs with near mass production efficiency … This preposition means that individualized or personalized goods are provided without the high cost (and, thus, price premiums) usually connected with (craft) customization. It is possible due to the capabilities of modern manufacturing technology, including flexible manufacturing systems and modular product structures, both of that reduce the trade-off between variety and productivity … Companies are therefore able to embrace both cost efficiency and a much closer reaction to customers' needs. (Berger et al. 2005: 2)

Lead-user theory considers an elite's potential to develop new product concepts or significant innovations within existing an existing concept, whereas mass customization has tended to focus on providing consumers with the opportunity to select their own design configurations from a given platform of choice. While this can entail 'building products to order', as in the case of the well-documented example of Dell Computers, there are a range of approaches that have been researched (see Salvador et al. 2009). Both von Hippel (2005) and Piller and Walcher (2006) have given attention in their research to the development of toolkits for user innovation 'as a technology that (1) allows users to design a novel product by trial-and-error experimentation and (2) delivers immediate (simulated) feedback on the potential outcome of their design ideas' (Piller et al. 2005: 2).

Reichwald and Piller (2006) differentiate between two forms of crowdsourcing: mass customization, which benefits the individual consumer by personalizing a product or service and open innovation, in which customers co-operate with the firm in broader ways to develop product innovations that can benefits a larger group of users. This latter concept is a vital theoretical framework for understanding these developments from the perspective of the firm. If Leadbeater's pro-am idea captures the nature of changes on the consumer/user side, open innovation encapsulates the radical shift taking place on the producer side. Open innovation provides us with a framework to integrate and account for many of the developments described above.

Henry Chesbrough (2003) argues that there is a paradigm shift underway in which firms are moving towards more open forms of innovation: 'In many industries today, the logic supporting an internally oriented, centralized approach to R&D has become obsolete. Useful knowledge is widespread in many industries and ideas must be used with alacrity if they are not to be lost. These factors create the new logic of Open Innovation, which embraces external ideas and knowledge in conjunction with internal R&D' (Chesbrough 2003: 177). Chesbrough contrasts the closed model of innovation at Xerox PARC with the open approach in evidence at IBM and Intel. The challenge is thus how to capture knowledge from outside the firm that can add value to its products, both upstream and downstream.

Following Chesbrough, other researchers have examined the value of online communities for companies committed to developing open innovation systems. According to Dahlander, Frederiksen and Rullani:

> Online communities, therefore, can constitute an important external source of innovation for those firms able to implement a constructive relationship with them. Individuals in these communities may not only be able to develop innovations that can be integrated into the firm, but also may come up with new perspectives on and ways of framing problems. The community may develop a shared and mutual understanding of what it is about, what in the new product design or features is valuable; it may create product/firm loyalty and establish among community participants a sense of belonging and meaning. (Dahlander et al. 2008: 118)

Nambisan and Sawhney (2007) propose a fourfold typology of open innovation strategies, which they describe as network-centric innovation. These are:

- The 'orchestra', which is controlled by a central decision-maker that brings together partners within a defined innovation space. An example that they cite is Boeing's development of the Dreamliner, in which the company assembled a consortium of global partners, each responsible for a specific aspect of innovation in the project.
- The 'creative bazaar' can be seen when one firm shops for innovation in a global bazaar of ideas and products. The corporate and intermediary models of crowdsourcing identified earlier fit into this category.
- The 'Jam Central' model is more community-led, group focused and improvisational, often without a dominant partner, sharing many of the characteristics of open-source communities.
- The 'MOD station' model is about modifying existing innovations and products in which a dominant firm builds a community of users to undertake innovation. The video game industry is an example of this.

Emergent models of open innovation are clearly evident, which involve 'the crowd' in a variety of different ways. Some arise from user communities and are allied to the ideologies of the open-source movement, while others are driven by the explicit strategies of companies seeking to gain vital competitive advantage. Ogawa and Piller argue that the notoriously high failure rates of newly launched products and the allied shortcomings of conventional market research are the key drivers for such consumer-focused methods. Consumer involvement in design and the use of on-line user communities can help to overcome many of the risks inherent in market research – a process that they term 'collective customer commitment'. As they argue: 'The beauty of collective consumer commitment is that innovative ideas can be explored at relatively little cost. If customers reject a particular design, it can easily be scrapped' (Ogawa and Piller 2006: 14).

The motivation for companies is clear – open innovation can reduce risk and enhance the range and quality of ideas that they have access to. But what is the motivation for the consumer? To date, little research has been conducted and there appears to be diverse motivations at work. Studies on motivation in open-source projects (Hars and Ou 2002; Bonaccorsi and Rossi 2004; Lakhani and Wolf 2005) seem to suggest that the primary motivation is the pleasure of creating. A study of participation on YouTube points to attention as a driver: 'The productivity exhibited in crowdsourcing exhibits a strong positive dependence on attention. Conversely, a lack of attention leads to a decrease in the number of videos uploaded and the consequent drop in productivity, which in many cases asymptotes to no uploads whatsoever' (Huberman et al. 2009: 8). A study of 651 participants in the iStockphoto site suggests a very different motivation: 'the crowd at iStockphoto is motivated by money and the opportunity to develop individual creative skills, not necessarily by the desire for peer recognition or the opportunity to build a network of friends and creative professionals' (Brabham 2009: 18).

Clearly there are different ideologies at work. Peer production is rooted in the principles of co-operation and the motivations of creativity for its own sake which is fundamental to hacker, DIY and 'crafter' cultures. However, perhaps increasingly pro-am communities regard themselves as equal competitors to professional creatives in fields such as design and thus seek financial rewards from their labours. The very nature of many crowdsourcing initiatives, which offer cash prizes and other financial incentives, is the driver here. This presents some acute issues and challenges for both corporate design users and the design professionals themselves, which are addressed below.

## ISSUES AND CHALLENGES

Working the crowd offers considerable advantages for companies and, in providing opportunities to contribute to design and innovation, user communities are also afforded benefits. But in this fast-changing and disruptive landscape of new technological possibility, new cultures of creation and new strategic developments, research is playing a game of catch-up on the implications of crowd-based innovation and the issues that arise from it. However, we can identify some emerging issues and challenges for the key partners involved: companies, design professionals and the wider community.

For companies, the fundamental challenge is to define an integrated strategic approach to the involvement of users and consumers in design processes that are built on the principles of open innovation in ways that are appropriate to the company's specific expertise, market position and future objectives. As we have seen above, there is no one single 'fits-all' strategy for open innovation and so an exploratory approach is required. Based on their case study of customer co-design at sports good manufacturer Adidas, Berger et al. (2005) identify four challenges that summarize their understanding of how such co-operation can develop effectively.

- The acquisition challenge refers to the organizational and technical infrastructure required by the firm in order to enhance the flow of information and knowledge from customers. It places a particular premium on the skills, motivation and expertise of those employees who work directly at the customer interface.
- The assimilation challenge concerns the company's internal processes, which allow the knowledge and information flowing from consumers to be integrated appropriately and productively with all the necessary internal systems and routines. In other words, how does the firm make maximum use of this customer knowledge and how does it re-engineer internal systems to enable this?

- The transformation challenge is that of using co-design to create a customer-focused organization strategically. A vital indicator here is that new value is created from co-design.
- The exploitation challenge involves the transformation of the co-design process into a sustainable business model that leverages existing and creates new competencies.

Such challenges apply equally to global players such as Adidas, as they do to small startups – indeed in the latter case perhaps more so. Threadless, for example, has built a sustainable and successful business by meeting these four challenges in an enterprise that was built from the ground up on the foundation of co-design. Whitla's research on crowdsourcing websites shows that smaller companies are currently the main users:

These firms often work with limited resources inside the firm and have limited budgets to spend on advertising agencies, public relation firms, graphic designers, photographers or whatever. These small firms can benefit greatly from access to a wide pool of skills available at reasonable prices through the technique of crowdsourcing. (Whitla 2009: 25)

However, for small companies to exploit this potential advantage fully, their assimilation challenge will be particularly acute, as they will need procedures and expertise to filter ideas and apply them internally (Hempel 2007).

Recent studies have emphasized the importance of well-designed tools to assist those consumers who wish to contribute to the design process, in particular TICs – or Toolkits for Idea Competitions. As one of these studies explains 'The performance of an idea competition may be significantly influenced by the design of the TIC's user interface, the procedure of idea formulation, features for collaborative idea creation and so on' (Franke and Piller 2006: 315). In itself, the company/customer interface and the support mechanisms provided to the customer to make a meaningful input to the design process becomes a critical and significant design challenge.

One final challenge for those companies embarking on any form of co-design strategy is related to the potential dangers of exploiting the customer. It is one thing to develop an online community of users that can cement a stronger relationship with a brand and quite another to fleece it of its creative ideas with little or no adequate remuneration. In the early days of crowdsourcing there was a certain novelty value for the consumer in such engagement but, with companies such as Apple offering generous terms to its iPhone app creators, there is a greater requirement for firms to reward on a fairer basis. As Whitla argues 'In today's business environment firms that engage in crowdsourcing activities will be required to be able to justify the social responsibility of their actions' (Whitla 2009: 12). Companies that ignore the social responsibility of fairness to their consumer/producers could rapidly lose brand commitment.

For designers and design managers, working the crowd presents an additional set of challenges. Kleeman et al. (2008: 22) nail this succinctly: 'For firms, outsourcing to the consumer carries a significant potential for increased profits, just as it puts regular employees at risk.' As a disruptive phenomenon, working the crowd will inevitably impact on the design labour market. Kleeman et al. describe the new consumer type emerging as the 'working consumer' ('arbeitender Kunde') and argue that it will increasingly take over parts of the value creation and production processes.

In September 2009, iStockphoto – a crowdsourcing provider of stock photography – announced that they would be sourcing logos. The reaction of one design blog the day after this announcement captures the feeling of many in the design profession:

> When it comes to crowdsourcing, the responsibility falls solely on designers to stand up and say NO. As long as there are thousands of designers submitting to these sites, they will continue to thrive. I personally vow to no longer associate myself with designers who undervalue our industry by allowing themselves to be taken advantage of as style-whores … and I encourage/challenge you to stand up, have a backbone and do the same. (The Donut Project, www.thedonutproject.com/2009/09/23/istockphoto-to-begin-selling-stock-logos/, accessed 24 September 2009)

Such strength of feeling is understandable but, aside from asserting the value and accountability of professionalism in design work, there is little impact that the design community can have on the rate of adoption of crowdsourcing. It will be worth examining the patterns of its use over the coming years and its impact on the professions, both in-house and consultancy based.

However, a more productive area to explore is that of how design will need to evolve to accommodate the needs of 'working consumers', whether they are sourced by the corporate sector or involved in peer production. The rise of strategic design and transformation design as broad areas of enquiry and practice suggests a way forward. 'Design, has historically focused on the "giving of form" whether two or three dimensional. Transformation design demands a shaping of behaviour – behaviour of systems, interactive platforms and people's roles and responsibilities' (Burns et al. 2006: 26).

The transformation of the processes of innovation and the tools and resources available to enable companies to 'work the crowd' and for the crowd to work for itself are perhaps the key design challenges of our age. Sanders and Stappers (2008) suggest that researchers and designers will need to learn how lead, guide, provide scaffolds and offer a clean slate to people who are at the various stages of creative empowerment. Fischer (2002) describes the challenge in terms of meta-design. The need here is to focus on the design of (1) the technical infrastructure that allows end-users to create, (2) a learning environment and work organization that allows a migration away from passive consumption and (3) the socio-technical environment in which people are recognized and rewarded by their contribution. The design challenge is therefore technological, educational and cultural/economic.

While 'professional' designers can and should rise to these challenges, they also lay within the domain of citizens to take responsibility for proposing and developing frameworks for co-creation that meet their cultural and political aspirations. This is a central issue concerning the politics of crowdsourcing, for this is an emerging phenomenon that has implications for power in society. As Kleeman has argued:

> Crowdsourcing gives consumers a new avenue of influence on corporate decision-making, at least indirectly through means such as recommending new designs and influencing public opinion. At the same time, consumers are themselves exposed to a new danger: the danger of being exploited by a corporation as a cheap supplier of valuable ideas stripped of control over their use. (Kleeman et al. 2008: 23)

Like the Internet itself, crowdsourcing has been driven by divergent ideologies and motivations. On the one hand, it has been driven by the market economy that recognizes in it the means to outsource creative work to non-professionals at a low cost and gain other advantages for the firm, such as enhanced brand loyalty. Left to its own devices it would strip mine consumers of their creative ideas and labours and reduce design management to a process of out-sourcing logistics. However, crowdsourcing is also driven by the open source and peer producing communities who see within it the potential for more inclusive and reciprocal economic relationships that are co-operatively organized and community based. Design management here becomes a richer process of transformation. This is not an 'either/or' scenario, but rather a set of contradictions and tensions that will be resolved through active engagement around these issues.

## CONCLUSIONS

Fundamentally, all of this is a major design challenge. The emphasis on design in the twentieth century was about enhancing value, creating desire, providing corporate clients with identity and proprietary innovation and involving people only as passive consumers. Design in the twenty-first century is something else. It is about enhancing engagement, creating community, providing clients and users with useful and useable tools and involving people as active participants. Certainly it remains an activity that is about developing and applying visual thinking at the highest levels and solving problems through making, iteration and prototyping. But these skills must now be applied in a very different context and in very different ways.

Table 31.1 pulls together some of the central issues raised in this chapter, indicating the relationships between a new economy, a new design and a new design management.

The emergent new economy is moving us beyond mass customization, placing demands on design to become more open, both as a process and as an outcome. The

**TABLE 31.1: The new design management**

| The New Economy | The New Design | The New Design Management |
|---|---|---|
| Going beyond mass customization | Making design open | Managing open and distributed design decision-making |
| Engaging users and consumers | Designing design tool kits | Managing the interfaces with co-designers |
| Rethinking innovation | Designing new systems of innovation | Emphasis on systems and applied design thinking rather than product |
| Rethinking professional roles | Design as enabler | Redefining professional roles and responsibilities |
| New economies of peer production | Design as an inclusive set of skills and knowledge base | Collaborative and cooperative models of management |
| Transformation of public and social services | Design for social innovation | Community-centred design practices |

challenge for design management is to manage a process of open and distributed de-sign decision making. This involves a profound challenge to existing business models and to ideas of intellectual property rights. As the new economy involves a shift from the 'producer–consumer' model to one where there is more engagement and partner-ship with users, then design's focus becomes more one of developing tool kits and interfaces for engagement. Developing brand engagement is less an issue of corpo-rate identity and the communication of values and more one of developing useable, relevant and practical tools. The design manager is therefore critically the manager of the interfaces with the wider community of co-designers. In rethinking innova-tion, the challenge is that of designing new systems of innovation as much as in-novative outcomes themselves. The management of this systems approach therefore becomes a priority. Professional roles (especially those of design professionals) have to be rethought and redefined. The designer's role becomes that of an enabler and the management issue here is to reconsider the roles, responsibilities, methods and tools that enable them to fulfil this new role. We can also already see the emergence of new economies based on peer production, requiring design to be more inclusive and management systems and approaches more co-operative and network based in their nature.

The transformation of public and social services is already under way in many major economies, often enabled by the processes described in this chapter. This of-fers an exciting and welcome new role for design in social innovation. Geoff Mulgan, former Director of Policy for Tony Blair and a key advocate of social innovation has described how 'the design world has started to engage with public services' (Mulgan

2007: 11). Significantly, most of the interesting initiatives have been led by independent design professionals working directly with communities. For example, Sarah Drummond explains how her My Police project uses social media as 'a democratic and non-agenda-setting way of engagement with the public, allowing them to voice their feelings and gain the service they want as a community' (Cairns 2010). Drummond is one of newly emergent generation of design-led social innovators who are 'working the crowd' for the social benefit of the crowd itself.

To 'orchestrate creative conversations', which is essential if the potential of crowds to work together more creatively and co-operatively is to be realized, design is playing to one of its core strengths. The power of visualization and design thinking to render complexity more understandable and to develop shared languages across disciplines and cultures is well proven. The challenge ahead is to develop and refine the new methods required to empower design as the critical and visionary force for positive change in the new age of the crowd.

## REFERENCES

Berger, C., Moslein, K., Piller, F. and Reichwald, R. (2005) 'Cooperation between Manufacturers, Retailers and Customers for User Co-design: Learning from Exploratory Research', *European Management Review* 1: 70–87.

Bødker, S., Ehn, P., Sjögren, D. and Sundblad, Y. (2000) 'Co-operative Design – Perspectives on 20 years with "the Scandinavian IT Design Model"', Proceedings of NordiCHI 2000.

Bonaccorsi, A. and Rossi, C. (2004) 'Altruistic Individuals, Selfish Firms? The Structure of Motivation in Open Source Software', *First Monday* 9(1), www.uic.edu/htbin/cgiwrap/bin/ojs/index.php/fm/ (accessed 5 April 2011).

Brabham, D. (2008) 'Crowdsourcing as a Model for Problem Solving: An Introduction and Cases', *Convergence: The International Journal of Research into New Media Technologies* 14(1): 75–90.

Brabham, D. (2009) 'Moving the Crowd at iStockphoto: The Composition of the Crowd and Motivations for Participation in a Crowdsourcing Application', *First Monday* 14(1): 1–19.

Burns, C., Cottam, H., Vanstone, C. and Winhall, J. (2006) *Red Paper 02: Transformation Design*. London: Design Council.

Cairns, G. (2010) 'Improving the Police Experience', *Guardian*, 28 April.

Chesbrough, H. (2003) *Open Innovation: the New Imperative for Creating and Profiting from Technology*. Boston, MA: Harvard Business School Press.

Cho, A. and Clery, D. (2009) 'International Year of Astronomy: Astronomy Hits the Big Time', *Science* 323(5912): 332–5.

Cross, N. (2006) *Designerly Ways of Knowing*. London: Springer.

Dahlander, L., Frederiksen, L. and Rullani, F. (2008) 'Online Communities and Open Innovation', *Industry and Innovation* 15(2): 115–23.

Fischer, G. (2002) 'Beyond "Couch Potatoes": From Consumers to Designers and Active Contributors', Proceedings of the Third Asian Pacific Computer and Human Interaction (APCHI '98), IEEE Computer Society, Washington, DC, USA.

Franke, N. and Piller, F. (2006) 'Value Creation by Toolkits for User Innovation and Design: The Case of the Watch Market', *Journal of Product Innovation Management* 21: 401–15.

Hall, T. and Bannon, L. (2005) 'Co-operative Design of Children's Interaction in Museums: A Case Study in the Hunt Museum', *CoDesign* 1(3): 187–218.

Hars, A. and Ou, S. (2002) 'Working for Free? Motivations for Participating in Open Source Projects', *International Journal of Electronic Commerce* 6: 25–39.

Hempel, J. (2007) 'Tapping the Wisdom of the Crowd', *Business Week,* 17 January.

Howe, J. (2006) 'The Rise of Crowdsourcing', *Wired* 14(6), www.wired.com/wired/archive/14.06/crowds.html (accessed 5 April 2011).

Howe, J. (2008) *Crowdsourcing: How the Power of the Crowd Is Driving the Future of Business.* London: Random House Business Books.

Huberman, B.. Romero, D. and Wu, F. (2009) 'Crowdsourcing, Attention and Productivity,' version of paper submitted for the 2009 World Wide Web Conference (Madrid), http://arxiv.org/abs/0809.3030 (accessed 5 April 2011).

Kittur, A., Chi, E. and Suh, B. (2008) 'Crowdsourcing for Usability: Using Micro-Task Markets for Rapid, Remote and Low-Cost User Measurements', Proceedings of the Twenty-sixth Annual SIGCHI Conference on Human Factors in Computing Systems, pp. 453–6.

Kleeman, F., Voß, G.G. and Rieder, K. (2008) 'Un(der)paid Innovators: The Commercial Utilization of Consumer Work through Crowdsourcing', *Science, Technology and Innovation Studies* 4(1): 5–26.

Kleinsmann, M. and Valkenburg, R. (2008) 'Barriers and Enablers for Creating Shared Understanding in Co-design Projects', *Design Studies* 29(4): 369–86.

Lakhani, K. and Wolf, R. (2005) 'Why Hackers Do What They Do: Understanding Motivation and Effort in Free/Open Source Software Projects', in J. Feller, B. Fitzgerald, S. Hissam and K. Lakhani (eds) *Perspectives on Free and Open Source Software.* Cambridge, MA: MIT Press.

Leadbeater, C. (2006) *The User Innovation Revolution: How Business Can Unlock the Value of Customers' Ideas.* London: National Consumer Council.

Leadbeater, C. (2008) *We Think.* London: Profile Books.

Leadbeater, C. (2009) *Original Modern Manchester's Journey to Innovation and Growth.* Provocation 11. London: NESTA.

Leadbeater, C. and Miller, P. (2004) *The Pro-Am Revolution How Enthusiasts are Changing our Economy and Society.* London: Demos.

Lévy, P. (2005) 'Collective Intelligence, A Civilisation: Towards a Method of Positive Interpretation', *International Journal of Politics* 18: 189–98.

Marx, K. and Engels, F. (1998) *The German Ideology: Including Theses on Feuerbach and an Introduction to the Critique of Political Economy.* London: Prometheus Books.

Mulgan, G. (2007) *Ready or Not? Taking Innovation in the Public Sector Seriously.* Provocation 03. London: NESTA.

Muller, M. (2002) 'Participatory Design: The Third Space in HCI', in J. Jacko and A. Sears (eds) *Handbook of Human Computer Interaction.* Mahwah, NJ: Erlbaum.

Näkki, P. and Virtanen, T. (2007) 'Utilising Social Media Tools in User-centred Design', CHI 2007 Workshop Supporting Non-professional Users in the New Media Landscape. San José, CA.

Nambisan, S. and Sawhney, M. (2007) *The Global Brain: Your Roadmap for Innovating Faster and Smarter in a Networked World*. Indianapolis: Wharton School Publishing.

Ogawa, S. and Piller, F. (2006) 'Reducing the Risks of New Product Development', *Sloan Management Review* 47(2): 65–72.

Piller, F., Berger, C., Moeslein, K. and Reichwald, R. (2004) 'Co-Designing the Customer Interface: Learning from Exploratory Research', Working paper, TUM Business School, Department of General and Industrial Management, Technische Universitaet Muenchen.

Piller, F. and Ihl, C. (2009) *Open Innovation with Customers: Foundations, Competences and International Trends*. Aachen: Technology and Innovation Management Group RWTH Aachen University.

Piller, F., Schaller, C. and Walcher, D. (2004) 'Customers as Co-Designers: A Framework for Open Innovation', Working paper, TUM Business School, Department of General and Industrial Management, Technische Universitaet Muenchen.

Piller, F., Schubert, P., Koch, M. and Moslein, K. (2005) 'Overcoming Mass Confusion: Collaborative Customer Co-design in Online Communities', *Journal of Computer-Mediated Communication* 10(4), http://jcmc.indiana.edu/vol10/issue4/piller.html (accessed 24 September 2009).

Piller, F. and Walcher, D. (2006) 'Toolkits for Idea Competitions: A Novel Method to Integrate Users in New Product Development', *R&D Management* 36(3): 307–18.

Reichwald, R. and Piller, F. (2006) *Interaktive Wertschöpfung. Open Innovation, Individualisierung und neue Formen der Arbeitsteilung*. Wiesbaden: Gabler.

Salvador, F., Piller, F. and Holan, P. (2009) 'Cracking the Code of Mass Customization', *Sloan Management Review* 50(3): 71–8.

Sanders, E. and Stappers, P. (2008) 'Co-creation and the New Landscapes of Design', *CoDesign* 4(1): 5–18.

Sanoff, H. (2006) 'Multiple Views of Participatory Design', *METU Journal of the Faculty of Architecture* 23(2): 131–43.

Smith, A. (1993) *An Inquiry into the Nature and Causes of the Wealth of Nations*. Ed. Sutherland, K. Oxford: Oxford University Press.

Surowiecki, J. (2004) *The Wisdom of Crowds: Why the Many are Smarter than the Few*. London: Abacus Books.

Toker, Z. (2007) 'Recent Trends in Community Design: The Eminence of Participation', *Design Studies* 28(3): 309–23.

Von Hippel, E. (2005) *Democratizing Innovation*. Cambridge, MA: MIT Press.

Wainwright, H. and Elliot, D. (1982) *The Lucas Plan: A New Trade Unionism in the Making*. London: Allison & Busby.

Wakeford, N. (ed.) (2004) *Innovation through People-centred Design – Lessons from the USA*. London: Department of Trade and Industry.

Whitla, P. (2009) 'Crowdsourcing and its Application in Marketing Activities', *Contemporary Management Research* 5(1): 15–28.

# 32

# On Managing as Designing

RICHARD J. BOLAND, JR.

## SETTING THE STAGE

The phrase 'managing as designing' was coined by Fred Collopy and I in 2002 as a way of signalling the need for change in the curriculum of business schools. MBA programmes have become a commodity, reflecting a uniform mindset in business schools – a mindset that sees managing as decision making and has been developing in business schools for over five decades. Seeing managing as designing is the explicit recognition that management is not value free, that it is deeply implicated in shaping and reshaping the social, economic and technological world we live in. As decision makers, managers can and do hide behind the claim that they are responsible for making objective, rational decisions that fulfil their obligation to maximize a company's shareholder wealth. Maximizing shareholder wealth requires them to make decisions that involve difficult trade-offs in selecting among alternative courses of action. These are hard decisions to make and they are highly skilled at the analysis of the facts of the situation that is required to make them.

## THE MANAGER AS DECISION MAKER

We object to the mindset behind a 'managing as decision making' view for two major reasons. First, for philosophical reasons, it is a dangerous and self-defeating mindset for both managers and the organizations they serve, which I will discuss below. Second, for practical reasons, the evidence of its failure is all around us – and is most dramatically seen in the collapse of the global financial system and the continuing exposures of high-profile corporate fraud.

Philosophically, viewing managers as decision makers portrays them as essentially passive. It portrays them as deciding what to do in a situation that is presented to them already formulated as a certain kind of problem with a known set of alternative actions that could be taken. Business schools teach them the quantitative analysis techniques they need in order to make that decision. Managing as designing reflects a different mindset – one in which managers are not objective or value free and are instead seeking to change the situation they are in, making it more desirable for all its stakeholders. Instead of being objective in selecting among an existing set of alternative choices, managers should be seeking betterment – for their organizations and society – by re-shaping how problems are understood and by inventing new alternatives to choose from. If they are not seeking betterment and inventing better alternatives than what is 'simply given' in a situation, they are *de facto* accepting the status quo and, if they accept the world as it is, then the world has no real need for managers. Bureaucrats would do just fine leading our organizations. To a large extent, bureaucrats are what business schools have trained their graduates to become.

To be sure, managers are not stereotypic bureaucrats who merely execute rules. They are MBA bureaucrats, with mighty analytical skills and decision techniques that grow more powerful, alluring and intoxicating every graduating class. These analytical skills do have merit and, for limited purposes, do perform the required calculations to answer correctly a narrow question. If the model of risk within their decision technique is a fair representation of the decision situation that they have defined, all is well. But, of course, that condition is only rarely met. The analytic and decision-making techniques are deeply flawed at their heart. For example, Herbert Simon, in his classic *Sciences of the Artificial*, identified a glaring failure of the most widely used decision model, which calculates the present value of a future event (Simon 1996). Known as the discounted net present value model, it is used to evaluate flows of cash over time, to compare returns on alternative investments and make a wide range of investment and planning decisions. But, as Simon points out, this widely used technique suffers from a fundamental logical flaw. Simply put, it values the future as worth less than the present. This devaluing of the future is overtly applied to money, so it seems to be an objective assessment that merely reflects a compounding of expected interest rates. But covertly, it applies to the full effect of the actions we take today and the consequences they have on non-monetary aspects of society, including our children and their children and all the future generations to come.

This is a very upside down view of how humans actually want to value their children and future generations. You may always have thought that it was worth making sacrifices for your children, that you act today so that they can have a better life than you. If you do feel that way and especially if you act that way, you are acting in contradiction to the decision-making models used by bright young MBAs to tell companies what is the appropriate action in a given circumstance.

The problem of valuing the future less than the present is just one rather small problem with the 'managing as decision making' mindset that guides the business school education of our future leaders. More pernicious by far are the philosophical assumptions about the world and our responsibility for acting in it that goes along with the management as decision-making mindset. Because managers as decision makers are depicted as passive recipients of a problem of a certain type, it is as if they are merely coming to a fork in the road and must choose which way to go. Apart from the occasional 'human value' or 'judgment' issues that defy our most powerful analytic decision techniques, we really wouldn't need managers at all. Further, by depicting the world as simply given, they bear no responsibility for having constructed it. There job is to be clear eyed in describing the situation as it is, not to be self-reflective and worried about how their own past actions have brought it into being.

## THE UNREALITY OF REALITY

Fortunately, the mindset of 'managing as decision making', which has shaped business school curriculums, does not reflect the reality of managing. That is why business schools are accused of a 'rigour/relevance gap'. Even though there are some high-level math modellers who have success with gambling in self-constructed markets, there are very few managers who live in that highly analytic world. Most managers are rarely, if ever, presented with a decision situation in which they simply choose among alternatives. Instead, they more often find themselves in an alien situation, in which the state of affairs is profoundly uncertain, the risk cannot be calculated, the possible ways forward are not evident and an understanding of the situation, its implications and the alternative actions that could be taken have to be invented before any decision could be made. These alien situations are known as 'wicked problems' and they are everywhere (cf. Rittel and Webber 1973).

Nonetheless, inventing situations and alternative actions is a troubling idea for business school students and faculty. Their toolkit of analytic techniques and decision making routines is premised on the existence of an objectively defined situation that is being dealt with rationally. If managers' decisions are often, if not always, thoroughly subjective and the decision situations they engage with are composed of issues, values, actors, conditions, environments, alternative courses of action and potential consequences that are invented by the managers themselves, then their claims for objectivity are obviously false. Their further claims of deserving large rewards for amazing performance are also false or worse.

The global economy is now reeling from the consequences of the decision-making pretensions of financial managers. Those financial managers used analytic models to turn mortgage loans made to unqualified homebuyers into collateralized debt instruments (CDI) rated as AAA investments. They also created insurance policies or credit default swaps, for betting on the default of CDIs. Those same firms then placed credit default swap wagers against the CDIs they had created and sold as solid investments. All the while, apparently unrecognized by regulators, those bets by the financial 'geniuses' were increasing the likelihood of defaults and their cascading consequences through the global financial system. The pretensions of 'objectivity' and of making rational and objective decisions on the part of financial managers has led us into a mess with no clear way out.

Most recently, we see the same financial managers have engaged in similar non-arm's-length transactions to structure debt instruments for Greece – instruments that reduce the visibility of the country's true debt. At the same time, those managers created markets for betting on the country's default and used them to place bets against Greece. Once again, we see the financial manager pretending to be objective and rational decision makers but all the while constructing the financial system that they are taking to be simply given as part of their objective environment. In both cases, the only financial gains to be had are from the financial managers' commissions and fees from selling the initial instruments and from their bets against the success of those instruments as they ride the stream of transactions up and down. From the viewpoint of managing as decision making, this enormous stream of revenue and its associated bonuses in reward of amazing performance may seem like brilliant business but, from a managing as designing viewpoint, it is nothing to be proud of or to be rewarded for.

The point of using this most dramatic and catastrophic situation as an example is to highlight the bankruptcy of the 'manager as decision maker' mindset that dominates business school curriculums and the self-images of their students. The image of the objective decision maker hides the reality of the deep personal involvement of each actor as a designer of the very system they operate within. This is true from the most micro level of the team they work with, the projects they work on, the organizations they work for and the 'markets' they work in. From top to bottom, the world of work is a self-constructed, self-perpetuating world that managers are designing as they act within it. A failure to realize the deeply self-referential nature of our social and economic systems is the undesirable, long-term consequence of the manager as decision maker mindset.

## THE MANAGER AS DESIGNER

The managing as designing mindset is a call to see that what most managers do, most of the time, is design. Entrepreneurs are only the most visible example of managers as designers. If we take off the false glasses and stop viewing managers as decision makers, we see them as designers at every level of every organization, in every industrial, service, professional and non-profit sector. Fundamentally, the financial managers who have brought down the world's economy did it through designing, not deciding. Why is it that they did not see that or that we do not see that?

One reason, as indicated in many ways above, is the longing to be seen as rational and objective. The manager as decision maker mindset is only one instance of that burning desire, which characterizes modernity. We need to overcome the fear of being labelled 'subjective' and 'relativist' and we need to open ourselves to the pervasiveness of designing in all human action, especially in managing. In a sense, managing as designing is an existentialist voice in management education, calling for us to recognize the freedom and choice that is our responsibility as authentic persons. It is a voice that reminds us we cannot escape from the responsibility of being human and exercising our will in shaping ourselves and our world.

Lou Pondy had a profound influence on me as a scholar. He was a colleague and mentor when I was a young assistant professor at the University of Illinois at Urbana Champaign from 1976 to his death in 1987. He had a wonderful way of getting to the heart of decision making and contrasting it with design. He argued that 'to decide' denotes a 'cutting away', as in a deciduous tree, which loses its leaves each Fall. A decision, then, is what is left after the cutting away of what is not wanted – it is thus a limiting of what could be, a closing off of other opportunities. Designing, in contrast, is an opening up of possibilities – an abductive leap to what could be.

In the domain of science, decision is the last moment of the theorize–hypothesize–test cycle: the moment when the scientist tests the null and either rejects the hypothesis or continues to consider that the theory being tested is correct. Design, in contrast, is the first two stages of the cycle, in which an alternative way of understanding is proposed as a theory of the phenomenon and an implication of that theory is hypothesized, along with a method to test it. The first two stages emphasize the scientist's deeply subjective involvement in creating knowledge, while the last stage allows the scientist to become objective and separated from the theory and hypothesis being tested.

Obviously, all three stages of the theorize-hypothesize-test cycle are necessary for science to proceed and the same is true for the process of management. Managing is a process of inquiry that is based on professional claims to knowledge about organizing and controlling an enterprise. Like the scientist, the manager develops and maintains that professional knowledge through inquiry – he or she has (or should have) a theory that leads to hypothesizing and testing – and that process of inquiry requires all three steps of the cycle. The manager is necessarily both a designer and a decider.

Managing as designing is not a call for abandoning the decision making role of managers, but for recognizing that a significant and I would argue determining, role of managers as designers has been systematically overlooked in management education. In short, we need whole persons as managers, but have been training business students to be only partial persons.

## Breaking the Bounds

In a decision-making role, managers are bounded in multiple ways. For one thing, managers are bound by the number of alternative actions that they are expected to choose among. The set of alternatives being considered reflects a particular framing of the problem and that is another way in which they are bound. The framing of the problem is analogous to the initial theorize and hypothesize stages of the scientist's inquiry process and is where the scientist can be truly open to new possibilities. Once a problem is framed, however and a choice is being made, the possibilities for openness are limited to the elaboration and refinement of details. This is precisely the wrong place to start a design effort, but it is where the manager as decision maker too often does. It leads to a false sense of accomplishment, akin to rearranging the deck chairs on a floundering ship.

Breaking the bounds of the 'managing as decision-making' mindset is not easy. It requires courage to challenge the labelling of the situation, the statement of the problem and the feeling that one must move forward in making the decision. The manager who stands up ands says, 'this is the wrong way to think about this situation' or 'we need to develop new alternatives' is not a welcome member of the decision-making team. That kind of challenge is taken as a threat, not only to the individual or group that has stated the problem and its alternatives but also to the timeline for making a choice. Time is a most important resource for the manager as decision maker. 'We don't have the luxury of time' or 'The time for planning is over, now we must decide' are the kinds of responses the manager as designer will receive. It's a bit like saying, 'we don't have time to think' and is very hard to overcome. It's like trying to stop a moving train and is not unlike the problem faced by the whistle blower.

Both the whistle blower and the manager as designer are stepping back from the rush of time and declaring that something is fundamentally wrong with how the organization is operating. In both cases the organization's managers are not being fully human and responsible and someone is standing up to say, 'no, enough, this can't go on.' For the whistle blower, the thing that is wrong is tied to ethical or legal failings. For the manager as designer, the thing that is wrong is mindset that is self-limiting, but may possibly lead to ethical or legal failings. Either way, taking risks to stand up against what is fundamentally wrong is the issue at stake and courage is required to do that.

## REFERENCES

Rittel, H. and Webber, M. (1973) 'Dilemmas in a General Theory of Planning', *Policy Sciences* 4: 155–69.

Simon, H. (1996) *The Sciences of the Artificial*. 3rd edn. Cambridge, MA: MIT Press.

# Conclusions: Design Management and Beyond

RACHEL COOPER AND SABINE JUNGINGER

In the introduction to this Handbook we focused on the origins of design management and the discourse in the field. We identified three paradigms that have influenced the ways in which we look at the relationship between design and management and the activities of designing and managing: design practice, design management and design capability. Looking at the chapters in this book we find that all three paradigms currently shape the practices, theories and methods of the field. In hindsight, it is interesting that the Fulbright review of design management in the UK, which took place in the early 1980s (see general introduction), fed a debate as to whether design management needed to develop a sound research base and theoretical models in order to gain credibility as a discipline within education, especially the management curriculum. The recent focus of management on design 'thinking' is one of the indicators that the work conducted over the past decades in design 'management' has still to achieve its key objectives: to establish design as a practice, process and as an attitude in organizations. The focus of design management on the management side of things, rather than on design and designing, might have contributed to this situation. Instead of looking at issues of management from a design perspective, too often the lens was turned on design from a management view.

However in reviewing the chapters presented here, we are delighted to observe a growing confidence in design and a sense of optimism about its role in management and within the organization. In this final chapter, we reflect on what our contributors have been saying and consider the future for design management within the context of challenges confronting us today. We have grouped our findings in three sections to differentiate between current observations, implications for research and for the development of the field.

## CURRENT OBSERVATION IN DESIGN MANAGEMENT

In our reading of the contributions to this handbook we see four issues that deserve discussion.

1. *A Return to Emergent Design Processes.* In what amounts to one of the clearest indications that design management is feeling confident about its design origins, there is a strong reversal of opinion as to the usefulness and applicability of stage gate processes as guides in product development. Many businesses and organizations want to know what design means, how it can help them and how they can employ design in a meaningful and cost-effective manner. The trouble is that, as the study by Lockwood shows, design is just not something that lends itself to linear sequential processes that can be simplified, for example through a stage gate process. They once promised to align unruly design activities into linear and sequential processes compatible with dominant management decision-making models but they are now considered to be constraints that prevent design from unfolding in full, restricting its capacity for emergence – a necessity for radical innovation. Yet, anxiety about non-linear and uncertain approaches in management persists. This is not passing unnoticed by major design consultancies, one of which (IDEO) has now developed a computer simulation program, *the design thinker,* targeted at management educators and management students new to design thinking. For practicing designers and design educators alike, such simulations raise some concern because real life and real experiences help to form a design attitude and design understanding and develop design skills and managers/ management students cannot forego the individual engagement in the complex and messy processes and methods that constitute designing. Research findings by our authors illustrate the pitfalls of these attempts and of simplifying and sterilizing the activities of design (cf. Lockwood, Part III). Guilfoyle's statement that design is anarchic remains – organizational concern is now how to harness anarchy without destroying it. This relates to the development of the design capability and the return of in-house design.

2. *The Return of In-house Design.* Organizations that strive for service innovation have to engage in deep systematic inquiries into their own structures, people, practices and beliefs. The knowledge and insights generated in this process are highly sensitive and directly linked to potential competitive advantages. Von Stamm (Part III) argues not for an in-house designer but for the need for in-house design capability. Borja de Mozota (Part III) takes a resource-based approach to understand design management and considers design thinking a resource that is difficult to imitate. This means that competitors have a difficult time in copying the success of a company. It also demands willingness from external consultants to learn from and engage with an organization's implicit and occasionally explicit design approach. This is an interesting situation for

design consultancies and the design profession. When the challenge involves organizational transformation and cultural change, external consultants can bring in the necessary design leadership to develop and instil a design attitude. But unlike in the early days of design consulting practices, this knowledge transfer and skill building are *the* actual deliverables. We have reached a stage where the tangible product (that which can be touched and experienced) is indeed the organically emerging by-product of the intangible product in development, namely the organizational transformation (through design). Kyung-won Chung and Yu-Jin Kim (Part III) see the reason for the rise of the in-house designer in the growing involvement of designers in strategic matters.

3. *Working in Tandem: Design Leadership and Design Champions.* Design leadership signifies a further development in design management and describes a new set of skills and responsibilities. It is a distinct role from that of the design champion, although the design champion, too, typically assumes a leading position endowed with authority in an organization (Gorb 1989). Design leadership requires a person to have a design attitude (cf. Black 1975; Michlewski 2009) whereas this is not a requirement for an organizational design champion. A design champion does not need to understand the inner workings of ongoing design activities or their elements. The role of a design champion is to provide the setting in which design capabilities can be developed, refined and applied. How important design champions are can be seen in the study by Ravasi and Stigliani (Part III). Many of the difficulties the design professionals report in their survey can be directly traced back to the absence of design champions in SMEs. However, their study also serves to highlight design leadership as weak aspect amongst many design consultancies. How else can we explain the fact that the participating designers view themselves more like victims of the non-design-educated business clients? Design leadership is clearly a role to be developed and resourced to enable organizations to drive forward and embed an organizational wide design capability. What skills and knowledge will future design leaders have to have?

4. *Implementation of Design Management Goes West: Design Education in Management Rises in the East.* It is striking that Part II on design management education is the most international in its consideration of design management. Many countries are currently discovering design management and seeking to embed design management practices in their national educational contexts. They are approaching design from many different angles and thus their interpretation and need for design management differs as well. This growth in interest is seen in the East, where design, especially in education, is recognized as adding value to the manufacturing base, where the role of brand development in differentiation is connected to the management of design and design strategy, yet the new challenges of sustainability, zero carbon and ageing populations are providing a further stimulus to create a generation of design manager to drive responses through the organization. However, when it comes to actual design practice

and design management implementation, the balance in terms of research and methods tips towards traditionally Western countries, where the thirty years of research and education provide the trajectory of knowledge. Part III of this handbook, which concerns implementation matters, is a testament to this situation.

So we have identified a move to embrace design, not to attempt to mould it into a management look alike and to embed design capability, through clear and skilled design leadership and to implement this capability based on research. However there are some gaps in research that we need to address. These are discussed below.

## RESEARCH TOPICS AND METHODS IN DESIGN MANAGEMENT

Taking the broad perspective as we have done in this book has highlighted a number of issues and topics that remain to be addressed in the field of design management research and education:

### *Design Management History and Knowledge*

Design management remains a field that prefers to concern itself with the here and now and with the future, in particular the future of products and services. The links to the origins and traditions of design management are rather weak. Working on this handbook, we increasingly found this lack of connection and reflection problematic; questions and answers tend to be repetitive and often reinvented. We see a need for a fully researched history of design management that looks both into the origins of design and of management and how they interconnect. The developments and interpretations in both of these fields have shaped and continue to shape the field of design management today. Although we have tried to build a bridge to early key works, a handbook simply cannot do justice to all the historic developments. The early works here clearly signal issues and methods that are fundamental in design management, while others are context specific. Our theoretical models must capture these and our design management practice recognize and test them.

Although we anticipated that our contributors would be most familiar with the work and research in their respective countries, we were amazed to find that much research sticks to national boundaries. This has at least two consequences from our perspective: First, it means that important work in other countries may not be acknowledged and secondly that the pace of advance in design management is hampered, as it takes a long time for important work, new concepts and new practices to become widely known beyond their regional or national borders.

### Few Explorations of Design Management in Small and Medium-sized Enterprises

There continues to be a need for research into how design can support small and medium-sized enterprises (SME). Due to the size and structure of many of these companies, a small design project can imply a huge potential for organization change, but also a huge potential risk. What this means is that designers who engage with SMEs are particularly well advised to familiarize themselves with organizational concepts relating to change and resistance to change. That only a few designers have developed this capability is one of the conclusions one can draw from the study by Ravasi and Stigliani (Part III). One cannot help but detect a certain victim attitude amongst the design consultancies of 'being misunderstood'. Frans Joziasse (Part III) puts his finger on the designers' wound and challenges them to develop methods and tools for design leadership that are more suitable for SMEs.

We cannot help but notice the tendency of design management experts to turn to large companies to explain their own theories or to claim these examples as evidence. Nobody denies that there are good lessons to be learned and insights to be gained by studying an outstanding example. But there are costs for focusing exclusively on already salient organizations. For one, there exist only a handful of iconic companies but there are a multitude of organizations and businesses to whom design management does, could or should apply. The situations, aims, products, services and markets of small and medium-sized businesses often differ significantly from those of large companies, let alone those that have already established certain design management practices. As Borja de Mozota points out in her chapter, the larger a company, the greater the likelihood that design management is already recognized to be a valuable asset. But it is the SMEs, as Frans Joziasse points out, that could benefit most from best practices in design leadership. In many ways, Josiazze's reflection complements the study conducted by Ravasi and Stigliani. Ravasi and Stigliani's study focused on the problems design consultancies experience when working with SMEs, but Joziasse takes the opposite stance and challenges designers to develop methods and tools in design leadership that SMEs can understand and apply. If design leadership is something design education has overlooked so far this then points to opportunities to develop new approaches to design leadership and especially design leadership in relation to SMEs.

### Blinded by the Light: About the Difficulty of Keeping a Critical Distance from Successful Companies

From AEG to Olivetti, from IBM to Sony, big and globally leading, iconic companies have appealed to design management researchers because this was the place where design managers were often highly visible. In their search for understanding design methods and design practices, it remains a popular exercise to take the leading companies as the 'one best model'. From AEG to Olivetti, from IBM to Sony to Starbucks, there was always one company that served to demonstrate that a design attitude contributes to the bottom line and to the unique market position of an organization. However, in light of the changes and shifts in the foci of design management practice and research, we now, for example, pursue questions of service design. Service design is being hailed as essential to businesses. What is the role of user pathways in this context? Improvements in the user experience continue to be one of the most reliable sources for organizational change and innovation. Reorienting the organization around people (Buchanan 2004), in turn, requires a design attitude throughout the organization as it treats the organization as a gestalt (Yoo et al. 2006). Perhaps we should look at less iconic companies to understand this process as well as the perceived 'successful' ones.

## Apple: A Worm Devouring Design?

It is not clear at what point one can attest to a research obsession but the fact that one in two chapters in this book refers to Apple Inc., the inventor of the iPod, iPhone, iTunes and iPad, might be an indication. Apple offers rich insights and material for the study and understanding of design management-related questions and issues. Yet few authors seem willing to ask hard questions that could produce rewarding and new knowledge. Instead, too many are satisfied with merely pointing to the company or its products, foregoing any deeper analysis or reflection. There is no doubt that Apple has achieved an overall design gestalt that encompasses its tangible products and its brand – but what about design attitudes within the organization? How far does the design attitude reach beyond the product development? And is this even important? Steve Jobs fills both the role of the design champion and the design leader. This, too, raises interesting questions that are worthwhile pursuing from a design management perspective. An in-depth critical study of design in Apple remains to be achieved.

   Aside from these academic concerns, practising designers gain from being able to explain the strengths and weaknesses of salient design companies. If designers use Apple to communicate to a business client what designers can do and how they add value, they should not act surprised if all their clients wants from them is 'an Apple'. In other words, pointing to an existing product seldom leads people to develop a design attitude that is conducive to emergence and exploration. It would be an irony of fate if Apple turns into a new design paradigm that restricts the role and purpose of design in the organization. As design management researchers, we can prevent the apple turning into the worm.

## Research Methods in Design Management

The chapters demonstrate the broad range of research methods used to explore the links between design and management, the role of design in management and the role of management in design. Case studies remain a popular tool for analysis. Yet there is still a need for design case studies that enhance our understanding of how products come into being and how this development process pushes and pulls the organization in unfamiliar ways into unfamiliar territory. Such case studies will also facilitate communication between designers and managers in the context of organization design. Action research is also a missed opportunity to contribute to change through design management practice and reflect upon it. Embedding design management researchers using ethnographic methods will provide the richness of data that will enable the field to strengthen our understanding of design management in relationship to organization behaviour and operations management.

## DESIGN MANAGEMENT AND BEYOND

Design management is dead. Long live design management.

As this book illustrates, design management discourse weaves around design practice, design management and design capacity whilst branding, product development, the consumer experience and organizational change remain relevant to design management. The real common denominator is the idea that designing requires managing and that managing also involves designing. Yet we are all witness to the failures of managements around us. Be it at our universities, in our client relationships or in our own businesses, we understand the need for structure and routines, processes and quality assurance. All of these aspects have to be managed but things already have to be in place before they can get managed. Rightly, people associate management with the tasks of monitoring, controlling, evaluating, troubleshooting, the motivation of employees and the allocation of resources. Yet design management ought to learn from the shortcomings of management that are now being revealed in various forms – the focus on short- versus long-term gains, risk versus innovation, markets versus society.

Design management, as a conscious area of research and practice, is still rather a young field. Early scholars in this growing field have focused on identifying and establishing rules and guidelines. The British Standards are an exemplary product of this early phase. Early work has also contributed to a base of design management literature and knowledge that has reinforced the need for tools and methods to guide and structure various design approaches in a business context. At the same time, this early era has produced some valuable tools and methods that are still useful. However, as in other small and emerging fields, the focus of establishing a useful and valid academic platform within the design community must mean that we do not overlook some important developments that are occurring simultaneously outside of the design and design management community. Design is recognized as playing a wider role in businesses and organizations in general, so we must also be prepared to illustrate how it can also contribute to wider global challenges with which we are now familiar, such as climate change, demographic change, economic failure and social breakdown.

How do we place design management within the domain of open/collaborative ICT supported innovation? Or when new models of global design and local manufacture change the design, manufacture, service paradigm? How do we use design management to investigate and promote sustainability and resilience, resilience of our infrastructure, our organizations and our communities in facing risks of breakdown in our energy supply or other services and resources, whether by act of nature or violence? How do we use design management to support the delivery of healthcare, education and social service in the developed and developing world? How do we develop design policy in organizations, nationally and internationally?

These are truly grand challenges, but despite being in a young field we believe that we are ready for the challenges ahead, whilst never forgetting our past. In the words of Bill Hannon, the founder of the Design Management Institute who retired in 2010 (Tobin 2010): 'We should never forget our roots. It's incredible to see how far design has come since then. Most of us didn't have a business background as we came out of design school. I felt we needed the tools to deal with an entirely different culture. We were doing design thinking all those years but didn't have a word to describe it.'

We agree that we should never forget our roots. And we should be proud of the achievements since then. But let us not forget that while design is imbued with a renewed sense of importance and relevance in its dealings with management, business and organizations and society, much work remains ahead of us.

## REFERENCES

Black, M. (1975) 'The Designer and Manager Syndrome', in T. Schutte (ed.) *The Art of Design Management – The Tiffany-Wharton Lectures on Corporate Design Management*. New York: Tiffany, pp. 41–55.

Buchanan, R. (2004) 'Management and Design: Interaction Pathways in Organizational Life', in R. Boland and F. Collopy (eds) *Managing as Designing*. Stanford, CA: Stanford University Press, pp. 54–63.

Gorb, P. (1989) *Design Management Journal*, vol. 1. Boston, MA: Design Management Institute.

Michlewski, K. (2009) 'Uncovering Design Attitude: Inside the Culture of Designers' *Organization Studies* 29: 373–92.

Tobin, J. (2010) 'A Man Ahead of His Time: Bill Hannon and the Founding of DMI', *Viewpoints* (June), www.dmi.org/dmi/html/publications/news/viewpoints/nv_vp_bh.htm (accessed 20 October 2010).

Yoo, Y., Boland, R., Jr. and Lyytinen, K. (2006) 'From Organization Design to Organization Designing' *Organization Science* 17(2): 215–29.

# INDEX

399452